A Research Strategy for Ocean-based Carbon Dioxide Removal and Sequestration

Committee on A Research Strategy for Ocean-based
Carbon Dioxide Removal and Sequestration

Ocean Studies Board

Division on Earth and Life Studies

A Consensus Study Report of

The National Academies of
SCIENCES · ENGINEERING · MEDICINE

THE NATIONAL ACADEMIES PRESS
Washington, DC
www.nap.edu

THE NATIONAL ACADEMIES PRESS **500 Fifth Street, NW** **Washington, DC 20001**

This activity was supported by contracts between the National Academy of Sciences and ClimateWorks Foundation, Contract No. 10004979. Any opinions, findings, conclusions, or recommendations expressed in this publication do not necessarily reflect the views of any organization or agency that provided support for the project.

International Standard Book Number-13: 978-0-309-08761-2
International Standard Book Number-10: 0-309-08761-9
Digital Object Identifier: https://doi.org/10.17226/26278
Library of Congress Control Number: 2022933129

Additional copies of this publication are available from the National Academies Press, 500 Fifth Street, NW, Keck 360, Washington, DC 20001; (800) 624-6242 or (202) 334-3313; http://www.nap.edu.

Printed in the United States of America

Suggested citation: National Academies of Sciences, Engineering, and Medicine. 2022. *A Research Strategy for Ocean-based Carbon Dioxide Removal and Sequestration*. Washington, DC: The National Academies Press. https://doi.org/10.17226/26278.

The National Academies of

SCIENCES • ENGINEERING • MEDICINE

The **National Academy of Sciences** was established in 1863 by an Act of Congress, signed by President Lincoln, as a private, nongovernmental institution to advise the nation on issues related to science and technology. Members are elected by their peers for outstanding contributions to research. Dr. Marcia McNutt is president.

The **National Academy of Engineering** was established in 1964 under the charter of the National Academy of Sciences to bring the practices of engineering to advising the nation. Members are elected by their peers for extraordinary contributions to engineering. Dr. John L. Anderson is president.

The **National Academy of Medicine** (formerly the Institute of Medicine) was established in 1970 under the charter of the National Academy of Sciences to advise the nation on medical and health issues. Members are elected by their peers for distinguished contributions to medicine and health. Dr. Victor J. Dzau is president.

The three Academies work together as the **National Academies of Sciences, Engineering, and Medicine** to provide independent, objective analysis and advice to the nation and conduct other activities to solve complex problems and inform public policy decisions. The National Academies also encourage education and research, recognize outstanding contributions to knowledge, and increase public understanding in matters of science, engineering, and medicine.

Learn more about the National Academies of Sciences, Engineering, and Medicine at **www.nationalacademies.org**.

The National Academies of

SCIENCES • ENGINEERING • MEDICINE

Consensus Study Reports published by the National Academies of Sciences, Engineering, and Medicine document the evidence-based consensus on the study's statement of task by an authoring committee of experts. Reports typically include findings, conclusions, and recommendations based on information gathered by the committee and the committee's deliberations. Each report has been subjected to a rigorous and independent peer-review process and it represents the position of the National Academies on the statement of task.

Proceedings published by the National Academies of Sciences, Engineering, and Medicine chronicle the presentations and discussions at a workshop, symposium, or other event convened by the National Academies. The statements and opinions contained in proceedings are those of the participants and are not endorsed by other participants, the planning committee, or the National Academies.

For information about other products and activities of the National Academies, please visit www.nationalacademies.org/about/whatwedo.

COMMITTEE ON A RESEARCH STRATEGY FOR OCEAN-BASED CARBON DIOXIDE REMOVAL AND SEQUESTRATION

Scott C. Doney, *Chair,* University of Virginia, Charlottesville
Holly Buck, University at Buffalo, NY
Ken Buesseler, Woods Hole Oceanographic Institution, Woods Hole, MA
M. Debora Iglesias-Rodriguez, University of California, Santa Barbara
Kathryn Moran, University of Victoria, BC
Andreas Oschlies, GEOMAR Helmholtz Centre for Ocean Research Kiel, Germany
Phil Renforth, Heriot-Watt University, Edinburgh, UK
Joe Roman, University of Vermont, Burlington
Gaurav N. Sant, University of California, Los Angeles
David A. Siegel, University of California, Santa Barbara
Romany Webb, Columbia Law School, New York, NY
Angelicque White, University of Hawai'i at Manoa, Honolulu

Staff

Kelly Oskvig, Senior Program Officer, Ocean Studies Board
Bridget McGovern, Research Associate, Ocean Studies Board
Trent Cummings, Senior Program Assistant, Ocean Studies Board, until July 2021
Elizabeth Costa, Program Assistant, Ocean Studies Board

OCEAN STUDIES BOARD

Larry A. Mayer (NAE), *Outgoing Chair*, University of New Hampshire, Durham
Claudia Benitez-Nelson, *Incoming Chair,* University of South Carolina, Columbia
Mark Abbott, Woods Hole Oceanographic Institution, Woods Hole, Massachusetts
Carol Arnosti, University of North Carolina, Chapel Hill
Lisa Campbell, Duke University, Durham, North Carolina
Thomas S. Chance, ASV Global, LLC (ret.), Broussard, Louisiana
Daniel Costa, University of California, Santa Cruz
John Delaney, University of Washington (ret.), Seattle
Scott Glenn, Rutgers University, New Brunswick, New Jersey
Patrick Heimbach, University of Texas, Austin
Marcia Isakson, University of Texas, Austin
Lekelia Jenkins, Arizona State University, Tempe
Nancy Knowlton (NAS), Smithsonian Institution (ret.), Washington, District of Columbia
Anthony MacDonald, Monmouth University, West Long Branch, New Jersey
Thomas Miller, University of Maryland, Solomons
S. Bradley Moran, University of Alaska, Fairbanks
Ruth M. Perry, Shell Exploration & Production Company, Houston, Texas
James Sanchirico, University of California, Davis
Mark J. Spalding, The Ocean Foundation, Washington, District of Columbia
Richard Spinrad, Oregon State University, Corvallis
Robert S. Winokur, Michigan Tech Research Institute, Silver Spring, Maryland

OSB Staff Members

Susan Roberts, Director
Stacee Karras, Senior Program Officer
Kelly Oskvig, Senior Program Officer
Emily Twigg, Senior Program Officer
Vanessa Constant, Associate Program Officer
Megan May, Associate Program Officer
Alexandra Skrivanek, Associate Program Officer
Bridget McGovern, Research Associate
Shelly-Ann Freeland, Financial Business Partner
Thanh Nguyen, Financial Business Partner
Trent Cummings, Senior Program Assistant, until July 2021
Kenza Sidi-Ali-Cherif, Program Assistant
Grace Callahan, Program Assistant
Elizabeth Costa, Program Assistant

Preface

Over the past several hundred years, society's expanding consumption of fossil fuels and extensive alteration of the terrestrial biosphere has led to a dramatic rise in atmospheric levels of carbon dioxide (CO_2) and other greenhouse gases. The resulting global climate change is one of the most pressing issues facing society today. Slowing human CO_2 emissions is particularly challenging because fossil fuel use is embedded widely in our modern energy system and economy. Thus, a broad net is being cast searching for a portfolio of solutions to decarbonize the economy and perhaps even actively remove and safely sequester CO_2 away from the atmosphere.

More than a half century ago, Revelle and Suess[1] wrote a pioneering study on the ocean's role in removing from the atmosphere excess CO_2 due to human emissions. Their paper was published at a time when there was quite limited scientific information on the ocean carbon system. It was published just months before the start of the iconic atmosphere CO_2 time series by David Keeling in 1957 at Mauna Loa, Hawai'i, that provides one of the most robust constraints on the fate of human CO_2 emissions. Based on decades of subsequent ocean science and carbon cycle research, modern estimates clearly indicate that natural ocean processes act to remove from the atmosphere about a quarter of human CO_2 emissions from fossil fuel consumption and deforestation. Thus, the ocean already provides an invaluable service slowing the atmospheric growth of CO_2 and associated climate change, though at the cost of rising levels of ocean acidification.

The predominant long-term fate of excess CO_2 from human emissions is to end up in the ocean over centuries to millennia. This raises the question of whether society could (and should) attempt to accelerate ocean processes that remove and store CO_2 away from the atmosphere. Numerous approaches for deliberate ocean carbon dioxide removal (CDR), ranging across biological and geochemical methods to more industrial techniques, have been proposed by scientists, engineers, and technologists. As described in the body of this report, there remain crucial unresolved questions regarding many aspects of ocean CDR, and this report provides an overview of our current state of understanding and a possible research path forward to resolve these major knowledge gaps.

[1] Revelle, R., and H.E. Suess, 1957: Carbon dioxide exchange between atmosphere and ocean and the question of an increase of atmospheric CO2 during the past decades, Tellus, 9(1), 18-27, https://doi.org/10.1111/j.2153-3490.1957.tb01849.x.

This is a report on a research agenda to better inform future societal decisions on ocean CDR; the Committee is not advocating either for or against possible future ocean CDR deployments, and the Committee recognizes that ocean CDR would, at best, complement the role of climate mitigation approaches including decarbonization.

This report builds heavily on previous National Academies of Sciences, Engineering, and Medicine studies, in particular the rationale and framing for research on CDR provided in the 2015 report on *Climate Intervention: Carbon Dioxide Removal and Reliable Sequestration*. The ocean CDR report here also adds to the more terrestrial focus of the 2019 report on *Negative Emissions Technologies and Reliable Sequestration: A Research Agenda*.

The report benefited greatly from many insightful presentations given by external speakers and the participants at the Committee's public meetings and workshop. Because of COVID-19, the Committee's work was completed under the unusual conditions of virtual-only meetings, a challenge compounded by a relatively short time span for the report. I want to extend a special thanks to the Committee members and National Academies staff for their dedication, energy, and thoughtful discussion and contributions over the past year.

Scott Doney
Chair, Committee on A Research Strategy for
Ocean Carbon Dioxide Removal and Sequestration

Acknowledgments

This Consensus Study Report was reviewed in draft form by individuals chosen for their diverse perspectives and technical expertise. The purpose of this independent review is to provide candid and critical comments that will assist the National Academies of Sciences, Engineering, and Medicine in making each published report as sound as possible and to ensure that it meets the institutional standards for quality, objectivity, evidence, and responsiveness to the study charge. The review comments and draft manuscript remain confidential to protect the integrity of the deliberative process.

We thank the following individuals for their review of this report:

Trisha Atwood, Utah State University, Logan
Miranda Boettcher, German Institute for International and Security Affairs, Berlin, Germany
Philip Boyd, University of Tasmania, Hobart, Australia
Wil Burns, Northwestern University, Evanston, IL
Ken Caldeira, Carnegie Institution for Science, Washington, D.C.
Ik Kyo Chung, Pusan National University, Busan, South Korea
Sarah Cooley, Ocean Conservancy, Washington, D.C.
Matthew Eisaman, Stony Brook University, NY
Winston Ho, The Ohio State University, Columbus
Bo Barker Jørgensen, Aarhus University, Denmark
Ricardo Letelier, Oregon State University, Corvallis
Shelly D. Minteer, University of Utah, Salt Lake City
Benjamin Saenz, Biota.Earth, Berkely, CA

Although the reviewers listed above provided many constructive comments and suggestions, they were not asked to endorse the conclusions or recommendations of this report nor did they see the final draft before its release. The review of this report was overseen by **Mark Barteau,** Texas

A&M University, and **Jim Yoder,** Woods Hole Oceanographic Institution. They were responsible for making certain that an independent examination of this report was carried out in accordance with the standards of the National Academies and that all review comments were carefully considered. Responsibility for the final content rests entirely with the authoring Committee and the National Academies.

During the study process, we had a number of speakers both at committee meetings and at a series of workshops. We thank the following individuals for their contributions during the study process: Brad Ack, Jess Adkins, Mark Preston Aragones, Trisha B. Atwood, David Babson, Lennart Bach, Philip Boyd, Ellen Briggs, Wil Burns, Ken Caldeira, Fei Chai, Francisco Chavez, Bill Collins, Sarah Cooley, Emily Cox, Andrew Dickson, Carlos M. Duarte, Matthew Eisaman, Julio Friedmann, Halley E. Froehlich, Oliver Geden, Dwight Gledhill, Olavur Gregersen, Nicolas Gruber, Jill Hamilton, Barbara Haya, Stephanie Henson, Joseph Hezir, Baerbel Hoenisch, K. John Holmes, Anna-Maria Hubert, Xabier Irigoien, Nick Kamenos, David P. Keller, David Koweek, Dorte Krause-Jensen, Tim Kruger, Ricardo Letelier, Catherine Lovelock, Niall MacDowell, Filip Meysman, Juan Moreno-Cruz, David P. Morrow, Ryan Orbuch, Andy Pershing, Albert J. Plueddemann, Greg Rau, Miles Richardson, Ros Rickaby, Kate Ricke, Andy Ridgwell, Ulf Riebesell, Joellen Russell, Christopher L. Sabine, Terre Satterfield, Raymond Schmitt, Gyami Shrestha, Lisa Suatoni, Shuchi Talati, Chris Vivian, Brian von Herzen, Marc von Keitz, George Waldbusser, Heather Willauer, Phillip Williamson, Richard Zeebe, and Robert Zeller.

Contents

Summary

As of 2021, atmospheric carbon dioxide (CO_2) levels have reached historically unprecedented levels, higher than at any time in the past 800,000 years. The increase is incontrovertibly due to anthropogenic CO_2 emissions from activities such as fossil fuel burning, agriculture, and historical land-use change. The current level of human emissions greatly exceeds the ability of nature to remove CO_2—simply reducing the levels of human emissions may not be enough to stabilize the climate. Carbon dioxide removal (CDR), sometimes referred to as negative emissions technologies, may prove valuable, in conjunction with reduced emissions to meet the global goal of limiting warming to well below 2°C, comparable to preindustrial levels, as established by the Paris Agreement.[1]

The 2015 National Academies report, *Climate Intervention: Carbon Dioxide Removal and Reliable Sequestration*, concluded that, to contribute to climate change mitigation, CDR approaches would need to be scaled up massively and that it is critical to begin research now to increase the viability and affordability of existing or new approaches to CDR. In response, the National Academies released a report in 2019 to provide a research agenda for advancing CDR and, specifically, for assessing the benefits, risks, and sustainable scale potential for a variety of land- and coastal-based CDR approaches. The study found that, to meet climate goals, some form of CDR will likely be needed to remove roughly 10 Gt CO_2/yr by mid-century and 20 Gt CO_2/yr by the end of the century. To help meet that goal, four land-based CDR approaches are ready for large-scale deployment: afforestation/reforestation, changes in forest management, uptake and storage by agricultural soils, and bioenergy with carbon capture and storage, based on the potential to remove carbon at costs below \$100/t CO_2. The 2019 report did not examine the more global ocean-based approaches but did recognize the potential for ocean-based CDR and the need for a research strategy to explore these options.

To address this gap in understanding and the need for further exploration into CDR options that could feasibly contribute to a larger climate mitigation strategy, with sponsorship from the ClimateWorks Foundation, the National Academies convened the Committee on a Research Strategy

[1] See https://unfccc.int/process-and-meetings/the-paris-agreement/the-paris-agreement.

for Ocean-Based Carbon Dioxide Removal and Sequestration. Specifically, this committee was assembled to develop a research agenda to assess the benefits, risks, and potential for responsible scale-up of a range of six specific ecosystem-based and technological ocean-based CDR (or ocean CDR) approaches, as defined by the sponsor. The resulting research agenda is meant to provide an improved and unbiased knowledge base for the public, stakeholders, and policy makers to make informed decisions on the next steps for ocean CDR, not to lock in or advocate for any particular approach. The committee's Statement of Task is presented in Box S.1. The committee's deliberations and report writing was informed by review of scientific and social science literature and by a series of public workshops and presentations, drawing expertise from the academic, governmental, and nongovernmental communities.

THE OCEAN AS A STRATEGY

The ocean covers 70 percent of Earth's surface; it includes much of the global capacity for natural carbon sequestration, and it may be possible to enhance that capacity through implementation of ocean-based CDR approaches. The ocean holds great potential for uptake and longer-term sequestration of anthropogenic CO_2 for several reasons: (1) the ocean acts as a large natural reservoir for CO_2, holding roughly 50 times as much inorganic carbon as the preindustrial atmosphere; (2) the ocean already removes a substantial fraction of the excess atmospheric CO_2 resulting from human emissions; and (3) a number of physical, geochemical, and biological processes are known to influence air–sea CO_2 gas exchange and ocean carbon storage.

By acting to remove CO_2 from the atmosphere and upper ocean and then store the excess carbon either in marine or geological reservoirs for some period of time, ocean CDR approaches could complement CO_2 emission reductions and contribute to the portfolio of climate response strategies needed to limit climate change and surface ocean acidification over coming decades and centuries. While rapid and extensive decarbonization and emissions abatement of other greenhouse gases in the United States and global economies are the primary action required to meet international climate goals, ocean-based and other CDR approaches could help balance difficult-to-mitigate human CO_2 emissions and contribute to mid-century to late-century net-zero CO_2 emission targets.

It is critical that ocean CDR approaches be assessed against the consequences of no action and as one component of a broad and integrated climate mitigation strategy. Without substantial decarbonization, emissions abatement, and potential options such as CDR, atmospheric CO_2 growth will continue unabated with associated rising impacts from climate change and ocean acidification, putting marine ecosystems at risk.

TECHNOLOGIES CONSIDERED

As directed by the Statement of Task (Box S.1), the committee examined six groups of ocean-based CDR approaches, depicted in Figure S.1:

- **Nutrient fertilization** (Chapter 3): Addition of micronutrients (e.g., iron) and/or macronutrients (e.g., phosphorus or nitrogen) to the surface ocean may in some settings increase photosynthesis by marine phytoplankton and can thus enhance uptake of CO_2 and transfer of organic carbon to the deep sea where it can be sequestered for timescales of a century or longer. As such, nutrient fertilization essentially locally enhances the natural ocean biological carbon pump using energy from the sun, and in the case of iron, relatively small amounts are needed.
- **Artificial upwelling and downwelling** (Chapter 4): Artificial upwelling is a process whereby water from depths that are generally cooler and more nutrient and CO_2 rich than

BOX S.1
Statement of Task

With the goal of reducing atmospheric carbon dioxide, an ad hoc committee will conduct a study exclusively focused on carbon dioxide removal and sequestration conducted in coastal and open ocean waters to:

A. Identify the most urgent unanswered scientific and technical questions, as well as questions surrounding governance, needed to: (i) assess the benefits, risks, and potential scale for carbon dioxide removal and sequestration approaches; and (ii) increase the viability of responsible carbon dioxide removal and sequestration;
B. Define the essential components of a research and development program and specific steps that would be required to answer these questions;
C. Estimate the costs and potential environmental impacts of such a research and development program to the extent possible in the timeframe of the study.
D. Recommend ways to implement such a research and development program that could be used by public or private organizations.

The carbon dioxide removal approaches to be examined include:

- Iron, nitrogen, or phosphorus fertilization
- Artificial upwelling and downwelling
- Seaweed cultivation
- Recovery of ocean and coastal ecosystems, including large marine organisms
- Ocean alkalinity enhancement
- Electrochemical approaches.

surface waters is pumped into the surface ocean. Artificial upwelling has been suggested as a means to generate increased localized primary production and ultimately export production and net CO_2 removal. Artificial downwelling is the downward transport of surface water; this activity has been suggested as a mechanism to counteract eutrophication and hypoxia in coastal regions by increasing ventilation below the pycnocline and as a means to carry carbon into the deep ocean.

- **Seaweed cultivation** (Chapter 5): The process of producing macrophyte organic carbon biomass via photosynthesis and transporting that carbon into a carbon reservoir removes CO_2 from the upper ocean. Large-scale farming of macrophytes (seaweed) can act as a CDR approach by transporting organic carbon to the deep sea or into sediments.
- **Recovery of ocean and coastal ecosystems** (Chapter 6): CDR and sequestration through protection and restoration of coastal ecosystems, such as kelp forests and free-floating *Sargassum,* and the recovery of fishes, whales, and other animals in the oceans.
- **Ocean alkalinity enhancement** (Chapter 7): Chemical alteration of seawater chemistry via addition of alkalinity through various mechanisms including enhanced mineral weathering and electrochemical or thermal reactions releasing alkalinity to the ocean, with the ultimate aim of removing CO_2 from the atmosphere.
- **Electrochemical approaches** (Chapter 8): Removal of CO_2 or enhancement of the storage capacity of CO_2 in seawater (e.g., in the form of ions or mineral carbonates) by enhancing its acidity or alkalinity, respectively. These approaches exploit the pH-dependent solubility of CO_2 by passage of an electric current through water, which by inducing water splitting

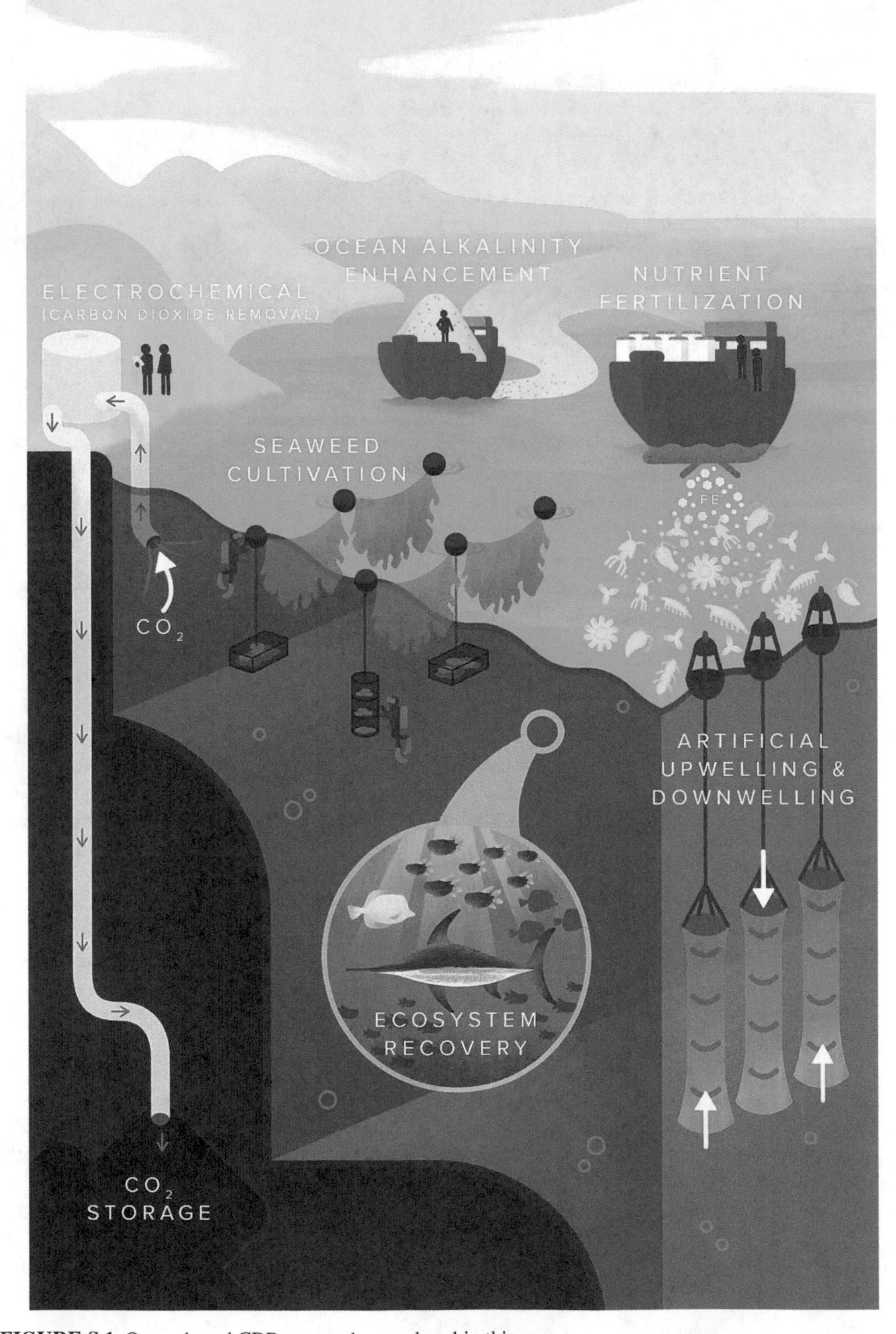

FIGURE S.1 Ocean-based CDR approaches explored in this report.

("electrolysis"), changes its pH in a confined reaction environment. As one example, ocean alkalinity enhancement may be accomplished by electrochemical approaches.

The assessment of these six ocean-based approaches is a representative sampling of ocean-based CDR and is not intended to be an exhaustive list. The application of the recommendations developed within the report can be extended to ocean CDR approaches broadly.

CDR POTENTIAL

To assess the potential of each of the six ocean CDR approaches as a viable path forward in a larger climate mitigation strategy, the committee used a variety of information sources including a review of the scientific literature and a series of public meetings held in the virtual setting with presentations by more than 65 experts from academic, governmental, and nongovernmental communities (see Appendix B for a list of experts invited to speak to the committee) to understand stakeholder interest, and to explore the current state of knowledge, potential, and limitations of ocean CDR approaches.

Each of the ocean-based CDR approaches was evaluated against a common set of criteria, where feasible, developed by the committee based on specific elements included in the Statement of Task and from a review of previous planning and synthesis documents on CDR. The criteria investigated include knowledge base, efficacy, durability, scale, monitoring and verification, viability and barriers, and governance and social dimensions.

The approaches were evaluated and given a ranking of low, medium, or high, along with a level of certainty, where appropriate. The committee's evaluation is summarized in Table S.1. Across all approaches, knowledge gaps remain in determining carbon sequestration efficacy, scaling, and durability, as well as environmental and social impacts and costs.

Common Challenges of Ocean CDR

Knowledge: The knowledge base is inadequate, based in many cases only on laboratory-scale experiments, conceptual theory and/or numerical models and needs to be expanded to better understand risks and benefits to responsibly scale up any of the ocean-based CDR approaches.

Governance: Social and regulatory acceptability is likely to be a barrier to many ocean CDR approaches, particularly ones requiring industrial infrastructure. There will be both project-specific and approach-specific social, political, and regulatory discussions, as well as contestation around the role of CDR broadly. Field-scale trials are likely to be a site of wider societal debate around decarbonization and climate response strategies.

Unknown environmental and social impacts: All ocean-based CDR approaches will modify the marine environment in some way, with both intended and unintended impacts. However, the knowledge base is weak on the unintended impacts and the consequences of both intended and unintended CDR impacts on marine ecosystems and coastal human communities.

Monitoring and verification: Monitoring and verification activities are essential to quantify the efficacy and the durability of carbon storage of ocean-based CDR approaches and to identify environmental and social impacts. Potential synergies may exist with other ocean and environmental or climate observing systems. Substantial challenges remain, however, particularly for observing impacts on marine organisms and the resulting implications for marine ecosystems as well as documenting regional- to global-scale impacts on ocean carbon storage.

TABLE S.1 Summary of Ocean CDR Scale-Up Potential

	Ocean Nutrient Fertilization	Artificial Upwelling/ Downwelling	Seaweed Cultivation	Ecosystem Recovery	Ocean Alkalinity Enhancement	Electrochemical Processes
Knowledge base What is known about the system (low, mostly theoretical, few in situ experiments; medium, lab and some fieldwork, few carbon dioxide removal (CDR) publications; high, multiple in situ studies, growing body of literature)	**Medium–High** Considerable experience relative to any other ocean CDR approach with strong science on phytoplankton growth in response to iron, less experience on fate of carbon and unintended consequences. Natural iron-rich analogs provide valuable insight on larger temporal and spatial scales.	**Low–Medium** Various technologies have been demonstrated for artificial upwelling (AU), although primarily in coastal regimes for short duration. Uncertainty is high and confidence is low for CDR efficacy due to upwelling of CO_2, which may counteract any stimulation of the biological carbon pump (BCP).	**Medium–High** Science of macrophyte biology and ecology is mature; many mariculture facilities are in place globally. Less is known about the fate of macrophyte organic carbon and methods for transport to deep ocean or sediments.	**Low–Medium** There is abundant evidence that marine ecosystems can uptake large amounts of carbon and that anthropogenic impacts are widespread, but quantifying the collective impact of these changes and the CDR benefits of reversing them is complex and difficult.	**Low–Medium** Seawater CO_2 system and alkalinity thermodynamics are well understood. Need for empirical data on alkalinity enhancement; currently, knowledge is based on modeling work. Uncertainty is high for possible impacts.	**Low–Medium** Processes are based on well-understood chemistry with a long history of commercial deployment, but is yet to be adapted for CO_2 removal by ocean alkalinity enhancement (OAE) beyond benchtop scale.
Efficacy What is the confidence level that this approach will remove atmospheric CO_2 and lead to net increase in ocean carbon storage (low, medium, high)	**Medium–High Confidence** BCP known to work and productivity enhancement evident. Natural systems have higher rates of carbon sequestration in response to iron but low efficiencies seen thus far would limit effectiveness for CDR.	**Low Confidence** Upwelling of deep water also brings a source of CO_2 that can be exchanged with the atmosphere. Modeling studies generally predict that large-scale AU would not be effective for CDR.	**Medium Confidence** The growth and sequestration of seaweed crops should lead to net CDR. Uncertainties about how much existing net primary production (NPP) and carbon export downstream would be reduced due to large-scale farming.	**Low–Medium Confidence** Given the diversity of approaches and ecosystems, CDR efficacy is likely to vary considerably. Kelp forest restoration, marine protected areas, fisheries management, and restoring marine vertebrate carbon are promising tools.	**High Confidence** Need to conduct field deployments to assess CDR, alterations of ocean chemistry (carbon but also metals), how organic matter can impact aggregation, etc.	**High Confidence** Monitoring within an enclosed engineered system, CO_2 stored either as increased alkalinity, solid carbonate, or aqueous CO_2 species. Additionality possible with the utilization of by-products to reduce carbon intensity.

Durability Will it remove CO_2 durably away from surface ocean and atmosphere (low, <10 years; medium, >10 years and <100 years; high, >100 years), and what is the confidence (low, medium, high)	**Medium** 10–100 years Depends highly on location and BCP efficiencies, with some fraction of carbon flux recycled faster or at shallower ocean depths; however, some carbon will reach the deep ocean with >100-year horizons for return of excess CO_2 to surface ocean.	**Low–Medium** <10–100 years As with ocean iron fertilization (OIF), dependent on the efficiency of the BCP to transport carbon to deep ocean.	**Medium–High** >10–100 years Dependent on whether the sequestered biomass is conveyed to appropriate sites (e.g., deep ocean with slow return time of waters to surface ocean).	**Medium** 10–100 years The durability of ecosystem recovery ranges from biomass in macroalgae to deep-sea whale falls expected to last >100 years.	**Medium–High** >100 years Processes for removing added alkalinity from seawater generally quite slow; durability not dependent simply on return time of waters with excess CO_2 to ocean surface.	**Medium–High** >100 years Dynamics similar to OAE.
Scalability Potential scalability at some future date with global-scale implementation (low, <0.1 Gt CO_2/yr; medium, >0.1 Gt CO_2/yr and <1.0 Gt CO_2/yr; high, >1.0 Gt CO_2/yr), and what is the confidence level (low, medium, high)	**Medium–High** Potential C removal >0.1–1.0 Gt CO_2/yr (medium confidence) Large areas of ocean have high-nutrient, low-chlorophyll conditions suitable to sequester >1 Gt CO_2/yr. Co-limitation of macronutrients and ecological impacts at large scales are likely. Low-nutrient, low-chlorophyll areas have not been explored to increase areas of possible deployment. (Medium confidence based on 13 field experiments.)	**Medium** Potential C removal >0.1 Gt CO_2/yr and <1.0 Gt CO_2/yr (low confidence) Could be coupled with aquaculture efforts. Would require pilot trials to test materials durability for open ocean and assess CDR potential. Current model predictions would require deployment of tens of millions to hundreds of millions of pumps to enhance carbon sequestration. (Low confidence that this large-scale deployment would lead to permanent and durable CDR.)	**Medium** Potential C removal >0.1 Gt CO_2/yr and <1.0 Gt CO_2/yr (medium confidence) Farms need to be many million hectares, which creates many logistic and cost issues. Uncertainties about nutrient availability and durability of sequestration, seasonality will limit sites, etc.	**Low–Medium** Potential C removal <0.1–1.0 Gt CO_2/yr (low–medium confidence) Given the widespread degradation of much of the coastal ocean, there are plenty of opportunities to restore ecosystems and depleted species. However, ecosystems and trophic interactions are complex and changing and research will be necessary to explore upper limits.	**Medium–High** Potential C removal >0.1–1.0 Gt CO_2/yr (medium confidence) Potential for sequestering >1 Gt CO_2/yr if applied globally. High uncertainty coming from potential aggregation and export to depth of added minerals and unintended chemical impacts of alkalinity addition.	**Medium–High** Potential C removal >0.1–1.0 Gt CO_2/yr (medium confidence) Energy and water requirements may limit scale. For climate relevancy, the scale will be double to an order of magnitude greater than the current chlor-alkali industry.

continued

TABLE S.1 Continued

	Ocean Nutrient Fertilization	Artificial Upwelling/ Downwelling	Seaweed Cultivation	Ecosystem Recovery	Ocean Alkalinity Enhancement	Electrochemical Processes
Environmental risk Intended and unintended undesirable consequences at scale (unknown, low, medium, high), and what is the confidence level (low, medium, high)	**Medium** (low–medium confidence) Intended environmental impacts increase NPP and carbon sequestration due to changes in surface ocean biology. If effective, there are deep-ocean impacts and concern for undesirable geochemical and ecological consequences. Impacts at scale uncertain.	**Medium–High** (low confidence) Similar impacts to OIF but upwelling also affects the ocean's density field and sea-surface temperature and brings likely ecological shifts due to bringing colder, inorganic carbon- and nutrient-rich waters to surface.	**Medium–High** (low confidence) Environmental impacts are potentially detrimental especially on local scales where seaweeds are farmed (i.e., nutrient removal due to farming will reduce NPP, carbon export, and trophic transfers) and in the deep ocean where the biomass is sequestered (leading to increases in acidification, hypoxia, eutrophication, and organic carbon inputs). The scale and nature of these impacts are highly uncertain.	**Low** (medium–high confidence) Environmental impacts would be generally viewed as positive. Restoration efforts are intended to provide measurable benefits to biodiversity across a diversity of marine ecosystems and taxa.	**Medium** (low confidence) Possible toxic effect of nickel and other leachates of olivine on biota, bio-optical impacts, removal of particles by grazers, unknown responses to increased alkalinity on functional diversity and community composition. Effects also from expanded mining activities (on land) on local pollution, CO_2 emissions.	**Medium–High** (low confidence) Impact on the ocean is possibly constrained to the point of effluent discharge. Poorly-known possible ecosystem impacts similar to alkalinity enhancement. Excess acid (or gases, particularly chlorine) will need to be treated and safely disposed. Provision of sufficient electrical power will likely have remote impacts.

Social considerations Encompass use conflicts, governance-readiness, opportunities for livelihoods, etc.	Potential conflicts with other uses of high seas and protections; downstream effects from displaced nutrients will need to be considered; legal uncertainties; potential for public acceptability and governance challenges (i.e., perception of "dumping").	Potential conflicts with other uses (shipping, marine protected areas, fishing, recreation); potential for public acceptability and governance challenges (i.e., perception of dumping).	Possibility for jobs and livelihoods in seaweed cultivation; potential conflicts with other marine uses. Downstream effects from displaced nutrients will need to be considered.	Trade-offs in marine uses to enhance ecosystem protection and recovery. Social and governance challenges may be less significant than with other approaches.	Expansion of mining production, with public health and economic implications; general public's potential for public acceptability and governance challenges (e.g., if perceived as "dumping").	Similar to OAE and to any industrial site. Substantial electrical power demand may generate social impacts.
Co-benefits How significant are the co-benefits as compared to the main goal of CDR and how confident is that assessment	**Medium** (low confidence) Enhanced fisheries possible but not shown and difficult to attribute. Seawater dimethyl sulfide increase seen in some field studies that could enhance climate cooling impacts. Surface ocean decrease in ocean acidity possible.	**Medium–High** (low confidence) May be used as a tool in coordination with localized enhancement of aquaculture and fisheries.	**Medium–High** (medium confidence) Placing cultivation facilities near fish or shellfish aquaculture facilities could help alleviate environmental damages from these activities. Bio-fuels also possible.	**High** (medium–high confidence) Enhanced biodiversity conservation and the restoration of many ecological functions and ecosystem services damaged by human activities. Existence, spiritual, and other non-use values. Potential to enhance marine stewardship and tourism.	**Medium** (low confidence) Mitigation of ocean acidification; positive impact on fisheries.	**Medium–High** (medium confidence) Mitigation of ocean acidification; production of H_2, Cl_2, silica.

continued

TABLE S.1 Continued

	Ocean Nutrient Fertilization	Artificial Upwelling/ Downwelling	Seaweed Cultivation	Ecosystem Recovery	Ocean Alkalinity Enhancement	Electrochemical Processes
Cost of scale-up Estimated costs in dollars per metric ton CO_2 for future deployment at scale; does not include all of monitoring and verification costs needed for smaller deployments during R&D phases (low, <$50/t CO_2; medium, ~$100/t CO_2; high, >>$150/t CO_2) and confidence in estimate (low, medium, high)	**Low** <$50/t CO_2 (low–medium confidence) Deployment of <$25/t CO_2 sequestered for deployment at scale is possible, but needs to be demonstrated at scale	**Medium–High**. >$100–$150/t CO_2 (low confidence) Development of a robust monitoring program is the likely largest cost and would be of similar magnitude as OIF. Materials costs for pump assembly could be moderate for large-scale persistent deployments. Estimates for a kilometer-scale deployment are in the tens of million dollars.	**Medium** ~$100/t CO_2 (medium confidence) Costs should be less than $100/t CO_2. No direct energy used to fix CO_2.	**Low** <$50/t CO_2 (medium confidence) Varies, but direct costs would largely be for management and opportunity costs for restricting uses of marine species and the environment. No direct energy used.	**Medium–High** >$100–$150/t CO_2 (low–medium confidence) Cost estimates range between tens of dollars and $160/t CO_2. Need for expansion of mining, transportation, and ocean transport fleet.	**High** >$150/t CO_2 (medium confidence) Gross current estimates $150–$2,500/t CO_2 removed. With further R&D, it may be possible to reduce this to <$100/t CO_2.
Cost and challenges of carbon accounting Relative cost and scientific challenge associated with transparent and quantifiable carbon tracking (low, medium, high)	**Medium** Challenges tracking additional local carbon sequestration and impacts on carbon fluxes outside of boundaries of CDR application (additionality).	**High** Local and additionality monitoring needed for carbon accounting similar to OIF.	**Low–Medium** The amount of harvested and sequestered carbon will be known. However, an accounting of the carbon cycle impacts of the displaced nutrients will be required (additionality).	**High** Monitoring net effect on carbon sequestration is challenging.	**Low–Medium** Accounting more difficult for addition of minerals and non-equilibrated addition of alkalinity, than equilibrated addition.	**Low–Medium**

Cost of environmental monitoring Need to track impacts beyond carbon cycle on marine ecosystems (low, medium, high)	**Medium** (medium–high confidence) All CDR will require monitoring for intended and unintended consequences both locally and downstream of CDR site, and these monitoring costs may be substantial fraction of overall costs during R&D and demonstration-scale field projects. This cost of monitoring for ecosystem recovery may be lower.					
Additional resources needed Relative low, medium, high to primary costs of scale-up	**Low–Medium** Cost of material: iron is low and energy is sunlight.	**Medium–High** Materials, deployment, and potential recovery costs.	**Medium** Farms will require large amounts of ocean (many million hectares) to achieve CDR at scale.	**Low** Most recovery efforts will likely require few materials and little energy, though enforcement could be an issue. Active restoration of kelp and other ecosystems would require more resources.	**Medium–High** Adaptation and likely expansion of existing fleet for deployment; infrastructure for storage at ports. Infrastructure support for expansion of mineral extraction, processing, transportation, and deployment.	**Medium–High** High energy requirements (1–2.5 MWh/t CO_2 removed) and build-out of industrial CDR.

Cost: Accurate estimation of the cost of a CDR approach at low technological readiness is challenging, and costs presented come with considerable uncertainty. It is typical for early-stage assessments to underestimate costs, and for that reason some recommend the inclusion of capital cost contingencies over 100 percent (effectively doubling the calculated capital cost). Cost discovery will be an important feature of a research strategy that aims to investigate approaches through increasing technology readiness.

RESEARCH RECOMMENDATIONS

Expanded research including field research is needed to assess ocean-based CDR techniques' potential efficacy in removing and sequestering excess carbon away from the atmosphere and the permanence or durability of the sequestered carbon on timescales relevant to societal policy decisions. Research is also needed to identify and quantify environmental impacts, risks, benefits, and co-benefits as well as other factors governing possible decisions on deployment such as technological readiness, development timelines, energy and resource needs, economic costs, and potential social, policy, legal, and regulatory considerations. Additionally, research on ocean CDR would greatly benefit from targeted studies on the interactions and trade-offs between ocean CDR, terrestrial CDR, greenhouse gas abatement and mitigation, and climate adaptation, including the potential of mitigation deterrence.

The specific research needed to advance understanding of ocean CDR is listed in Tables S.2 and S.3, including associated time-frame and cost estimates. Table S.2 summarizes the foundational research identified in Chapters 2 and 9 as research priorities common across ocean CDR approaches including potential social, policy, legal, and regulatory considerations. The research included in Table S.2 is meant to inform the framework for any future ocean-based CDR effort. Table S.3 then summarizes research needed to better understand the feasibility of that particular approach, with bolded text indicating priority. Additional details on the research activities listed in Table S.2 and S.3 can be found in the corresponding chapters.

Early research findings might indicate a low viability for particular approaches. The research agenda below is to be adaptive, meaning that decisions on future investments in research activities will need to take into account new findings on the efficacy and durability of a technique, whether the social and environmental impacts outweigh benefits or face social and governance challenges. Generally speaking, showstoppers can be anticipated for some approaches from factors both internal and external to the research. Internal showstoppers include findings that indicate that the viability is so low as to not warrant further research investments. There may also be external showstoppers to the research, such as lack of social license or governance challenges that preclude further investigation. A conceptual diagram depicting how the research program could start and evolve is shown in Figure S.2.

The research needs, Tables S.2 and S.3, are presented within the context of adhering to Recommendations 1, 2, and 3 (below). Those research needs shown in bold in Table S.3 are identified as priorities for taking the next steps to advance understanding of that particular approach while the elements in Table S.2 lay the framework for ocean CDR broadly. Recommendation 1 includes elements that should be included in any ocean CDR research program. Recommendation 2 prescribes components common to implementation of any ocean CDR research program, and Recommendation 3 defines priorities for any ocean CDR research program. Recommendations 1, 2, and 3 are broadly applicable to any ocean-based CDR approach; they are not limited to the six approaches explored in this report.

Recommendation 1: Ocean CDR Research Program Goals. To inform future societal decisions on a broad climate response mitigation portfolio, a research program for ocean CDR should

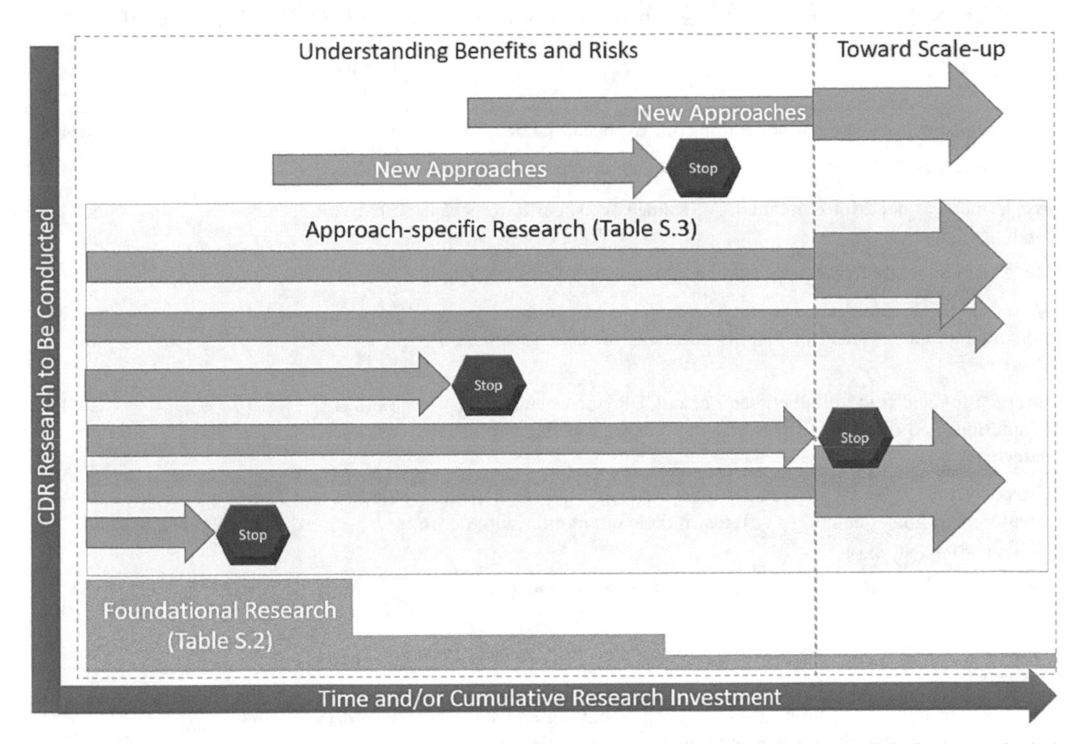

FIGURE S.2 Conceptual timeline of ocean-based CDR research based on Tables S.2 and S.3. Stops included on the diagram represent possible internal and external showstoppers or barriers to a particular approach.

be implemented, in parallel across multiple approaches, to address current knowledge gaps. The research program should not advocate for or lock in future ocean CDR deployments but rather provide an improved and unbiased knowledge base for the public, stakeholders, and policy makers. Funding for this research could come from both the public and private sectors, and collaboration between the two is encouraged. The integrated research program should include the following elements:

1. Assessment of whether the approach removes atmospheric CO_2, in net, and the durability of the CDR, as a primary goal.
2. Assessment of intended and unintended environmental impacts beyond CDR.
3. Assessment of social and livelihood impacts, examining both potential harms and benefits.
4. Integration of research on social, legal, regulatory, policy, and economic questions relevant to ocean CDR research and possible future deployment with the natural science, engineering, and technological aspects.
5. Systematic examination of the biophysical and social interactions, synergies, and tensions between ocean CDR, terrestrial CDR, mitigation, and adaptation.

Common Components

No single research framework will be adequate for all CDR approaches within a comprehensive research strategy, because knowledge base and readiness levels differ substantially. There are, however, several common components that are relevant to research into any ocean CDR approach.

TABLE S.2 Foundational Research Priorities Common to All Ocean-Based CDR

	Estimated Budget	Duration (years)	Total Cost
Model international governance framework for ocean CDR research	\$2M–3M/yr	2–4	\$4M–\$12M
Application of domestic laws to ocean CDR research	\$1M/yr	1–2	\$1M–\$2M
Assessment of need for domestic legal framework specific to ocean CDR	\$1M/yr	2–4	\$2M–\$4M
Development of domestic legal framework specific to ocean CDR			
Mixed-methods, multi-sited research to understand community priorities and assessment of benefits and risks for ocean CDR as a strategy	\$5M/yr	4	\$20M
Interactions and trade-offs between ocean CDR, terrestrial CDR, adaptation, and mitigation, including the potential of mitigation deterrence	\$2M/yr	4	\$8M
Cross-sectoral research analyzing food system, energy, sustainable development goals, and other systems in their interaction with ocean CDR approaches	\$1M/yr	4	\$4M
Capacity-building research fellowship for diverse early-career scholars in ocean CDR	\$1.5M/yr	2	\$3M
Transparent, publicly accessible system for monitoring impacts from projects	\$0.25M/yr	4	\$1M
Research on how user communities (companies buying and selling CDR, nongovernmental organizations, practitioners, policy makers) view and use monitoring data, including certification	\$0.5M/yr	4	\$2M
Analysis of policy mechanisms and innovation pathways, including the economics of scale-up	\$1–2M/yr	2	\$2M–\$4M
Development of standardized environmental monitoring and carbon accounting methods for ocean CDR	\$0.2M/yr	3	\$0.6M
Development of a coordinated research infrastructure to promote transparent research	\$2M/yr	3–4	\$6M–\$8M
Development of a publicly accessible data management strategy for ocean CDR research	\$2–3M/yr	2	\$4M–\$6M
Development of a coordinated plan for science communication and public engagement of ocean CDR research in the context of decarbonization and climate response	\$5M/yr	10	\$50M
Development of a common code of conduct for ocean CDR research	\$1M/yr	2	\$2M
Total Estimated Research Budget (Assumes all 6 CDR approaches moving ahead)	**\$29M/yr**	**2–10**	**\$125M**

TABLE S.3 Research Needed to Advance Understanding of Each Ocean CDR Approach (Bold type identifies priorities for taking the next step to advance understanding of each particular approach; more details on each research need provided in individual chapters)

	Estimated Budget	Duration (years)	Total Budget
Ocean Fertilization			
Carbon sequestration delivery and bioavailability	**$5M/yr**	**5**	**$25M**
Tracking carbon sequestration	**$3M/yr**	**5**	**$15M**
In-field experiments, >100 t Fe and >1,000 km^2 initial patch size followed over annual cycles	**$25M/yr**	**10**	**$250M**
Monitoring carbon and ecological shifts	$10M/yr	10	$100M
Experimental planning and extrapolation to global scales	$5M/yr	10	$50M
Total Estimated Research Budget	$48M/yr	5–10	$440M
Estimated Budget of Research Priorities	**$33M/yr**	**5–10**	**$290M**
Artificial Upwelling and Downwelling			
Technological readiness: Limited and controlled open-ocean trials to determine durability and operability of artificial upwelling technologies (~100 pumps tested in various conditions)	**$5M/yr**	**5**	**$25M**
Feasibility studies	$1M/yr	1	$1M
Tracking carbon sequestration	$3M/yr	5	$15M
Modeling of carbon sequestration based on achievable upwelling velocities and known stoichiometry of deep-water sources. Parallel mesocosm and laboratory experiments to assess potential biological responses to deep water of varying sources.	$5M/yr	5	$25M
Planning and implementation of demonstration-scale in situ experimentation (>1 year, >1,000 km) in region-sited-based input from modeling and preliminary experiments	$25M/yr	10	$250M
Monitoring carbon and ecological shifts	$10M/yr	10	$100M
Experimental planning and extrapolation to global scales (early for planning and later for impact assessments)	$5M/yr	10	$50M
Total Estimated Research Budget	$54/yr	5–10	$466M
Estimated Budget of Research Priorities	**$5M/yr**	**5–10**	**$25M**
Seaweed Cultivation			
Technologies for efficient large-scale farming and harvesting of seaweed biomass	**$15M/yr**	**10**	**$150M**
Engineering studies focused on the conveying of harvested biomass to durable oceanic reservoir with minimal losses of carbon	**$2M/yr**	**10**	**$20M**
Assessment of long-term fates of seaweed biomass and by-products	**$5M/yr**	**5**	**$25M**
Implementation and deployment of a demonstration-scale seaweed cultivation and sequestration system	$10M/yr	10	$100M
Validation and monitoring of the CDR performance of a demonstration-scale seaweed cultivation and sequestration system	$5M/yr	10	$50M
Evaluation of environmental impacts of large-scale seaweed farming and sequestration	**$4M/yr**	**10**	**$40M**
Total Estimated Research Budget	$41M/yr	5–10	$385M
Estimated Budget of Research Priorities	**$26M/yr**	**5**	**$235M**

continued

TABLE S.3 Continued

Ecosystem Recovery			
Restoration ecology and carbon	**$8M/yr**	**5**	**$40M**
Marine protected areas: Do ecosystem-level protection and restoration scale for marine CDR?	**$8M/yr**	**10**	**$80M**
Macroalgae: Carbon measurements, global range, and levers of protection	**$5M/yr**	**10**	**$50M**
Benthic communities: disturbance and restoration	$5M/yr	5	$25M
Marine animals and CO_2 removal	**$5M/yr**	**10**	**$50M**
Animal nutrient-cycling	$5M/yr	5	$25M
Commercial fisheries and marine carbon	$5M/yr	5	$25M
Total Estimated Research Budget	$41M/yr	5–10	$295M
Estimated Budget of Research Priorities	**$26M/yr**	**5–10**	**$220M**
Ocean Alkalinity Enhancement			
Research and development to explore and improve the technical feasibility/and readiness level of ocean alkalinity enhancement approaches (including the development of pilot-scale facilities)	$10M/yr	5	$50M
Laboratory and mesocosm experiments to explore impacts on physiology and functionality of organisms/communities	**$10M/yr**	**5**	**$50M**
Field experiments	**$15M/yr**	**5–10**	**$75M–$150M**
Research into the development of appropriate monitoring and accounting schemes, covering CDR potential and possible side effects	$10	5–10	$50M–$100M
Total Estimated Research Budget	$45M/yr	5–10	$180M–$350M
Estimated Budget of Research Priorities	**$25M/yr**	**5–10**	**$125–$200M**
Electrochemical Processes			
Demonstration projects including CDR verification and environmental monitoring	**$30M/yr**	**5**	**$150M**
Development and assessment of novel and improved electrode and membrane materials	**$10M/yr**	**5**	**$50M**
Assessment of environmental impact and acid management strategies	$7.5M/yr	10	$75M
Coupling whole rock dissolution to electrochemical reactors and systems	**$7.5M/yr**	**10**	**$75M**
Development of hybrid approaches	**$7.5M/yr**	**10**	**$75M**
Resource mapping and pathway assessment	$10M/yr	5	$50M
Total Estimated Research Budget	$72.5M/yr	5–10	$475M
Estimated Budget of Research Priorities	**$55M/yr**	**5–10**	**$350M**

Recommendation 2: Common Components of an Ocean CDR Research Program. Implementation of the research program in Recommendation 1 should include several key common components:

1. The development and adherence to a common research code of conduct that emphasizes transparency and open public data access, verification of carbon sequestration, monitoring for intended and unintended environmental and other impacts, and stakeholder and public engagement.
2. Full consideration of, and compliance with, permitting and other regulatory requirements. Regulatory agencies should establish clear processes and criteria for permitting ocean CDR research, with input from funding entities and other stakeholders.
3. Co-production of knowledge and design of experiments with communities, Indigenous collaborators, and other key stakeholders.
4. Promotion of international cooperation in scientific research and issues relating to the governance of ocean CDR research, through prioritizing international research collaborations and enhancement of international oversight of projects (e.g., by establishing an independent expert review board with international representation).
5. Capacity building among researchers in the United States and other countries, including fellowships for early-career researchers in climate-vulnerable communities and underrepresented groups, including from Indigenous populations and the Global South.

Research Priorities

Based on the present state of knowledge, there are substantial uncertainties in all of the ocean CDR approaches evaluated in this report. **The best approach for reducing knowledge gaps will involve a diversified research investment strategy that includes both crosscutting, common components (Table S.2) and coordination across multiple individual CDR approaches (Table S.3) in parallel (Figure S.2).** The development of a robust research portfolio will reflect a balance among several factors: common elements and infrastructure versus targeted studies on specific approaches; biotic versus abiotic CDR approaches; and more established versus emerging CDR approaches.

Crosscutting foundational research priorities listed in Table S.2 include research on international governance and the domestic legal framework of ocean CDR research. Other priorities include the development of a common code of conduct for ocean CDR research and coordinated research infrastructure including components on standardized environmental monitoring and carbon accounting methods, publicly accessible data management, and science communication and public engagement.

The research priorities in Table S.3 for each of the four biotic ocean CDR approaches differ based on the current knowledge base, extent of previous research, and distinctions in the underlying biological processes. Evaluation of research needs across CDR approaches is more challenging, suggesting some investment in all methods; however, a first-order attempt at prioritization can be constructed based on current knowledge. **Among the biotic approaches, research on ocean iron fertilization and seaweed cultivation offer the greatest opportunities for evaluating the viability of possible biotic ocean CDR approaches; research on the potential CDR and sequestration permanence for ecosystem recovery would also be beneficial in the context of ongoing marine conservation efforts.**

For abiotic ocean CDR approaches, the research agenda (Table S.3) will be most impactful if it combines a thorough understanding of potential environmental impacts alongside technology development and upscaling efforts. Based on present understanding, there is considerable CDR potential for ocean alkalinity enhancement, which spans a number of approaches including, but not restricted to, ocean liming, accelerated rock weathering, and electrochemical methods for alkalinity enhancement, among others. Next steps for alkalinity enhancement research offer large opportunities for closing knowledge gaps but include the complexity of undertaking large-scale experimentation to assess whole ecosystem responses across the range of technologies and approaches for increasing alkalinity. Therefore, **among the abiotic approaches, research on ocean alkalinity enhancement, including electrochemical alkalinity enhancement, has priority over electrochemical approaches that only seek to achieve CDR from seawater (also known as carbon dioxide stripping).**

Recommendation 3: Ocean CDR Research Program Priorities. A research program should move forward integrating studies, in parallel, on multiple aspects of different ocean CDR approaches, recognizing the different stages of the knowledge base and technological readiness of specific ocean CDR approaches. Priorities for the research program should include development of

1. Overarching implementation plan for the next decade adhering to the crosscutting strategy elements in Recommendation 1 and incorporating from its onset the common research components in Recommendation 2 and Table S.2. Progress on these common research components is essential to achieve a foundation for all other recommended research.
2. Tailored implementation planning for specific ocean CDR approaches focused on reducing critical knowledge gaps by moving sequentially from laboratory-scale to pilot-scale field experiments, as appropriate, with adequate environmental and social risk reduction measures and transparent decision-making processes (priority components bolded in Table S.3).
3. Common framework for intercomparing the viability of ocean CDR approaches with each other and with other climate response measures using standard criteria for efficacy, permanence, costs, environmental and social impacts, and governance and social dimensions.
4. Research framework including program-wide components for experimental planning and public engagement, monitoring and verification (carbon accounting), and open publicly accessible data management.
5. Strategy and implementation for engaging and communicating with stakeholders, policy makers, and publics.
6. Research agenda that emphasizes advancing understanding of ocean fertilization, seaweed cultivation, and ocean alkalinity enhancement.

CONCLUDING REMARKS

Ocean CDR approaches are already being discussed widely, and in some cases promoted, by scientists, nongovernmental organizations, and entrepreneurs as potential climate response strategies. At present, society and policy makers lack sufficient knowledge to fully evaluate ocean CDR outcomes and weigh the trade-offs with other climate response approaches, including climate adaptation and emissions mitigation, and with environmental and sustainable development goals. Research on ocean CDR, therefore, is needed to decide whether or not society moves ahead with deployment, and to assess at what scales and locations the consequences of ocean CDR would be acceptable.

1

Introduction

1.1 HUMAN PERTURBATIONS TO THE GLOBAL CARBON CYCLE

As of 2021, atmospheric carbon dioxide (CO_2) levels reached historically unprecedented levels, nearly 50 percent higher than preindustrial values only two centuries ago (Figure 1.1; GML, 2021). Present atmospheric CO_2 levels are higher than at any time in the past 800,000 years and likely several million years. The cause of the CO_2 increase over the 19th, 20th, and early 21st centuries is clearly and incontrovertibly identified as human activities including fossil-fuel burning, agriculture, and historical land-use change including deforestation (Figure 1.2; Friedlingstein et al., 2020). Human CO_2 emissions over the most recent decade were close to 35 billion tons of CO_2 per year (10^9 t CO_2/yr = 1 Gt CO_2/yr).[1] The atmospheric accumulation of CO_2 would be even higher if not for land and ocean carbon sinks that together currently remove more than half the amount of human emissions from the atmosphere; however, these natural sinks may become less effective in a future high-CO_2, warmer world. The current level of human emissions greatly exceeds the ability of nature to remove CO_2, and a reduction on the order of 90 percent in human emissions is required to stabilize atmospheric CO_2 at some specified level, and approximately net-zero human CO_2 emissions is needed to stabilize climate because of inertia in the Earth system. In a number of climate scenarios, a period of net negative human CO_2 emissions occurs later in this century to compensate for other greenhouse gases (GHGs) and to address overshoots in atmospheric CO_2. An overall climate change response strategy will include climate mitigation approaches to reduce CO_2 and other GHG emissions. Carbon dioxide removal (CDR) approaches (Box 1.1) could be used in conjunction with emissions abatement to compensate for positive human emissions of CO_2 or

---

[1] For consistency, the unit metric tons (Mt) of CO_2 is used through most of this report when referring to CO_2 removal from the atmosphere. One billion Mt of CO_2 is equivalent to 0.128 parts per million (ppm) in global average atmosphere CO_2 mixing ratio, where, for reference, the 2021 CO_2 mixing ratio is approximately 415 ppm. But the reader should be aware that some sources in literature, policy documents, and press releases use a range of different units (e.g., 1 Gt CO_2 = 1015 g CO_2 = 1 Pg CO_2), and sometimes mass of carbon, C, is used rather than mass of CO_2 (1 Gt CO_2 = 0.273 Gt C). Ocean biological and geophysical carbon fluxes and stocks, for example, are often reported in carbon units rather than CO_2 units, reflecting the multiple chemical forms of carbon in biomass pools and in inorganic carbon dissolved in seawater.

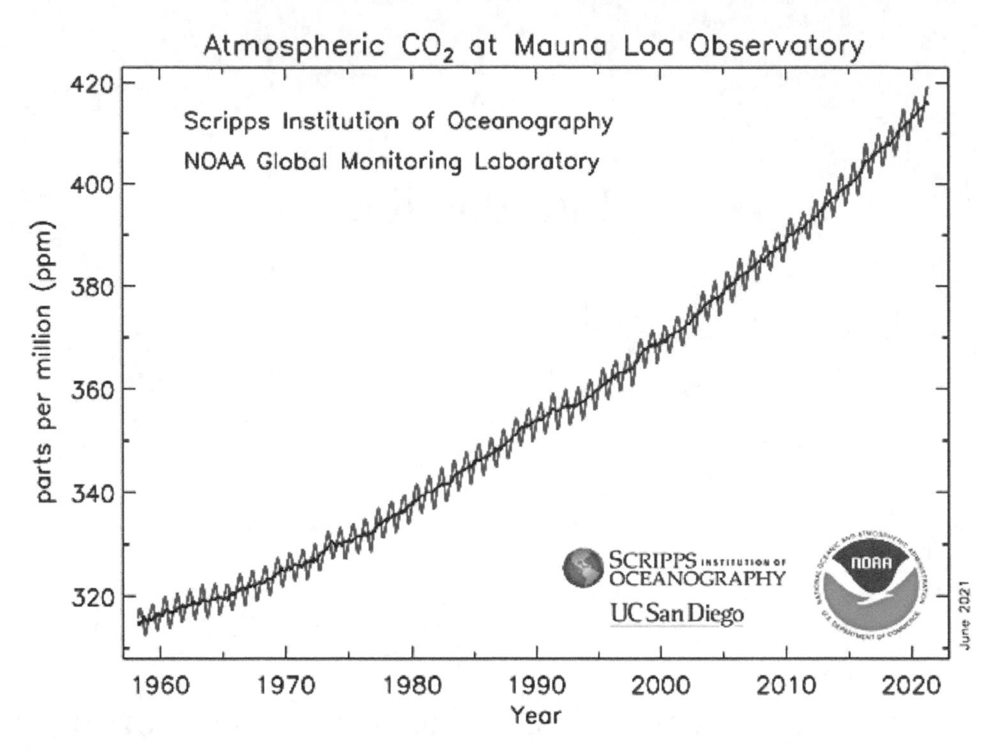

FIGURE 1.1 Time history of atmosphere CO_2 mixing ratio from the Mauna Loa, Hawai'i Observatory, the longest available instrumental record. The seasonally adjusted atmospheric CO_2 mixing ratio in mid-2021 exceeded 415 ppm compared to a preindustrial mixing ratio of 280 ppm measured from ice cores. SOURCE: GML, NOAA, 2021.

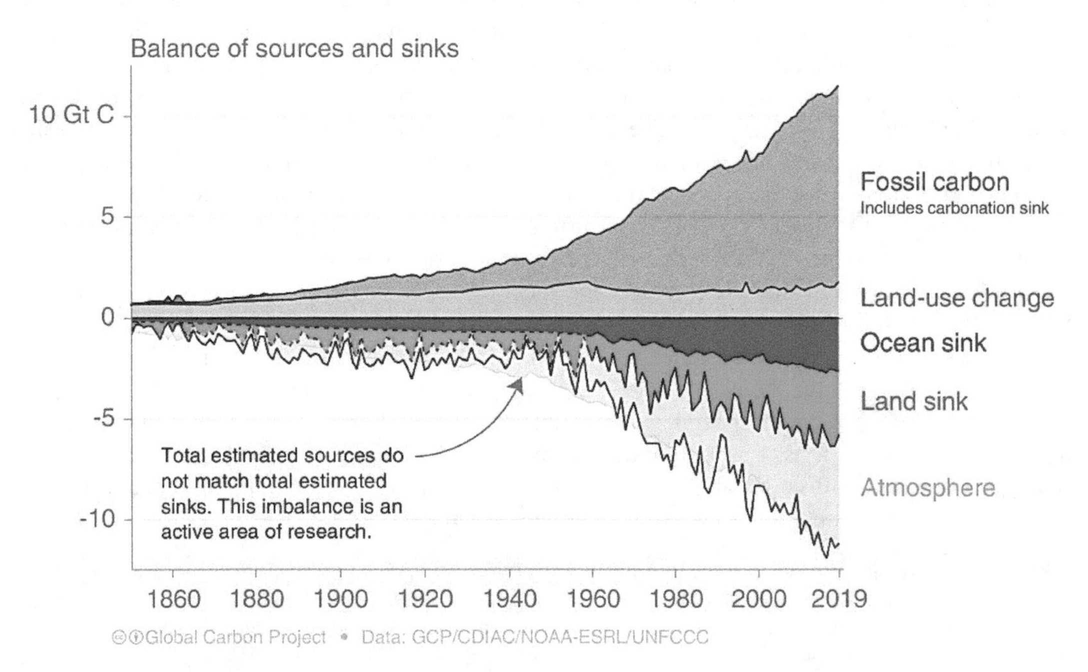

FIGURE 1.2 Reconstruction over time of the industrial period of human carbon emissions to the atmosphere from fossil fuel use and land-use change (positive fluxes), ocean and land sinks (negative fluxes), and atmospheric accumulation. SOURCES: (a) Friedlingstein et al., 2020; (b) Global Carbon Project, 2020; (a) and (b) licensed under Creative Commons CC BY 4.0.

BOX 1.1
Carbon Dioxide Removal Approaches

Carbon dioxide removal (CDR) approaches span a wide range of biotic and abiotic methodologies, but all involve some step for removing or capturing CO_2 from the atmosphere or some reservoir in close contact with the atmosphere (e.g., surface ocean) and then storing or sequestering that CO_2 in some other reservoir to ensure limited release back to the atmosphere for some period of time (NASEM, 2019). Proposed approaches include protecting and enhancing terrestrial and marine ecosystems that remove and store carbon for long periods of time to more technological methods such as direct air capture that would directly remove CO_2 from the atmosphere via industrial chemical methods and then store that CO_2 in a geological reservoir. Intermediate approaches have also been proposed, such as bioenergy with carbon capture and storage (Hanssen et al., 2020). The focus of this report is on an assessment of a selection of ocean-related CDR approaches that would indirectly remove CO_2 from the atmosphere by lowering or redistributing CO_2 in the ocean water column.

contribute to net negative CO_2 emissions; this would require the durable storage of the removed carbon in some reservoir(s) away from the atmosphere for a sufficiently long period of time, typically taken as decades to centuries.

The rising level of atmospheric CO_2 is a major global concern because CO_2 is a key heat trapping gas, or GHG (USGCRP, 2017; IPCC, 2021), and elevated CO_2 levels are a major factor driving observed anthropogenic climate change that has already increased global average surface temperature by 1.12°C from preindustrial levels (NCEI, 2020) as seen in Figure 1.3. Although CO_2 is only one of several GHGs, contributing about 74 percent of the present total radiative imbalance leading to global warming (WRI, 2020; see also National Oceanic and Atmospheric Administration [NOAA] Earth System Research Laboratories[2]), the contribution of CO_2 to overall anthropogenic warming will likely grow in the future because of the long lifetime of excess CO_2 in the atmosphere (multidecadal and longer), ocean, and land biosphere system, compared with other GHGs (e.g., methane)[3] (IPCC, 2021). While this report concentrates on removal of excess atmospheric CO_2, comprehensive climate mitigation strategies incorporate approaches to reduce human emissions of all GHGs and may even explore deliberate removal of gas beyond CO_2 such as methane.

The many impacts of climate change on managed and natural ecosystems and across human society are well documented in the scientific literature and in national and international assessment reports (e.g., USGCRP, 2018). In addition to climate change and associated ocean warming (Laufkötter et al., 2020) and decline in subsurface oxygen levels (Breitburg et al., 2018), marine ecosystems are also experiencing changes in seawater chemistry, termed ocean acidification, associated with the ocean uptake of excess CO_2 (Pershing et al., 2018). A wide range of marine organisms including shellfish and corals appear to be sensitive to ocean acidification, and the impacts extend to coastal human communities reliant on marine resources such as wild-caught fisheries, aquaculture, and marine tourism and recreation (Doney et al., 2020). Ocean acidification and climate change are closely linked because of a common underlying causal factor: large human CO_2 emissions to the atmosphere. CDR approaches for climate mitigation may also improve ocean acidification conditions, at least for some parts of the surface ocean.

The ocean holds great potential for uptake and longer-term sequestration of anthropogenic CO_2 for several reasons: (1) the ocean acts as a large natural reservoir for CO_2, holding roughly

[2] See https://www.esrl.noaa.gov/.

[3] This statement should not be taken to suggest that other GHG emissions are unimportant or do not need to be reduced.

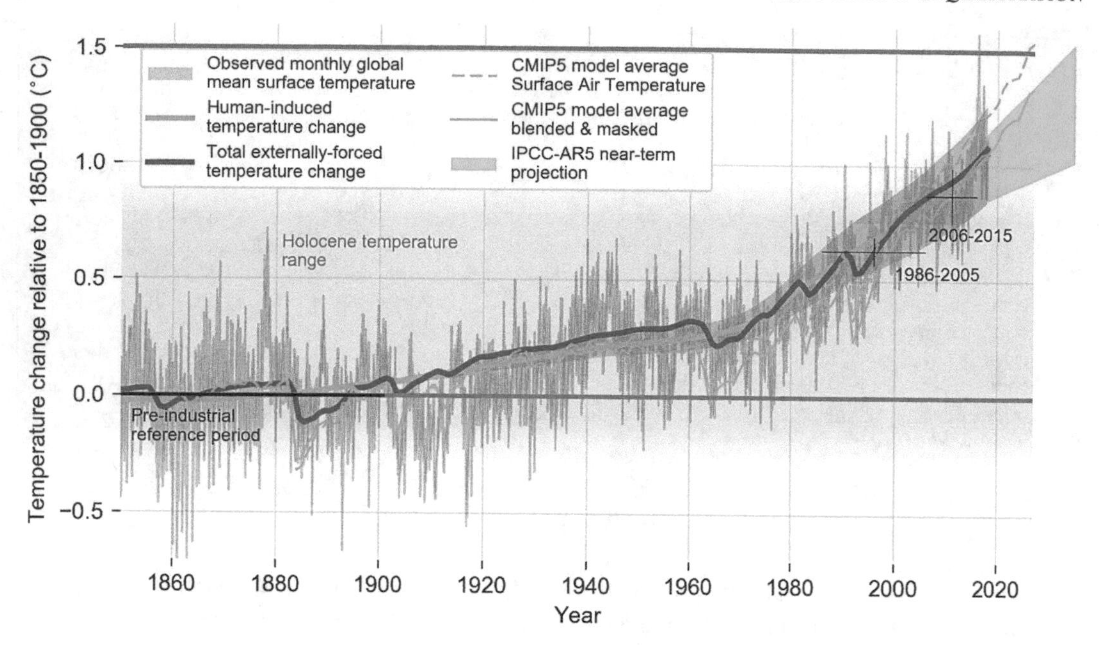

FIGURE 1.3 Historical global mean surface temperature. SOURCE: IPCC SR1.5, Figure 1.2, Allen et al., 2018.

50 times as much inorganic carbon as the preindustrial atmosphere; (2) the ocean already removes a substantial fraction of the excess atmospheric CO_2 from human emissions; and (3) a number of physical, geochemical, and biological processes are known to influence air–sea CO_2 gas exchange and ocean carbon storage. The ocean covers 70 percent of Earth's surface; it includes much of the global capacity for natural carbon sequestration, and it may be possible to enhance that capacity through implementation of an ocean-based CDR strategy.

1.2 CLIMATE MITIGATION, DECARBONIZATION, AND CARBON DIOXIDE REMOVAL

Without deliberate action to reduce human CO_2 emissions, continued rapid CO_2 accumulation in the atmosphere increases the expected magnitude of climate change and ocean acidification (UNEP, 2017; IPCC, 2021). Therefore, mitigation efforts to curb future climate change focus heavily on reducing the emissions of CO_2 (and other GHGs), which could potentially utilize CDR. On timescales of a few years, the atmosphere is relatively well mixed, and the global mean atmospheric CO_2 trend effectively reflects global net CO_2 fluxes (sources minus sinks). Reducing CO_2 emissions through abatement and removing CO_2 via CDR, therefore, are approximately equivalent on these timescales from the perspective of the atmospheric CO_2 budget, especially for the next several decades when emissions will remain well above net zero. The symmetry in the response of atmospheric CO_2 to emissions and removals may break down if there is a long delay (decades) in implementing CDR, such as may occur in atmospheric CO_2 overshoot scenarios, where land and ocean carbon processes may decrease the effectiveness of CDR (Zickfeld et al., 2021).

A climate target is needed to frame the amount and timing for emissions abatement and CDR. The international Paris Climate Agreement provides one such framing used widely in Intergovernmental Panel on Climate Change (IPCC) reports and the scientific literature. The Paris Agreement calls for limiting global warming to well below 2°C, preferably to 1.5°C, compared to preindustrial levels. In model simulations, meeting the Paris target requires reaching net-zero CO_2 emissions globally well before the end of the century, with earlier mid-century net-zero targets for developed nations (NRC, 2015a; NASEM, 2019). Multiple abatement pathways will need to be pursued to rapidly decarbonize the U.S. and global economies including efforts to increase energy efficiency, switch to non-CO_2–emitting energy sources including renewables, capturing and sequestering CO_2 from point sources in geological reservoirs, and reducing CO_2 (and other GHG) emissions from land use (NASEM, 2021a).

While reaching net zero from current CO_2 emissions is primarily an effort in decarbonization and emissions abatement, essentially all of the climate mitigation scenarios that meet the Paris Agreement target include some form and amount of CDR (also commonly referred to as negative emission technologies or NETs) to balance residual CO_2 emissions from difficult-to-decarbonize sectors of economies (e.g., long-distance transportation, cement and steel production), to provide time for the development and implementation of different decarbonization approaches, and to compensate for short-term overshoots in emissions and atmospheric CO_2 and GHG levels. The reliance on, or even need for, CDR approaches varies considerably depending on the climate and socioeconomic scenarios and integrated assessment models used to evaluate human emissions and barriers to decarbonization. However, some form of CDR is common in climate/socioeconomic scenarios that attempt to limit global warming to well below 2°C, preferably to 1.5°C, compared to preindustrial levels in line with the international Paris Agreement (Figure 1.4; Allen et al., 2018; Fuhrman et al., 2019; Canadell et al., 2021; IPCC, 2021). The remaining carbon budget, or allowable cumulative future net human CO_2 emissions, is a particularly useful framing for assessing requirements for decarbonization and CDR. At current human CO_2 emission rates, the remaining carbon budgets would be expended in a little more than a decade to a few decades to stay below the Paris Agreement climate targets. The low-climate-warming scenarios used in the IPCC (2021) report reflect integrated assessment model simulations that often require substantial CDR, on the order of 10 Gt CO_2/yr or more by mid- to late-century, to stay within the remaining carbon budget constraints or to address atmospheric CO_2 overshoot where additional CDR is required to generate periods of net-negative human CO_2 later this century (Fricko et al., 2017; Riahi et al., 2017).

Historical and future global warming levels from now until mid-century (2050) approximately scale with cumulative CO_2 emissions, and CO_2 emissions reduction and CDR targets can be framed in terms of the remaining carbon budgets to meet specified warming targets (IPCC, 2021). For example, the estimated remaining carbon budget for a 1.5°C target is only about 300–900 Gt CO_2, a relatively small amount compared to current emissions of about 35 Gt CO_2/yr and the historical (1850 to 2019) cumulative human CO_2 emissions of 2,390 ± 240 Gt CO_2 (IPCC, 2021). Aggressive efforts to reduce emissions of methane and other GHGs would expand the remaining carbon budget but would need to be done in concert with decarbonization and CDR to stabilize global climate. This point is brought home well in a quote from the IPCC AR6 report:

> Emission pathways that limit globally averaged warming to 1.5°C or 2°C by the year 2100 assume the use of CDR approaches in combination with emission reductions to follow net negative CO_2 emissions trajectories in the second half of this century. For instance, in SR1.5, all analyzed pathways limiting warming to 1.5°C by 2100 with no or limited overshoot include the use of CDR . . . Affordable and environmentally and socially acceptable CDR options at scale well before 2050 are an important element of 1.5°C-consistent pathways. (IPCC, 2021, pp. 4-81–4-82)

There is considerable uncertainty, however, in the cumulative CDR requirement as noted in the IPCC (2018) *Global Warming of 1.5°C* report, where on the order of 100–1,000 Gt CO_2 of CDR over the 21st century is projected to meet the 1.5°C target with limited or no overshoot.

As illustrated in Figure 1.4, the potential scale for peak CDR demand varies greatly from zero to a few Gt CO_2/yr up to tens of Gt CO_2/yr. The range in CDR demand reflects the climate warm-

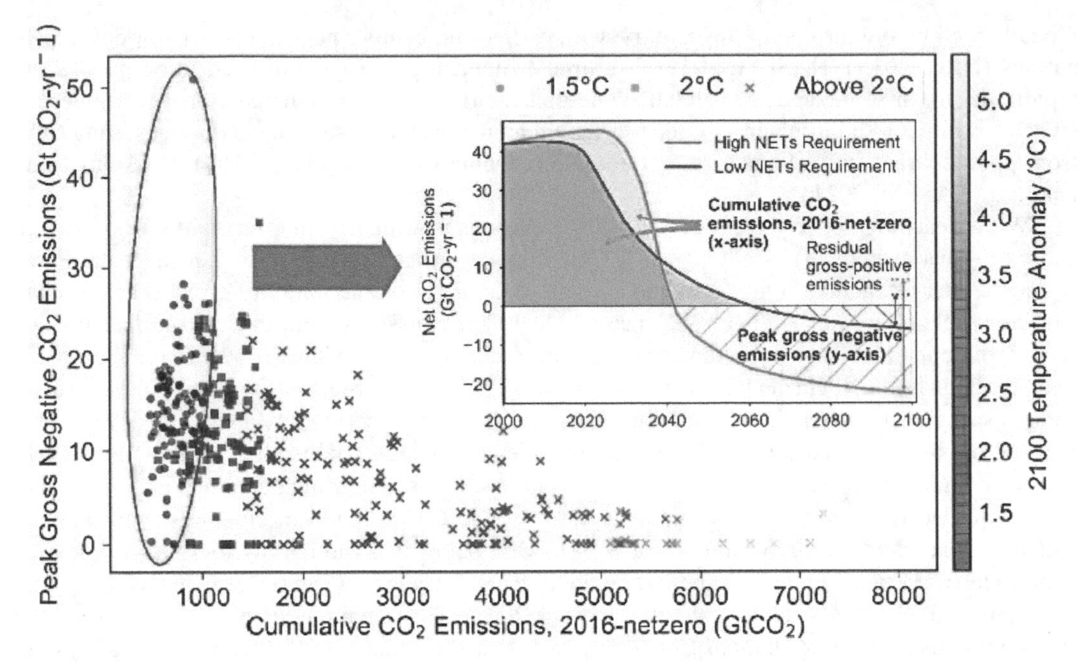

FIGURE 1.4 Integrated assessment model scenarios consistent with limiting end-of-century warming to 1.5°C (circles) or 2°C (squares) above preindustrial levels have exceedingly small remaining cumulative CO_2 emission budgets (from 2016 to year of net-zero emissions), and nearly all require significant future CO_2 removal from the atmosphere using NETs. As illustrated in the inset, for a given temperature target (e.g., those trajectories that achieve 1.5°C indicated by the red circle), future NET requirements are governed both by the magnitude of the GHG pulse emitted previously as well as by residual gross-positive emissions from those sectors of the economy that are recalcitrant to decarbonization once a climate policy is implemented (e.g., air travel). Increased cumulative emissions after 2016 before reaching net zero generally corresponds to increased peak future NETs deployment and associated impacts. SOURCE: Copyright © 2019 Fuhrman, McJeon, Doney, Shobe and Clarens.

ing target (a lower 1.5°C target results in more CDR demand than less aggressive warming targets such as 2° or 3°C), as well as assumptions about socioeconomic pathways, climate sensitivity, and the response of land and ocean carbon sinks to climate change. Even larger amounts of CDR would be required if the goal is not simply climate stabilization but rather to shift climate back toward preindustrial conditions as has been proposed by some as an even more challenging and not well-agreed-upon possible future objective to occur after climate stabilization has been achieved. The required annual scale of CDR is thus comparable to the amount of CO_2 that is absorbed by the global ocean currently, ~9 Gt CO_2/yr (Friedlingstein et al., 2020). Significantly, the required CDR scale is a substantial fraction of current fossil fuel CO_2 emissions, ~35 Gt CO_2/yr (Friedlingstein et al., 2020) and is also larger than the largest manufacturing industries that is cement production (Andrew, 2018; Cao et al., 2020; IEA, 2020). It is important, however, to keep in mind that these CDR estimates reflect model estimates using only a subset of possible approaches, typically land-based afforestation/reforestation and bioenergy with carbon capture and sequestration, and adding CDR approaches with different land, resource, energy, and cost constraints can result in a different estimates of total CDR, emissions abatement, and residual human CO_2 emissions for the same climate target (e.g., recent studies adding direct air capture CDR, Fuhrman et al., 2020). Studies highlight the challenge of accomplishing the projected level of CDR with land-based approaches alone, particularly for reforestation and bioenergy with carbon capture and sequestration CDR methods that require substantial amounts of land with impacts on food and water supplies, energy use, and fertilizer demand (Fuhrman et al., 2020). A CDR portfolio approach that includes less land-intensive methods, such as direct air capture, (Fuhrman et al., 2021) along with ocean-based methods may be more appropriate.

A CDR objective of removing and durably storing tens of Gt CO_2/yr by mid-century will likely be quite challenging to achieve. Today, global CO_2 sequestration activities accomplish <0.1 Gt CO_2/yr storage (Liu et al., 2018; Page et al., 2020; Townsend and Gillepsie, 2020). IEA (2021a) estimates that the global capacity of carbon capture, utilization, and storage facilities for CO_2 capture in 2020 was only about 40 Mt CO_2/yr. At present, industrial-scale CDR approaches are even smaller scale than pilot and demonstration plants; for example, a new direct air capture facility in Iceland would, at fully planned capacity, remove 4 kt CO_2/yr (Gertner, 2021). While ocean-based CDR is an important direction for achieving large-scale CDR, the scale of the challenge is daunting. For context, consider the scale of one of the largest manufacturing sectors, the cement industry, which produces around 4.5 Gt/yr of cement clinker, resulting in the downstream production of more than 20 Gt/yr of concrete (i.e., a mixture of cement, stone, sand, and water). The required rapid ramp-up of CDR scale implied by integrated assessment model scenarios involves the creation of a new sector, de novo, that is of a size similar to the cement/concrete sector, albeit in 30 years. The construction and commissioning of large capital facilities needed for many CDR approaches would require time, even with expected advances along technology learning curves. For example, to plan, permit, build, and commission a cement plant that produces ~1 Mt/yr of cement clinker requires at least on the order of 4 years. Growing policy or market demand for CDR, alone, does not guarantee success in reaching adequate scale for CDR. Substantial investments likely would be needed at multiple stages of innovation for a technology, as described in Nemet et al. (2018), addressing factors on both the supply side (e.g., research and development, demonstrations, and scale-up) and the demand side (e.g., demand pull, niche markets, and public acceptance). For this sectoral CDR scaling, it is furthermore important to consider the operational cost reductions that could be achieved, in time, and the amount of capital investment that is needed to stand up a CDR industry. Investments in CDR sectors will also depend on overall demand and price for CDR.

End-to-end carbon removal, ranging from carbon capture to geological sequestration (e.g., from point sources or the atmosphere), currently costs in the vicinity of ~\$70 to \$700/t CO_2. It has been suggested that it is necessary that carbon removal from the atmosphere be achieved at a net cost less than \$100 (net present value based on 2021 dollars) (Budinis et al., 2018; Pilorgé et al., 2020; IEA, 2021b). The NASEM (2019) CDR report adopted a value of <\$100/t CO_2 as a rough guide for "economical" CDR approaches, and for consistency this report uses the same cutoff, acknowledging that more expensive CDR approaches may also be considered because of other factors such as low resource demands and co-benefits. The requisite level of cost reduction, beyond technology improvements, requires a variety of actions including achieving (1) economies of scale, (2) massive replicability in manufacturing and deploying technology components and capital assets, and (3) the use of abundant, cost-effective, and accessible materials and components in technological systems. What remains prerequisite, however, are clear and consistent approaches for environmental and construction permitting, which limits overhead costs and restricts escalations of the overnight cost of construction.

Yet another important aspect related to the deployment of CDR technologies, however, implies the integration of carbon management solutions with existing industrial operations. For ocean-based CDR this could include coupling with marine aquaculture, shipping, and transportation systems, and connecting with and learning from coastal industrial operations such as desalination and chemical production. Some shore-based industrial facilities, for example, are already supplied with seawater intakes that could be utilized in some ocean-based CDR approaches; these intakes can account for a substantial fraction of the cost of capital construction. Such synergistic integration of engineering with existing technologies (e.g., particularly for electrochemical approaches) may provide opportunities to accelerate demonstration projects and scaling up of some ocean-based CDR technologies. Further, some electrochemical CDR approaches produce hydrogen gas as a by-product, and assuming that the initial electrical energy source for the CDR comes from low-CO_2 sources, the hydrogen gas could be used as a clean fuel (or temporary energy storage) to reduce overall costs and carbon intensity of the CDR approach or other industrial activity.

1.3 SEAWATER CO_2 AND CARBONATE SYSTEM CHEMISTRY

The addition or removal of CO_2 from the ocean alters the acid-base chemistry of seawater. CO_2 gas dissolved in seawater, aqueous CO_2 (CO_2 (aq)), can react with water to form carbonic acid (H_2CO_3), a weak acid:

$$CO_2\ (aq) + H_2O \leftrightarrow H_2CO_3$$

The partial pressure of CO_2 gas, pCO_2, varies proportionally with the concentration of CO_2 (aq) and is also influenced by temperature and salinity. The hydration reaction of CO_2 is relatively rapid and for most applications can be assumed to be at equilibrium. At seawater pH (~8), H_2CO_3 decomposes into bicarbonate (HCO_3^-), an inorganic carbon ion, and a hydrogen ion (H^+):

$$H_2CO_3 \leftrightarrow HCO_3^- + H^+$$

Similarly, a bicarbonate ion can decompose into a carbonate ion (CO_3^{2-}):

$$HCO_3^- \leftrightarrow CO_3^{2-} + H^+$$

The seawater inorganic carbon system acid-base reactions are also in equilibrium as a function of temperature, salinity, and pressure. The dissolved inorganic carbon (DIC) is the sum of aqueous CO_2, H_2CO_3, HCO_3^-, and CO_3^{2-}, with HCO_3^- dominating at seawater pH. The addition of CO_2 increases seawater DIC, and the resulting production of H^+ ions increases the acidity (lowers the pH) of seawater, where pH is defined on a logarithmic scale from the H^+ ion concentration, pH = $-\log_{10} [H^+]$. The seawater concentration of DIC is much higher than that in freshwater because of the high seawater alkalinity, a measure of the acid buffering capacity and a reflection of the balance of inorganic ions from rock weathering and other processes.

The addition of CO_2 gas to seawater from either physical or biological processes (e.g., respiration of organic matter) increases the DIC concentration but does not affect alkalinity. Because the CO_2 hydration reaction produces a weak acid H_2CO_3, CO_2 addition also makes the seawater more acidic (lowers pH) and shifts the partitioning of the inorganic carbon ions that make up DIC, increasing CO_2 and HCO_3^- and lowering CO_3^{2-}. Increasing seawater alkalinity, for example, by adding a base (e.g., sodium hydroxide [NaOH]), shifts the inorganic ion partitioning in the opposite sense, lowering CO_2 and increasing pH and CO_3^{2-}. The dissolution of calcium carbonate ($CaCO_3$) minerals into seawater increases both DIC and alkalinity, with the alkalinity increasing to twice that of DIC, and also results in an increase in the pH and CO_3^{2-}.

Air–sea CO_2 flux is controlled thermodynamically by the difference between the partial pressure of CO_2 (pCO_2) between the surface ocean and atmosphere. Thus, the addition or removal of CO_2 and alkalinity can affect air–sea exchange by altering CO_2 (aq) and pCO_2. For example, the formation of organic matter by photosynthetic organisms in the surface ocean involves the uptake of CO_2, which acts to lower seawater pCO_2 and enhance the downward flux of CO_2 from the atmosphere into the surface ocean. The kinetics of air–sea CO_2 gas exchange are relatively slow because the gas CO_2 (aq) is a small fraction of the large seawater DIC reservoir, and the inorganic carbon ions do not directly exchange with the atmosphere. Typical gas exchange equilibration timescales for the surface ocean are on the order of a year, and surface water pCO_2 can exhibit larger differences (disequilibrium) from atmospheric pCO_2.

1.4 OCEAN CARBON CYCLE AND OCEAN ANTHROPOGENIC CO_2 UPTAKE

The ocean geographic patterns and seasonal cycle of air–sea CO_2 vary substantially because of the interplay of biological and physical processes, resulting in ocean regions with both outgassing and ingassing from the atmosphere. Rising atmospheric CO_2 levels from human emissions shift the balance toward enhanced downward CO_2 flux from the atmosphere and ocean, and the anthropogenic CO_2 perturbation flux overlays the natural, preindustrial patterns. Globally the current net air–sea flux of anthropogenic CO_2 is roughly a quarter of emissions for fossil fuel consumption or about 9 Gt CO_2/yr (Friedlingstein et al., 2020), after accounting for the effects of river carbon inputs. The rate of anthropogenic CO_2 uptake is constrained by multiple observational approaches including global surveys of air–sea CO_2 flux, temporal changes in the ocean inventory of DIC, proxy methods based on other transient tracers such as chlorofluorocarbons, and numerical ocean models. The rate of ocean uptake of anthropogenic CO_2 is primarily controlled by physical circulation and the rate at which surface waters are exchanged into the thermocline and deep ocean.

The ocean uptake of anthropogenic CO_2 is a relatively small perturbation that occurs on top of the large natural background cycling and storage associated with the marine carbon system. The detection and quantification of anthropogenic CO_2 uptake because of rising atmospheric CO_2 has been a decades-long challenge for marine chemists and oceanographers, involving a combination of work to improve the accuracy and precision of seawater pCO_2, DIC, and alkalinity measurements and extensive field surveys of surface- and deep-ocean chemical and physical properties and their change over time. Detection and attribution of large-scale changes in ocean carbon storage due to ocean CDR approaches will have similar challenges. For comparison the cumulative ocean uptake of anthropogenic CO_2 from 1850 to 2019 is estimated to be about 591 Gt CO_2 or equivalently 161 Gt C (Friedlingstein et al., 2020), while the natural stock or reservoir of ocean inorganic carbon is about 38,000 Gt C (Figure 1.5). The large background ocean carbon inventory reflects a number of factors, in particular, the large seawater alkalinity that results in large DIC concentrations in equilibrium with any particular atmospheric CO_2 level and the transfer of carbon from the surface ocean–atmosphere reservoirs to the deep ocean by the long-term action of the biological carbon pump (Figure 1.6).

The biological pump consists of two complementary components: biological production of organic matter and the formation of biominerals (i.e., shells and skeletons made from $CaCO_3$). These processes in the surface ocean result in the surface uptake of DIC and export of carbon to depth via several physical and biological pathways (e.g., gravitational particle sinking, physical mixing of organic carbon into the mid-waters, and active biological transport via vertical migration). The majority of the exported organic carbon and biomineral $CaCO_3$ is respired (or remineralized for $CaCO_3$) back into DIC in the upper 1,000 meters of water column, and respiration and remineralization continue in the deep-water column and at the sediment surface, with only a small fraction (<1 percent for organic carbon, 10–20 percent for inorganic carbon; Andersson, 2014) buried in marine sediments. Ocean physical circulation then returns the respired and remineralized DIC and associated nutrients to the surface ocean on timescales of years to centuries.

Over time the biological pump acts to lower surface DIC and increase deep-ocean carbon storage, generating a vertical gradient of DIC in the ocean, until a steady state is reached where the downward flux of organic matter and carbonate is balanced by the upward flux of excess DIC (Sarmiento and Gruber, 2013). On the timescale of the ocean overturning circulation of centuries to

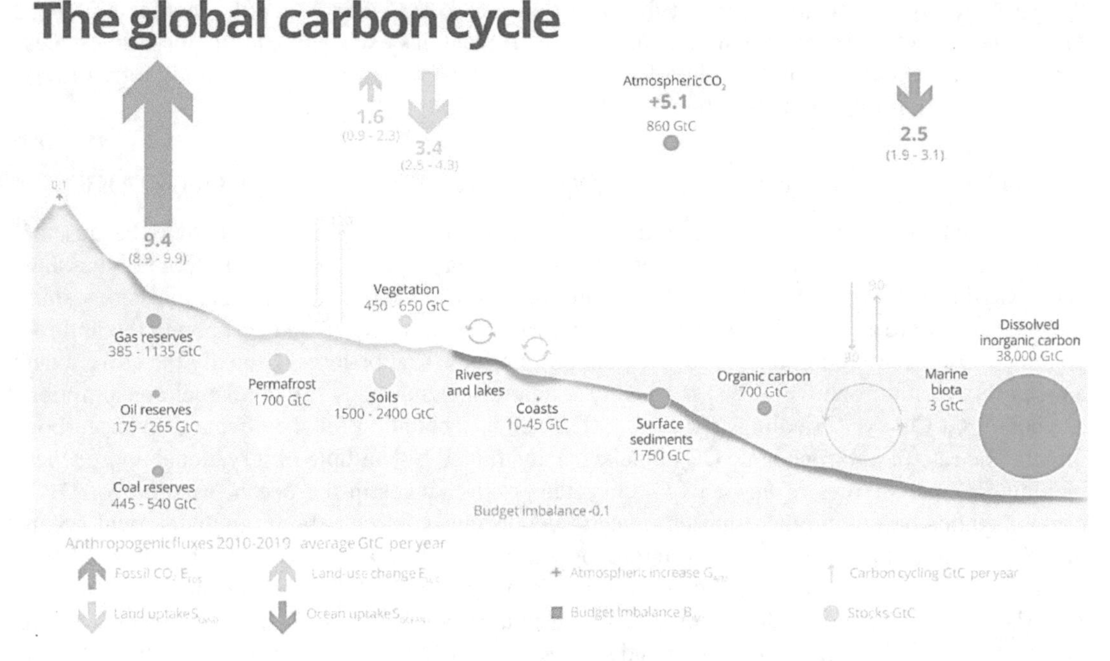

FIGURE 1.5 Schematic of the overall perturbation of the global carbon cycle (Gt C/yr) caused by human activities, averaged globally for the decade 2010–2019. Large arrows indicate direction of perturbations in surface-atmosphere fluxes, and the plus sign (+) for atmosphere CO_2 indicates the growth in the atmospheric carbon reservoir. The anthropogenic perturbation occurring on top of an active carbon cycle, with fluxes (Gt C) and stocks (Gt C/yr) represented in the background. SOURCE: Friedlingstein et al., 2020, licensed under Creative Commons CC BY 4.0.

a millenia, the biological pump acts to maintain a much lower atmospheric CO_2 level than would occur in its absence. Variations in ocean circulation and the biological pump are also indicated as the most likely cause of large variations of ~100 ppm in atmospheric CO_2 documented in ice-core records for the past ~900,000. Cold glacial periods exhibited lower atmospheric CO_2 than warm interglacial periods including preindustrial conditions, with shifts between the two states occurring on timescales of 10,000 years or more. The effect of the biological pump on drawing down surface pCO_2 depends on the ratio of organic to $CaCO_3$ production and export because the effect of the alkalinity decline from $CaCO_3$ formation increases surface water pCO_2, opposing the effect of organic matter production and declining DIC. Ocean carbon storage is also sensitive to the fraction of sinking organic matter that reaches the deep sea prior to being respired (often termed the remineralization length scale) and to the extent of biological nutrient utilization in high-latitude surface ocean, especially in the Southern Ocean. The preindustrial ocean appears to have been in a steady state, with only small variations in atmospheric CO_2, for hundreds to thousands of years prior to the sharp growth of atmospheric CO_2 following the industrial revolution in the 1800s (see Figure 1.7).

From the perspective of anthropogenic CO_2 uptake by the ocean, the biological pump only contributes to the extent that it has been perturbed away from this steady state (Broecker, 1991), and even then many processes that may alter the rate of the biological pump have a relatively small impact on net ocean carbon storage because there is partial to nearly complete cancelation between the changes in the biological export flux and compensating changes in physical transport of DIC. For example, enhanced upwelling of nutrient-rich subsurface water can enhance biological productivity and export flux, but the physical upwelling supplying the extra nutrients also brings up excess

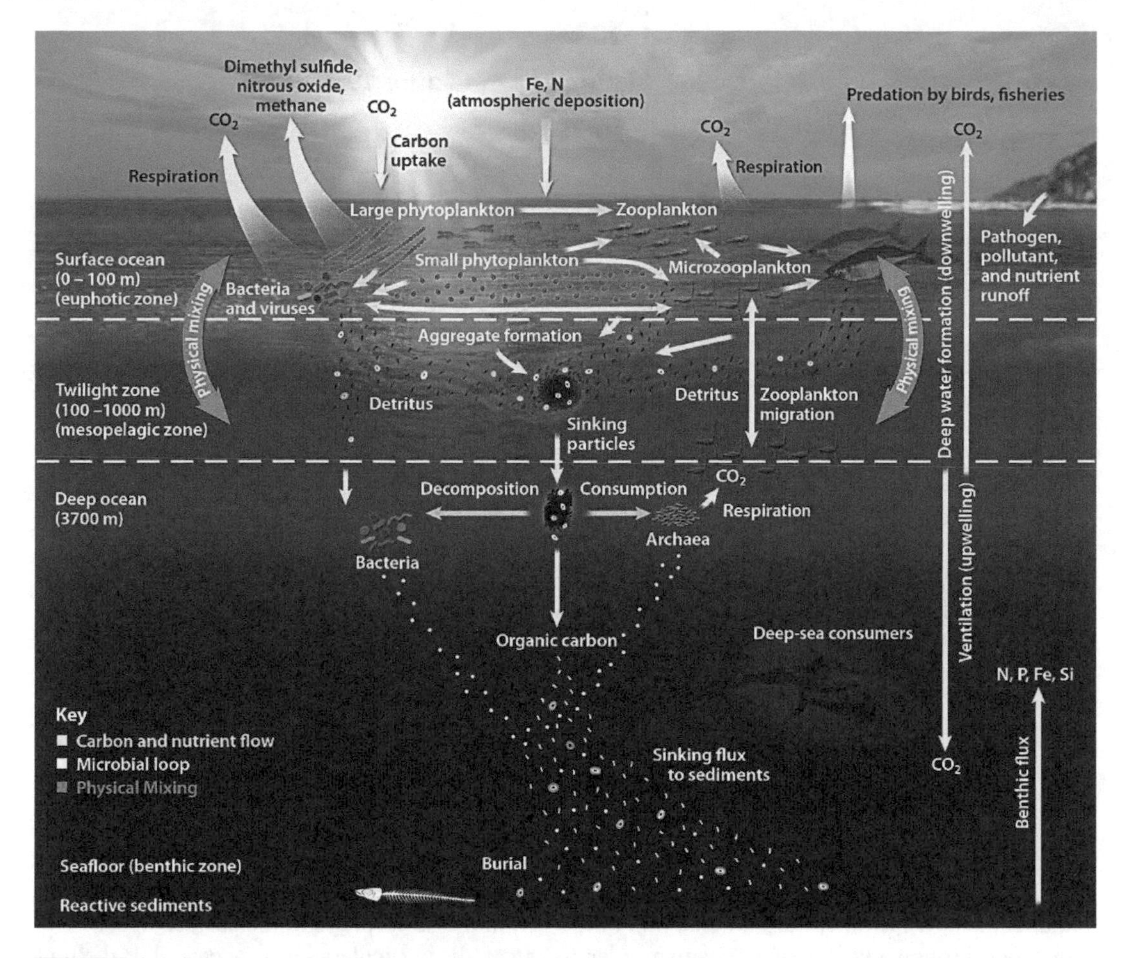

FIGURE 1.6 Schematic of the ocean carbon cycle illustrating carbon and nutrient flows (yellow), food web and microbial loop processes (white/light blue), and physical mixing (orange). SOURCE: Office of Biological and Environmental Research of the U.S. Department of Energy Office of Science.

DIC; both the nutrients and excess DIC come from prior respiration of organic matter in the subsurface ocean. Substantial biological perturbations in global-scale ocean carbon storage require either shifts in the depth patterns of export and remineralization, alteration in the fraction of nutrients in the deep ocean supplied by respiration versus physical transport from the surface ocean, primarily the Southern Ocean with abundant surface macronutrients, or decoupling of carbon–nutrient relationships (Boyd and Doney, 2003; Sarmiento and Gruber, 2013). As documented in the 2021 IPCC report (Canadell et al., 2021), at least so far, there is only weak evidence of detectable changes in the global-scale ocean biological pump affecting net ocean carbon storage due to climate change or ocean acidification on a large scale and low confidence in our understanding of the magnitude and sign of ocean biological feedbacks to CO_2 storage and climate; the ocean uptake in anthropogenic CO_2 is attributed almost wholly to physicochemical processes and trends in human CO_2 emissions. However, given the large carbon fluxes associated with the biological carbon pump, with 5–12 Gt C/yr leaving the surface ocean annually (Siegel et al., 2014), relatively small variations in the function of the biological pump (e.g., carbon-to-nutrient ratios in organic matter; organic-to-inorganic carbon ratios; fraction of organic matter reaching the deep sea) have the potential to modify the vertical partitioning of ocean DIC, ocean carbon storage, and atmospheric CO_2 level.

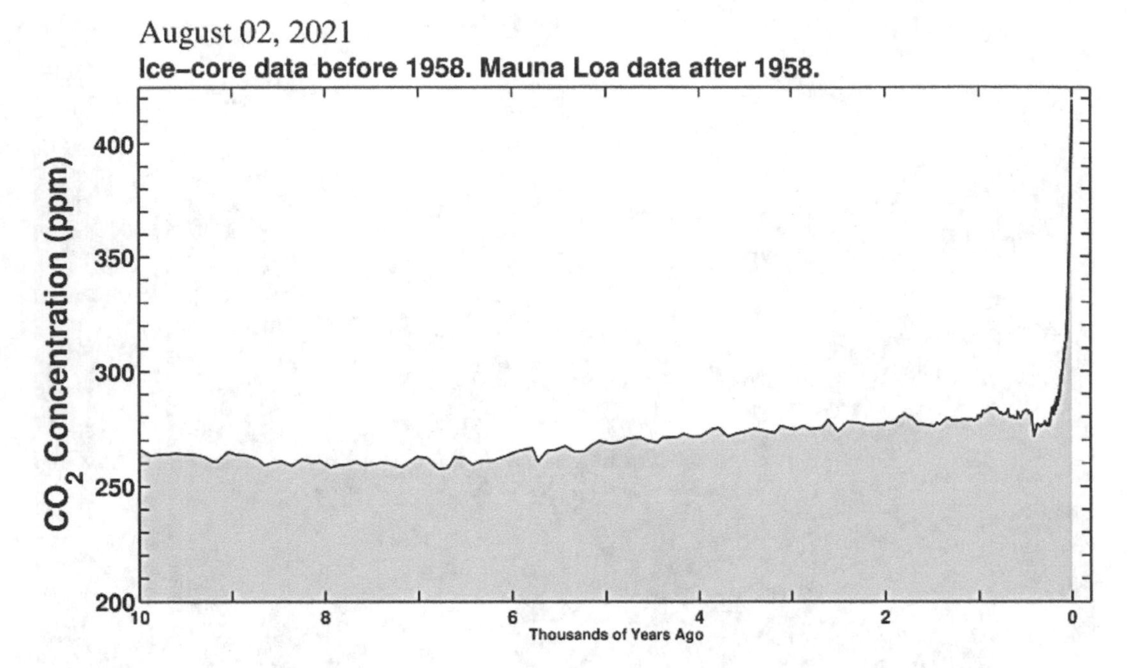

FIGURE 1.7 The Keeling Curve. Credit: Scripps Institution of Oceanography.

1.5 OCEAN-BASED CARBON DIOXIDE REMOVAL

The objective of any CDR approach is to remove excess CO_2 from the atmosphere and store or sequester this carbon in some other reservoir away from the atmosphere for some time period, typically decades or longer. For ocean CDR, the removal from the atmosphere is indirect via an enhancement of the downward air–sea flux of CO_2 from the atmosphere to the surface ocean. This can occur through a variety of mechanisms, depending on the particulars of the ocean CDR method, including increasing the alkalinity and thus the DIC holding capacity of surface seawater; removing CO_2 from seawater for storage in some nonmarine or geological reservoir and thus creating a CO_2 or pCO_2 deficit in surface waters; removing surface CO_2 by increasing the storage of organic carbon in biomass, detritus, and dissolved organic carbon pools; directly injecting CO_2 into the deep ocean; or enhancing the biological transport of organic carbon from the surface ocean to the deep sea. As detailed in subsequent chapters, current scientific understanding of these ocean CDR approaches is insufficient to inform societal decision making and also differs substantially across the range of possible approaches (Box 1.2).

The timescale of carbon sequestration driven by ocean CDR will depend on the location and form of the excess carbon. Relevant questions include the permanence of changes in seawater DIC holding capacity, timescales for conversion of excess organic matter back to DIC, water column physical transport pathways of excess DIC back to the surface, and the leakage rate to the water column or atmosphere of geological or sediment sequestration. Ocean circulation pathways and rates are key to sequestration timescales for CDR approaches that deposit carbon in the water column or at the seafloor. The ocean thermocline covering roughly the upper 1,000 meters of the water column exhibits relatively rapid ventilation timescales of years to a few decades, and carbon must be transported into the deep sea (depth > ~1,000 meters) to achieve century-long sequestration times (Siegel et al., 2021a). This can be problematic for CDR approaches that enhance the ocean biological carbon pump because typically only a small fraction of sinking organic matter passes 1,000 meters, the remainder being respired back to CO_2 in the upper ocean and thermocline.

Here we consider ocean CDR techniques with sufficiently long sequestration permanence or durability to contribute to the portfolio of climate mitigation approaches in development to reduce atmospheric CO_2 this century and beyond and to buy time in the short term for deployment of other mitigation approaches. Although there is no uniform agreement in the literature or policy discussions on a specific sequestration threshold, the committee focused on methods that could potentially deliver durable CO_2 sequestration on timescales of several decades to a century or longer. Sequestration shorter than a decade is likely too short to be an effective policy tool. No sequestration method is foolproof, and probabilistic approaches will be warranted, along with monitoring, to evaluate the expected risk of CO_2 release back to the surface ocean and atmosphere over time.

The efficacy of CDR methods typically is evaluated as the near-term (weeks to months) removal of CO_2 from the atmosphere. Simply shifting carbon from one ocean carbon reservoir to another or to a geological reservoir is insufficient for climate mitigation if one cannot demonstrate the actual removal of CO_2 from the atmosphere. On longer timescales, the effectiveness of any CDR technique, land- or ocean-based, depends on the response of the full Earth system that will tend to dampen the atmospheric CO_2 response (Canadell et al., 2021). Lowering atmospheric CO_2 by any form of CDR reduces the growth rate of atmospheric CO_2 and thus slows the physicochemical uptake of anthropogenic CO_2 by the ocean, even in some cases possibly causing a small CO_2 outgassing. The same is true for human CO_2 emissions, where the growth rate in the atmospheric CO_2 inventory is only slightly less than half of human emissions (the airborne fraction) because of land and ocean carbon sinks.

The durability of ocean-based CDR must also take into consideration the impacts of ongoing and future climate change and ocean acidification. Ocean acidification and elevated CO_2 reduce the buffer capacity of seawater, lowering the effectiveness of CO_2 uptake. Ocean warming, for example, is expected to decrease CO_2 solubility, increase vertical stratification in the ocean, and alter ocean circulation patterns and marine ecosystem dynamics. In model simulations, climate carbon-cycle feedbacks reduce ocean CO_2 uptake somewhat, but the dominant factor governing the magnitude of the ocean sink remains strongly dependent on the CO_2 emissions scenario.

Ocean CDR approaches must also be assessed against the consequences of no action. Without substantial decarbonization, emissions abatement, and potential options such as CDR, atmospheric CO_2 growth will continue unabated with associated rising impacts from climate change and ocean acidification. Marine ecosystems and ocean resources are vulnerable to both climate change and ocean acidification (Pershing et al., 2018), and these impacts should be considered when evaluating the environmental impacts of ocean CDR. Even excluding deliberate ocean CDR, the ocean will continue to act as a sink for anthropogenic CO_2 because of physicochemical uptake; this process will continue into the future even if other mitigation options stabilize atmospheric CO_2 at some elevated level. In fact, over time the ocean will naturally sequester a larger and larger fraction of anthropogenic CO_2 (Archer et al., 2009) with the rate controlled on decadal to millennial timescales by ocean physical circulation and overturning and on longer timescales by adjustments in the marine cycling of $CaCO_3$ and the rate of marine $CaCO_3$ sedimentation (Figure 1.8).

1.6 ORIGIN AND PURPOSE OF THE STUDY

In 2013, the Board on Atmospheric Sciences and Climate convened the Committee on Geoengineering Climate: Technical Evaluation and Discussion of Impacts. The committee produced two reports, one on CDR, *Climate Intervention: Carbon Dioxide Removal and Reliable Sequestration* (NRC, 2015a), and a second on solar radiation management (NRC, 2015b). Since the publication of the CDR report, interest in developing strategies for carbon sequestration has increased in concert with the increasing recognition of the potential need to employ CDR to prevent the more dire consequences associated with past and current GHG emissions. For CDR approaches to meaning-

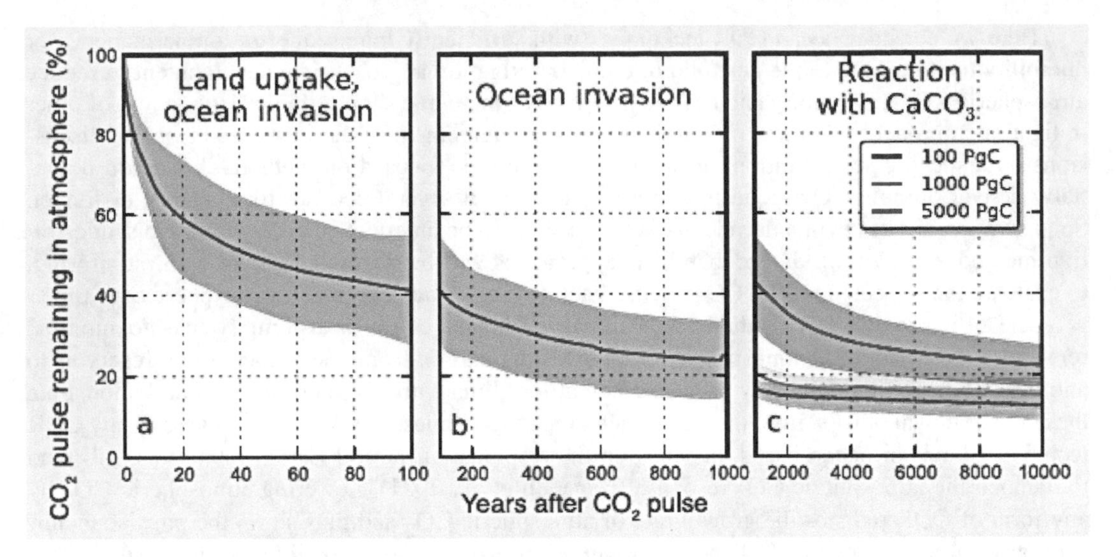

FIGURE 1.8 Percentage of emitted CO_2 remaining in the atmosphere in response to an idealized instantaneous CO_2 pulse emitted to the atmosphere in year 0 as calculated by a range of coupled climate–carbon cycle models. (a and b) Multimodel mean (blue line) and the uncertainty interval (±2 standard deviations, shading) simulated during 1,000 years following the instantaneous pulse of 100 petagrams (Pg) C (Joos et al., 2013). (c) A mean of models with oceanic and terrestrial carbon components and a maximum range of these models (shading) for instantaneous CO_2 pulse in year 0 of 100 Pg C (100 Gt C) (blue), 1,000 Pg C (orange), and 5,000 Pg C (red line) on a time interval up to 10,000 years (Archer et al., 2009). Text at the top of the panels indicates the dominant processes that remove the excess of CO_2 emitted in the atmosphere on successive timescales. SOURCE: Box 6.1, Figure 1 from Ciais et al., 2013.

fully contribute to a portfolio of responses to climate change, they need to "occur at a truly massive scale" (NRC, 2015a). It will be challenging to develop technologies to remove significant amounts of carbon from the atmosphere at a scale and cost that can be adopted in time to meet global targets for limiting warming "to well below 2 degrees Celsius, while pursuing efforts to limit the increase to 1.5 degrees" (Conference of Parties to the U.N. Framework Convention on Climate Change, December 2015).

The NRC (2015a) CDR report advised that "if carbon dioxide removal technologies are to be viable, it is critical now to embark on a research program to lower the technical barriers to efficacy and affordability while remaining open to new ideas, approaches, and synergies." In 2019, the National Academies published a report that advances this goal, *Negative Emissions Technologies and Reliable Sequestration: A Research Agenda* (NASEM, 2019). The study found that, to meet climate goals, some form of CDR will likely be needed to remove roughly 10 Gt CO_2/yr by mid-century and 20 Gt CO_2/yr by the end of the century. To help meet that goal, four land-based CDR approaches were ready for large-scale deployment: afforestation/reforestation, changes in forest management, uptake and storage by agricultural soils, and bioenergy with carbon capture and storage, based on the potential to remove carbon at costs below \$100/t CO_2. The study included a detailed research agenda to assess the benefits, risks, and sustainable scale potential for those four land-based approaches to CDR. The committee also examined approaches described as coastal blue carbon, limited to nearshore coastal land management strategies (e.g., seagrasses and wetlands), concluding that the potential for removing carbon is lower than other approaches but continued research is warranted to understand how future uptake of carbon may be affected by climate change and coastal management practices.

BOX 1.2
Research Progress and Scientific Confidence

In an ideal situation, research investments lead over time to more accumulated knowledge that reduces scientific uncertainties and increases confidence in scientific conclusions in a roughly linear fashion. Following past anecdotal experiences with new research fields, however, the path linking the volume of research to scientific confidence may be nonlinear or even nonmonotonic, with peaks and valleys over time (Busch et al., 2015) (Figure 1.9). In some cases, early research may show promising results that do not hold up to further studies as more studies highlight previously unforeseen complexities; the engagement of a larger scientific community also brings in diverse new perspectives and backgrounds to the field. Thus, an initial high confidence level early in a field's development may be followed by a period of reduced scientific confidence. Only after more extensive research and deeper understanding may a robust measure of confidence (and corresponding uncertainties) emerge. This is relevant to developing a research strategy for ocean CDR because of the striking differences in maturity across different approaches.

For example, current knowledge for several approaches (ocean alkalinity enhancement, electrochemical, and artificial upwelling) is based primarily on theoretical considerations and modeling, analogs to natural processes, and simple laboratory and/or small-scale tests. Understanding for some aspects of these approaches is relatively low, particularly for potential unintended environmental consequences, but optimism is high among some proponents that responsible CDR can be developed at a scale that benefits society. This optimism may or may not hold up with the further and more comprehensive research over time that is likely required to properly characterize the actual underlying confidence level needed to inform societal decisions. The evolution of scientific confidence over time can be an issue when communicating with the public, policy makers, and stakeholders.

In contrast, at least for ocean iron fertilization, a longer period of study and open-ocean fieldwork has led to a substantial understanding of the upper-ocean biological responses to iron fertilization and effects on ocean CO_2 uptake, at least over the time span of the experiments. At the same time, the added scientific understanding has led to a decrease in acceptance of ocean iron fertilization by some groups, even within the scientific community, given the complexity of what is known now to be still large uncertainties regarding the intended and unintended consequences at scale (e.g., Chisholm et al., 2001; Johnson and Karl, 2002).

The important point of the schematic is that CDR approaches may follow different learning curves over time, and some care may be required in interpreting early results. The scientific confidence regarding the viability of different CDR approaches may change with time as further research expands scientific understanding, requiring flexibility in the design of any research program and periodic review of priorities as the science evolves.

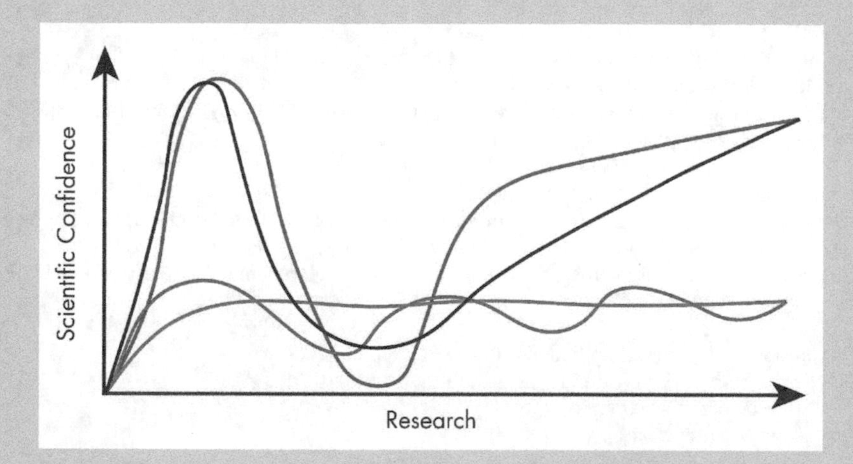

FIGURE 1.9 Potential learning curves over time for ocean CDR approaches.

The 2019 report did not examine the more global ocean-based approaches but did recognize the potential for ocean-based CDR and the need for a research strategy to explore these options. To address this gap in understanding and the need for further exploration into CDR options that could feasibly contribute to a larger climate mitigation strategy, the National Academies convened the Committee on A Research Strategy for Ocean-Based Carbon Dioxide Removal and Sequestration. Specifically, this committee was assembled to develop a research agenda to assess the benefits, risks, and potential for responsible scale-up of a range of ecosystem-based and technological ocean-based CDR approaches. The committee's Statement of Task is presented in Box 1.3. The six approaches in the Statement of Task, as defined by the study sponsor, are representative of the range of proposed ocean-based CDR approaches; they should not be taken as a comprehensive list of all ocean-based approaches. Additionally, this report does not repeat the work on blue carbon in vegetated coastal ecosystems covered in the 2019 National Academies report.

The intended audience for this report is wide ranging, including those interested in incorporating ocean-based CDR as part of a larger climate mitigation strategy. The committee's task (Box 1.3) and focus was on identifying research and development needs within the ocean-based CDR space that could supply information to decision makers considering next steps involved in the scale-up of promising ocean-based CDR solutions.

Funding for the study came from the ClimateWorks Foundation, a nonprofit organization serving as a philanthropic platform for advancing climate solutions. As part of ClimateWorks Ocean CDR Portfolio, this task included examination of six groups of ocean-based CDR approaches, to identify key scientific and technological questions, including questions surrounding governance and societal dimensions that could increase the viability of responsible use of the ocean as a mechanism for carbon removal from Earth's atmosphere.

BOX 1.3
Statement of Task

With the goal of reducing atmospheric carbon dioxide, an ad hoc committee will conduct a study exclusively focused on carbon dioxide removal (CDR) and sequestration conducted in coastal and open ocean waters to:

A. Identify the most urgent unanswered scientific and technical questions, as well as questions surrounding governance, needed to: (i) assess the benefits, risks, and potential scale for carbon dioxide removal and sequestration approaches; and (ii) increase the viability of responsible carbon dioxide removal and sequestration;
B. Define the essential components of a research and development program and specific steps that would be required to answer these questions;
C. Estimate the costs and potential environmental impacts of such a research and development program to the extent possible in the timeframe of the study.
D. Recommend ways to implement such a research and development program that could be used by public or private organizations.

The carbon dioxide removal approaches to be examined include:

- Iron, nitrogen, or phosphorus fertilization
- Artificial upwelling and downwelling
- Seaweed cultivation
- Recovery of ocean and coastal ecosystems, including large marine organisms
- Ocean alkalinity enhancement
- Electrochemical ocean CDR approaches.

1.7 STUDY APPROACH AND FRAMEWORK FOR ASSESSMENT

The study is organized around the six groups of ocean-based CDR approaches identified in the Statement of Task: nutrient fertilization, artificial upwelling and downwelling, seaweed cultivation, ecosystem recovery, alkalinity enhancement, and electrochemical approaches, illustrated in Figure 1.10. Chapter 2 of the report covers a series of crosscutting issues—legal, regulatory, and governance issues, social dimensions and justice considerations, and economic and funding considerations—foundational to all ocean-based CDR approaches. Chapter 2 also includes a subsection on common ocean monitoring requirements that will be needed for both CDR verification and assessment of environmental impacts. Chapters 3–8 then document the six ocean-based CDR groups followed by a synthesis chapter (Chapter 9):

- **Nutrient fertilization** (Chapter 3): Addition of micronutrients (e.g., iron) and/or macronutrients (e.g., phosphorus or nitrogen) to the surface ocean may in some settings increase photosynthesis by marine phytoplankton, and can thus enhance uptake of CO_2 and transfer of organic carbon to the deep sea where it can be sequestered for timescales of a century or longer. As such, nutrient fertilization essentially locally enhances the natural ocean biological carbon pump using energy from the sun, and in case of iron, relatively small amounts of iron are needed.
- **Artificial upwelling and downwelling** (Chapter 4): A process where water from depths generally cooler and more nutrient and carbon dioxide rich than surface waters is pumped into the surface ocean. Artificial upwelling has been suggested as a means to increase localized primary production and ultimately export production and net CDR. Artificial downwelling is the downward transport of surface water; this activity has been suggested as a mechanism to counteract eutrophication and hypoxia in coastal regions by increasing ventilation below the pycnocline and as a means to carry carbon into the deep ocean.
- **Seaweed cultivation** (Chapter 5): The process of producing macrophyte organic carbon biomass via photosynthesis and transporting that carbon into a carbon reservoir removes CO_2 from the upper ocean. Large-scale farming of macrophytes (seaweed) can act as a CDR approach by transporting organic carbon to the deep sea or into sediments.
- **Recovery of ocean and coastal ecosystems** (Chapter 6): CDR and sequestration through protection and restoration of coastal ecosystems, such as kelp forests and free-floating *Sargassum,* and the recovery of fishes, whales, and other animals in the oceans.
- **Ocean alkalinity enhancement (OAE)** (Chapter 7): Chemical alteration of seawater chemistry via addition of alkalinity through various mechanisms including enhanced mineral weathering and electrochemical or thermal reactions releasing alkalinity to the ocean, with the ultimate aim of removing CO_2 from the atmosphere.
- **Electrochemical approaches** (Chapter 8): Removal of CO_2 or enhancement of the storage capacity of CO_2 in seawater (e.g., in the form of ions or mineral carbonates) by enhancing its acidity or alkalinity, respectively. These approaches exploit the pH-dependent solubility of CO_2 by passage of an electric current through water, which by inducing water splitting ("electrolysis") changes its pH in a confined reaction environment. As one example, OAE may be accomplished by electrochemical approaches.

For clarity in the report, we refer to carbon dioxide removal or CDR as intentional efforts to remove CO_2 from the atmosphere and store or sequester that carbon in some reservoir isolated from the atmosphere for some extended period of time, typically multiple decades or longer. Natural biotic and abiotic processes also act to sequester carbon away from the atmosphere—the atmosphere holds only a fraction of the amount of carbon in the ocean and land biosphere, let alone

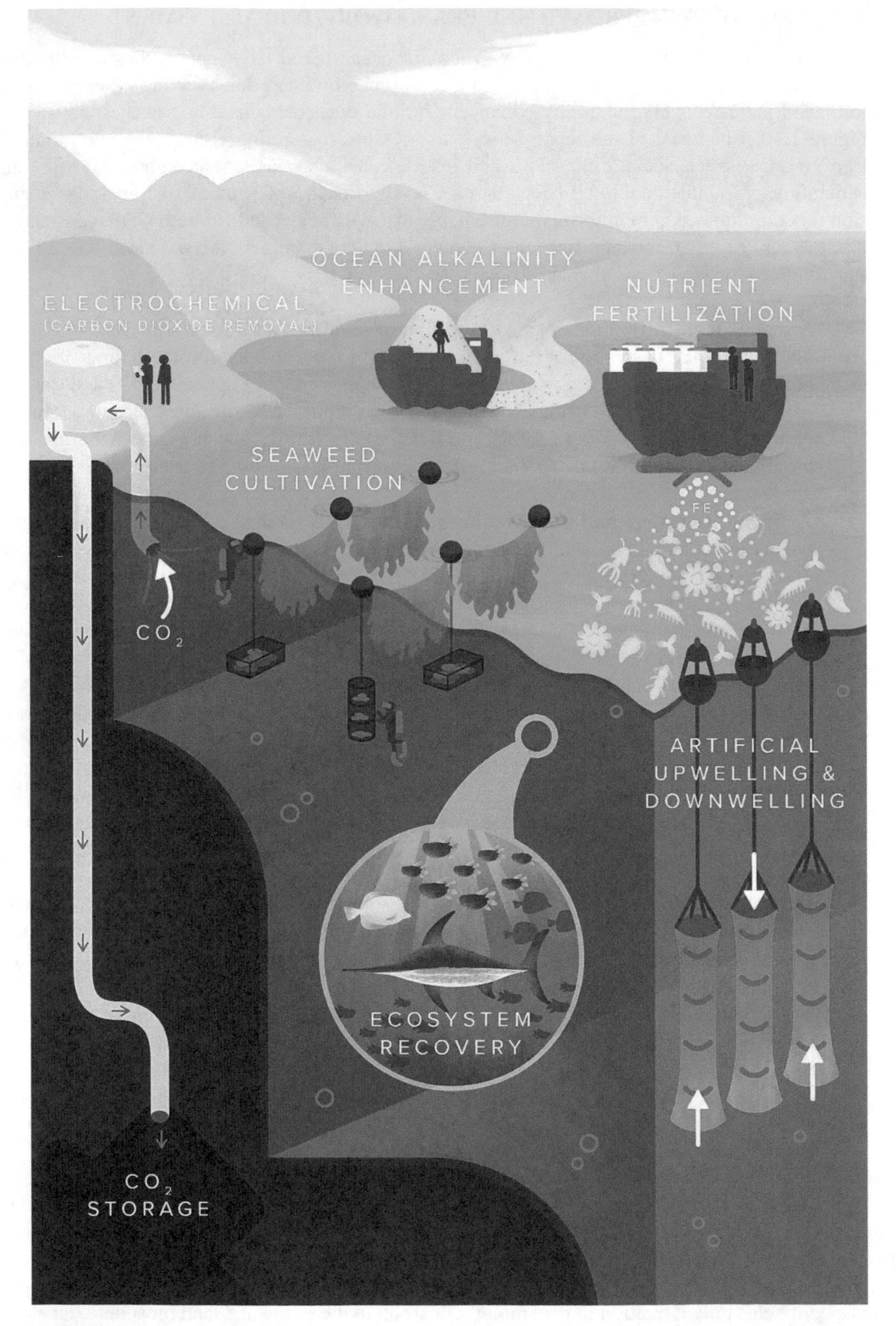

FIGURE 1.10 Ocean-based CDR approaches considered in this study.

more slowly evolving geological reservoirs. Where possible, we attempt to keep distinct deliberate human actions that enhance carbon storage away from the atmosphere; they are often perturbations on the much larger natural carbon fluxes and storage reservoirs.

The committee used a variety of information sources to inform and enrich deliberations and conduct their assessment, including a review of the scientific literature and a series of public meetings held in the virtual setting including four workshops and two additional public sessions. More than 65 experts from academic, governmental, and nongovernmental communities (see Appendix B for a list of experts invited to speak to the committee) were invited to present to the committee to assist the committee in better understanding stakeholder interest and exploring the current state of knowledge, potential, and limitations of ocean-CDR approaches. Workshop and meeting programs were developed to encourage discussion from diverse perspectives on ocean CDR feasibility and included presentations, made publically available, as well as moderated panel discussions incorporating questions from the committee and the online audience.

Each of the six groups of ocean-based CDR approaches was evaluated against a common set of criteria, where feasible. The criteria were developed by the committee based on specific elements included in the Statement of Task and from a review of previous planning and synthesis documents on CDR (e.g., NRC, 2015a; GESAMP, 2019; NASEM, 2019). The criteria were also used as prompts for invited speakers for the committee's public sessions. These criteria together inform discussion on the *viability* (or *feasibility*) of responsible CDR and sequestration as highlighted in the Statement of Task. The criteria investigated include the following:

Knowledge base: What is the current state of scientific and technical understanding and readiness? How much of current understanding is based on theory and models and laboratory-scale experiments versus contained field experiments (i.e., mesocosms or similar approaches) and uncontained ocean perturbation experiments? What are the main knowledge gaps, and what are the uncertainties and/or confidence in this knowledge?

Efficacy: Can effective CDR from the atmosphere be demonstrated? Does the approach meet additionality? That is, on the system-level scale, what is the expected net CDR from the atmosphere and are there any compensating climate mitigation effects such as release of other GHGs? What, if any, downstream effects will occur and how does this influence efficacy?

Durability or permanence: Where is the excess carbon stored? On what timescale(s) will the carbon be released back into the atmosphere? What are the risk factors, both natural and social, associated with CO_2 release? Until widely accepted methods are developed to equate varied durability terms, longer and more durable storage terms have greater value.

Monitoring and verification: What are the monitoring and verification activities needed to quantify CDR efficacy (carbon accounting of the CDR from atmosphere, the increase in carbon stored in some non-atmosphere reservoir, and timescale of loss of sequestered carbon back to atmosphere)? Similarly, what are the monitoring needs to identify environmental and social impacts? Are there potential synergies with other ocean and environmental/climate observing systems?

Scale: What is the potential scale of the CDR technique, in terms of annual CDR, at partial up to full deployment? Are there geographic constraints on efficacy and total scale? Is there information on the temporal ramp-up rate to deploy the approach at scale? To facilitate comparisons across methods, a nominal annual scale of 0.1 Gt CO_2/yr is used. While smaller than the possible total CDR demand of up to tens of Gt CO_2/yr, the nominal scale may be sufficient to contribute to a portfolio of CDR approaches.

Viability and barriers: What is the potential viability of the CDR approach for deployment, taking into consideration a full suite of technical, scientific, economic, safety, and sociopolitical factors? What are the possible environmental and social impacts of the CDR approach, considering both intended and unintended consequences? Are the impacts localized to the marine environment, or do they extend into coastal and terrestrial regions? Are there possible co-benefits of the CDR approach, or is the CDR approach a co-benefit for some other environmental or conservation goal? What are the costs of the CDR approach ($/t CO_2) including the CO_2 removal/sequestration and the monitoring and verification costs for carbon accounting and environmental or social impacts? What are the energy, resource, infrastructure, land, and ocean-space requirements for the CDR approach?

Governance and social dimensions: What is the governance landscape for research on and possible future deployment of the CDR approach? Here governance is defined broadly to mean the legal, policy, and social context in which activities relating to ocean-based CDR and sequestration take place. It encompasses the laws and rules applying to activities, as well as the policies, processes, and institutions by which decisions about activities are made, including the role of various stakeholders and the public in decision-making. What are the social dimensions and environmental justice issues associated with the CDR approach?

Research and Development (R&D) opportunities: What are the R&D opportunities for the CDR approach over the next decade with the objective that research investments in the near term should better inform societal decisions in the future about potential deployment or not of a CDR approach? How can CDR research programs be framed in terms of "responsible innovation," defined as "taking care of the future through collective stewardship of science and innovation in the present" (Stilgoe et al., 2013)? Are there best practices for CDR research that include transparency, adequate monitoring (for accounting and for environmental and social impacts) that limit any potential negative impacts of the research, and include engagement of coastal communities and the public? Are there research opportunities for expanding knowledge by moving research from modeling and laboratory scale to carefully constructed field experiments? Will CDR research have co-benefits of improving ocean science understanding? What are the possible funding mechanisms for CDR research?

2

Crosscutting Considerations on Ocean-based CDR R&D

This chapter addresses several crosscutting considerations that are relevant to all ocean carbon dioxide removal (CDR) techniques. It begins with a discussion of the existing international and domestic legal frameworks for ocean CDR research and deployment. That is followed by a discussion of the social dimensions of ocean CDR, including issues relating to public and community acceptance, environmental and climate justice considerations, and the political dynamics of ocean CDR. Finally, the chapter discusses other factors affecting the viability of ocean CDR, including monitoring and verification and funding. The chapter concludes with a discussion and summary of research needed to address these foundational, cross-cutting considerations.

2.1 LEGAL AND REGULATORY LANDSCAPE

The current legal framework for ocean CDR is highly fragmented, in large part due to the shared nature of the oceans. Around 60 percent of the oceans comprise so-called international waters, which are not under the authority or control of any one country, but rather open to use by all in accordance with international law. Coastal countries and, in some cases, their administrative divisions, share authority over the remainder of the oceans. As such, depending on where they occur, ocean CDR projects may be subject to various international and/or domestic laws.

At the international level and domestically in the United States, there is no single, comprehensive legal framework specific to ocean CDR research or deployment. Although there has been an attempt to regulate certain ocean CDR techniques—most notably, nutrient fertilization—under existing international agreements, there remain significant gaps in the international legal framework.

Notwithstanding the lack of international and domestic law specifically governing ocean CDR research and deployment, projects could be subject to a variety of general environmental and other laws. Because those laws were developed to regulate other activities, there is often uncertainty as to how they will apply to ocean CDR research and deployment. Further research is needed both to resolve unanswered questions about the application of existing law to ocean CDR projects and to develop new model governance frameworks for such projects.

Developing a clear and consistent legal framework for ocean CDR is essential to facilitate research and (if deemed appropriate) full-scale deployment, while also ensuring that projects are conducted in a safe and environmentally sound manner. Having appropriate legal safeguards in place is vital to minimize the risk of negative environmental and other outcomes and should help to promote greater confidence in ocean CDR among investors, policy makers, and other stakeholders. It is, however, important to avoid imposing inappropriate or overly strict requirements that could unnecessarily hinder ocean CDR research and deployment. Having clearly defined requirements should simplify the permitting of projects and reduce uncertainties and risks for project developers.

Jurisdiction over the Oceans

The extent of countries' jurisdiction over the oceans is defined by international law as set out in the 1982 United Nations Convention on the Law of the Sea (UNCLOS). Although the United States is not a party to UNCLOS, it recognizes many of its provisions (including those discussed in this subsection) as forming part of customary international law, and thus abides by them.

UNCLOS distinguishes the oceans from countries' internal waters (see Figure 2.1). The dividing line between the two is known as the baseline and is normally the low-water line along the relevant country's coast.[1] Waters situated landward of the baseline are internal waters over which the country has full sovereign rights.[2] Ocean waters, situated beyond the baseline, are divided into several zones, each of which has a different legal status (see Table 2.1).

The U.S. Territorial Sea and Exclusive Economic Zone (EEZ) are shown in Figure 2.2. Jurisdiction over the U.S. territorial sea is shared among the coastal states and territories and the federal government. Each coastal state has primary jurisdiction over areas extending 3 nautical miles from

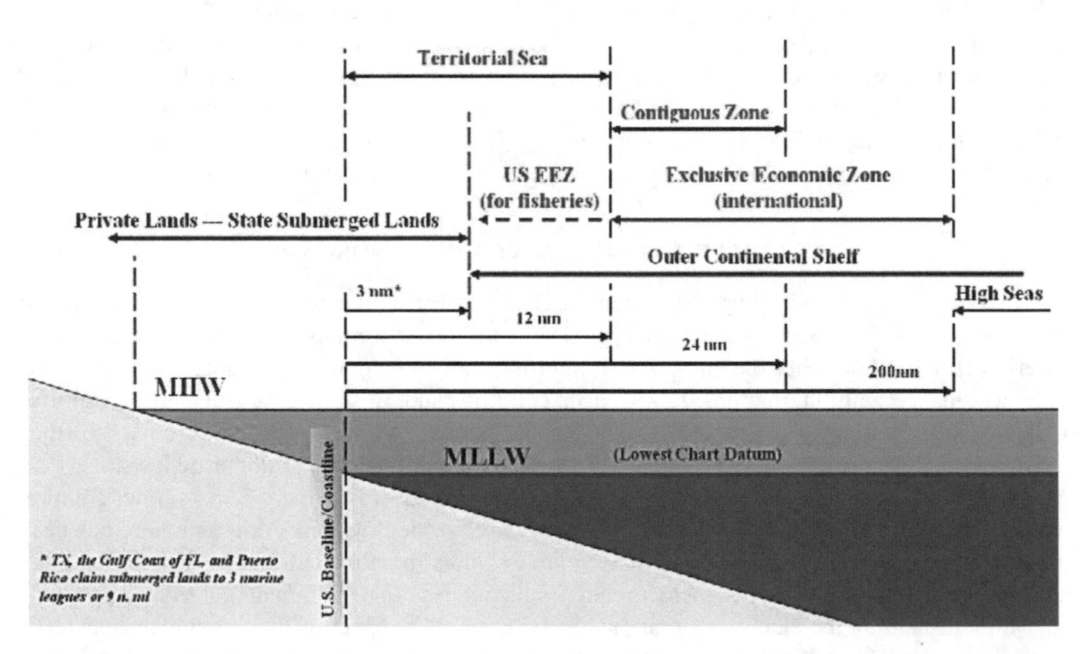

FIGURE 2.1 Maritime zones identified in United Nations Convention on the Law of the Sea. SOURCE: NOAA.

[1] Art. 5, United Nations Convention on the Law of the Sea, Dec. 10, 1982, 1833 U.N.T.S. 397 (hereinafter "UNCLOS"). See also art. 7, UNCLOS (providing for the use of "straight baselines" in some circumstances).

[2] Art. 8, UNCLOS.

TABLE 2.1 Zonal Jurisdictions in Ocean Waters

Zone	Location	Status
Territorial sea	0 to 12 nautical miles from the baseline	Part of the sovereign territory of the coastal country.[a]
Contiguous zone	12 to 24 nautical miles from the baseline	Country has authority to prevent and punish infringement of its customs, fiscal, immigration, or sanitary laws and regulations.[b]
Exclusive economic zone (EEZ)	12 to 200 nautical miles from the baseline	Country has sovereign rights to explore for, exploit, conserve, and manage natural resources and perform other activities for the economic exploitation of the zone, and jurisdiction over artificial islands and other structures, marine scientific research, and the protection and preservation of the marine environment.[c]
Continental shelf	12 to 200 nautical miles from the baseline or the outer edge of the continental margin (subject to certain limits)	Country has sovereign rights to explore and exploit natural resources in the continental shelf.[d]
High seas	Areas not included in the above categories	No country has sovereign rights. Open to use by all countries.[e]

[a] Art. 2-3, United Nations Convention on the Law of the Sea, Dec. 10, 1982, 1833 U.N.T.S. 397 (hereinafter "UNCLOS").
[b] Art. 33, UNCLOS.
[c] Art. 55-57, UNCLOS.
[d] Art. 76-78, UNCLOS.
[e] Art. 86-87, UNCLOS.

its coastline, except in parts of the Gulf of Mexico, where the jurisdiction of Texas and Florida extends 9 nautical miles from the coast.[3] Puerto Rico's jurisdiction also extends 9 nautical miles from the coast, while other territories only have jurisdiction over areas within 3 nautical miles of the coast.[4] (Areas under the primary jurisdiction of states or territories are referred to as "state waters.")

Local governments have limited jurisdiction in state waters in some areas. Additionally, the federal government retains some regulatory authority in state waters (e.g., to regulate commerce, navigation, national defense, and international affairs).[5] The federal government also has exclusive authority over federal waters, which extend beyond state waters, up to 200 nautical miles from the baseline.

Some Native American tribes have rights to fish in U.S. state and federal waters and co-manage fishery resources with state and federal governments.[6] U.S. courts have held that tribal fishing rights create an implied duty on the part of state and federal governments to avoid damage to fish habitat.[7] Federal agencies are required to consult with tribal officials before taking any action that will "have substantial direct effects on one or more Indian tribes."[8] The National Oceanic and Atmospheric Administration (NOAA) has issued guidelines for conducting such consultations.[9]

[3] 43 U.S.C. §§ 1301 & 1312; U.S. v. Louisiana, 100 S. Ct. 1618 (1980), 420 U.S. 529 (1975), 394 U.S. 11 (1969), 389 U.S. 155 (1967), 363 U.S. 1 (1960), 339 U.S. 699 (1950).

[4] 48 U.S.C. §§ 749 & 1705.

[5] 43 U.S.C. § 1314.

[6] See e.g., Treaty with the Dwamish, Suquamish, etc. (commonly known as the Treaty of Point Elliot), Art. 5, Jan. 22, 1855, 12 Stat. 927.

[7] United States v. Washington, 853 F.3d 946 (9th Cir. 2017), *cert. granted,* 138 S. Ct. 735 (2018).

[8] Executive Order No. 13175, 65 Fed. Reg. 65249 (2000).

[9] See http://www.legislative.noaa.gov/policybriefs/NOAA%20Tribal%20consultation%20handbook%206%2013.11%20final.pdf.

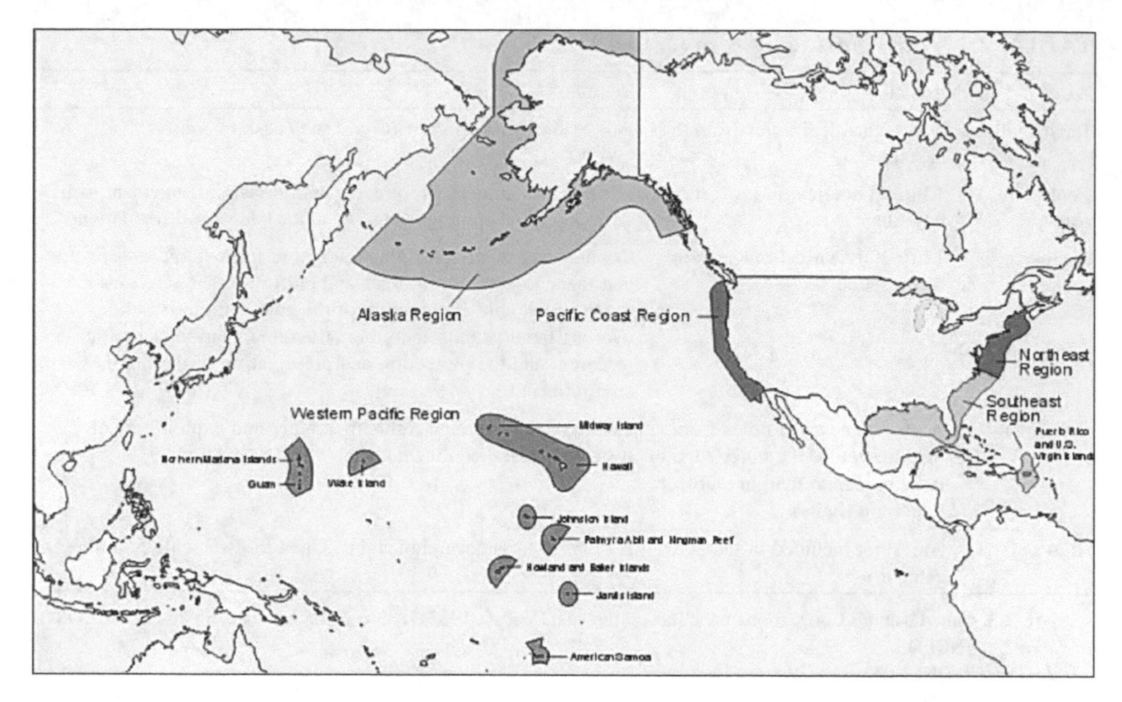

FIGURE 2.2 U.S. territorial sea and exclusive economic zone. SOURCE: NOAA (2021).

International Law Relevant to Ocean CDR

Ocean-based activities are governed by a large body of international law, comprising both international agreements that specific countries have consented to be bound by and customary rules that establish universal legal standards that are binding on all countries. This body of international law was developed to deal with issues such as ocean access, marine pollution, and fisheries management. At the time they were negotiated, none of the international ocean agreements was intended to regulate ocean CDR research or deployment. However, many of the agreements include provisions that could apply to in situ testing, and/or full-scale deployment, of one or more ocean CDR techniques. The parties to one agreement—the London Protocol—have adopted an amendment that is intended to establish a specific regulatory framework for so-called marine geoengineering activities that involve the addition of materials to the oceans (e.g., nutrient fertilization). That amendment has not yet taken effect, however.

Previous studies have considered the application of existing international law to projects involving research into, or full-scale deployment of, various ocean CDR techniques (e.g., Freestone and Rayfuse 2008; Abate and Greenlee, 2009; Verlaan, 2009; Proelss, 2012; Proelss and Hong, 2012; Kuokkanen and Yamineva, 2013; Scott, 2013; Reynolds, 2015, 2018a, 2018b; McGee et al., 2017; Brent et al., 2018; Brent et al., 2019; GESAMP, 2019; Webb et al., 2021). The studies have generally concluded that existing international law is poorly suited to regulating ocean CDR. Studies have identified various gaps and shortcomings in the existing international legal framework and highlighted challenges that may arise from its application to ocean CDR. Some have also recommended principles to guide the development of new international governance frameworks for ocean CDR (e.g., Abate and Greenlee, 2010; McGee et al., 2017).

This section summarizes the prior research on the application of existing international law to ocean CDR. It is important, at the outset, to note the limited effect of some of the international laws discussed. While all countries are generally bound by the rules of customary international law, international agreements are only binding on countries that are party to them. Several of the international agreements discussed below have a relatively small number of parties and therefore limited application. Moreover, international agreements and customary international law generally do not impose binding obligations on private actors (e.g., individuals and corporations). However, to comply with their international legal obligations, countries may be required to adopt domestic laws that apply to private actors under their jurisdiction. This would include private actors engaging in ocean CDR activities:

- in the relevant country's territorial sea or EEZ, and
- in other areas, where activities are performed:
    - using vessels registered or "flagged" in the relevant country; or
    - in some cases, using materials that were loaded onto a vessel in the relevant country.

Note also that some international agreements establish different rules for ocean CDR research versus full-scale deployment. Where that is the case, it is noted below. Many agreements do not, however, expressly distinguish between research and deployment.

Relevant Principles of Customary International Law

Several rules of customary international law could apply to research into, and full-scale deployment of, ocean CDR techniques. Previous studies (e.g., Reynolds, 2015; Brent et al., 2019) have concluded that ocean CDR and other geoengineering activities could trigger the so-called "no harm" rule of customary international law. Under the no harm rule, countries have a "responsibility to ensure that activities within their jurisdiction or control do not cause damage to the environment of other[s]" or the global commons (including the high seas).[10] The rule imposes a "due diligence" obligation on countries to "do the utmost" to avoid or minimize transboundary environmental harm, including by adopting and enforcing domestic laws to control potentially harmful activities.[11] Researchers (e.g., Brent et al., 2019; Webb et al., 2021) have concluded that, to fulfil their obligation, countries may need to establish domestic laws respecting ocean CDR.

Countries also have a procedural obligation under customary international law to assess whether projects under their jurisdiction are at risk of causing significant transboundary environmental harm (e.g., to other states' territory or the high seas).[12] While there is no agreed definition of what constitutes "significant" harm, the International Law Commission has interpreted the term as requiring damage that is "more than detectable, but need not be at the level of serious or substantial."[13] Past research (e.g., Brent et al., 2019) has identified various factors relevant to assessing the risk of harm from ocean CDR, including the sensitivity of the area likely to be affected and the nature, scale, and permanence of the effects. Ultimately, however, the assessment will need to be undertaken on a case-by-case basis by the country under whose jurisdiction the activity occurs.

---

[10] Declaration of the United Nations Conference on Environment and Development, Principle 2, UN Doc A/CONF.151/26/ Rev. 1, June 3-14, 1992.

[11] Responsibilities and Obligations of States Sponsoring Persons and Entities with Respect to Activities in the Area, Advisory Opinion, Int'l Tribunal for the Law of the Sea, Case No. 17, 110-116 (Feb. 2011).

[12] Certain Activities Carried Out by Nicaragua in the Border Area (Costa Rica v. Nicaragua) & Construction of a Road in Costa Rica Along the San Juan River (Nicaragua v. Costa Rica) (International Court of Justice, General List Nos 150 and 152, 16 Dec. 2015).

[13] International Law Commission, Draft Articles on Prevention of Transboundary Harm from Hazardous Activities, with Commentaries 152 (2001), https://legal.un.org/ilc/texts/instruments/english/commentaries/9_7_2001.pdf.

Additional international law obligations apply where the initial assessment indicates that a project presents significant risks. A more comprehensive environmental impact assessment (EIA) must be conducted for risky projects and, where the EIA confirms the potential for significant transboundary environmental damage, those potentially affected must be notified and consulted with.[14] However, international law does not dictate the conduct of the EIAs or consultations, giving countries broad discretion to determine how to comply.

Relevant International Agreements

UNFCCC, Kyoto Protocol, and Paris Agreement

The United Nations Framework Convention on Climate Change (UNFCCC) was adopted in May 1992 and entered into force in March 1994. As of August 2021, 196 countries, including all United Nations member states, and the European Union were party to the UNFCCC.[15] A subset of parties subsequently agreed to the Kyoto Protocol, which was adopted in December 1997 and entered into February 2005, and the Paris Agreement, which was adopted in December 2015 and entered into force in November 2016. The United States never became a party to the Kyoto Protocol, but is a party to the Paris Agreement.[16]

Past studies (e.g., Proelss, 2012; Reynolds, 2015; Craik and Burns, 2016; Brent et al., 2019) have concluded that the UNFCCC, Kyoto Protocol, and Paris Agreement implicitly approve the use of CDR techniques to mitigate climate change. The overarching goal of the UNFCCC is to stabilize atmospheric greenhouse gas concentrations at a level that will "prevent dangerous anthropogenic interference with the climate system."[17] To that end, the UNFCCC requires each developed country party to take steps to limit its greenhouse gas emissions and "protect[ ] and enhanc[e] its greenhouse gas sinks,"[18] which could be achieved through ocean CDR. The UNFCCC defines the term "sink" broadly to include "any process, activity or mechanism which removes a greenhouse gas, an aerosol or a precursor of a greenhouse gas from the atmosphere."[19] The definition is not limited to naturally occurring techniques and would encompass human interventions.

The Kyoto Protocol similarly requires developed country parties to protect and enhance greenhouse gas sinks and to conduct research into, and adopt policies to promote the use of, "carbon dioxide sequestration techniques."[20] That term, although not defined in the Kyoto Protocol, could include ocean CDR techniques that result in the storage of CO_2 in the marine environment.

---

[14] Responsibilities and Obligations of States Sponsoring Persons and Entities with respect to Activities in the Area, Advisory Opinion, Int'l Tribunal for the Law of the Sea, Case No. 17, 145-149 (Feb. 2011); Certain Activities Carried Out by Nicaragua in the Border Area (Costa Rica v. Nicaragua), Judgement, ICJ Rep. 2015, 665 at 706-707 (Dec. 2015).

[15] See https://treaties.un.org/Pages/ViewDetailsIII.aspx?src=IND&mtdsg_no=XXVII7&chapter=27&Temp=mtdsg3&clang=_en.

[16] The United States adopted the Paris Agreement on September 3, 2015. On November 4, 2019, the United States notified the United Nations Secretary General of its intent to withdraw from the Paris Agreement. Under the terms of the Paris Agreement, the withdrawal took effect 1 year later on November 4, 2020. The United States rejoined the Paris Agreement on January 20, 2021. See https://www.un.org/sg/en/content/sg/note-correspondents/2017-08-04/note-correspondents-paris-climate-agreement; https://www.google.com/url?q=https://treaties.un.org/doc/Publication/CN/2019/CN.575.2019-Eng.pdf&sa=D&source=editors&ust=1628009662970000&usg=AOvVaw3poBcPe4P3tUXvNd5YuBp9; https://www.state.gov/the-united-states-officially-rejoins-the-paris-agreement/.

[17] Art. 2, United Nations Framework Convention on Climate Change, May 9, 1992, S. Treaty Doc No. 102-38, 1771 U.N.T.S. 107 (hereinafter UNFCCC).

[18] Art. 4(2)(a), UNFCCC.

[19] Art. 1, UNFCCC.

[20] Art. 2(1)(a)(ii) & (iv), Kyoto Protocol to the United Nations Framework Convention on Climate Change, Dec. 10, 1997, 2303 U.N.T.S. 148.

Finally, the Paris Agreement requires all parties, including the United States, to take steps to mitigate climate change, with the objective of limiting global warming to "well below" 2°C, and ideally 1.5°C, above preindustrial levels.[21] Under the Paris Agreement, parties "aim to reach global peaking of greenhouse gas emissions as soon as possible" and "to achieve a balance between anthropogenic emissions by sources and removal by sinks" in the second half of the century.[22] The Paris Agreement thus anticipates that parties may mitigate climate change both by reducing anthropogenic emissions and increasing removals by sinks. The Agreement expressly states that parties "should take action to conserve and enhance, as appropriate, sinks and reservoirs of greenhouse gases."[23] Ocean CDR could, at least in some circumstances, be viewed as a way of enhancing sinks.

Each party to the Paris Agreement determines the extent to which, and how, it will contribute to the achievement of the Agreement's goals and communicates that information in its "nationally determined contribution" (NDC).[24] One recent study (Gallo et al., 2017) found that 27 parties included coastal carbon sequestration techniques (also known as blue carbon) in their initial NDCs. Others (e.g., Craik and Burns, 2016; Brent et al., 2019) have concluded that parties could, consistent with the terms of the Paris Agreement, incorporate ocean CDR into their NDCs. The Paris Agreement does not, however, expressly require parties to engage in ocean CDR techniques or establish specific rules for their use.

Convention on Biological Diversity

The Convention on Biological Diversity (CBD) was adopted in June 1992 and entered into force in December 1993. As of August 2021, there were 196 parties to the CBD, giving it near global coverage.[25] Notably, however, the United States is not a party to the CBD.

The CBD aims to promote "the conservation of biological diversity, [and] the sustainable use of its components." Article 3 of the CBD reiterates the customary international law obligation of countries to avoid transboundary environmental harm. Under Article 7, parties must, as far as possible and appropriate, identify potentially harmful activities and monitor their effects. Article 14 requires parties to implement procedures to conduct EIAs of proposed activities that are likely to have significant adverse effects on biological diversity and allow for public participation in the assessment process. Where an activity is likely to have transboundary effects, parties must notify, and consult with, the potentially affected countries before it occurs.[26] Parties must have in place arrangements for responding to activities that present a "grave and imminent danger" to biological diversity and, if the danger is transboundary, immediately notify potentially affected countries and take action to prevent or minimize the danger.[27]

The parties to the CBD have adopted a series of nonbinding decisions specifically addressing ocean fertilization and other so-called "geoengineering" activities. Decision IX/16, adopted in October 2008, recommends that "ocean fertilization activities" be avoided "until there is an adequate scientific basis on which to justify such activities, including assessing associated risks, and a global, transparent and effective control and regulatory mechanism is in place for these activities."[28] Decision XI/16 incorporates an exception for "small scale research studies within coastal waters," which

[21] Art. 2(1)(a) & 4, Paris Agreement, Dec. 12, 2015, U/N. Doc. FCCC/CP/2015/L.9/Rev/1 (hereinafter Paris Agreement).

[22] Art. 4(1), Paris Agreement.

[23] Art. 5, Paris Agreement.

[24] Art. 3 & 4, Paris Agreement.

[25] See https://www.cbd.int/information/parties.shtml.

[26] Art. 14, Convention on Biological Diversity, June 5, 1992, 1760 U.N.T.S. 79, 143 (hereinafter CBD).

[27] Art. 14, CBD.

[28] Para. C(4), Report of the Conference of the Parties to the Convention on Biological Diversity on the Work of its Ninth Meeting, Decision IX/16 on Biodiversity and Climate Change, Oct. 9, 2008 (hereinafter Decision IX/16).

may be authorized "if justified by the need to gather specific scientific data, and should also be subject to a thorough prior assessment of the potential impacts of the research studies on the marine environment, and be strictly controlled, and not be used for generating and selling carbon offsets or any other commercial purposes."[29]

In October 2010, the parties to the CBD adopted Decision X/33, which reiterates that ocean fertilization activities should be "addressed in accordance with decision IX/16."[30] Decision X/33 also deals more broadly with "geoengineering activities," which were initially defined to include "any technologies that deliberately reduce solar insolation or increase carbon sequestration from the atmosphere on a large scale that may affect biodiversity."[31] The decision recommends that parties and other governments:

> [e]nsure, . . . in the absence of science based, global, transparent and effective control and regulatory mechanisms for geo-engineering, and in accordance with the precautionary approach and Article 14 of the Convention, that no climate-related geo-engineering activities[ ] that may affect biodiversity take place, until there is an adequate scientific basis on which to justify such activities and appropriate consideration of the associated risks for the environment and biodiversity and associated social, economic and cultural impacts, with the exception of small scale scientific research studies that would be conducted in a controlled setting . . . if they are justified by the need to gather specific scientific data and are subject to a thorough prior assessment of the potential impacts on the environment."[32]

The parties reaffirmed the above recommendation in October 2012 in Decision XI/20[33] and again in the December 2016 Decision XIII/4.[34] In Decision XI/20, the parties also adopted a broader definition of "geoengineering," which includes:

> (a) Any technologies that deliberately reduce solar insolation or increase carbon sequestration from the atmosphere on a large scale and that may affect biodiversity . . .
> (b) Deliberate intervention in the planetary environment of a nature and scale intended to counteract anthropogenic climate change and/or its impacts . . .
> (c) Deliberate large-scale manipulation of the planetary environment . . .
> (d) Technological efforts to stabilize the climate system by direct intervention in the energy balance of the Earth for reducing global warming.[35]

Webb et al. (2021, p.19) concluded that this definition would encompass ocean CDR projects "undertaken for the purpose of mitigating climate change." However, they and others (e.g., Sugiyama and Sugiyama, 2010; Bodansky, 2011; Reynolds, 2018b; Brent et al., 2019) note that the practical effect of Decision X/33 is limited because it is nonbinding and uses soft language. While one nongovernmental organization (NGO)—the ETC Group—has argued that Decision X/33 creates a "de facto moratorium" on geoengineering that arguably overstates its legal effect (ETC Group, 2010). The decision expressly allows geoengineering research projects meeting specified criteria and, as noted, the prohibition on nonresearch projects is not legally binding.

[29] Para. C(4), Decision IX/16.

[30] Para. 8(w), Report of the Conference of the Parties to the Convention on Biological Diversity on the Work of its Tenth Meeting, Decision X/33 on Biodiversity and Climate Change, Oct. 29, 2010 (hereinafter "Decision X/33").

[31] Note 3, Decision X/33.

[32] Para 8(w), Decision X/33.

[33] Para. 1, Report of the Conference of the Parties to the Convention on Biological Diversity on the Work of its Eleventh Meeting, Decision XI/20 on Climate-Related Geoengineering, Dec. 5, 2012 (hereinafter Decision XI/20).

[34] Preamble, Report of the Conference of the Parties to the Convention on Biological Diversity on the Work of Its Thirteenth Meeting, Decision XIII/4, Dec. 10, 2016 (hereinafter Decision XIII/4).

[35] Para. 5, Decision XI/20.

United Nations Convention on the Law of the Sea

UNCLOS was adopted in December 1982 and entered into force in November 1994. A separate Agreement for the Implementation of the Provisions of UNCLOS Relating to the Conservation and Management of Straddling Fish Stocks and Highly Migratory Fish Stocks (Straddling Fish Stocks Agreement) was adopted in August 1995 and entered into force in November 2001. As of August 2021, there were 168 parties to UNCLOS, and 91 parties to the Straddling Fish Stocks Agreement.[36] The United States is a party to the Straddling Fish Stocks Agreement only.

Ocean CDR research projects may be subject to Part XIII of UNCLOS, which deals with "marine scientific research." Although UNCLOS does not define what constitutes "marine scientific research," the term is commonly interpreted to encompass any "scientific investigation . . . concerned with the marine environment," including the water column and seabed. Researchers (e.g., Proelss and Hong, 2012; Brent et al., 2019, p. 19) have concluded that projects aimed at demonstrating or testing ocean CDR techniques would qualify if conducted "in situ" in the ocean.

Part XIII of UNCLOS recognizes the right of each country to conduct marine scientific research within its own territorial sea and EEZ, within the terrestrial sea and EEZ of another country with that country's consent, and on the high seas.[37] The right to conduct marine scientific research is, however, subject to countries' general duty under UNCLOS to protect and preserve the marine environment (discussed below). Marine scientific research must be conducted "exclusively for peaceful purposes," in accordance with "appropriate scientific methods," and must not "unjustifiably interfere with other legitimate uses" of the oceans.[38]

Countries wanting to conduct marine scientific research in the EEZ or on the continental shelf of another country must provide the host country with detailed information about the nature and objectives of the project, precisely where and when it will occur, and the activities and equipment to be used.[39] The host country must be given the opportunity to participate in the project and, if requested, access to the research data, samples, and results.[40] The research results must also be made available internationally.[41] Brent et al. (2019) have argued that these reporting requirements could help promote transparency in ocean CDR research. It is, however, important to note that the requirements will only apply to a subset of ocean CDR research projects—that is, those that are conducted by one country in the EEZ or on the continental shelf of a second country.

Part XII of UNCLOS, dealing with "Protection and Preservation of the Marine Environment," includes several provisions that could affect both research and full-scale ocean CDR projects. Article 193 recognizes that countries have a "sovereign right to exploit their natural resources." At least one study (Reynolds, 2018a) has concluded that ocean CDR may be viewed as a means of exploiting natural resources, specifically the ocean's ability to absorb CO_2, and thus within countries' sovereign rights.

Countries must exercise their sovereign rights in accordance with international law, including their obligation, under customary international law, to avoid significant transboundary environmental harm. Under Articles 192 and 193 of UNCLOS, countries also have a general obligation to protect and preserve the marine environment, and must exercise their sovereign rights in accordance with that obligation. UNCLOS includes several provisions requiring countries to take steps to con-

[36] See https://www.un.org/Depts/los/reference_files/chronological_lists_of_ratifications.htm#Agreement%20relating%20to%20the%20implementation%20of%20Part%20XI%20of%20the%20Convention.

[37] Art. 245 & 246, UNCLOS.

[38] Art. 240, UNCLOS.

[39] Art. 248, UNCLOS.

[40] Art. 249, UNCLOS.

[41] Art. 249, UNCLOS.

trol marine pollution[42] and monitor and mitigate its effects.[43] Similarly, the Straddling Fish Stocks Agreement requires parties (including the United States) to minimize pollution and its impacts, particularly on endangered fish and nonfish species.[44] The term "pollution" is defined broadly to mean:

> the introduction by man, directly or indirectly, of substances or energy into the marine environment, including estuaries, which results or is likely to result in such deleterious effects as harm to living resources and marine life, hazards to human health, hindrance to marine activities, including fishing and other legitimate uses of the sea, impairment of quality for use of the sea water and reduction of amenities.

Several researchers (e.g., Boyle, 2012; Reynolds, 2015, 2018b; Marshall, 2017) have argued that this definition could encompass CO_2 in the marine environment. Ocean CDR techniques that remove CO_2 from the marine environment could, therefore, be viewed as a form of pollution control. However, others (e.g., Brent et al., 2019; Webb et al., 2021) argue that ocean CDR techniques involving the addition of materials to ocean waters, such as ocean iron fertilization and ocean alkalinity enhancement, could themselves be considered pollution of the marine environment. Article 195 of UNCLOS requires parties, when taking steps to control pollution, to avoid merely transforming one type of pollution into another. That could have implications for projects that remove CO_2, which may be considered a form of pollution, from ocean waters by adding other materials, which may also constitute pollutants, into the water.

Researchers (e.g., Reynolds, 2018a,b; Webb et al., 2021) have recommended a case-by-case assessment of ocean CDR projects. Where a project is found to involve pollution of the marine environment, the country under whose jurisdiction it occurs will need to comply with various requirements imposed under UNCLOS. Among other things, the party must notify affected countries and competent international authorities and study and document the effects of the project.[45]

UNCLOS provides that countries that fail to fulfil their "international obligations concerning the protection and preservation of the marine environment . . . shall be liable in accordance with international law."[46] Disputes may be referred to the International Tribunal for the Law of the Sea, the International Court of Justice, or a specially constituted arbitral tribunal.[47] Where a country is found to have breached its international obligations, it must cease the offending conduct (if it is continuing), "offer appropriate assurances and guarantees of non-repetition," and "make full reparation" for any damage caused to others.[48]

London Convention and Protocol

The Convention on the Prevention of Marine Pollution by Dumping of Wastes and Other Matter (London Convention) was adopted in November 1972 and entered into force in August 1975. A protocol to the London Convention (the London Protocol) was adopted in November 1996 and entered into force in March 2006. The London Protocol will replace the Convention once ratified by all contracting parties. Until that occurs, the two instruments operate concurrently. Countries that are party only to the London Convention are bound solely by that instrument, whereas those that have ratified both are subject to the London Protocol. As of August 2021, there were 87 parties

[42] Art. 194, 196 & 210-212, UNCLOS.
[43] Art. 198, 199, 200, & 204-206, UNCLOS.
[44] Art. 5, Straddling Fish Stocks Agreement.
[45] Art. 194, 196, 198, 202-209 & 211-212, UNCLOS.
[46] Art. 235(1), UNCLOS.
[47] Art. 287, UNCLOS. See also Annex VII & VIII, UNCLOS.
[48] Resolution Adopted by the United Nations General Assembly, Responsibility of States for Internationally Wrongful Acts, A/RES/56/83 (Jan. 28, 2002).

to the London Convention, and 53 parties to the London Protocol.[49] The United States is a party only to the London Convention.[50]

Countries that are party to the London Convention and/or London Protocol must adopt domestic laws to control the dumping of waste and other matter in the oceans.[51] Both the London Convention and London Protocol define "dumping" to include the "deliberate disposal of waste or other matter at sea from vessels, aircraft, platforms, or other man-made structures."[52] The definition expressly excludes "the placement of matter [in the sea] for a purpose other than mere disposal," where "such placement is not contrary to the aims of" the London Convention or Protocol.[53]

Parties to the London Convention must prohibit the dumping of eight "blacklisted" substances (identified in Annex I to the Convention and listed in Table 2.2) but can permit the dumping of other substances.[54] In contrast, parties to the London Protocol must prohibit the dumping of all substances, except eight "whitelisted" substances (identified in Annex I to the Protocol and listed in Table 2.2), which may be dumped with a permit.[55]

Some ocean CDR techniques, including ocean alkalinity enhancement, nutrient fertilization, and seaweed cultivation (in some cases), may involve "dumping" within the terms of the London Convention and Protocol (Scott, 2013; Webb et al., 2021). Whether a party to the London Conven-

TABLE 2.2 Blacklisted and Whitelisted Substances as Identified in Annex I to the London Protocol

Blacklisted Substances Under the London Convention[a]	Whitelisted Substances Under the London Protocol[b]
1. Organohalogen compounds	1. Dredged material
2. Mercury and mercury compounds	2. Sewage sludge
3. Cadmium and cadmium compounds	3. Fish waste and material resulting from industrial fish processing operations
4. Persistent plastics and other persistent synthetic materials	4. Vessels, platforms, and other manmade structures
5. Crude oil, petroleum, refined petroleum products, distillate residues, and oil wastes	5. Inert, inorganic geological material
6. Radioactive matter (except that containing de minimis levels of radioactivity)	6. Organic material of natural origin
7. Materials produced for biological and chemical warfare	7. Bulk items "comprising iron, steel, concrete or similarly unharmful materials" (subject to some limitations)
8. Industrial waste generated by manufacturing or processing operations	8. Carbon dioxide streams from carbon dioxide capture processes for sequestration.

[a]Materials containing substances 1 through 5 as "trace contaminants" only are not blacklisted. Materials containing any of the above substances, except radioactive matter, are not blacklisted if they "are rapidly rendered harmless by the physical, chemical or biological processes in the sea" and do not "make edible marine organisms unpalatable" or "endanger human health or that of domestic animals."

[b]The listed materials cease to be whitelisted if they contain levels of radioactivity greater than "de minimis concentrations."

[49] See https://www.imo.org/en/OurWork/Environment/Pages/London-Convention-Protocol.aspx.

[50] See https://www.epa.gov/ocean-dumping/ocean-dumping-international-treaties#:~:text=The%20United%20States%20ratified%20the,Parties%20to%20the%20London%20Convention.

[51] Art. IV, Convention on the Prevention of Marine Pollution by Dumping of Wastes and Other Matter, Dec. 29, 1972 (hereinafter London Convention); Art. 4, Protocol to the Convention on the Prevention of Marine Pollution by Dumping of Wastes and Other Matters, Nov. 7, 1996 (hereinafter London Protocol).

[52] Art. III(a), London Convention; Art. 1(4.1), London Protocol.

[53] Art. III(b), London Convention; Art. 1(4.2), London Protocol.

[54] Art. IV, London Convention.

[55] Art. 4, London Protocol.

tion and/or Protocol can permit such activities will, therefore, depend on the nature of the substances to be dumped. Substances blacklisted under the London Convention include "[p]ersistent plastics and other persistent synthetic materials," such as "netting and ropes," which could possibly be dumped in some ocean CDR projects (e.g., seaweed cultivation; Webb et al., 2021). Most ocean CDR projects are, however, unlikely to involve the dumping of blacklisted substances and thus could be permitted by parties to the London Convention (Freestone and Rayfuse, 2008; Scott, 2013; Webb et al., 2021). In contrast, many ocean CDR projects involving dumping likely could not be permitted by parties to the London Protocol, because the materials used are not whitelisted under the Protocol (Freestone and Rayfuse, 2008; Scott, 2013; Webb et al., 2021). One possible exception is seaweed cultivation as "organic material of natural origin," which is whitelisted under the London Protocol. (There is, however, some uncertainty as to whether the sinking of cultivated seaweed even constitutes "dumping" and is thus covered by the London Convention; see Chapter 5).

In 2008, the parties to the London Convention and London Protocol adopted a nonbinding resolution (LC-LP.1, 2008), in which they agreed that the instruments apply to ocean fertilization projects "undertaken by humans with the principal intention of stimulating primary production in the oceans" (except conventional aquaculture and mariculture and other projects related to the creation of artificial reefs).[56]

Resolution LC-LP.1 (2008) specifies when ocean fertilization projects should be considered "dumping" for the purposes of the London Convention and Protocol. The resolution draws a distinction between research and other (nonresearch) ocean fertilization projects. According to the resolution, projects involving "legitimate scientific research . . . should be regarded as [involving the] placement of matter for a purpose other than mere disposal."[57] As such, research projects will fall outside the definition of dumping, provided they are not contrary to the aims of the London Convention or Protocol. (As discussed above, the definition of "dumping" in the London Convention and Protocol excludes the "placement of matter for a purpose other than mere disposal," where such placement is not contrary to the aims of the Convention or Protocol.)

Resolution LC-LP.1 calls for a case-by-case assessment of research proposals.[58] An Assessment Framework for Scientific Research Involving Ocean Fertilization was adopted in October 2010.[59] The 2010 framework provides for a two-stage assessment process, beginning with an "initial assessment" to consider whether the project "has proper scientific attributes and qualifies as "legitimate scientific research," followed by an "environmental assessment" to evaluate its potential effects on the marine environment and measures to mitigate those effects.[60]

The 2010 framework envisages that the initial and environmental assessments will be conducted by the country in whose jurisdiction the project will take place. According to the framework, countries "should" establish processes for consulting with "all stakeholders," including other potentially affected countries. Following consultation, and based on the initial and environmental assessments, the country with jurisdiction over the project must determine whether or not it is contrary to the aims of the London Convention/Protocol. The framework states that countries "should" only conclude that a project is not contrary to the aims of the London Convention/Protocol if "conditions are in place to ensure that, as far as practicable, environmental disturbance would be minimized, and the scientific benefits maximized."[61] The framework recommends that action be taken to "manage and mitigate risks" and states that this may be achieved by imposing "temporal restrictions (e.g.,

[56] Art. 1-2 & note 3, Resolution LC-LP.1 (2008) on the Regulation of Ocean Fertilization, Oct. 31, 2008.

[57] Art. 3, Resolution LC-LP.1 (2008).

[58] Art. 4-5, Resolution LC-LP.1 (2008).

[59] Resolution LC-LP.2 (2010) on the Assessment Framework for Scientific Research Involving Ocean Fertilization, Oct. 14, 2010.

[60] Annex 6, Resolution LC-LP.2 (2010).

[61] Annex 6, Resolution LC-LP.2 (2010).

during certain oceanographic conditions or biologically important times for species of concern), spatial restrictions (e.g., proximity to areas of special concern and value), and delivery restrictions (e.g., substances, tracers, amounts, repetition)" on projects.[62] Additionally, according to the framework, projects should be carefully monitored and a contingency plan developed to enable prompt response (including "cessation of fertilization activities") if environmental impacts are more severe than anticipated.[63] (Note that these recommendations do not apply to projects that are classified as "dumping" and permitted under the London Convention or Protocol.)

Resolution LC-LP.1 (2008) declares that nonresearch ocean fertilization projects "should be considered as contrary to the aims of the Convention and Protocol" and thus "should not be allowed." That directive is not legally binding, however. Past studies (e.g., Scott, 2013; Webb et al., 2021) have concluded that parties to the London Convention, including the United States, could issue permits authorizing nonresearch ocean fertilization projects that do not involve the discharge of any blacklisted substance.

Building on Resolution LC-LP.1 (2008), in October 2013, the parties to the London Protocol agreed to amend that instrument to establish a new regulatory framework for "marine geoengineering" defined as:

> a deliberate intervention in the marine environment to manipulate natural processes, including to counteract anthropogenic climate change and/or its impacts, and that has the potential to result in deleterious effects, especially where those effects may be widespread, long lasting or severe.[64]

The 2013 amendment provides that:

> Contracting Parties shall not allow the placement of matter into the sea from vessels, aircraft, platforms or other man-made structures at sea for marine geoengineering activities listed in annex 4, unless the listing provides that the activity or the subcategory of an activity may be authorized under a permit.[65]

Annex 4 currently only lists ocean fertilization (as defined above). In the future, Annex 4 could be expanded to include other ocean CDR techniques, which involve the addition of materials to the oceans. The Joint Group of Experts on the Scientific Aspects of Marine Environmental Protection (GESAMP) has established a working group to "[p]rovide advice to the London Parties to assist them in identifying those marine geoengineering techniques that might be sensible to consider for listing in" Annex 4.[66] Researchers (e.g., Brent et al., 2019; Webb et al., 2021) have concluded that ocean alkalinity enhancement and seaweed cultivation could be included in Annex 4. However, according to Brent et al. (2019), artificial upwelling and downwelling are unlikely to qualify for inclusion because they "involve[ ] the transfer of water/nutrients from one part of the ocean to another, rather than the introduction of new matter." On this view, artificial upwelling and downwelling would not constitute "dumping" for the purposes of the London Convention or Protocol, and thus not be subject to those instruments or the 2013 amendment.

In its current form, Annex 4 prohibits the issuance of permits for ocean fertilization projects, except those involving legitimate scientific research.[67] The process set out in the 2010 assessment framework is to be used to determine whether ocean fertilization projects qualify as legitimate sci-

---

[62] Annex 6, Resolution LC-LP.2 (2010).

[63] Annex 6, Resolution LC-LP.2 (2010).

[64] Annex 1, art. 1, Resolution LP.4(8), Amendment to the 1996 Protocol to the Convention on the Prevention of Marine Pollution by Dumping of Wastes and Other Matter, 1972 to Regulate Marine Geoengineering, Oct. 18, 2013.

[65] Annex 1, art. 1, Resolution LP.4(8).

[66] See http://www.gesamp.org/work/groups/41.

[67] Annex 4, art. 1.2 & 1.3, Resolution LP.4(8).

entific research that can be permitted.[68] The 2013 amendment also includes a general assessment framework that may be used for other types of marine geoengineering activities if or when they are listed in Annex 4.[69]

The 2013 amendment had not yet entered into force as of August 2021. Under the terms of the London Protocol, to enter into force, amendments must be ratified by at least two-thirds of the parties to the Protocol.[70] As of August 2021, just 6 parties (Estonia, Finland, Germany, the Netherlands, Norway, and the United Kingdom), out of 53, had ratified the 2013 amendment, which is well below the two-thirds threshold (IMO, 2021, p. 566). Even if the two-thirds threshold is met, the amendment will only take effect for countries that are party to the London Protocol and have ratified the amendment. The amendment will not affect the United States and other countries that are party only to the London Convention.

Other Relevant International Agreements

A range of other international agreements could apply to ocean CDR activities in some circumstances. Examples include the Convention on the Prohibition of Military or Any Other Hostile Use of Environmental Modification Techniques, the Convention on Environmental Impact Assessment in a Transboundary Context, the Convention Concerning the Protection of World Cultural and Natural Heritage, and the Convention on the Conservation of Antarctic Marine Living Resources.

Additionally, in June 2015, the United Nations General Assembly agreed to develop a new agreement under UNCLOS on the conservation and sustainable use of marine biodiversity in areas beyond national jurisdiction (commonly referred to as the "Biodiversity Beyond National Jurisdiction" or "BBNJ" Agreement). Brent et al. (2019) suggested that the new agreement could incorporate rules relating to ocean CDR.

Domestic U.S. Law Relevant to Ocean CDR

There are currently no domestic U.S. laws specifically targeting ocean CDR. However, depending on the ocean CDR technique employed, projects could be subject to various general U.S. environmental and other laws. For example, several coastal states have general aquaculture laws, which could apply to some seaweed cultivation projects.

The application of existing U.S. law to ocean CDR has been the subject of little research (see, e.g., Janasie and Nichols, 2018; Webb, 2020; Prall, 2021; Webb et al., 2021). One ongoing project, led by researchers at Columbia University, is examining the application of U.S. environmental law to several ocean CDR techniques. To date, however, the researchers have only published an analysis of laws applicable to ocean alkalinity enhancement and seaweed cultivation (Webb et al., 2021). Other studies have examined the U.S. laws governing the sub-seabed storage of CO_2 that could occur in some ocean CDR projects (e.g., involving electrochemical engineering) (Webb and Gerrard, 2018, 2019). All of the studies to date have focused primarily on the application of federal environmental law to ocean CDR. There has been little analysis of potentially applicable state and local laws, implications for tribal rights, liability, and other issues.

The U.S. laws applicable to ocean CDR research and deployment will depend on where projects occur. Near-shore projects occurring within state waters (i.e., up to 3, or in the Gulf of Mexico, 9 nautical miles from shore) may be subject to U.S. federal, state, and/or local laws. Only federal law will apply to projects that occur entirely within federal waters (i.e., beyond state waters, up to

[68] Preamble, para. 3 & Annex 4, art. 1.3, Resolution LP.4(8). See also Resolution LC-LP.2 (2010) on the Assessment Framework for Scientific Research Involving Ocean Fertilization, Oct. 14, 2010.

[69] Annex 5, Resolution LP.4(8).

[70] Art. 21, London Protocol.

200 nautical miles from shore). Some projects (e.g., certain types of ocean alkalinity enhancement) may necessitate onshore activities (e.g., mining) that are subject to different laws.

A full review of all U.S. federal, state, and local laws potentially applicable to ocean CDR is not attempted here. However, in Table 2.3, we provide a non-exhaustive list of key federal environmental laws that could have implications for the use of different ocean CDR techniques. Those laws can be divided into five broad categories as follows:

- *Environmental Review Laws:* The National Environmental Policy Act (NEPA) and state equivalents may apply where ocean CDR projects are undertaken, approved, or supported by a federal or state government entity. Briefly, NEPA requires preparation of an environmental impact statement (EIS) for any major project undertaken, approved, or supported by a federal agency that "significantly affect[s] the quality of the human environment" (42 U.S.C. § 4332(2)(C)). The EIS must include an analysis of the natural, economic, social, and cultural resource effects of the project and alternatives (42 U.S.C. § 4332(c); 49 C.F.R. Part 1502). It must be developed with public input, and state and federal agencies may also be required to consult with Native American tribes (40 C.F.R. Part 1503).
- *Species Protection Laws:* Ocean CDR projects affecting marine species or their habitats may implicate various federal laws. One example is the Endangered Species Act (ESA), which requires each federal agency to "insure that any action authorized, funded, or carried out by [it] . . . is not likely to jeopardize the continued existence of any endangered species or threatened species" (16 U.S.C. § 1536(a)(2)). To that end, agencies must consult with the Fish and Wildlife Service about any action that could affect terrestrial species (including coastal species such as sea otters and polar bears), and with the National Marine Fisheries Service (NMFS) about any action that could affect marine species (16 U.S.C. § 1536(a)(2)). Consultation with NMFS is also required where a federal agency action could harm "essential fish habitat" designated under the Magnuson-Stevens Fishery Conservation and Management Act (MSFCMA) (16 U.S.C. § 1855(b)(2)). The ESA and Marine Mammal Protection Act (MMPA) also prohibit government and private actors from killing, harming, or otherwise "taking" endangered species and marine mammals, respectively (16 U.S.C. §§ 1538(a)(1)(B)-(C) & 1372(a)). Regional fisheries councils established under the MSFCMA develop fisheries management plans that are designed to restore depleted stocks and set annual catch limits to prevent overfishing (16 U.S.C. § 1852).
- *Coastal and Ocean Management Laws:* The Coastal Zone Management Act (CZMA) requires federal agency activities that have coastal effects to be consistent, to the maximum extent practicable, with any applicable state coastal management plan (16 U.S.C. § 1456(c)(1)-(2)). Prior to undertaking any activities with coastal effects, the relevant federal agency must consult with affected states to ensure consistency (16 U.S.C. § 1456(c)(1)(C); 15 C.F.R. § 930.34). This requirement would be triggered where a federal agency undertakes or authorizes an ocean CDR project that could affect land or water use or natural resources in state waters or adjacent shorelands (16 U.S.C. § 1456(c)(1)(A); 15 C.F.R. § 930.31). Additional requirements would apply where ocean CDR projects are conducted in, or affect, areas designated as marine sanctuaries under the National Marine Sanctuaries Act (NMSA). Permits are required to stop or anchor vessels, submerge grappling, suction, and other devices, install seabed cables, and perform certain other activities within marine sanctuaries (15 C.F.R. §§ 922.61 & 922.62). It is unlawful to "destroy, cause the loss of, or injure" any living or nonliving resource that contributes to the conservation, recreational, ecological, historical, scientific, cultural, archaeological, educational, or esthetic value of a marine sanctuary (16 U.S.C. §§ 1432(8) & 1436(1)).

- *Ocean Dumping Laws:* Ocean CDR projects that involve the discharge of materials into ocean waters may be regulated under the Clean Water Act (CWA) or Marine Protection, Research, and Sanctuaries Act (MPRSA). The CWA applies to the discharge of certain materials, classified as "dredge or fill" materials or "pollutants" (including "rock"), within 3 nautical miles of the U.S. coast (33 U.S.C. §§ 1311(a), 1342, 1344, & 1362). The MPRSA applies to discharges of any material from a vessel, aircraft, or manmade structure within 12 nautical miles from the coast and in other areas where the materials dumped were transported from the United States or on a U.S.-registered vessel or aircraft (33 U.S.C. §§ 1402 & 1411-1413). Both the CWA and MPRSA require permits for discharges (33 U.S.C. §§ 1342, 1344, & 1412-1413).
- *Seabed Use Laws:* Use of the seabed underlying state waters (e.g., to anchor structures) is regulated by the relevant coastal state and typically requires a lease or other authorization therefrom. Authorization from the federal government is required to use the seabed underlying federal waters (known as the outer continental shelf). The Outer Continental Shelf Lands Act (OCSLA) authorizes the Bureau of Ocean Energy Management, within the Department of the Interior, to issue leases for energy and mineral development and related activities on the outer continental shelf (43 U.S.C. § 1337). Currently, however, there is no framework for leasing the outer continental shelf for other purposes (e.g., ocean CDR). Structures in both federal and state waters may also require approval from the U.S. Army Corps of Engineers under the Rivers and Harbors Act (RHA, 33 U.S.C. § 403) and the U.S. Coast Guard under the Aids to Navigation Program (33 C.F.R. Part 64).

The application of U.S. law to ocean CDR projects will differ depending on precisely where each project takes place and the precise activities involved. Most ocean CDR techniques involve some active intervention in the marine environment, for example, the installation of structures (e.g., pipes) or discharge of materials (e.g., iron) in ocean waters. One exception is ecosystem recovery, which may be achieved through more passive approaches, such as changes in fisheries and ocean management. Those changes may, in some circumstances, further the goals of existing domestic environmental laws (see Chapter 6). As such, implementing ecosystem recovery-based approaches may be simpler, from a legal perspective, than pursuing other ocean CDR techniques. **Techniques that involve installing structures or discharging materials into ocean waters could be subject to numerous permitting and other legal requirements.** Different requirements will apply to different techniques. Initial research focused on seaweed cultivation and ocean alkalinity enhancement indicates that projects will often require multiple federal and state permits (Webb et al., 2021). **In many cases, there are no established permitting processes, leading to significant uncertainty as to how projects will be treated. Formulating permitting processes early on could help to facilitate research and, if deemed appropriate, full-scale deployment of ocean CDR techniques.** Prior to any research or deployment, extensive consultation will generally be required with affected communities, Native American tribes, government bodies, and other stakeholders.

Summary

Establishing a robust legal framework for ocean CDR is essential to ensure that research and (if deemed appropriate) deployment is conducted in a safe and responsible manner that minimizes the risk of negative environmental and other outcomes. There is currently no single, comprehensive legal framework for ocean CDR research or deployment, either internationally or in the United States. At the international level, while steps have been taken to regulate certain ocean CDR techniques—most notably, ocean fertilization—under existing international agreements, significant uncertainty and gaps remain. Domestically, in the United States, initial studies suggest that a range of general environmental and other laws could apply to ocean CDR research and deployment. Those laws were, however, developed to regulate other activities and may be poorly suited to ocean CDR. Further study is needed to evaluate the full range of U.S. laws that could apply to different ocean CDR techniques and explore possible reforms to strengthen the legal framework to ensure that it appropriately balances the need for further research to improve understanding of ocean CDR techniques against the potential risks of such research, and put in place appropriate safeguards to prevent or minimize negative environmental and other outcomes.

2.2 SOCIAL DIMENSIONS AND JUSTICE CONSIDERATIONS

Ocean CDR has a number of social dimensions—definitions to help describe these dimensions are included in Box 2.1. These include public and community acceptance of ocean CDR, the social and economic impacts of developing new industries, the social relations that those new industries and practices will involve (i.e., between workers and companies, between communities, between members of households as men's and women's work and roles are affected by these new industries), the political dynamics of ocean CDR, and the social implications from the environmental impacts of ocean CDR. In other words, **the social dimensions and justice considerations of ocean CDR are broader than "social acceptance" and will need to be researched and addressed if ocean CDR is to be supported, effective, and just.**

However, while it is possible to map out potential social dimensions of ocean CDR, the empirical evidence base for making strong claims about how they will manifest is constrained because the deployment of large-scale CDR is in the future. Marine carbon removal approaches in particular are emerging and at an early stage of technological readiness. Aside from ocean fertilization, there are very few studies of the social dimensions of marine CDR (Bertram and Merk, 2020; Cox et al., 2021).

Crucially, the social implications do not inhere in the technologies; they are influenced by the particulars of deployment and policy. It is challenging to identify and quantify benefits or risks for technologies or practices in the abstract, because many of them emerge from the ways in which they are deployed. This implies that social science research anticipating the social dimensions of ocean CDR will need to be place specific and multisited. But because policies influencing ocean CDR will be developed at state, national, and international levels, addressing the social dynamics will also require multiscalar research that can link national and international actions with place-specific implications, as well as understand how place-specific developments influence policy.

Insights for Ocean CDR from Analog Activities in the Marine Space

Examining other ocean activities can suggest considerations for the social dynamics of ocean CDR. Activities that include utilization of oceans for cultivation (aquaculture) or environmental outcomes (conservation, blue carbon, and other ecosystem services) are relevant to techniques such as marine kelp sequestration and ecosystem recovery. Other industrial uses of the sea, such as mineral extraction (deep-sea mining, oil and gas extraction), renewable energy, or offshore carbon capture storage (CCS) will be relevant to anticipating the social dynamics of techniques perceived as industrial, or those that involve geologic sequestration.

How applicable these analogs are, and whether and how people understand marine CDR through them, is a key research question. People make sense of new developments based on pre-existing knowledge structures that are seen as related (Koschinsky et al., 2018). For example, for deep-sea mining, which is relatively unknown to publics, it is uncertain whether it will be anchored to terrestrial mining, oil or gas extraction, fracking, etc.; it would be connected to different images

TABLE 2.3 Application of Laws to U.S. Ocean Zonal Jurisdictions

Ocean CDR Technique						
Location	Ecosystem Recovery	Seaweed Cultivation	Nutrient Fertilization	Artificial Upwelling/ Downwelling	Ocean Alkalinity Enhancement	Electrochemical Engineering
U.S. state waters	NEPA[a] CZMA[a,c] ESA[e] MMPA[f] NMSA[g] MSFCMA[a,h]	RHA[b] MPRSA[d] NEPA[a] CZMA[a,c] ESA[e] MMPA[f] NMSA[g] MSFCMA[a,h]	MPRSA CWA NEPA[a] CZMA[a,c] ESA[e] MMPA[f] NMSA[g] MSFCMA[a,h]	RHA[b] CZMA[a,c] ESA[e] MMPA[f] NMSA[g] MSFCMA[a,h] MPRSA[d]	MPRSA CWA NEPA[a] CZMA[a,c] ESA[e] MMPA[f] NMSA[g] MSFCMA[a,h]	RHA[b] NEPA[a] CZMA[a,c] ESA[e] MMPA[f] NMSA[g] MSFCMA[a,h]
U.S. federal waters (within territorial sea)	NEPA[a] CZMA[a,c] ESA[e] MMPA[f] NMSA[g] MSFCMA[a,h]	OCSLA[i] MPRSA[d] NEPA[a] CZMA[a,c] ESA[e] MMPA[f] NMSA[g] MSFCMA[a,h]	MPRSA NEPA[a] CZMA[a,c] ESA[e] MMPA[f] NMSA[g] MSFCMA[a,h]	OCSLA[i] CZMA[a,c] ESA[e] MMPA[f] NMSA[g] MSFCMA[a,h] MPRSA[d]	MPRSA NEPA[a] CZMA[a,c] ESA[e] MMPA[f] NMSA[g] MSFCMA[a,h]	OCSLA[i] NEPA[a] CZMA[a,c] ESA[e] MMPA[f] NMSA[g] MSFCMA[a,h]
U.S. federal waters (within EEZ)	NEPA[a] CZMA[a,c] ESA[e] MMPA[f] NMSA[g] MSFCMA[a,h]	OCSLA[i] MPRSA[d,j] NEPA[a] CZMA[a,c] ESA[e] MMPA[f] NMSA[g] MSFCMA [a,h]	MPRSA NEPA[a] CZMA[a,c] ESA[e] MMPA[f] NMSA[g] MSFCMA[a,h]	OCSLA[i] CZMA[a,c] ESA[e] MMPA[f] NMSA[g] MSFCMA[a,h] MPRSA[d,j]	MPRSA[j] NEPA[a] CZMA[a,c] ESA[e] MMPA[f] NMSA[g] MSFCMA[a,h]	OCSLA[i] NEPA[a] CZMA[a,c] ESA[e] MMPA[f] NMSA[g] MSFCMA[a,h]

High seas	ESA[e] MMPA[f]	MPRSA[d,j] ESA[e,k] MMPA[f,k]	MPRSA[j] ESA[e,k] MMPA[f,k]	ESA[e,k] MMPA[f,k] MPRSA[d,j]	MPRSA[j] ESA[e,k] MMPA[f,k]	ESA[e,k] MMPA[f,k]

[a] If project is undertaken, authorized, or funded by the federal government.
[b] If structures installed in connection with project obstruct navigation.
[c] If project could affect land or water use or natural resources in state waters.
[d] If project involves the discharge of any material into ocean waters.
[e] If project could affect endangered or threatened species or their critical habitat.
[f] If project could affect marine mammals.
[g] If project is conducted, or could affect resources, in a marine sanctuary.
[h] If project could affect waters or submerged land designated as essential fish habitat.
[i] If project involves use of the seabed.
[j] If vessels registered, or loaded in, the United States are used to discharge material.
[k] If project is performed by a person subject to the jurisdiction of the United States.

BOX 2.1
Definitions of Key Terminology

Social acceptance is a concept with various sociocultural, political, market, and community dimensions (Wüstenhagen et al., 2007; Wolsink, 2019; Batel, 2020; Devine-Wright and Wiersma, 2020). It refers to the broadest, most general level of acceptance by the public, key stakeholders, and policy makers.

Community acceptance has to do with how specific, local stakeholders in communities and local governments respond to the technology (Wüstenhagen et al., 2007; Devine-Wright and Wiersma, 2020).

Stakeholders are defined most generally as people who have an interest in or are affected by a decision (NOAA, 2015). Stakeholders can be organizations, groups, or individuals who have a particular interest in a project, sector, or system, whether that interest be business, cultural, or personal; stakeholders can be identified through a stakeholder analysis process. Stakeholders may be from the private sector, non-governmental organizations, or the public sector.

Social license, or social license to operate (SLO), refers to community sanctioning and tacit acceptance of operations (Koschinsky et al., 2018). SLO is a concept that emerged in the 1990s in relation to the mining industry (van Bets et al., 2016), but then became an object of scholarly study. Social license is based on relationships between communities, operators, and governments; it can be conceptualized across scales (i.e., local SLO as well as national SLO may be relevant); and it is a dynamic condition and may be revoked (Gough et al., 2018).

Social support is distinct from both social acceptance and social license; a community may not contest an intervention, but this does not mean it enjoys broad or continuing support. Support may come and go. As a study of offshore carbon capture and storage noted, if benefits are shared unequally or do not manifest, this can build up into opposition (Mabon et al., 2014).

Social impacts are changes caused by interventionist activity and can be positive or negative, or direct and indirect (Koschinsky et al., 2018). Social impacts could involve changes in employment, access to resources or services, and changes in social relations, in terms of between groups or within family structures (Franks et al., 2011; Koschinsky et al., 2018). Social impacts are complex and cumulative; they may go across the life cycle of a project and change as the project moves through phases, from identification to construction to closure. Different communities may have different experiences and perceptions of the social impacts of the same intervention (Koschinsky et al., 2018). Social impacts can also be different at different scales of technology deployment, and they can occur at sites distant from the actual activity. Social impact assessments attempt to identify social impacts (Vanclay, 2019).

and knowledges (Koschinsky et al., 2018). Research has shown that CDR techniques associated with CCS on land can be associated with fracking (Cox et al., 2021). In other words, for each marine CDR technique, there may be top-of-mind associations through which people “read” CDR, which may differ between individuals or communities. However, what those analogs are has not been studied.

The literature on social acceptance and social impacts of offshore renewable energy, offshore CCS, deep-sea extraction, aquaculture, blue carbon and payments for marine ecosystem services, and marine conservation suggests the following seven takeaways that are relevant for marine CDR:

1. Marine activities are not necessarily more acceptable than terrestrial activities. Social acceptance for ocean activities has unique challenges, especially in terms of defining who are legitimate stakeholders for offshore projects.

Social acceptance for ocean industries and practices brings challenges in terms of specifying a community, property rights, and so-called "not in my backyard" or NIMBYism (Soma and Haggett, 2015). NIMBYism is the objection to something undesirable being built or situated in one's neighborhood (Merriam-Webster, n.d.). For example, when the company Nautilus Minerals wanted to engage in deep-sea mining off of Papua New Guinea, some scholars observed that it had to create a community, and that creating a public to engage with means the results are unstable and partial (Filer and Gabriel, 2018; Childs, 2019). Nautilus established a new concept called "coastal area of benefit" (CAB) based on an artificial sense of spatial boundaries, which was critiqued by communities outside the area who were denied access to its material benefits (Childs, 2019). More populous communities were defined as outside the CAB, which reduced the financial burden to the company (Childs, 2019). These challenges with defining affected communities will apply to some marine CDR techniques.

Public rights at sea imply that people feel a sense of ownership over natural resources such as seascapes, even if they do not literally own them (Haggett, 2008; Soma and Haggett, 2015). Even when there is strong sociopolitical acceptance of technologies in the abstract, there may be local opposition to specific proposals. With offshore wind, this has to do with the visual impacts of big projects on the character of the seascape, impacts on tourism, and concerns about decision-making and justice (Devine-Wright and Wiersma, 2020). Spatial proximity has been found to be a factor in support of offshore wind, with attitudes becoming favorable at a greater distance (Krueger et al., 2011; Devine-Wright and Wiersma, 2020). This may also be the case with many ocean CDR projects, but again, this has not been studied.

However, studies in renewable energy have moved away from NIMBYism as an explanation for rejection, which is also relevant offshore (Soma and Haggett, 2015). Moreover, research with stakeholders and publics in Scotland has challenged a narrative that offshore CCS would be more acceptable than onshore CCS (Mabon et al., 2014). More generally, some analysts have argued that ocean CDR would face greater public acceptability challenges than terrestrial CDR, since the ocean is perceived as fragile, critical for human life, emotionally valuable, and difficult to experiment upon in a controlled way (Cox et al., 2021). This is worth keeping in mind to the extent that marine CDR might be proposed as a "solution" for social opposition to CDR deployments on land.

2. Carbon removal will be only one of many factors driving change in marine environments, and its social dimensions need to be assessed in the context of blue growth and marine conservation goals to maximize co-benefits and avoid unintended harms to other goals.

The wider social context of ocean CDR is that of aspirationally transitioning to a sustainable blue economy (Claudet et al., 2020), on the one hand, and that of a "blue acceleration" of competing interest for ocean food, material, and space on the other hand (Jouffray et al., 2020). Ocean CDR needs to be understood in this wider context, because these twin conversations about the future of the ocean—of not only how to save or restore the ocean and creating sustainable ocean practices, but also how ocean space is being industrialized and increasingly under pressure as a new frontier to exploit—will shape how communities and policy makers around the world view ocean CDR. For example, in a study of offshore CCS, for publics and stakeholders, CO_2 storage was only one of many factors driving change in the marine and coastal environment, with more concerns about offshore wind than CCS, as well as concerns about ocean acidification and extreme weather (Mabon et al., 2014). What happens in these other domains will interact with the social dynamics of ocean CDR.

Ocean CDR must avoid conflicting with other environmental aims. The literature on payment for ecosystem services emphasizes avoiding incentives that reward maximizing payments for carbon at the loss of another service, such as incentivizing fast-growing mangrove stands to maximize carbon credits (Lau, 2013). This has led to a discussion of bundling or stacking ecosystem services (Lau, 2013). Research on multitrophic aquaculture or mariculture also emphasizes holistic frameworks that can optimize for local food consumptions or livelihoods rather than for overall profits (Cisneros-Montemayor et al., 2019). Synergies are being explored between sectors, such as mariculture and offshore wind farms for farming bivalves and algae (Cisneros-Montemayor et al., 2019); assessments of marine CDR could evaluate it along these existing developments.

Notably, the acceptability of ocean CDR could be constrained by missteps in both the blue economy space and the terrestrial CDR/climate tech space. A report on social license in the Blue Economy for the World Oceans Council noted that the loss of social license to operate in one sector could impact the level of societal trust in the broader Blue Economy concept and lead to concerns about "blue-washing" (Voyer and van Leeuwen, 2018).

3. Perceptions of "naturalness" are important in terrestrial CDR, and could affect the acceptability of ocean CDR techniques as well as the scale at which they are deployed.

With terrestrial CDR, approaches perceived as natural are appraised by publics as more favorable (Merk et al., 2019; Wolske et al., 2019). This seems to hold true for ocean CDR, though evidence is limited (Bertram and Merk, 2020). In some early studies, ocean fertilization was appraised more negatively than land-based CDR, with higher perceived risks, and was perceived as an engineered rather than natural approach (Amelung and Funke, 2014; Bertram and Merk, 2020; Jobin and Siegrist, 2020).

Naturalness is socially constructed, and the ocean may be perceived as a special natural environment. The sea has its own sense of place distinct from the mainland (Gee, 2019; Devine-Wright and Wiersma, 2020). In a dialogue about seafloor exploration and mining, communities, NGOs, and marine users perceived the marine environment as more sensitive and fragile than the terrestrial environment (Mason et al., 2010). Beliefs about the sea or ocean have been identified as important in understanding how people view offshore wind (Bidwell, 2017). What some literature in renewable energy has found to be important is "place-technology fit," with place having attributes of meaning and attachment (Devine-Wright and Wiersma, 2020). Whether the place fits the technology can be related to what else is going on there; that is, in a study of offshore wind in Guernsey, it was found that certain areas were seen to be more pristine and disfavorable for wind, but areas near industrial sites were seen as more acceptable (Devine-Wright and Wiersma, 2020).

Scale may be a key parameter in terms of naturalness. For example, macroalgal farming may be natural, but in a study of social license for commercial seaweed farming in Scotland and France, acceptability was inversely related to the scale of the industry and the area occupied by the farms (Billing et al., 2021). As one respondent put it, it is important to stay at the local level; interviewees critiqued high-level European strategies for developing "blue gold" that they felt did not account for local impacts such as site abandonment (which could leave structures in the sea), introduction of invasive species, and seaweed washing ashore (Billing et al., 2021). Issues with large-scale aquaculture were projected onto the seaweed industry. Small-scale seaweed cultivation, however, was described as simple and organic, and based on relationships and trust, contrasting with large-scale cultivation, which was associated with technocracy (Billing et al., 2021).

More generally, understandings and representations of the ocean differ across cultures; for example, in Pacific island cultures it may be valued as a spiritual heritage and common good that is not distinct from the land (Koschinsky et al., 2018). So for example, when a deep-sea mining company wanting to operate in Papua New Guinea tried to mitigate concerns about fish stocks by explaining that ocean space was divided into three layers that did not mix, that was rejected by communities, who understand the ocean to be connected and inhabited by spiritual beings (Childs, 2020). Indigenous peoples around the world have their own conceptions of the sea, and coastal

Indigenous and First Nations peoples in North America have their own worldviews that influence their traditional marine management practices, for example, around treating nonhuman kin respectfully (Lepofsky and Caldwell, 2013). This means that there is not just one generic way perceptions of naturalness will impact ocean CDR, and it implies a place-specific approach to social science research on it. In other words, deliberative or qualitative methods would be needed to understand how people in various cultures understand the social and cultural dimensions of marine CDR, including how perceptions of the value and role of nature shape support or opposition to marine CDR.

4. Stakeholders and publics will be concerned about how to govern novel risks, as well as reversibility.

Ocean CDR proposals may be perceived as highly risky (Cox et al., 2021). Salient dimensions of risk include the degree of control that people have, how voluntary it is, how familiar the risk is, and how severe the consequences might be (Cox et al., 2021). There will be questions of how well equipped institutions are to manage these risks. In a study of offshore CCS, respondents wondered whether existing governance could deal with something where the effects are likely to be irreversible and uncertain across long periods of time and across complex three-dimensional volumes (Mabon et al., 2014). Arguments based upon the precautionary principle may evolve, as they did in deep-sea mining—it was difficult to prove in advance that a deep sea mine would not have negative impacts on ocean life that coastal communities valued (Filer and Gabriel, 2018). There are also questions about the reversibility of ocean CDR, here meaning not in terms of the permanence of the carbon sequestration, but rather referring to the ability to undo interventions and their unanticipated consequences (Bellamy et al., 2017).

5. Social benefits will be important for social acceptance.

The Blue Economy discourse often focuses on the contribution of ocean economic sectors to global gross domestic product, and the amount of jobs that can be provided ($1.5 trillion and 31 million jobs). However, there has been less attention to how these benefits are distributed, even though the social aspects are key to achieving sustainable management of the oceans (Cisneros-Montemayor et al., 2019).

Benefits could involve jobs, new funds for community priorities based on taxation or profits from developments, and so on. The definition is imprecise, and not enough attention has been paid to what concrete benefits might be. In the literature on payment for ecosystem services, benefits are often described in terms of incentives, including monetary or in-kind incentives (e.g., capacity building, training, infrastructure building, and codification of access rights) (Lau, 2013). The cultural context of the community is important for determining the incentives or benefits. Compensation can be one form of benefit, but deserves more research (Walker et al., 2014; Soma and Haggett, 2015). Public ownership of projects as a means of benefiting communities has also been identified as deserving more research (Soma and Haggett, 2015).

Co-benefits are sometimes discussed rather than benefits (i.e., the carbon sequestration is considered primary, and the social benefits are co-benefits). There are protocols to assess co-benefits: within the blue carbon sector, voluntary carbon markets have considered Climate, Community and Biodiversity Standards, and "Social Carbon" co-benefits (Cisneros-Montemayor et al., 2019). However, for communities, the co-benefits, rather than the carbon sequestration, may be the primary consideration.

Literature on sustainable coastal and marine management describes a tension: if funds are spent efficiently, more conservation could be incentivized; yet if the poor are providing these ecosystem services for the lowest payment, then the burden falls disproportionately on them, and they may not be able to refuse payments, meaning that the situation is less than voluntary (Lau, 2013). Paying for ecosystem services could also lock poor coastal communities into agreements that prevent them from using their resources more profitably in the future. These considerations are important for blue carbon and kelp aquaculture, but potentially apply broadly to other CDR techniques as well—the cheapest CDR per ton may not be the most equitable or socially beneficial.

6. Changes in access to common pool resources related to implementation of marine CDR are a concern that needs to be addressed.

Coastal and marine resources have traditionally been open access and an important source of livelihoods. Studies of payment for marine ecosystem services indicate that both formal and de facto changes in use and access rights will affect the communities, as well as the success of the intervention (Lau, 2013). With payment for ecosystem services, even if the goal is not to alleviate poverty, equity and poverty alleviation will need to be addressed in designing the schemes (Lau, 2013). Blue carbon projects have also been critiqued for pushing out traditional users (Cisneros-Montemayor et al., 2019). The social lessons from blue carbon projects might reasonably be the same for things such as kelp cultivation—incorporating livelihood aspects as part of project design, involving local community at all stages of planning and implementation, and considering the needs of local communities during development can ensure that cultivation or sequestration in one area does not lead to activities that would reduce carbon sequestration elsewhere (Wylie et al., 2016). An examination of coastal carbon sequestration projects noted that the protection and government management of natural resources can lead to traditional management systems being replaced, and communities losing their ability to change their management strategies in response to environmental change (Herr et al., 2019). Another analysis of coastal carbon, on mangrove ecosystems in the Philippines, critiqued the latest framing of mangroves under a new global framing of "blue carbon" as bearing technocratic and financialized ideals of sustainability, and argued that the need to consult and benefit local communities is widespread in discourse but rarely has clear implementations and strategies (Song et al., 2021). The concern is that coastal communities may lose their customary rights, and their interests will be marginalized at the demands of international priorities (Song et al., 2021).

7. Public engagement will be important for social acceptance and procedural justice in ocean CDR, though it is not a guarantee of these.

Public engagement involves a dialogue between scientists and nonscientists that attempts to involve the public in discussions about the direction and pace of technology development (Corner et al., 2012). Public engagement is often seen as lying on one end of a spectrum of public participation, with public informing or consulting on the other end, as a more limited or one-way form of participation that can be manipulative (Arnstein, 1969). Rationales for public engagement include normative rationales (dialogue is an important part of democracy; engaging the public on important decisions and new technologies is the right thing to do); substantive rationales (public engagement can improve the quality of the research); and instrumental rationales (it can increase legitimacy and trust) (Corner et al., 2012; Fiorino 2016).

Public engagement has been widely discussed in regard to science, to emerging technologies broadly, and to energy technologies more specifically. Public engagement is one factor in the persistence of social acceptance, and the timing, content, and processes of public engagement are important (Soma and Haggett, 2015). Public engagement is no guarantee of project development: consultations perceived as checkbox exercises could worsen or generate opposition (Soma and Haggett, 2015). Thin and consultative participatory engagements can result in participatory exhaustion or backlash (MacArthur, 2015). The literature emphasizes the importance of public engagement; generally speaking, though, some recent theoretical research has discussed the need to move away from fixed assumptions of what it means to participate and technocratic processes toward an understanding of participation that focuses on diverse collectives of participation, and trying to build a system in which multiple forms of public involvement can happen (Chilvers et al., 2018). There are many recommendations of best practices for public engagement and participation in decisions regarding the environment, some of which are discussed in reports such as the U.S. Environmental Protection Agency's (EPA's) *Public Participation Guide* (aimed at government agencies) (U.S. EPA, 2021) or the National Research Council's *Public Participation in Environmental Assessment and Decision Making* (NRC, 2008). When it comes to public engagement in research and development more specifically, it will be important to make sure the results of engagements feed back into research, involve a diversity of researchers, and allocate sufficient time and resources for the process.

Environmental Justice and Climate Justice

Environmental justice is a goal, a movement, and a field of research. Environmental justice as a movement began with groups concerned with civil rights, the environment, worker health and safety, Indigenous land rights, environmental racism, and more (Schlosberg and Collins, 2014). The definition of environmental justice used by the EPA is "the fair treatment and meaningful involvement of all people regardless of race, color, national origin, or income, with respect to the development, implementation, and enforcement of environmental laws, regulations, and policies."

However, environmental justice is a multidimensional concept, which includes distributive, procedural, reparative, intergenerational, and recognitive dimensions (Holifield, 2013). Distributive justice is concerned with the fair allocation of environmental risks or harms as well as the ability to access environmental benefits. Procedural justice involves participation in the decision-making processes around environmental risks and benefits. Corrective or reparative justice involves whether the restorative measures or penalties for environmental harms are fair. Intergenerational justice involves the fair treatment of future generations. Recognitive justice means that policies and programs meet the standard of fairly considering and representing the cultures, values, and situations of affected parties (Whyte, 2011). In the context of Native communities, tribal cultures may have their own conceptions of environmental justice that have existed before nontribal discussions (Whyte, 2011). Conceptions of justice may also include nonhumans. For example, a study of conceptions of justice in two Papua New Guinea fishing communities found that respondents articulated fish as subjects of justice; needing to rest, having a chance to breed, and so on (Lau et al., 2021).

Ocean carbon removal technologies will have different environmental justice implications based on how they are deployed, including the policies that support them and the actors and motivations driving them. For example, a macroalgae project led by a community that is compensating for its own hard-to-abate emissions would have different justice implications than one instituted by a company who is selling carbon removal credits to another company, even though the activity might look the same from a biophysical perspective, in terms of carbon flows. These two different deployments would have different distributive and procedural justice implications, in terms of where the benefits flow and how decisions are made.

Climate justice, which developed out of environmental justice discourse (Agyeman et al., 2016), emphasizes that climate change is not just a matter of warming, but of justice. Climate inequalities exist within and between nations, and climate change has disproportionate impacts on historically marginalized or underserved communities. Those who have contributed the least to the problem are bearing the greatest harms, including communities in the global South, Indigenous groups, and future generations. Historical responsibility approaches to climate justice are based on the polluter pays principle, while rights-based approaches emphasize the right to develop out of poverty before bearing the responsibility of mitigation (Schlosberg and Collins, 2014).

Climate justice considerations go beyond the impacts of particular approaches to the entire concept of carbon removal and its role in net-zero scenarios. Carbon markets have historically been seen by the environmental justice movement as giveaways to polluters at the expense of environmental justice communities (Schlosberg and Collins, 2014), and policies that allow continued pollution in one area with removal in another area will naturally be questioned. For example, civil society organizations have critiqued blue carbon for turning the carbon ecosystems into a commodity that legitimizes continued emissions elsewhere (Song et al., 2021). An analysis of public engagement with carbon removal found that respondents were concerned with "environmental dumping," or analogies with the dumping of polluting processes on poorer populations (McLaren et al., 2016), which may be a particular issue with marine carbon removal given the phenomenon of ocean or marine dumping. There is a question of who has to bear the burden of carbon removal, and who is enjoying liberties because of it. Scholars have analyzed fair-share emissions and carbon removal quotas (Pozo et al., 2020; Dooley et al., 2021). A particular concern is that offsetting via carbon removal could deprive poor nations and regions of "cheap" carbon removal options and make their path toward net zero harder while giving wealthy nations an easier path (Carton et al., 2021; Rogelj et al., 2021). There are intergenerational justice concerns with creating a temporal equivalence between emissions and removals that puts more burden on future generations (Hansen et al., 2017; Lawford-Smith and Currie, 2017; Carton et al., 2021).

When thinking about the justice implications of CDR, it is also valuable to weigh the counterfactual scenario of not having CDR available. Ethicists have also argued that the use of large-scale negative-emission technologies may be permissible due to the extreme harms that would result from failing to stabilize the climate (Lenzi, 2021).

Mitigation deterrence, or reducing or delaying mitigation, is another key climate justice issue that has been a long-standing concern of climate advocates (McLaren et al., 2016; Campbell-Arvai et al., 2017; Markusson et al., 2018). For example, if a CDR activity is perceived as being substitutable with mitigation, this can lead to mitigation deterrence; science is central to this, because it helps create new objects in which to invest (Markusson et al., 2018). CDR could also produce rebound effects, and the anticipated or imagined future availability of CDR could also delay emissions reductions (McLaren, 2020). Other scholars have pointed out that since policy makers can (and should) do both CDR and mitigation, and that framing the issue in terms of substitutable actions actually makes this substitution more likely, a risk-response feedback framework that assesses particular policy packages would be more fruitful than the framework of mitigation deterrence (Jebari et al., 2021). Regardless, the idea that carbon removal can delay cutting emissions and phasing out fossil fuels in a form of "nontransition" (Cox et al., 2020) has been cited as a key concern for publics.

Scale has emerged as a central issue in assessing the environmental and climate justice implications of CDR, in terms of both the technology and decision-making (Cox et al., 2018). Many technologies are relatively risk-free in the abstract or at small scale (Cox et al., 2018), but their social implications accrue at larger scales. Moreover, environmental justice impacts that are evident when examining local scales may be addressable only at regional or national scales (Buck, 2018), especially when thinking about complex supply chains or remote actors. There may also be demands on the global scale. Countries in the Global South may argue that in line with the principle of common but differentiated responsibilities and respective capabilities, CDR should be deployed by developed nations who should reach net-negative levels first (Mohan et al., 2021). **Environmental justice concerns are not "local issues" and climate justice concerns are not "global concerns"; a multiscalar framework is needed to research and understand them.**

Coastal Community Research and Engagement

The opportunities to advance knowledge and understanding of any ocean-based carbon removal solution are greatly enhanced when the barriers to participation are removed. Recognizing this, **it is critical that research and development activities incorporate equity, diversity, and inclusion with a particular focus on coastal communities, especially Indigenous communities that reside in their broadest sphere of influence (e.g., mining activities associated with alkalinity enhancements) and marginalized coastal communities** (e.g., Felthoven and Kasperski, 2013). There are two aspects of this engagement: (1) research conducted in communities should follow ethical protocols for engagement, and (2) efforts should be pursued to include community members in research activities.

Following ethical protocols for engagement means complying with both Institutional Research Board processes such as university researchers might undertake when using human subjects—even if the people doing the research are not associated with a university—as well as complying with ethical procedures that local groups may have set out. Many Indigenous jurisdictions have their own ethical review protocols that pertain to research activities in their territories. One resource for engaging with Indigenous communities on ocean research is the Ocean Frontier Institute's Indigenous Engagement Guide, published in 2021 to provide guidance on effectively and respectfully engaging and communicating with Indigenous communities (Ocean Frontier Institute and Dillon Consulting, 2021).

Including community members in research activities, or co-producing research with communities, is a growing focus in sustainability research broadly as well as coastal environmental research. This work is ongoing within the community of ocean observation monitoring as well. For the past 30 years, the global ocean observing community has gathered once a decade at the OceanObs conference, and at the 2019 conference, for the very first time, this gathering included Indigenous delegates from Canada, the United States including coastal states such as Hawaii, the South Pacific Islands, and New Zealand (Figure 2.3). An important outcome of the conference was the publication of the Coastal Indigenous Peoples' Declaration (Indigenous Delegates at OceanObs'19, 2019) calling on the ocean community to formally recognize the traditional knowledge of Indigenous people worldwide and the commitment to establish meaningful partnerships. Ocean CDR research can build on this growing recognition of the importance of meaningful partnerships in ocean monitoring and related fields.

FIGURE 2.3 OceanObs19 included 53 delegates from Indigenous communities representing Fiji, Samoa, Maori New Zealand, Hawaii, and Canada. Photo Credit: Ocean Networks Canada.

There are several approaches to achieve meaningful partnerships, but all should ensure that there is a full understanding of the approaches taken among the partners (Figure 2.4; Alexander et al., 2019).

Examples of test cases specific to ocean monitoring are described by Kaiser et al. (2019), one in Canada and one in New Zealand. Their recommendations for successful partnerships include practices for two-way knowledge sharing, proposal co-design, documented project plans, incorporation of educational resources, mutually agreed upon monitoring, and data and results sharing. Use of cross-cultural resources were recommended for any future ocean monitoring projects. These partnerships also have potential for greater impacts through a more robust knowledge of community needs now and going forward.

2.3 OTHER CROSSCUTTING CONSIDERATIONS

Considerations that are also important for advancing research on ocean-based carbon removal include monitoring and verification of carbon removed and other environmental impacts, valuation of added benefits including ecosystem services, the economics of different approaches, and policy mechanisms to support research and deployment (where deemed appropriate).

Monitoring for Environmental Impacts and Enhancements

Among the six approaches that are the focus of this study, four are location-specific solutions that will impact the local ecosystem—fertilization, enhanced upwelling and downwelling, alkalinity enhancement, and electrochemistry. Each of these will require tailored carbon accounting and environmental monitoring for the specific location.

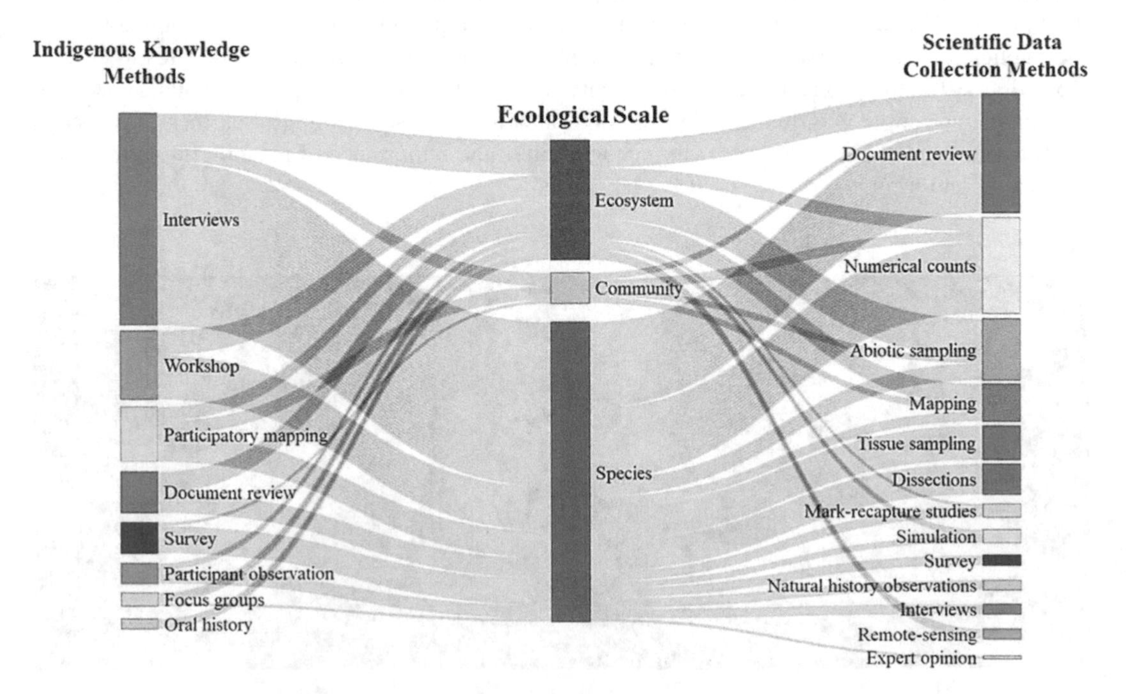

FIGURE 2.4 Relationship between Indigenous knowledge methods, ecological scale, and scientific data collection methods. SOURCE: Alexander et al., 2019. Licensed under Creative Commons CC BY 4.0.

For the two others—ecosystem recovery and seaweed cultivation—environmental monitoring for both negative and positive impacts would mimic the approaches described for ecosystem recovery in Chapter 6. The exception to this would be where seaweed cultivation is solely focused on growth for deepwater disposal. For this application, research is needed to understand the fate of placement of kelp mounds in deep-water environments.

The monitoring systems are summarized in Table 2.4 and described in more detail in the following text.

In the case of fertilization, the area of treatment is relatively large and thus would require a mix of monitoring systems. Satellite-based ocean color would be used to document the extent of treatment on the surface ocean. A suite of autonomous surface vehicles and water column gliders (the number would depend on the areal extent), outfitted with sensors (e.g., temperature, salinity, pressure, partial pressure of CO_2 [pCO_2], oxygen [O_2], nutrients, nitrate, pH, turbidity, etc.) would be deployed on and beneath the treated bloom area. At least one area outside of the bloom area, but in a similar water depth and ecosystem, could be monitored with a mooring hosting similar sensors. These sensor systems would deliver an important subset of essential ocean variables (see, e.g., Danovaro et al., 2020) to document the resulting bloom size and its fate in the water column on ocean physical and chemical properties, and sediment traps would be used to document the impact on the seafloor and the carbon sequestered. Ship expeditions, used to deploy these in situ suites of sensors, would also deploy biogeochemical Argo floats throughout the area of treatment and beyond, as well as neutrally buoyant sediment traps set to capture sediment over a range of water depths, including close to the seafloor.

Approaches and scales for the application of artificial upwelling and downwelling as a carbon removal solution are in nascent stages. For upwelling, CDR is through surface ocean fertilization and, hence, monitoring would align well with that of iron fertilization, described above. It would likely also require monitoring of CO_2 from the upwelled water that is released into the atmosphere prior to fertilization of the surface waters. A surface mooring with infrared, temperature, salinity, pCO_2, nutrients, and O_2 sensors would measure the CO_2 at the surface and in the atmosphere as well as the chemistry to understand the conditions for fertilization and pH changes.

Models have shown that most of the carbon removal from upwelling would be land based because of the cooling effects of the upwelling enhancing soil uptake of CO_2. This effect would also require additional monitoring of surrounding terrestrial soils.

Research for alkalinity enhancement is at a stage where the next step to advance knowledge is mesoscale experiments. Integration and development of configurations for monitoring would include in situ and robotic carbonate chemistry observation platforms and biogeochemical Argo flotillas to assess and quantify this solution's ability to durably store CO_2 in seawater. The full-cycle carbon accounting must also include any carbon-intensive energy sources used for the source rocks.

Electrochemical approaches can be characterized as an industrial plant solution where there is the ability to directly measure the tonnage of CO_2 extracted, making carbon accounting straightforward. Some of the approaches include releases of water solutions of differing chemistry, and so monitoring both the positive and negative impacts on the ecosystem would be important. Many of the monitoring approaches described in the monitoring section of Chapter 6 could be tailored to the industrial electrochemical coastal or ship-based settings.

Where seaweed cultivation's sole focus is farming kelp for deep-water disposal onto the seafloor, a monitoring program should be established in each of the ecosystem areas where kelp will be delivered to the seafloor. It is generally recognized that kelp detrital carbon is remineralized through grazing and microbial decomposition in shallow-water settings and grazing is significantly reduced in deep water, thus becoming a potential carbon sink. These same rates would also ideally be measured to assess the amount of carbon sequestered and its durability (e.g., to assess whether on the same or longer timescales as physical oceanographic estimates of deep-sea sequestration).

Experiments could be conducted in the kelp sequestration seafloor region using a combination of a single large mound (approximately production-scale size) and a series of small mounds deployed within mesh bags with varying mesh size openings that would provide a mechanism to isolate grazing by smaller colonizers to better understand individual species rates.

The rates of grazing and secondary colonization may also represent an ecosystem enhancement as a benthic succession, much like whale falls in the deep sea (Pedersen et al., 2021). Whale-fall monitoring to assess benthic successions has succeeded with the regular use of a seafloor lander equipped with a time-lapse video camera and an acoustic Doppler profiler (Aguzzi et al., 2018). This approach would be appropriate for monitoring a large kelp mound on the seafloor.

Monitoring for Certification

Scaling up negative emissions technologies will require certifiable metrics that document the amount of carbon removed, obtained from systems that accurately measure carbon removed and the durability of its removal. These data can then be used by established and trusted entities for certification, which in turn attracts financing, especially in the voluntary market.

An example that is easy to measure is electrochemical removal of CO_2 where the tonnage of CO_2 extracted is directly measured. The monitoring is transparent and readily certifiable.

Ocean alkalinity enhancement is more challenging because the certification would be based on foundational chemistry (well understood) combined with sensors and model results. This means that the monitoring approach used for verification and certificates must be incorporated into the design and implementation of more complex approaches. Fertilization and artificial upwelling and downwelling have similar certification challenges.

Seaweed cultivation has complexities too, especially if it is grown for multiple uses. If it is grown only for carbon sequestration in the deep sea, accounting of the carbon removal is described above in monitoring for environmental impacts and enhancements because of the dual purpose of the sensors and approaches needed. Seaweed cultivation for other purposes has the same types of certification issues as, for example, the forest industry has had in the past. Some of that carbon would go back into the atmosphere and the ocean, thus reducing its durability. Some could also be used to produce durable products. The carbon accounting would be faced with questions about additionality and carbon credits that must be accounted for and managed. Lessons learned from forestry and this sector's offset practices (see, e.g., Gifford, 2020) could be applied to seaweed cultivation to avoid creating an ocean CO_2 removal market that is viewed as fraudulent or illegitimate. Therefore it is essential to co-establish robust monitoring, accounting, and verification protocols. Interdisciplinary research can help determine what the protocols could be and how they can be robust.

Policy Support for Ocean CDR

Government policy will have a major bearing on the nature and scope of ocean CDR research and deployment (if any). In the United States, climate policy has traditionally focused on emission reductions, and CDR-specific policy remains nascent (Schenuit et al., 2021). In recent years, however, the U.S. Congress has shown an interest in CDR. The Energy Act of 2020, which was passed by Congress with bipartisan support and incorporated into the Fiscal Year 2021 omnibus spending bill, directed the Secretary of Energy to review CDR approaches (including approaches involving the "capture of carbon dioxide . . . from seawater") and recommend policy tools to advance their deployment. The Energy Act also appropriated funding for the establishment of a government "research, development, and demonstration program . . . to test, validate, or improve technologies and strategies to remove carbon dioxide from the atmosphere on a large scale."

While not framed specifically as a CDR policy, the 45Q tax credit carbon capture program, which was first adopted in 2009 and significantly expanded in 2018, provides financial incentives

TABLE 2.4 Environmental Monitoring Solutions of Ocean CDR Approaches

CDR Approach	Systems	Measurements	Capital Costs	Annual Operating Costs	Comments
Ecosystem recovery and seaweed cultivation	Marine protected area, Intergrated Ocean Observing System (IOOS)-type monitoring	Water quality, nutrients, biodiversity, sound, pollution; qualitative assessment of carbon removal	Existing regional systems; capital upgrades $0.5M/yr per region	$4M/yr per region	May require changing IOOS mandates
Ocean alkalinity enhancement	IOOS-type monitoring in the region(s) of treatment	Water quality, nutrients, biogeochemistry; modeling needed for volume of carbon removed, plus carbon intensity of mining and deploying minerals	Same as above, plus biogeochemical Argo (10 per region) = $9M and capital upgrades $0.5M/yr per region	$4M/yr per region, plus operating costs for floats (comms; personnel; shiptime) $1M/yr	May require changing IOOS mandates; determining CO_2 emissions from mining and deployment also required
Artificial upwelling and downwelling and nutrient fertilization	Satellite data; biogeo Argo; gliders; autonomous moorings for measuring change; autonomous moorings as control; sediment traps (seafloor and mid-water)—see Figure 3.5	Surface bloom extents; water properties, nutrients, dissolved oxygen, pCO_2, sediment accumulation	Existing ships, autonomous vehicles; (capital costs for autonomous moorings, sediment traps, biogeo Argo) = $5M	For the geographic area equivalent to the EXPORTS project, ship time $7M; data analyses, research AND modeling support $25M	The EXPORTS field campaign deployments conducted over a region of a couple of hundreds of km
Electrochemical Processes	Monitoring incorporated into the engineered processes		Costs are included in the costs for the system.	Costs are included in the costs for the system.	These solutions are engineered and controlled. Monitoring of the source of power for carbon accounting is important.

for direct air capture and storage of CO_2. Under the program, certain projects involving the geological sequestration of CO_2 captured at qualifying industrial facilities or directly from the ambient air (i.e., via direct air capture) are eligible for a tax credit for their first 12 years of operation. The credit is only for projects that capture CO_2 from the ambient air and sequester it onshore or offshore in sub-seabed geological formations within the U.S. territorial sea or EEZ. Other ocean-based CDR and storage projects do not qualify for the credit.

There is general agreement among climate-focused economists that scaling up CDR, including ocean-based approaches, will require significant government spending (Bednar et al., 2019) and internationally agreed-upon financial incentive policies (Honegger and Reiner, 2018). Some studies have recommended that governments establish CDR-specific policy goals and mechanisms that are separate from, but aligned with, other climate policies, for example, dealing with emissions reductions (e.g., Bellamy, 2018; Geden and Schenuit, 2020). For example, Geden and Schenuit (2020) have argued that countries' net-zero targets "should be explicitly divided into emission reduction targets and removal targets," and separate policy frameworks adopted for each. Bellamy (2018) concluded that further empirical research is needed to evaluate policy options for incentivizing CDR.

While it may seem premature to explore policy options to support ocean CDR, particularly given the early stage of development of many techniques, **delaying policy engagement could impact future CDR scale-up.** In this regard, Lomax et al. (2015) have warned that "excluding [CDR] from near-term policy attention would reduce any incentives for businesses and research organizations to expend effort and investment on advancement of [CDR], and to engage with policy to develop suitable support for [CDR]-oriented businesses." Lomax et al. (2015) argue that policy frameworks should "keep the [CDR] option open." However, it is equally important that policy not lock in future deployment of CDR or deter other actions to address climate change, particularly emission reductions. Past studies have recommended policy designs to reduce the risk of mitigation deterrence (McLaren et al., 2019; Geden and Schenuit, 2020).

CDR policy has been the subject of relatively little previous research. Some studies have explored the use of carbon pricing, credit, or similar market mechanisms to pay for CDR (e.g., Honegger and Reiner, 2018; Platt et al., 2018; Fajardy et al., 2019; Rickels et al., 2020). Fajardy et al. (2019) found that many existing carbon pricing schemes "only penalize [carbon dioxide] emissions and do not remunerate removal" and, in any event, existing carbon prices are generally too low to stimulate CDR deployment. They argue that, unless carbon pricing schemes change and carbon prices increase, some form of "negative emissions credit" will be needed to pay for CDR. This would require accurate and verifiable carbon accounting (discussed above). The complexities associated with, and current lack of standardization in, carbon accounting have been identified as potential barriers to the use of carbon pricing or credit mechanisms (Lomax et al., 2015). The use of such mechanisms could also raise environmental justice and other concerns that have not been fully explored in prior research.

Other policy instruments to support CDR could include direct government grants for research and development, tax credits similar to the existing 45Q program, and procurement and supply-chain standards that incorporate CDR (Friedman, 2019; Sivaram et al., 2020; Schenuit et al., 2021). One study has also recommended the adoption of policies tied to the "non-climate co-benefits" of CDR (Cox and Edwards, 2019). Further analysis and comparison of these and other policy options are needed. It will be particularly important to consider the social and distributional impacts of pursuing different policy options (Bellamy, 2018). In particular, consideration should be given to those who bear the risks and reap the benefits of ocean CDR technologies under different policies. The drivers of, and approaches to developing, robust and effective CDR policy also require further study.

Looking beyond government policy, ocean CDR research and deployment (if any) could be funded by the philanthropic community, or driven by the market. Market pull does not yet exist in this space, except for a small number of niche approaches, such as that developed by Stripe Climate. The Stripe Climate model enables online businesses to direct a portion of their revenues to supporting the scale-up of CDR technologies.[71] Notably, Stripe Climate does not put a price on CO_2, or use removals to generate carbon offsets.[72]

There is the possibility that the private voluntary carbon market could grow to consider ocean CDR through, for example, the Taskforce on Scaling Voluntary Carbon Markets. The Taskforce was launched by the UN Special Envoy for Climate Action and Finance and is sponsored by the Institute of International Finance. It has more than 250 member institutions, which represent buyers and sellers of carbon credits, standards setters, market infrastructure providers, and other interested bodies. Its goal is to grow voluntary carbon markets, including by identifying and addressing integrity and quality concerns.

Markets will be most helpful when one or more ocean CDR approaches reach a high level of technology readiness and scale. It is possible that Wright's law could apply. Pioneered by Theodore Wright in 1936, Wright's law has been and continues to be a framework for forecasting cost declines as a function of growth in production—for every cumulative doubling of units produced, costs will fall by a constant percentage. As costs decline, market interest in the direct purchase and operation of the solutions and/or purchase of carbon removal services using these solutions would grow, bolstered by financial incentives from governments.

2.4 ADDRESSING RESEARCH GAPS

Several key research gaps exist that are foundational to the forward movement and success of any ocean CDR approach. These research questions are described below, summarized in Table 2.5, and woven into the committee's recommendations in Chapter 9.

Legal Research Gaps

There are several key gaps in the existing body of research on the legal and regulatory landscape for ocean CDR. First, while many prior studies have discussed the application of existing international law to ocean CDR, most have been largely descriptive. Some studies have identified unresolved questions and highlighted potential challenges associated with the application of existing international law to ocean CDR. However, there has been comparatively little normative research, exploring what a "model" international legal framework would look like. Such a framework could provide the basis for development of a new international agreement governing ocean CDR research. Achieving broad acceptance of such an agreement could prove difficult, however. Past efforts to develop international rules for ocean CDR and related research (e.g., under the CBD and London Convention and Protocol) have primarily yielded nonbinding resolutions and decisions. One exception is the 2013 amendment to the London Protocol, but that has yet to take effect, having been ratified by just six countries. Nevertheless, developing a model international legal framework could help to inform future discussions, including the ongoing negotiations surrounding the BBNJ Agreement. The model framework could also provide useful guidance to the research community and support development of a code of conduct for research (see Chapter 9).

Whereas the treatment of ocean CDR projects under international law has been well studied, comparatively little research has explored the application of domestic law to such projects. Research to date has focused on only a subset of ocean CDR techniques and principally examined the application of federal environmental law thereto. The studies have been largely descriptive and have not

[71] See https://stripe.com/climate.

[72] See https://stripe.com/docs/climate/faqs.

examined in detail whether existing federal law is sufficient or appropriate to regulate ocean CDR (though some studies have highlighted uncertainties or challenges associated with the application of existing federal law). There has been no comprehensive review of all state, territory, and local laws applicable to ocean CDR projects and limited analysis of the potential tribal rights implications of such projects. The liability of project developers for environmental and other harms has also received little attention. Further research into the existing domestic legal framework is needed to determine whether it is sufficient and appropriate to regulate ocean CDR. While some studies have highlighted uncertainties or challenges associated with the application of existing domestic law, none has fully evaluated the need for, or utility of adopting, a new legal framework specific to ocean CDR or analyzed what such a framework should look like.

Research Gaps in Social Dimensions

When it comes to social dimensions, there are applicable insights from adjacent domains, but there is very little empirical research directly on ocean CDR. As for *what* should be researched, most social dimensions can be judged research gaps, but it is possible to make general observations about *how* the research should be done, in terms of research that is interdisciplinary, inclusive, multiscalar, and cross-sectoral. First, understanding the social dimensions of ocean CDR will require research that is interdisciplinary from the project outset, meaning that people from various disciplines are shaping the research questions and approaches. Second, ocean CDR also needs a more diverse research community and would benefit from support for early-career and established researchers from diverse backgrounds. In 2017, Blacks and Hispanics comprised just 1.5 percent and 3 percent of the occupations of "Earth scientists, geologists, and oceanographers," and the study *Global Change Research Needs and Opportunities for 2022-2031* points out that this lack of inclusion undermines the capacity of U.S. science to generate knowledge that is credible, relevant, and legitimate (NASEM, 2021b). Third, research should be multiscalar, in terms of understanding both site-specific considerations and national and international policies and how they shape each other; mixed qualitative and quantitative methods will be crucial for this.

While the limited amount of research on social dimensions means that most things are a research gap, we can specifically point to three important ways of approaching the needs. First, research on the social dimensions of ocean CDR would benefit from a cross-sectoral framework, meaning that ocean CDR should be considered in the context of food systems, energy systems and energy access, and so on. For example, how particular ocean CDR approaches would interact with local and global food systems is a research question that would benefit from social and biophysical scientists working together. An assessment of the relevant systems would be a logical first step. Second, another key area of research is understanding how different ocean CDR approaches would interact with the sustainable development goals. Third, it is also critical to understand how ocean CDR interacts with mitigation, adaptation, and terrestrial CDR, both biophysically and socially. A research program on the social dimensions of ocean CDR should include these three approaches.

Monitoring, Economics, and Policy Research Gaps

The research program should fund a transparent, publicly accessible system for monitoring impacts from projects. Research is also needed on how user communities view and use monitoring data and certification processes. This is important for designing robust certification schemes that are accessible and trusted by multiple user communities. Research should also be conducted on policy mechanisms and innovation pathways, including on the economics of scale-up. It is important to analyze not just what potential policy mechanisms exist, but also who is affected by different policies.

Research Agenda Costs

The research costs here are approximate and were compiled based on experiences of the committee and similar research agendas. They reflect what might be practically necessary for developing the required knowledge base to begin to scale up ocean CDR to climate-significant scales.

For example, the recent National Academies report on terrestrial CDR (NASEM, 2019) recommended $5 million per year for 10 years on social science research on cost-effective adaptative management of coastal blue carbon and the response of coastal land owners and managers to carbon removal and storage incentives, $1 million a year for 3 years for extension and outreach to forest landowners, $2 million a year for 3 years to study barriers to agricultural soil carbon adoption, $5 million a year for 10 years to study the social and environmental impacts of carbon mineralization, $1 million a year for 10 years on public engagement with geological sequestration, etc. The figures in this report are of similar scope, given a more compressed timescale. To further put this in context, the coastal carbon sequestration research agenda in the 2019 report recommended $1.16–$1.19 billion for research on coastal carbon sequestration, with the majority of that dedicated to an integrated network of coastal sites over 20 years; the social science component was $50 million (about 4 percent), and another $40 million was recommended for a publicly accessible data center. In this report, the recommended social science and governance research portion is about 5 percent of the total budget, similar to the blue carbon research recommendation in the 2019 report. It is also about 9 percent of the priority research items, recognizing that understanding the social feasibility and governance considerations is important for further investment in these approaches.

Another reference point is the National Academies report *Reflecting Sunlight: Recommendations for Solar Geoengineering Research and Governance* (NASEM, 2021c), which recommended spending $20 million to $40 million over 5 years (~20 percent of the total research budget) on research into "social dimensions," including "public engagement, political and economic dynamics, governance research, ethics and philosophy." This recommended spending on social science and governance activities reflects the understanding that addressing climate change with emerging technologies is a social and governance matter as much as a technical one.

TABLE 2.5 Research and Development Needs to Address Overarching Research Gaps

No.	Recommended Research	Question(s) Answered	Estimated Research Budget ($M/yr)	Time Frame (years)
2.1	Model international governance framework for ocean CDR research	How can the existing international governance framework for ocean CDR research be improved? Is there an alternative framework(s) that could better facilitate needed ocean CDR research while ensuring that research is conducted in an open, transparent, responsible, and environmentally and socially acceptable manner?	2–3	2–4
2.2	Application of domestic laws to ocean CDR research	What are the full range of domestic laws (federal, state, local, and tribal) applicable to each ocean CDR approach?	1	1–2 initially (and ongoing as needed)
2.3	Assessment of need for domestic legal framework specific to ocean CDR	What is the need for and utility of establishing a domestic legal framework specific to ocean generally or individual ocean CDR approaches?	1	2–4
	Development of domestic legal framework specific to ocean CDR	What does a "model" domestic legal framework for ocean CDR (either generally or by approach) look like? What should it require with respect to ex ante review of projects, stakeholder consultation, monitoring and verification, publication of data, etc.?		
2.4	Mixed-methods, multisited research to understand community priorities and assessment of benefits and risks for ocean CDR as a strategy	What are the potential harms and benefits of ocean CDR approaches, for livelihoods and for communities?	5	4
2.5	Interactions and trade-offs between ocean CDR, terrestrial CDR, adaptation, and mitigation, including the potential of mitigation deterrence	What are said interactions and trade-offs?	2	4
2.6	Cross-sectoral research analyzing food system, energy, sustainable development goals, and other systems in their interaction with ocean CDR approaches	What implications do ocean CDR techniques have for food systems? For energy production and access? For achieving the sustainable development goals?	1	4
2.7	Capacity-building research fellowship for diverse early-career scholars in ocean CDR	How can we build interdisciplinary, cross-sectoral, diverse expertise in ocean CDR?	1.5	2
2.8	Transparent, publicly accessible system for monitoring impacts from projects	How do public data on monitoring project impacts influence public perception and awareness of ocean CDR projects? How do they influence scientific research?	0.25	4

2.9	Research on how user communities (companies buying and selling CDR, nongovernmental organizations, practitioners, policy makers) view and use monitoring data, including certification	What are the strengths and weaknesses of various certification approaches? What makes certification robust and trustworthy?	0.5	4
2.10	Analysis of policy mechanisms and innovation pathways, including on the economics of scale-up	What are policy options for scaling ocean CDR? How do different pathways and policies for scaling up ocean CDR affect both societies and the outcomes of the CDR?	1–2	2
2.11	Development of standardized environmental monitoring and carbon accounting methods for ocean CDR		0.2	3

3

Nutrient Fertilization

3.1 OVERVIEW

Ocean-based carbon dioxide removal (ocean CDR) via nutrient fertilization refers to the addition of micronutrients (e.g., iron [Fe]) and/or macronutrients (e.g., phosphorus [P], nitrogen [N], silica [Si]) to the surface ocean with the deliberate intent to (1) increase photosynthesis by marine phytoplankton, and thus enhance uptake of carbon dioxide (CO_2) from surface waters, and to (2) enhance the transfer of the newly formed organic carbon to the deep sea away from the surface layer that is in immediate contact with the atmosphere. Step (1) can be accomplished wherever growth of phytoplankton is limited by nutrients, which is the case for some or all phytoplankton over large regions of the ocean, except eutrophic regions such as often found close to continental margins. Achievement of step (1) has been demonstrated for a number of fertilization experiments via in situ measurements and remote sensing of ocean color. There is larger scientific uncertainty about achieving step (2). Depending on the location, export depth, and remineralization rates of sinking particles, carbon can be sequestered for 100- to 1,000+-year timescales in the deep ocean. As such, nutrient fertilization aims to enhance locally the magnitude and efficiency of the natural ocean biological carbon pump (BCP; Figure 3.1) using energy from the sun and nutrients either from within the ocean (see Chapter 4) or from outside the ocean. In the case of fertilization with micronutrients such as Fe, relatively small amounts of iron may be needed relative to potential (C) C sequestration, whereas the amount (mass and volume) of nutrient required for fertilization with nitrogen, phosphorus, or silicate will be many orders of magnitude higher.

According to criteria described in Chapter 1, the committee's assessment of the potential for ocean nutrient fertilization as a CDR approach is discussed in Sections 3.1–3.5 and summarized in Section 3.6. The research needed to fill gaps in understanding of ocean fertilization, as an approach to durably removing atmospheric CO_2, is discussed and summarized in Section 3.7.

FIGURE 3.1 Schematic of the ocean BCP with an emphasis on C transport to depth and return times relevant to the timescale of C storage in the mid and deep ocean. SOURCE: Natalie Renier, Woods Hole Oceanographic Institution.

3.2 KNOWLEDGE BASE

Ultimately the BCP sets the vertical gradient in dissolved inorganic carbon (DIC) through the depletion of inorganic carbon in the surface waters due to incorporation into biomass during photosynthesis, and the net remineralization of organic forms of carbon below the euphotic zone. This gradient is maintained as the strength of the BCP balances the vertical components of ocean mixing that work to homogenize these gradients. Ultimately, the strength of the associated surface CO_2 depletion affects the partitioning of CO_2 between ocean and atmosphere (Takahashi, 2004). Models suggest that if we were to turn off the BCP globally, net atmospheric CO_2 levels would increase by 200 parts per million volume (ppmv) on timescales of many hundred years (Sarmiento and Toggweiler, 1984; Maier-Reimer et al., 1996). Likewise, if the depth of C remineralization on sinking particles were to deepen by 24 meters on average, atmospheric CO_2 could, at least on long timescales, decrease by 10–25 ppmv (Kwon et al., 2009) due to steeper vertical gradients of inorganic carbon and enhanced vertical exchange at shallower depths. As an ocean CDR approach, the main goal would thus be to strengthen, or increase, the net transport of organic carbon out of the euphotic zone, and thereby increase the efficiency of the BCP, thus decreasing the carbon content of the surface waters in contact with the atmosphere while boosting the fraction of carbon that is transported to the deep sea where it can be sequestered on timescales >100 years.

Iron Fertilization

Historically the evidence for a control on atmospheric CO_2, via changes in the supply of iron, comes from the geological record and the glacial-interglacial cycles and correlations between CO_2 as captured in ice cores and dust, a primary source of iron to the ocean (e.g., Martin et al., 1990). This correlation was popularized by John Martin in his famous quip first made at Woods Hole Oceanographic Institute in 1988: "Give me half a tanker of iron and I'll give you the next ice age." The implied potential for a high leverage in terms of a much higher mass of carbon removed per mass of iron applied spurred interest in a potentially efficient way to remove atmospheric CO_2 by oceanic Fe fertilization. What followed next were several fundamental experiments in the lab and in bottles at sea demonstrating this connection between Fe limitation and phytoplankton growth in high-nutrient, low-chlorophyll waters (HNLC) where phytoplankton growth was shown to be limited by iron rather than macronutrients (Martin and Fitzwater, 1988; Morel and Price, 2003). This was followed by more than a decade of purposeful open-ocean Fe addition experiments (Figure 3.2).

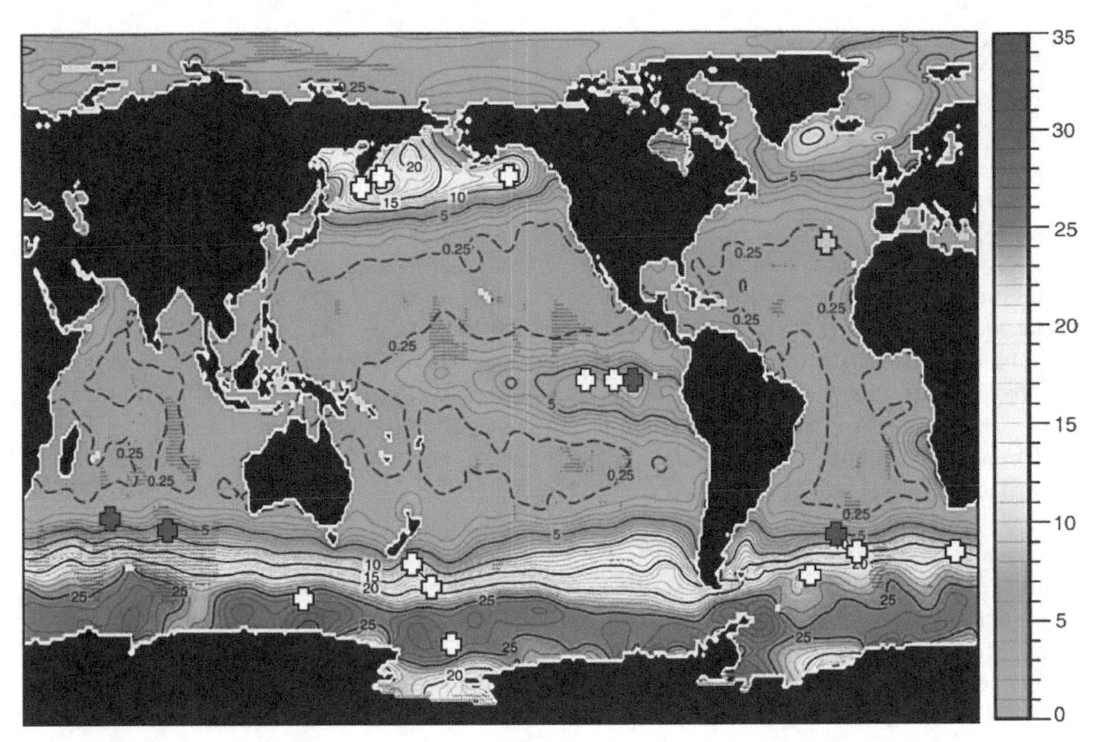

FIGURE 3.2 Annual surface mixed-layer nitrate concentrations in units of micromoles per liter with approximate site locations for artificial ocean iron fertilization (aOIF) experiments (white crosses), natural OIF studies (red crosses), and a study of Fe and P enrichment (green cross). SOURCE: Modified from Boyd et al., 2007, with addition of LOHAFEX aOIF site study in 2009 (Smetacek and Naqvi, 2010).

The goal of prior artificial ocean iron fertilization (aOIF) experiments was largely focused on assessment of the primary response to added iron, and not to track C sequestration and its impact on deeper ocean layers. Several manuscripts and reports have been written on the results of these aOIF studies (de Baar et al., 2005; Boyd et al., 2007; Yoon et al., 2018; GESAMP, 2019), and a consensus has been established that an increase in photosynthetic CO_2 uptake can generally be achieved. As an ocean CDR approach, this open-ocean testing of the impact of Fe enrichment puts this method far ahead of others in terms of the knowledge base. Consequently, it has also put this method at the forefront of public concerns regarding all forms of "geoengineering" and has led to many groups having already formed strong opinions for or against OIF. These social acceptance issues are often focused on OIF, yet this is only one ocean CDR approach, and many of the same acceptance issues would be common to at least all biotic ocean CDR approaches and in many cases abiotic ocean CDR as well, especially if deployed at scale (see Chapter 2). Also given that these early field experiments were conducted largely without international oversight, they prompted the establishment of guidelines for future ocean fertilization (OF) research under international agreements (see Section 3.4 and Chapter 2). However, at the time of writing, those guidelines were not legally binding.

Although the increase in photosynthetic CO_2 uptake (step i) via aOIF is well established, there is less consensus about the transfer and subsequent storage of carbon at depth (step ii). In summary, the two main questions that arise from deliberate aOIF experiments, and are common to all CDR approaches in this report, are: (a) Will it work (to remove carbon from the surface ocean and impact atmospheric CO_2 for some period of time?) and (b) What are the biogeochemical consequences (both intended and unintended)? These issues are explored in more detail below, particularly regarding use of aOIF for ocean CDR. We consider its efficacy and permanence, possible consequences when done at scale, and the ecological and geochemical impacts and future research directions. Note also that any attempt to deliberately alter the oceans' BCP will have consequences that should be considered relative to the status quo of doing nothing.

In addition to aOIF studies, natural systems with episodic or local high Fe delivery have improved the knowledge base for OIF. For example, a natural analog for natural nutrient fertilization is the atmospheric deposition of volcanic ash that leaches trace metals in seawater, generally promoting primary productivity (see Fisheries, below, and e.g., Duggen et al., 2007; Jones and Gislason, 2008; Hamme et al., 2010; Browning et al., 2014; Zhang et al., 2017). Study of Fe sources and impacts around islands in the Southern Ocean has also provided many clues as to the impacts of OIF at larger scales (see Export Efficiencies, below, and e.g., Blain et al., 2008; Pollard et al., 2009). Another source of nutrients to coastal environments is the deposition of ash from wildfires, a phenomenon that appears to be increasing in frequency and intensity as a result of anthropogenic perturbation (Jolly et al., 2015; Cattau et al., 2020). Only a few studies have considered the effects of fires on coastal marine ecosystems when increases in atmospheric deposition of metals or macronutrients are observed (Young and Jan, 1977; Sundarambal et al., 2010; Kelly et al., 2021). One example is an unusual bloom and coral reef die-off during 1997 in Indonesia that has been explained by Fe deposition into the surface ocean by nearby wildfires. Also, an unusual phytoplankton community composition in the Santa Barbara Channel (Kramer et al., 2020) appears to be the result of atmospheric deposition of ash leaching metals and carbon following the Thomas Fire in California in 2017 (Kelly et al., 2021).

Macronutrient Fertilization

The global carbon cycle, marine biogeochemistry, and Earth's climate are thought to have been affected by the supply of macronutrients from continental weathering and on timescales of tens of thousands to hundreds of thousand years. Compared to OIF, ocean macronutrient fertilization (OMF) has received less attention in the scientific community (but see Harrison, 2017). It has the obvious disadvantage of much larger amounts of material required per ton of carbon removed (see Costs and Energy, below). One possible advantage of OMF compared to OIF is the fact that low-nutrient, low-chlorophyll (LNLC) regions are easier to access than the Southern Ocean, the prime candidate region for OIF. Fertilization with inorganic nitrogen has been investigated and suggested as a CDR measure in N-limited LNLC regions (Lawrence, 2014) where sufficient phosphate is available. The few available cost estimates have been low (Jones and Young, 1997, estimate $20/t CO_2). While inorganic N fertilizer can, in principle, be fixed from the atmosphere, albeit at substantial energetic costs, a marine application of phosphate will have to consider that phosphate is a nonrenewable resource also needed in agricultural food production.

While the increase in photosynthetic CO_2 uptake (step i) is widely assumed uncontested, an unexpected decrease in chlorophyll biomass was observed in response to phosphate addition to the ultraoligotophic eastern Mediterranean, indicating that complex food web dynamics and ecosystem responses have to be carefully accounted for when making inferences on C fluxes induced by OF (Thingstad et al., 2005). It is not yet clear whether this is an issue for OMF and less so or not so for OIF. For OMF, the transfer of carbon to the deep ocean has received little attention and will face the same issues as for OIF. The following sections will therefore concentrate on OIF, for which a larger number of theoretical and experimental studies have been performed and the knowledge base is considered advanced compared to OMF.

3.3 EFFICACY

Export Efficiencies

Thirteen open-ocean aOIF studies were conducted between 1993 and 2009 by the oceanographic community as research experiments, resulting in a significant body of literature and several reviews comparing them (Boyd et al., 2007; de Baar et al., 2008; Yoon et al., 2018; GESAMP, 2019). In these field experiments, from 350 to 4,000 kilograms (kg) of iron was added in the form of Fe sulfate dissolved in acidic waters and released in the propeller wash of a moving ship more than 25 to 300 square kilometers (km^2) in one or multiple additions resulting in initial Fe concentrations between ≈1 to 4 nanomoles (nM). These experiments resulted in variable growth response (net primary productivity [NPP] increased by <400 to >1,700 milligrams (mg) C per square meter per day) and shifts in community structure, largely driven by the growth of diatoms of several types (Table 1 in Trick et al., 2010; Tables 2 and 4 in Yoon et al., 2018). Observations from ships extended from as short as 10 days to 30–40 days, and in most cases, the fate of the enhanced growth was not studied due to the limited time on site, the lack of appropriate sampling and measurement tools for particulate organic carbon (POC) fluxes, and in many cases, continued addition of iron that kept the bloom in progress.

To use these results to address the C sequestration efficiencies in response to iron, we need to consider not just the molar ratio of iron added to carbon incorporated into algal growth (C:Fe of 150,000:500,000; Sunda and Huntsman, 1995; de Baar et al., 2008), but also the ratio of iron to carbon that is exported. In summarizing early aOIF experiments in the Southern Ocean, Buesseler and Boyd (2003) noted that two studies—Southern Ocean Iron RElease Experiment (SOIREE) and EisenEx (Eisen is iron in German)—showed no increase in C export in the form of sinking POC to depth within 13 to 23 days after fertilization. A third aOIF study, Southern Ocean Iron Experiment (SOFeX)-South, showed a measurable increase in POC flux between the control and fertilized patch after 30 days, with a C:Fe molar export ratio of 8,000 at 100 meters. These authors noted that the aOIF observations were too short to determine the ultimate fate of the Fe-induced POC export, but these data did not support some of the more optimistic claims surrounding the low cost and small amount of iron needed for ocean CDR (Buesseler and Boyd, 2003). Using the same 100 meters boundary for POC export, de Baar et al. (2008) reported C:Fe export ratios ranging from 650 to 6,600 in three aOIF studies, including those reported by Buesseler and Boyd (2003). De Baar et al. (2008) attributed this relatively modest efficiency compared to algal growth needs, as being due to 75 percent of the added iron being rapidly associated with colloidal forms and subsequently quickly lost via scavenging and hence unavailable for algal growth.

In support of this proposed Fe loss mechanism, de Baar et al. (2008) summarized several natural OIF studies where there was a nearby island source of natural iron in the Southern Ocean resulting in long-standing, yet locally variable bloom and export responses (Blain et al., 2008; Pollard et al., 2009). During one of these, CROZET, C:Fe export ratios ranging from 5,400 to >60,000 were found. Even higher natural C:Fe export ratios were found off the Kerguelen Plateau (up to 174,000). De Baar et al. (2008) compared different estimates of the C:Fe efficiency made using several methods, from looking at POC determined by traps and radionuclide methods, to quantifying export by calculating upward diffusive fluxes of iron and calculating a C balance. Suffice it to say that natural OIF studies showed higher C export ratios in response to iron than aOIF. Presumably, in the natural system, the community response is more likely to reach a steady-state or at least seasonal balance between sources and losses and is less impacted by the episodic nature of aOIF experiments as conducted to this point. One area of research and development (R&D) would thus be looking at the forms of iron added, increasing the Fe-binding ligands in an attempt to minimize losses (see Research Agenda, below), varying the input from pulse to continuous, as well as extending observations to full growth cycle including the bloom demise (several months).

For OIF to sequester carbon from the atmosphere, we need to consider not only the C:Fe ratios leaving the surface, but also the extent to which carbon associated with sinking particles (or other pathways of the BCP) is attenuated with depth, as it is only with C transport below at least the depth of annual winter mixing that carbon can be considered sequestered in terms of a CDR approach (see discussion of durability, below). Few of the aOIF experiments had depth-resolved C export production (EP) measurements, but one that did, LOHAFEX, observed a factor of 8 decrease in POC flux between 100 and 450 meters (using neutrally buoyant sediment traps; Table 5 in Yoon et al., 2018). This is not dissimilar to the expected range in POC attenuation associated with sinking particles and the natural BCP. In a summary of shallow POC flux attenuation below the euphotic zone in the natural BCP, Buesseler et al. (2020) found that up to 90 percent of the sinking POC flux can be lost in the first 100 meters below the euphotic zone, though in some settings, essentially no attenuation can be measured in those first 100 meters, depths over which POC flux attenuation is typically the greatest. This flux attenuation is the result of combined processes that convert sinking forms of carbon to nonsinking forms, such as occurs with "sloppy feeding" by zooplankton on large organic aggregates, and by heterotrophic consumption of sinking particles and conversion to dissolved organic and inorganic carbon by resident zooplankton, microbes, and other animals in the mesopelagic.

If attenuation efficiencies can be controlled or altered during purposeful additions at sites where the communities are more likely to sequester carbon, such as after the sinking of intact diatom cells, then the effectiveness of CDR would be directly affected, or at least the amount of iron needed greatly reduced. Looking again at natural systems, this total loss of carbon starting with NPP and export out of the euphotic zone and transferred 100 meters below, varies from 1–50 percent (export efficiency - C export/NPP). So in estimating the effectiveness of OF for ocean CDR, there remains a large uncertainty in these factors, which determines costs and potential biogeochemical impacts below a purposeful event, as well as its permanence (see below). Also of importance, is that the depth of remineralization for carbon, iron, and other macronutrient remineralization will differ. For example, the depth of remineralization typically follows the order of P < N < C < biogenic silica, from shallowest to deepest, but little is known about the remineralization depth of sinking particulate iron (Lamborg et al., 2008), which is presumably shallower for biogenically incorporated iron and deeper for detrital iron, which would track more closely the lithogenic fraction of the particle flux. More recent studies confirm the importance of particle composition and type in regulating Fe remineralization (Bressac et al., 2019). R&D directed at measuring and purposefully changing these export ratios is needed and cannot be answered by these initial 13 aOIF experiments. As such, current cost estimates for OIF (see below) are limited by the variations in export ratios, but compared to other CDR methods, particularly abiotic ones, OIF would require only a small amount of iron to have a large impact on C sequestration.

Durability or Permanence of CDR

Similar to the permanence issue for land-based CDR, any ocean-based CDR is only as effective as its durability, or timescale over which carbon is removed from and then returned to the atmosphere. This would hold whether using a C capture and storage method, where CO_2 was deliberately injected into the deep ocean, or as discussed in this section, carried into the deep ocean via sinking organic matter, such as in response to stimulation of phytoplankton due to OF. Likewise, biotic methods that deliberately sink organic matter from macrophytes would face a similar issue with durability depending upon where and to what degree the carbon degraded during sinking (see Section 6.2 on Macroalgae). Here we use 100 years to define what is considered "durable" (or "permanent") C sequestration, similar to several land-based options such as enhanced management of forests. This sequestration time frame in the ocean is largely determined by depth and location and is set by the mixing and circulation properties of the ocean.

Primeau (2005) characterized a "first-passage time" as the time when a fluid element at depth in the ocean will make its first return contact with the surface ocean, and thus CO_2 would be able to leak back into the atmosphere. While the Atlantic Ocean in his model had generally younger water mass ages than the Pacific Ocean (difference by about a factor 2 in the deep basins), first-passage times were found to be more uniform over different latitudes and ocean basins. In a model-based analysis employing a steady-state assumption for ocean circulation, Primeau (2005) found that these times were generally greater than 200 years for depths below 500 meters and about 600 years at 2,000-meter depth. A more recent study that illustrates the global pattern versus depth over the 100-year time horizon for C injections can be found in a model by Siegel et al. (2021). The shallow retention times are quite short, with less than 50 percent of the carbon retained more than 100 years in large parts of the ocean if carbon is introduced above 200–500 meters, but carbon is largely retained in most areas when introduced below 1,000 meters, with retention times of centuries, except in the North Atlantic Gyre, along the Southern Ocean polar frontal regions, and in the Southern Indian Ocean east of Africa (Figure S2 in Siegel et al., 2021a).

Another way to consider the timescale for C sequestration is to consider the fraction of CO_2 retained given variations in the attenuation of sinking POC flux from the surface to the 100-year sequestration depth. Siegel et al. (2021) show this as a map of the fraction of carbon retained for 100 years that leaves the surface euphotic zone (Figure 3.3), using the Martin et al. (1987) POC attenuation power-law exponent b which is a best-fit parameter for POC flux versus depth between generally 100 and 1,000 meters. A larger b signifies faster POC flux attenuation and thus less carbon brought to depth where it is sequestered. For example, with the global average b of 0.8 for POC flux attenuation, around 30 percent of the carbon leaving the surface would reach a depth of >100-year sequestration (Figure 3.3, center). This is not surprising since it is well known that much of the sinking POC flux in the ocean is lost due to natural processes that remineralize carbon in mid-waters. Using a faster carbon attenuation ($b = 1.0$) results in less carbon being sequestered (Figure 3.3, right), whereas slower C attenuation ($b = 0.6$) results in many regions exceeding 50 percent retention over 100-year timescales or longer (Figure 3.3, left). In practice, b varies from >1 to <0.5 (Buesseler et al., 2020), but the response to aOIF has been the generation of diatom blooms that in the natural ocean are more often characterized by a lower b, hence the map showing b of 0.6 may be a better predictor of regional patterns of C sequestration for a surface source of fresh POC following aOIF. The issue of deliberately reducing or selecting for low POC attenuation efficiencies is an area of further research since the overall effectiveness of OF as a CDR approach will depend greatly on the fraction that reaches the deep ocean (see Export Efficiencies, above, and further discussion below).

A region particularly well suited for long C sequestration might be the Southern Ocean south of the biogeochemical divide separating the Antarctic from the sub-Antarctic (Marinov et al., 2006) where surface waters and sinking matter enter the deep cell, or "unproductive Southern Ocean circuit" according to Toggweiler et al. (2006) of the global overturning circulation (Ferrari et al., 2014). Besides reaching long first-passage times of the deep waters entering the deep overturning cell, a second advantage compared to other regions is that the supply of macronutrients originates from the shallower cell and is thus not affected by OIF-induced changes in macronutrients. Removal of macronutrients from waters south of the biogeochemical divide would also have less deleterious effects on biological productivity elsewhere. Such effects would otherwise lead to slow saturation of the OIF-induced global mean air–sea fluxes under continuous fertilization, whereas no such saturation and sequestration timescales exceeding 100 years are seen in a modeling study south of the biogeochemical divide (Sarmiento et al., 2010).

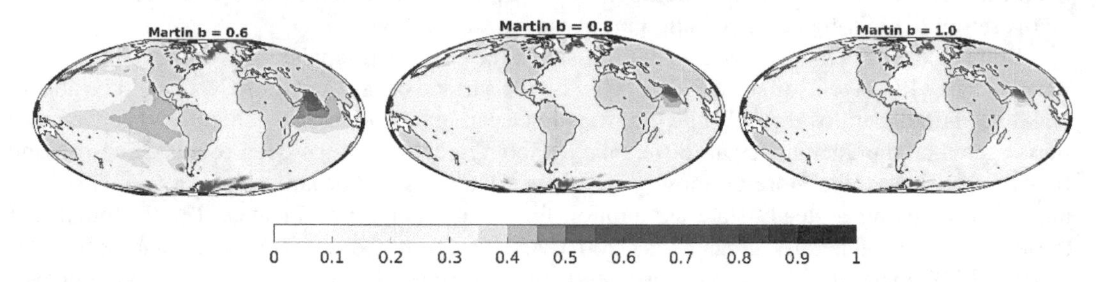

FIGURE 3.3 Color shows the fraction of CO_2 retention for 100 years or more, in response to a surface bloom. Values for *b* follow the parameterization by Martin et al. (1987) with 0.8 being a global average, and 1.0 and 0.6 indicating greater and lower particulate organic carbon flux attenuation, respectively. SOURCE: Adapted from Figure S3 in Siegel et al., 2021a, licensed by Creative Commons CC BY 4.0.

Of significance to ocean CDR is the need for (1) thorough measurements and models, to quantify the permanence for a given site, and (2) deliberately selecting sites and enhancing export efficiencies to optimize for maximal sequestration time. Enhanced C removal and efficient transport via the BCP will depend on the pathway that carbon takes to reach the deep sea. For example, the physical pumps that transport suspended POC or mix dissolved organic carbon below the surface will not lead to long-term sequestration in regions other than those where deep waters are formed. Neither will most active migrations, such as zooplankton diel vertical migration, because these processes are largely limited to <1,000 meters (Boyd et al., 2019). However, the gravitational settling of POC does reach the seafloor, with on average ≈10 percent of the carbon fixed via NPP in the euphotic zone reaching 1,000 meters (Martin et al., 1987) and of that, <1 percent is buried in ocean sediments. So it is the gradient in sinking POC remineralization that will set BCP sequestration efficiencies for a given location. These remineralization rates vary with the speed of particle settling and are likely modulated at least by temperature and oxygen (Devol and Hartnett, 2001; Van Mooy et al., 2002; Boscolo-Galazzo et al., 2021). These BCP efficiencies vary widely in natural settings (e.g., Buesseler et al., 2020), and could potentially be altered if one could select for formation of blooms with fast-sinking pellets (e.g., salps), carcasses, and/or sites with less microbial degradation (e.g., colder waters and low O_2). Ultimately, this return flow of carbon to the surface from the site of export sets the time frame for the permanence OF as a CDR approach, and this must be considered and compared relative to other CDR approaches.

Monitoring and Verification

Effective CDR requires C accounting that is transparent and verifiable and requires ways to monitor the ecosystem responses in the upper ocean where OF is applied and in the deep sea where carbon is intended to be sequestered. In the case of OF, the area of treatment for the demonstration projects could be relatively large (>1,000–10,000 km^2) with a timescale of several months to years. The broad synoptic images (swath widths ≥2,500 km), high spatial resolution (~1 km), and rapid resampling (nearly daily global coverage) make satellite ocean color observations well suited to document the enhancement of surface ocean productivity created via OF (e.g., Westberry et al., 2013). Previous OIF studies have used satellite maps of phytoplankton chlorophyll concentrations to map the extent of and changes in phytoplankton due to trace nutrient addition (e.g., Abraham et al., 2000; Westberry et al., 2013). Existing satellite data products can also be used to assess changes in the phytoplankton Fe stress either via changes in the chlorophyll-to-C ratio or the solar-stimulated chlorophyll fluorescence line height (Behrenfeld et al., 2005, 2009; Westberry et al., 2013; Xiu et al., 2014). In an exciting new development, the National Aeronautics and Space Administration's (NASA's) upcoming Plankton, Aerosol, Cloud, ocean Ecosystem (PACE) mission[1] will provide global, hyperspectral (5-nm resolution) observations of ocean color reflectance spectra (Werdell et al., 2019), improving the quantification of phytoplankton composition from satellite data (e.g., Uitz et al., 2015; Xi et al., 2017; Kramer et al., 2021). Merged satellite altimetry data products will also be useful for assessing the trajectory of surface water parcels and stirring of patches, which may in turn affect the efficiency by which Fe additions are utilized by the surface ocean (e.g., Abraham et al., 2000). Thus, satellite observations provide a suite of data products useful for monitoring and verifying the effectiveness of OIF for ocean CDR.

However, satellite data only measure surface ocean properties and do not provide estimates of C export and other necessary biogeochemical determinations. The biogeochemical Argo profiling float network (which would continue to operate after the start of the treatment) can be used to track surface productivity and variations in plankton biomass (Yang et al., 2021) as well as provide needed subsurface data. A suite of autonomous surface vehicles and water column gliders (the number would depend on the aerial extent), outfitted with abiotic sensors (temperature, salinity, pressure, pCO_2, O_2, bio-optics, and nutrients) would be deployed on and beneath the treated bloom area. These autonomous sensor platforms profile between the sea surface and ocean depths every few days, measuring water properties and relaying data via satellite. At least one area outside of the bloom area, but in a similar water depth and ecosystem, could be monitored as a control, with similar sensors and sampling systems. These sensor systems would deliver an important subset of essential ocean variables (see Danovaro et al., 2020) to document the resulting bloom size and its impact on ocean physical and chemical properties. To quantify organic C uptake and the strength and efficiency of the BCP, methods that sample and/or optically characterize particles are needed (McDonnell et al., 2015). To quantify the extent of C storage below known reference depths for C sequestration (see section above on permanence), sediment traps (e.g., Buesseler et al., 2008), bio-optics (e.g., Giering et al., 2020), or other radionuclide-based particle flux tracers (e.g., Waples et al., 2006) would be used to measure the POC and associated elemental fluxes to the deep ocean. Consideration in these monitoring strategies for assessment of sinking POC also needs to consider the horizontal displacement of surface particle sources and the eventual location of C sequestration in the deep sea, that is, consideration of particle source funnels (Siegel et al., 2008). Examples of several monitoring technologies are shown in Figure 3.4.

Management of deep ocean ecosystems used for ocean CDR requires establishment of benchmark conditions and monitoring regional-scale threats, such as marine heat waves and ocean deoxygenation, in addition to site-specific monitoring of the C removal. Ships of opportunity (e.g., Smith et al., 2019) could also be coordinated and engaged to corral the floats to remain in the regions of interest or reseed an area with floats and other monitoring systems. The ships themselves could also provide data and data products. Data and data products from decades-long "Line" surveys in some regions (e.g., Line P in the Northeast Pacific) provide a historic framework for understanding benchmarks and/or long-term ocean change in an HNLC area suitable for OIF (Wong et al., 1995; Timothy et al., 2013).

Given that the potential for the area to be fertilized is extremely large, DARPA's (Defense Advanced Research Projects Agency's) Ocean of Things,[2] currently under development, will be useful for monitoring impacts. Such systems are formed by an interconnected network of small, inexpensive, and potentially biodegradable sensors and floats. Their sensor suites include measuring sea-surface temperature over a large region, so that it can be mapped with sufficient density to better understand ocean currents and mixing. The premise is that this dense measurement network

[1] See https://pace.gsfc.nasa.gov.

[2] See https://oceanofthings.darpa.mil/.

FIGURE 3.4 Instruments for monitoring impacts of OF as a CDR approach include (1) profiling float with underwater vision profiler camera; (2) wind-driven surface autonomous vehicle; (3) neutrally buoyant sediment trap; (4) Argo profiling float; (5) PELAGRA sediment trap; (6) fleet of small MINIONS floats; (7) surface buoy for open-ocean mooring; (8) ocean glider; (9) deep-moored time-series sediment trap; (10) wave-powered surface autonomous vehicle. SOURCE: (1) David Luquet, (2) Beth Hamel, NOAA, (3) Alyson Santoro, NASA, (4) NASA's Earth Observatory, (5) Amala Mahadevan, NSF, (6) Melissa Omand, (7) D. Macintyre, NOAA, (8) NOAA, (9) MBARI, (10) PMEL, NOAA.

can be combined with remote sensing data and models to merge observations and assess the fate of carbon. The program is in development, so there is an opportunity to incorporate small sensors specific for OF monitoring, such as ocean MINIONS[3] a small, inexpensive isopycnal float with onboard sensors and an upward-looking camera that quantifies POC export associated with sinking particles (Melissa Omand, University of Rhode Island, ongoing personal communication).

To assess the potential positive or negative impacts on regional fisheries, information on trends in species diversity and fish abundances is a key metric. Imagery from fixed-observatory underwater cameras and regular ship-based video transect surveys could be used to develop data products on species diversity and abundance. Manual, machine vision, and crowdsourcing tools are several approaches that extract biological information from video and photo archives to reveal trends in species numbers (see, e.g., Matabos et al., 2017). Understanding impacts higher up the food chain would require sampling with traditional nets and use of new techniques such as eDNA methods (e.g., Closek et al., 2019) to catch trends that occur over multiple years of growth and adaptation to purposefully altered conditions.

Additionality and Downstream Effects

The aim of OIF is to stimulate photosynthesis, the production of biomass and uptake and redistribution of carbon. In this process, nutrients other than iron are taken up and will be redistributed as well. The larger the amount of carbon sequestered is, the larger will be the redistribution of nutri-

[3] See https://twilightzone.whoi.edu/work-impact/technology/minions/.

ents. This will have local and remote effects on nutrient fields and therefore on nutrient limitation, biological production, biological diversity, and, eventually, the marine BCP.

One effect that has received attention is the so-called nutrient robbing (Shepherd, 2009), whereby macronutrients utilized during aOIF-induced biological production are not available for biological production and associated C uptake elsewhere. Besides its ecological implications, this represents a nonlocal CO_2 leakage and presents difficulties for appropriate accounting of CO_2 sequestration achieved by aOIF (Oschlies et al., 2010a). This effect is of particular concern for aOIF in the Fe-limited surface waters of the tropical and subpolar North Pacific, where no deep waters with significant amounts of unutilized (preformed) nutrients are produced, and hence all macronutrients in the surface layer will essentially be used up anyway, and aOIF would predominantly lead to a relocation of the areas of biological production. The net CO_2 sequestration inferred from individual patchy fertilization experiments may thus be considerably overestimated (Aumont and Bopp, 2006).

Nutrient robbing would also occur for aOIF in the Southern Ocean, where surface waters tend to be replete upon subduction, thereby forming nonzero preformed nutrients. Model studies indicate that Southern Ocean OIF will lead to nutrients being trapped in the Southern Ocean and less nutrients will be exported to regions farther north, eventually leading to a reduction of biological production north of the Southern Ocean (Oschlies et al., 2010a). Relieving the Fe stress on diatoms via Southern Ocean OIF may also lead to changes in the Si:N ratio of the organic matter export and, consequently, to a change in silicic acid leakage from the Southern Ocean (Holzer et al., 2019). Nutrient robbing by Southern Ocean aOIF is likely to be accompanied by eventual reduction in biological production in much of the "world ocean" outside the fertilization region. Models suggest that these Southern Ocean nutrients currently fuel up to three-quarters of the biological production in the global ocean north of 30°S (Sarmiento et al., 2004; Marinov et al., 2006).

3.4 SCALABILITY

Like all ocean CDR approaches, models are used to assess the scale at which OF would affect the global carbon cycle. These models have focused on a particular region (e.g., Southern Ocean) or HNLC regions globally and often use the complete drawdown of surface ocean macronutrients to simulate enhanced primary production and the amount of potential C removal. There are important differences between models, however, including the extent to which deep C sequestration is considered versus shallow C export; whether nutrient co-limitation is included; the timescale of removal and reequilibration of CO_2 with the atmosphere; and, for example, ignoring the impact of OIF on LNLC regions via stimulation of N_2 fixation. These are just some examples of why the estimates thus far on the total scale of OIF alone range widely, from a fraction of a Gt C/yr to up to 3–5 Gt C/yr (Table 3.1), with a recent GESAMP (2019) report settling on 1 Gt C/yr (3.7 Gt CO_2) as the maximum theoretical potential. Practical consideration for engineering such large-scale deployments is also not considered. Deliberate alteration of ocean ecosystems to this extent would have many impacts and feedbacks not included in these models, but certainly the potential exists to augment the natural BCP of 5 to 12 Gt C/yr (C flux at the base of the euphotic zone; Siegel et al., 2014) by a Gt/yr or more.

One outcome of these models is that regional differences in ocean CDR capacity for OIF are large. For example, numerical models generally show a maximum C sequestration potential when OIF is applied to the entire Southern Ocean, the largest HNLC region of the world ocean, during the growing season when growth is not limited by light. This would lead to substantial net air–sea CO_2 fluxes (Aumont and Bopp, 2006). In contrast, OIF was found to have limited impact when applied in the equatorial Pacific (Gnanadesikan et al., 2003).

TABLE 3.1 Ocean Iron Fertilization Global Sequestration Potential

Source	Year	Gt C/yr	Comments
Aumont and Bopp	2006	1 to 2	106–227 Gt/C over 100 years provided in reference to five previous model studies, Table 1
Buesseler et al.	2008	0.2 to 0.3	Reported several hundred million for HNLC areas only
GESAMP report	2019	1	Maximum potential based on model predictions, Table 4.4
Strong et al.	2009	up to 1	Southern Ocean only—refers mostly to Zahariev et al., 2008
Cao and Caldeira	2010	up to 3	Deplete all surface PO_4 by 2100—Table 2—822 – 541 = 281 Gt C until 2100 (difference with and without OIF), so over 90 years about 3 Gt C/yr
Oschlies et al.	2010a	<1	Decadal to centennial timescale Southern Ocean OIF only, but global impacts considered, including downstream impacts and CO_2 backflux—Table 1
Keller et al.	2014	1 to 5	Southern Ocean only south of 30°, decreasing quickly from 5 to 1 if measured on centennial scales
Natural BCP	Various	5 to 12	Natural BCP for reference of euphotic zone C loss (e.g., Siegel et al., 2014)

Over longer timescales, a model applying OIF everywhere south of 30°S found that OIF-induced air–sea flux of CO_2 is largest during the first year, reaching 5 Gt C/yr in Keller et al. (2014), but drops to less than 2 Gt C/yr within 10 years and about 1 Gt C/yr on centennial timescales. The large C uptake during the first year can be explained by the large macronutrient reservoir that becomes accessible upon the relaxation of Fe limitation. Export of organic matter and subsequent remineralization at depth leads to trapping of much of these nutrients (and carbon) in the Southern Ocean. Upwelling of the nutrients and respired C trapped in the Southern Ocean offsets a substantial fraction of the OIF-mediated downward flux of POC, leading to a substantial return flux of respired CO_2 to the atmosphere and an atmospheric uptake efficiency, defined as the ratio of air–sea CO_2 flux to export production (Jin et al., 2008), of less than 0.5. The uptake efficiency is also affected by the lowering of atmospheric pCO_2 by successful OIF (or other CDR schemes). This will shift the CO_2 air–sea partial pressure difference toward a net efflux of CO_2 from the ocean to the atmosphere. A similar efflux might also occur for the net C flux between the terrestrial biosphere, where photosynthetic CO_2 uptake is often stimulated by elevated atmospheric CO_2 concentrations. The compensating effect of such effluxes due to CDR-induced changes in atmospheric CO_2 changes from a few percent in the first year of CDR operation to about 10 percent on decadal and 50 percent on centennial timescales (Oschlies, 2009).

3.5 VIABILITY AND BARRIERS

General Considerations

The intention of OF is to stimulate photosynthesis and the production of organic matter. Similar to any biological CDR method, this intentional perturbation of natural ecosystems will change species composition, food web structure, and biodiversity, and will generate winners and losers until a new ecosystem is established.

Fertilization-induced enhancement of biological production will also lead to enhanced remineralization and oxygen consumption. While oxygen levels will thus decline below fertilization areas, the trapping of nutrients is expected to lead to a decline in biological production and eventually oxygen consumption in other regions of the world ocean. For Southern Ocean OIF, the volume of low-oxygen waters located in the tropical oceans may thus even shrink despite a global decline in the marine oxygen inventory (Oschlies et al., 2010a).

The remineralization part of the nitrogen cycle also involves nitrification, during which nitrous oxide (N_2O) is produced. A second pathway for enhanced production of N_2O is associated with anaerobic remineralization in low-oxygen environments that may expand in response to OF. Detailed understanding of the rates of N_2O production and possible consumption is still lacking, but direct measurements during the aOIF SOIREE measured increased N_2O emissions from the fertilized patch that would offset 6 percent to 12 percent of the OIF-induced CO_2 uptake (Law and Ling, 2001). Similar offsets were inferred from models (Jin and Gruber, 2003; Oschlies et al., 2010a). It is important that this offsetting of enhanced biological CO_2 uptake by N_2O produced from enhanced remineralization will likely occur for any biological marine CDR scheme as well as for biological terrestrial CDR schemes.

Another non-CO_2 greenhouse gas (GHG) that has been observed in nine aOIFs is production of dimethyl sulfide (DMS), which can lead to the formation of cloud condensation nuclei above the ocean and thus provide additional positive co-benefit in terms of reducing global temperatures (e.g., Law, 2008). But field results are variable, with larger DMS increases seen in the Southern Ocean versus the North Pacific.

In summary, a number of trace gases could be affected by OIF, not only N_2O and DMS, but various halocarbons, methane (CH_4), and isoprene (see Figure 1 in Law, 2008), and accounting for their positive and negative feedbacks on climate needs to be included in research studies of OIF as an ocean CDR approach.

Harmful Algal Blooms

All aOIFs are intended to produce changes to community composition as a consequence of adding iron. Of concern to many has been the possible increase in the abundance of *Pseudonitzschia*, a diatom genus known to produce the harmful neurotoxin domoic acid (DA) (Silver et al., 2010; Trick et al., 2010). This unintended consequence is often put forward in the public media as a reason not to continue with OIF as an ocean CDR approach (Allsopp et al., 2007; Harris, 2012; Tollefson, 2017). Looking more closely, there are few data to support this concern based upon direct measurements of DA, including studies by Marchetti et al. (2009), who did not detect increased DA production by *Pseudonitzschia* in response to the aOIF Subarctic Ecosystem Response to Iron Enrichment Study (SERIES) in the northeastern Pacific. Trick et al. (2010) point out that 6 of the 11 aOIFs produced increases in *Pseudonitzschia* abundances, so roughly half of the experiments had the potential to cause unintended harmful algal blooms (HABs). However, Trick et al. (2010) measured DA/cell in natural conditions, from a single northeastern Pacific profile (Ocean Station Papa) and saw little difference in the DA quota for incubations of Fe- and non-Fe–enriched cells ($3–4 \times 10^{-6}$ picograms [pg] DA/cell). But using these estimates of low DA per cell and assumptions about transfer to the DA in water (not measured), they postulate a possible toxigenic response to an aOIF deployment, if conducted at a scale 100 to 1,000 times larger than any aOIF experiment thus far. In other words, with enough *Pseudonitzschia*, one might see a harmful response.

In another study to consider DA production, Silver et al. (2010) measured DA in stored cells from two natural settings and two aOIF studies (Southern Ocean Iron Experiment–South and FeExII). The results show highly variable DA per cell and only elevated DA in two settings, the natural northwestern Pacific (K2 site DA/cell = 0.9 ± 07 pg/cell for four samples) and in the aOIF site SOFeX-S (0.9 ± 0.2 for two samples). Values at these two sites were much higher than after the aOIF experiment FeExII (0.04 ± 0.02 for two samples) with values equally low as at a natural site in the Gulf of Alaska (0.03 ± 0.07 for four samples). The point is that based upon these 10 samples, there is no clear evidence of additional DA per cell after aOIF relative to natural systems. It is only when the abundance of *Pseudonitzschia* increases in response to OIF that there may be conditions where harmful responses are possible. Silver et al. (2010) noted that "neurotoxin impacts at higher trophic levels, well known in shelf and coastal regions, have not yet been reported in open ocean systems." They conclude that caution is warranted, but as with any ocean CDR approach, there will be unintended consequences that will be important to study, and thus be able to predict, if one were to move from aOIF research to large-scale implementation.

Co-benefits

Fisheries

OF has increasingly been proposed as a method for fisheries enhancement, in addition to, or in place of using it as a method for ocean CDR. At the most basic level, enhancements to the base of the food chain should lead to increases in fish stocks, at least if other variables remain similar. This concept was put forward early in the framework of reducing global hunger, based upon the addition of nitrogen to the ocean to increase production and thus conversion to seafood for human consumption (Jones and Young, 1997). In part, given the controversy surrounding aOIF and lack of a C credit market, commercial interest has shifted in several recent cases to this "ocean seeding" idea. For example, in 2012, the Haida Salmon Restoration Corporation (HSRC) asked the Haida Nation village of Old Massett in British Columbia to fund a commercial venture to deliberately release 100 tons of iron off Haida Gwaii as a means to enhance the local salmon fishery. Controversy remains about the legality of this effort (Tollefson, 2012; Wilson, 2013), and it was also lacking in the public release of data or peer-reviewed studies documenting the impacts. While after-the-fact study of remote sensing images and plankton sampling did document a bloom within the study area (Batten and Gower, 2014; Xiu et al., 2014), no links could be made to enhanced fisheries.

We are thus left with no evidence on the potential positive or negative impacts on fisheries of the 2012 event, though follow-on proposals for aOIF have been put forward with the specific goal to enhance the local fisheries. In one such case, Oceanos[4] is proposing an Fe addition in the Humboldt Current area in the territorial waters off Peru. Whereas links further up the food chain may be impossible or at least difficult to demonstrate with commercial-scale OIF, natural OIF events may provide clues as to the possible link between OIF and fisheries enhancements.

One of the best-documented natural OIF events that has been tied to fisheries is the 2008 Kasatochi volcanic eruption off the Aleutian archipelago. Hamme et al. (2010) documented a large-scale biogeochemical response of a doubling of surface chlorophyll over an area of 1.5 to 2×10^6 km^2 and an observed decrease in surface pCO_2 by 30 µatm (8 percent) and increase in pH at Ocean Station Papa from 8.08 to 8.13. This productivity enhancement and decrease in CO_2 was attributed to the response to the addition of iron from Kasatochi and resulted in what they estimate to be a C export event on the order of 0.01 Gt C (0.04 Gt CO_2). Olgun et al. (2013) further studied the release of iron from the volcanic ash and supported their findings that enough iron would have been added to support the enhanced productivity seen by remote sensing.

The link between the Kasatochi event and enhanced fisheries, however, remains controversial. Parsons and Whitney (2012) were the first to suggest that the volcanic iron induced a massive diatom bloom in the Gulf of Alaska that enhanced the food supply for adolescent sockeye salmon,

[4] See http://oceaneos.org/.

leading to one of the strongest sockeye returns on record in 2010 for the Fraser River. Olgun et al. (2013) noted that this diatom bloom could support a larger zooplankton copepod food source for these juvenile salmon. From the timing of the bloom and magnitude of the response, they thought it was "very plausible" that the eruption enhanced salmon survival, though they point to several other factors that can affect ocean survival of salmon. McKinnell (2013) looked more broadly at Sockeye salmon spawning success in the Fraser River and challenged whether the volcanic event was the cause. They present a case that the survival was unremarkable in the historical record for the Fraser River, and that several other factors refute this idea, such as that the region with the anomalous chlorophyll enhancement is not where the juveniles migrate, and that no other salmon from that feeding region had unexpectedly high returns.

It is not surprising that a link between short-term OIF enrichments and fisheries are hard to document given the episodic supply of iron during a volcanic eruption and the subsequent enhancement of fish stocks years later. That decoupling in time and the wide range of processes that impact fisheries will make it difficult to attribute a positive fisheries co-benefit to a local event or sustained regional OF. And finally, if the goal of OF is enhanced fisheries for human consumption, then the ocean CDR benefits decline as carbon is returned to the atmosphere via respiration of food supply on land.

Ocean Acidification

In terms of other co-benefits, if the consequences of OF are a reduction in surface ocean DIC, as seen in 11 of the 13 aOIF experiments (Yoon et al., 2018), this would result in a pH increase and thus a decrease in surface ocean acidification (OA), at least temporarily during the drawdown period. Since OA is considered to be detrimental to carbonate-producing marine life in particular (Doney et al., 2020), this would be a co-benefit by maintaining or increasing pH over scenarios without OF. Using a simple ocean model, Cao and Caldeira (2010) predict that the impact of OIF given an extreme scenario of complete surface phosphate removal would reduce atmospheric CO_2 by 130 ppm, but increase surface ocean pH by only 0.06, relative to the same emissions without OIF.

Interestingly Cao and Caldeira (2010) further emphasize that this surface OA decrease, or co-benefit, would be accompanied by further acidifying the deep ocean, as also expected for other approaches of moving atmospheric CO_2 into the deep ocean (Reith et al., 2019). This could have negative impacts on the growth of deep-sea corals, as well as the metabolic processes of deep-sea biota in general (e.g., Siebel and Walsh, 2001). In effect, OF might be a co-benefit for surface corals and shell fisheries, but a shift to less favorable conditions in the deep ocean, similar to the arguments for potential negative impacts of mid-water and deep-ocean oxygen decreases in response to OF (see General Considerations). Oschlies et al. (2010a) inferred a pH decline by more than 0.1 units over large parts of the mid-depth Southern Ocean after simulated multidecadal OIF.

Lacking longer or larger-scale aOIF studies to directly examine deep-ocean impacts, studies of natural OF systems may provide another line of evidence regarding the impacts of OIF on deep-sea biota. When comparing a naturally Fe-enriched setting versus nearby controls off the Crozet Plateau in the Southern Ocean, the Fe-induced increased supply of organic carbon to the seafloor led to greater densities and biomass of deep-sea animals (Wolff et al., 2011). In fact, a similarity in deep-sea ecology in the Fe-enriched site and the productive northeastern Atlantic was noted, with the suggestion that aOIF could similarly increase the benthic biomass and species composition.

No matter what the impact of OF on the deep sea, it should be noted that what deliberate and large-scale OF would do is essentially speed up the natural processes that are already happening, under any current scenario of enhanced CO_2 in the atmosphere. For example, in one emissions scenario, 40 percent of fossil-fuel CO_2 would be stored in the ocean with OIF by 2100 versus 27 percent without OIF (Cao and Caldeira, 2010).

Cost and Energy

Ocean fertilization approaches leverage mass ratios between nutrients and organic carbon (Table 3.2) such that the costs of raw materials could be relatively low when normalized to the mass of CO_2 removed. The costs in Table 3.2 do not include other parts of the supply chain (i.e., transport, loading, and addition to the ocean) or monitoring. It is clear that for macronutrient fertilization (N, Si, P), the amount of macronutrient added would be much greater than OIF, and hence the raw material costs are greater as reflected in the cost per ton of CO_2 removed. Similarly, to stimulate CO_2 removal at a meaningful scale (~1 Gt CO_2/yr), then the N and P production would be equivalent to 30 percent to 40 percent of current markets, whereas OIF would consume <0.1 percent of the current Fe market. Thus the challenge of obtaining iron and its land-based impacts from mining would be far smaller than for other macronutrients and ocean CDR methods such as alkalinity enhancement that have far greater material needs and hence impacts on land (see Chapter 7). Furthermore, phosphorus is a nonrenewable resource and its use in large-scale OF would compete with its use for agriculture.

The deployment costs for spreading nutrients in the ocean is also relatively low, especially in the case of iron where relatively small amounts are needed. For instance, the HSRC in its 2012 project chartered a fishing vessel to put 100 tons of iron a few hundred kilometers off the coast of Haida Gwaii at a reported cost of $2.5M (e.g., Biello, 2012). Early estimates for OIF that include both materials and delivery were as low as $2/t C ($0.5/t CO_2; Markels and Barber, 2002).

It is clear that costs for OF, however, are very sensitive to (1) the efficiency of nutrients added to stimulated C removal and (2) the ratio between carbon removed and that which is permanently sequestered at depth. On the basis of different Fe:C_{seq} efficiencies, Boyd (2008) illustrates that the cost per ton for OIF can vary from <$3/t C_{seq} to >$300/t C (<$1 to >$80/t CO_2, Figure 3.5), providing a best estimate based on aOIF experiments of $30–300/t C ($8–80 /t CO_2). Other estimates at a larger scale suggest that the costs of aOIF could be as low as <$10/t C (Harvey, 2008; Renforth et al., 2013). The HSRC project mentioned earlier did not include monitoring and verification of C storage, but using a modest Fe:C ratio of 1,000, one can estimate a cost of $25/t C sequestered. A key research question is therefore to better predict and quantify these Fe:C_{seq} ratios, which will be

TABLE 3.2 Ratios of Nutrients to CO_2 Removed and Market Pricing and Production Comparisons

Nutrient	Ratio of Theoretical Maximum Carbon Dioxide Removed (t CO_2/t)[a]	Market Price of Material ($/t element)	Market Price Normalized to CO_2 ($/t CO_2)[b]	Percent of Total Annual Production to Facilitate 1 Gt CO_2
Nitrogen	21	1,000 (ammonium nitrate)[c]	48	~34
Silicon (diatoms)	11	300–1,700[d] (e.g., Si fume) 1–5 (silicate rock)	27–155, possibly <1 for silicate rock	—
Phosphorus	150	300[e] (phosphate rock)	2	~40
Iron	780–78,000	350 (65% iron ore)[f]	<0.4	<0.1

[a] Expressed as t CO_2 for comparability (assuming a molar ratio of C:Si:P:Fe of 106:15:16:1:0.1-0.001 [Brzezinski, 2004] to derive maximum removal rates per ton of nutrient added).

[b] These represent material costs per theoretical ton of CO_2 removed only, not the levelized cost of net C removal (see below for more on costs).

[c] See https://www.intratec.us/chemical-markets/ammonium-nitrate-price.

[d] See http://www.microsilica-fume.com/silica-fume-price-per-ton.html.

[e] See https://www.indexmundi.com/commodities/?commodity=rock-phosphate&months=60.

[f] See https://tradingeconomics.com/commodity/iron-ore.

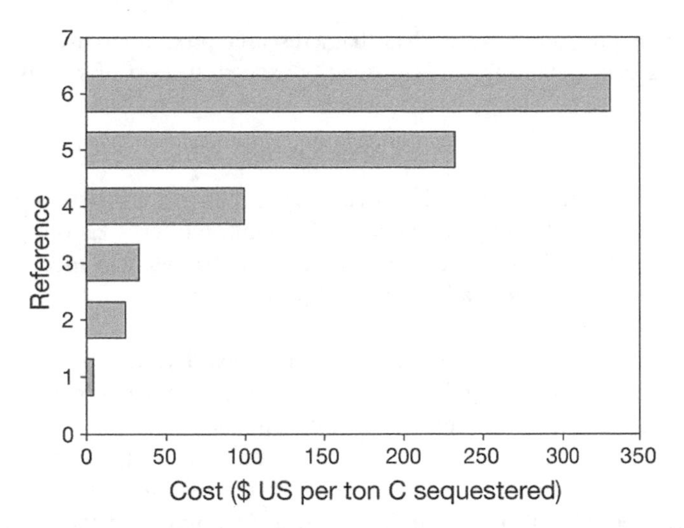

FIGURE 3.5 Estimates of the cost of C sequestration by OIF. These six estimates differ largely because of varying ratios of Fe:C sequestered as detailed in the original references and discussed above in Export Efficiencies. Note that costs based on CO_2 sequestration would be 3.7 times lower. SOURCE: Boyd, 2008.

set largely by both the bioavailability of the added iron and the extent of shallow remineralization of sinking POC flux that is stimulated in response to the Fe addition.

As with any ocean CDR approach, monitoring intended and unintended consequences to include changes in geochemistry beyond carbon and changes in ocean ecology requires additional costs that are rarely quantified. Perhaps the best way to estimate these costs is to scale costs based on prior research studies and aOIF field programs. These field studies looked at changes beyond the C balance, including shifts in plankton productivity and community structure; the consequences of other limiting macronutrients; and in the waters below the in situ Fe addition and in control sites, changes to oxygen, N_2O, CH_4, DMS; and other potential consequences, such as noted above, and including the presence of HABs. No official accounting is available, but budgets for SOFeX, for example, the last U.S. experiments in 2002 were on the order of $10M for two aOIF field deployments.

A more recent cost accounting for a research study of the BCP can be taken from NASA's EXport Processes in the Ocean from Remote Sensing (EXPORTS) program.[5] In the North Atlantic in 2021, EXPORTS used three ships and several of the latest autonomous platforms similar to that which would be needed for new in situ demonstration projects (see Monitoring and Verification). Although EXPORTS did not add iron, it looked at the fate of carbon in natural settings and ecological and community structure, with shipboard measurements largely over the course of 1 month. Using this more recent field experiment as an example, costs ran into $15M to $20M per field site, which included measurements of physical, biological, and geochemical processes, including C stocks and rates, such as C uptake, export, and remineralization. This was still of relatively modest duration and size (1 month, 10,000 km^2) but it was conducted as a Lagrangian time series much like would be needed for CDR studies. The cost would be similar to scale appropriately for more comprehensive, CDR-focused aOIF science experiments and pilot-scale demonstration projects. No one is suggesting that an operational CDR approach would take the same level of detail and hence cost

[5] See https://oceanexports.org/.

as much as a research program, but it is clear that a research program to track the consequences and fate of C, including ecological impacts, will readily exceed the cost of simply deploying the iron.

Governance

The legal framework for ocean CDR is discussed in Chapter 2. Many of the international and domestic laws discussed in that chapter could apply to nutrient fertilization.

Nutrient fertilization is the only ocean CDR technique for which a specific governance framework has been adopted at the international level. Specifically:

- In 2008, the parties to the Convention on Biological Diversity (CBD) adopted a nonbinding decision recommending that governments take a "precautionary approach" and refrain from engaging in nutrient fertilization, except for "small scale research studies within coastal waters" (Para. C4, 2010). The decision further states that small-scale research studies may only be authorized "if justified by the need to gather specific scientific data," and should be subject to "thorough" review and "strictly controlled." This was reaffirmed in a separate decision issued in 2010 (Para. 8w, 2010).
- Also in 2008, the parties to the London Convention and Protocol adopted a nonbinding resolution stating that nutrient fertilization projects "should not be allowed," unless they are undertaken for the purposes of "legitimate scientific research" (Art. 3-5, Resolution LC-LP.1, 2008). The parties subsequently adopted a framework for reviewing research proposals to evaluate their "scientific attributes" and potential environmental impacts (Resolution LC-LP.2, 2010).
- In 2013, the parties to the London Protocol agreed to an amendment that effectively prohibits nutrient fertilization, except for research purposes (Resolution LP.4(8), 2013).

Note that neither the 2008 decision under the CBD nor the 2008 resolution under the London Convention and Protocol is legally binding. The 2013 London Protocol amendment will become legally binding once ratified by two-thirds of the parties to the Protocol, but that has not yet occurred and appears unlikely in the near future. Even if the ratification threshold is met, only parties to the London Protocol that have ratified the amendment would be bound by it. Notably, the amendment would not bind the United States, which is only a party to the London Convention and not the London Protocol (see Chapter 2).

Researchers (e.g., Webb et al., 2021) have concluded that the United States could, at least in some circumstances, undertake or authorize ocean CDR projects involving the addition of materials to ocean waters (presumably including nutrient fertilization) without violating the terms of the London Convention. Projects would, however, have to comply with applicable domestic law. Table 2.3, in Chapter 2, identifies key U.S. federal environmental laws that could potentially apply to nutrient fertilization. No detailed research has been conducted on the application of those and other U.S. laws. We understand that such research is being conducted as part of an ongoing project, led by Columbia University researchers, but they had not published their findings at the time of writing.

Ocean CDR is likely to face social acceptance challenges given its divisive history. With open-ocean activities, stakeholder analysis and defining relevant publics can be challenging. Moreover, if OF receives social license on a local or community scale, it may not receive social license globally; conversely, global social license will not translate to community-scale acceptance. One way scientists can provide the basis for robust societal debate is to provide research that aligns with open-science principles. For the research to assess OF as a CDR approach, there would need to be transparency in planning and open access to all data; results disseminated at open meetings and in peer-reviewed journals in a timely manner; full compliance with international laws; study of intended and unintended ecological effects in both surface and subsurface waters; and assessment of impacts beyond the study area and extrapolation to global scales. If academic, private, and government scientists are involved in collective experiments, there needs to be the ability to maintain independence in their ability to report data and interpretations thereof. Finally, if research activities are funded by commercial interests, there is even a greater need for a clear and transparent code of conduct to ensure that results are considered unbiased and accepted by the public.

3.6 SUMMARY OF CARBON DIOXIDE REMOVAL POTENTIAL

The criteria for assessing the potential for ocean nutrient fertilization as a feasible approach to ocean CDR, described in Sections 3.2–3.5, is summarized in Table 3.3.

3.7 RESEARCH AGENDA

While OF, and OIF in particular, has a longer history of scientific study than all other ocean CDR approaches, these studies were not intended as a test of the feasibility and cost of OIF for large-scale CDR and climate mitigation, or to fully assess environmental impacts at deployment scales. Modeling studies, on the other hand, often focused on the sequestration potential, environmental impacts, and, sometimes, cost estimates of large-scale deployment. Efforts to bridge local experimental scales and global modeling scales (e.g., Aumont and Bopp, 2006) should be encouraged to help maximize the information gained. The earlier OIF studies do serve as a pilot-scale work that can be used to pose several key questions that would be answered with additional laboratory, field, and modeling studies as part of a portfolio of ocean CDR research activities. These research questions can be grouped broadly by the ones on "will it work" related to C sequestration effectiveness and "what are the intended and unintended consequences" related to changes to ocean ecosystems that are an intended part of responsible ocean CDR of any type.

These pilot studies taught us that aOIF experiments would need to be significantly longer and larger than earlier ones that used 0.3–4 tons of iron (II) sulfate ($FeSO_4$) and covered 25–300 km^2 with ship-based observations lasting 10–40 days. A demonstration-scale aOIF field study might need to add up to 100–1,000 tons of iron (using planes, or autonomous surface vehicles), cover up to 1 million km^2 (1 percent of HNLC waters), and last for at least an entire growth season with multiyear follow-up. This would be a scale similar to the Kasatochi volcanic eruption in the Gulf of Alaska (see Fisheries) that caused no permanent harm, but was of a size that it could be readily tracked and pH and CO_2 impacts could be measured, and it provided a regional C loss out of the surface of 0.01–0.1 Gt C (0.04–0.4 Gt CO_2) (Hamme et al., 2010; Longman et al., 2020).

If these demonstration projects were conducted in different HNLC settings (Southern Ocean, Gulf of Alaska, Equatorial Pacific) and LNLC as well, one could document and reach a predictive understanding of the differing ecological and biogeochemical responses. Documenting the CDR impact and understanding and minimizing any long-term ecological damage would be key to the success of any aOIF demonstration project. Several recent large-scale studies of the BCP, such as NASA EXPORTS, can be used to estimate costs ($15M–$25M per site) and duration (3–5 years) of any one such field study, resulting in a demonstration-scale research program and its synthesis and modeling thereof, with a total budget surpassing $200M over 10 years (see summary of research costs in Table 3.4).

On the path to such demonstration projects there is also work to be done in labs, mesocosms, and on smaller scales. In addition, model improvements to better capture the cycling of not just carbon but also iron (e.g., Black et al., 2020) are needed, with careful attention to permanence issues and downstream impact that may only or best be captured by models, and include realistic export and all pathways of the BCP. Observations and models of naturally enriched OIF settings (islands

TABLE 3.3 CDR Potential of Ocean Nutrient Fertilization

Knowledge base What is known about the system (low, mostly theoretical, few in situ experiments; medium, lab and some fieldwork, few carbon dioxide removal (CDR) publications; high, multiple in situ studies, growing body of literature)	**Medium–High** Considerable experience relative to any other ocean CDR approach with strong science on phytoplankton growth in response to iron, less experience on fate of carbon and unintended consequences. Natural Fe-rich analogs provide valuable insight on larger temporal and spatial scales.
Efficacy What is the confidence level that this approach will remove atmospheric CO_2 and lead to net increase in ocean carbon storage (low, medium, high)	**Medium–High Confidence** Biological carbon pump (BCP) known to work and productivity enhancement evident. Natural systems have higher rates of carbon sequestration in response to iron but low efficiencies seen thus far would limit effectiveness for CDR.
Durability Will it remove CO_2 durably away from surface ocean and atmosphere (low, <10 years; medium, >10 years and <100 years; high, >100 years), and what is the confidence (low, medium, high)	**Medium** 10–100 years Depends highly on location and BCP efficiencies, with some fraction of carbon flux recycled faster or at shallower ocean depths; however, some carbon will reach the deep ocean with >100-year horizons for return of excess CO_2 to surface ocean.
Scalability What is the potential scalability at some future date with global-scale implementation (low, <0.1 Gt CO_2/yr; medium, >0.1 Gt CO_2/yr and <1.0 Gt CO_2/yr; high, >1.0 Gt CO_2/yr), and what is the confidence level (low, medium, high)	**Medium–High** Potential C removal >0.1–1.0 Gt CO_2/yr (medium confidence) Large areas of ocean have high-nutrient, low-chlorophyll conditions suitable to sequester >1 Gt CO_2/yr. Co-limitation of macronutrients and ecological impacts at large scales are likely. Low-nutrient, low-chlorophyll areas have not been explored to increase areas of possible deployment. (Medium confidence based on 13 field experiments).
Environmental risk Intended and unintended undesirable consequences at scale (unknown, low, medium, high), and what is the confidence level (low, medium, high)	**Medium** (low to medium confidence) Intended environmental impacts increase net primary production and carbon sequestration due to changes in surface ocean biology. If effective, there are deep-ocean impacts and concern for undesirable geochemical and ecological consequences. Impacts at scale uncertain.
Social considerations Encompass use conflicts, governance-readiness, opportunities for livelihoods, etc.	Potential conflicts with other uses of high seas and protections; downstream effects from displaced nutrients will need to be considered; legal uncertainties; potential for public acceptability and governance challenges (i.e., perception of "dumping").
Co-benefits How significant are the co-benefits as compared to the main goal of CDR and how confident is that assessment	**Medium** (low confidence) Enhanced fisheries possible but not shown and difficult to attribute. Seawater dimethyl sulfide increase seen in some field studies that could enhance climate cooling impacts. Surface ocean decrease in ocean acidity possible.

TABLE 3.3 Continued

Cost of scale-up Estimated costs in dollars per metric ton CO_2 for future deployment at scale; does not include all of monitoring and verification costs needed for smaller deployments during R&D phases (low, <\$50/t CO_2; medium, ~\$100/t CO_2; high, >>\$150/t CO_2) and confidence in estimate (low, medium, high)	**Low** <\$50/t CO_2 (low–medium confidence) Deployment of <\$25/t CO_2 sequestered for deployment at scale are possible, but need to be demonstrated at scale
Cost and challenges of carbon accounting Relative cost and scientific challenge associated with transparent and quantifiable carbon tracking (low, medium, high)	**Medium** Challenges tracking additional local carbon sequestration and impacts on carbon fluxes outside of boundaries of CDR application (additionality).
Cost of environmental monitoring Need to track impacts beyond carbon cycle on marine ecosystems (low, medium, high)	**Medium** (medium–high confidence) All CDR will require monitoring for intended and unintended consequences both locally and downstream of CDR site, and these monitoring costs may be substantial fraction of overall costs during R&D and demonstration-scale field projects.
Additional resources needed Relative low, medium, high to primary costs of scale-up	**Low–Medium** Cost of material: iron is low and energy is sunlight.

and volcanic events primarily in HNLC regions) have also proved useful in gaining an understanding of sequestration efficiencies and longer-term biological responses and should continue. Consideration of co-benefits (e.g., fisheries) and other non-CO_2 GHGs (N_2O, DMS, CH_4, and O_2) would also be needed. Finally, the technology needed to monitor OF is growing, but investment in new designs of autonomous vehicles with biogeochemical sensors, such as on bioARGO floats and gliders, and new optical/camera systems and particle collectors would allow for better tracking of C and other responses. On the research side, new ways to query the genetic shifts in the marine food web could be quite informative. All of this work would require resources on the scale of the current Ocean Observatories Initiative nodes, or Long-Term Ecological Research sites, and/or be put in place as enhancements to the decadal surveys already under way in ocean sciences.

Also necessary are research activities into the social costs and public acceptance of the deliberate manipulation of the ocean commons. While the legal framework under which this research could be conducted has been set in place by the London Convention and London Protocol, it has not been tested, and there are several unresolved questions about its application. Thus we do not know whether the current international agreements would work to allow research but limit unwanted practices in terms of study design, transparency, claims of C credits, and ecosystem enhancements or detriments. Again, these social and legal issues would only grow in importance with any large-scale ocean CDR deployments and will affect the ability to move forward with funding and permitting, even for research. As noted above, while some of these activities are small scale and can be done by individuals, many of the outstanding research questions will require an emphasis on demonstration-scale studies that are larger and longer than done previously. The outcome of studies on these demonstration scales is essential if we are to deploy any or a combination of different ocean- and land-based CDR approaches. Finally, studies are needed to better define costs and benefits of OF so that we can reliably predict the consequences and scales over which the benefits outweigh the costs relative to doing nothing, and against other land- and ocean-based CDR approaches.

In summary, some specific examples of research needs include:

1. **C sequestration efficiencies.** This is a key factor in setting impact on C storage and hence climate. These efficiencies are set by ability for a given amount of nutrient (iron) to enhance C sequestration. New topics for studies would include the following: Can we increase bioavailability of iron and reduce removal by Fe scavenging and thus enhance phytoplankton growth? Can we enhance C and Fe export to the deeper ocean to increase permanence? How can we observe or estimate or model sequestration times in a manner robust enough for accounting purposes? What level of robustness or reliability is required? Are there better methods to track carbon and added iron? Can we engineer designs for improved Fe delivery at larger scales? How do we optimize the deployment to increase production and export or durability (location, season, duration, and continuous versus pulsed delivery)? What are the consequences of multiple nutrient and other limitations to C sequestration responses? How do we manipulate conditions to get food web response that maximizes C sequestration (fast and efficient sinking, low grazing)? How do we improve our monitoring technologies to track consequences of OF to not only carbon, but also full biogeochemical responses and through the food chain?
2. **Ecological responses.** In addition to the intended additional C sequestration, unintended and unexpected consequences have occurred that we need to know more about: What would the impact of OF be on planktonic food webs? Would fisheries be enhanced? What are impacts on higher marine trophic levels and how would one recognize them? Would responses include HABs that threaten open-ocean or coastal systems? Would production of other GHGs enhance (DMS) or reduce (N_2O) the climate impact of CDR? Would downstream ecosystems be limited by intended macronutrient removal? What are responses to OF in low-nutrient settings such as the rates if N_2 fixation? What are consequences to geochemical conditions in the deep ocean that may alter deep-ocean ecosystems (O_2, changing DIC or pH)?
3. **Social acceptance, governance, and deployment costs.** For OF, some of the pressing advancements needed are the following: Are the London Convention and London Protocol sufficient to regulate research and demonstration-scale experiments (possibly), and eventual larger-scale deployments with potential downstream impacts (not likely)? What is the best code of conduct that should be followed for research on OF (and other CDR approaches)? Would OF research be considered acceptable and reversible by a society that is experiencing climate change consequences very differently (i.e., benefits of OF may be separated greatly in time and space from negative consequences of climate change)? If OF was included in C removal markets or platforms, where and to whom would the benefits go, and what would the risks be? What could be the harms and benefits of different policy models for OF deployment?

Many of these research agendas will need to include modeling, whether it is of the localized field experiments or global ecosystem–biogeochemical models to assess the long-term and remote consequences. These models will need to include the cycling of nutrients, including iron. However at present there is low confidence in model projections of Fe distribution and fluxes (Tagliabue et al., 2016; Black et al., 2020). Full earth systems models will be needed to link changes in ocean physics and biogeochemistry to atmospheric CO_2 and climate (e.g., Bonan and Doney, 2018). High enough resolution will be needed to include at least mesoscale physical interactions. Long-term models are needed to assess the full consequences of downstream impacts. Finally, the need for experimental and observational data to validate and verify models will be essential if we are to use models to extrapolate to scales that are not readily measurable in ocean sciences.

3.8 SUMMARY

For the purposes of reducing atmospheric CO_2, this chapter outlines the state of our knowledge (Table 3.3) and key remaining questions that need responsible and transparent study to advance OF research, in particular OIF (Table 3.4). Given that OIF mimics natural systems, much can be learned from studying Fe-enriched "hot spots" near islands or after volcanic events. But studying natural systems is not sufficient to predict outcomes of deliberate OIF as a CDR approach. Unlike many other CDR methods, an international framework has been proposed for evaluating demonstration projects, but it is not legally binding, and there remain many unresolved questions and gaps in the governance framework. Future projects could be 10–100 times larger than prior aOIF experiments, adding hundreds to thousands of tons of iron and resulting in blooms of 10^5–10^6 km^2 that could be tracked and studied for longer than a single annual growth cycle. The potential for net C sequestration of OIF is large enough (Gt C sequestered for >100 yr) and Fe needs are small enough—0.1 percent of annual Fe ore production is 10^6 t/yr, which could lead to 1 Gt C/yr sequestration (3.7 Gt CO_2) with Fe:C_{seq} efficiencies of only 1:1000—to warrant additional study. This biotic approach has relatively high scalability and low costs for deployment, though challenges would include verifiable C accounting and, as for most ocean CDR at scale, careful monitoring of intended and unexpected ecological effects up and down the food chain.

Even if the costs or impacts prove unacceptable for large-scale deployment globally, many companies are already suggesting OF as a way to enhance fisheries, and so having these studies in place could help to inform regulation of the scale and locations over which OF may be allowed or not. It is therefore important to conduct these studies as a basis of evidence for policy makers to contain entrepreneurs and other organizations that do not choose to follow international standards, or plan, organize, and report results in a transparent manner that upholds scientific standards and complies with international protocols. The relatively low cost of entry to initiate an Fe-induced bloom, \$1M–\$2M for hundreds of tons of iron and a small ship, make OIF an approach that does not require huge investments, making it prone to misbehavior by individuals or small organizations or companies. If done well, OF may be an imperfect action done for a good purpose, such as for fisheries enhancement or CDR, but if done poorly or outside of regulated and transparent studies, it has the potential to leave a legacy of unknown and possibly unacceptable impacts. Thus this investment in OF research is warranted whether one believes that it can work on large scales for CDR, or if one simply wants to regulate misuse of the global ocean commons.

TABLE 3.4 Research and Development Needs on Ocean Fertilization

#	Recommended Research	Question Answered	Environmental Impact of Research	Social Impacts of Research	Estimated Research Budget ($M/yr)	Time Frame (yr)
3.1	**C sequestration delivery and bioavailability**	**Can we increase bioavailability of Fe, and ease of delivery, and should delivery be as pulse or continuous for increasing CDR?**	**Modest because mostly laboratory, mesocosm, and modeling studies are needed**	**Modest because studies are mostly shore based**	5 Improvements in Fe:C ratio have a major impact on ultimate deployment, so it is advisable to start immediately	**5**
3.2	**Tracking C sequestration**	**How can we track enhanced C fluxes? Are there new methods for tracking carbon or Fe?**	**Likely done as part of larger field experiments**	**See field experiments**	**3** **New methods for tracking carbon and iron from surface to depth are needed**	**5**
3.3	**In field experiments using more than 100 t Fe over a 1,000 km^2 or greater initial patch size, followed over annual cycles**	**What are CDR efficiencies at scale and what are the intended and unintended ecological impacts?**	**Modest** **Regional impacts during Fe addition period and some concerns beyond test boundaries. May also reduce the effects of acidification in upper ocean. If effective for CDR, impacts are expected on deep-ocean geochemistry. Observations and models are needed.**	**Modest/High** **Early public concern with OIF for ocean geoengineering due to possible unknown ecological shifts, i.e. harmful algal blooms and co-production of other greenhouse gases. Recent emphasis on co-benefit of enhanced fisheries is yet to be verified.**	**25** **Research needs to measure all possible geochemical, physical, and ecological impacts to gauge effectiveness and impacts at scale (costs as noted in text are based upon smaller and shorter aOIF and biological carbon pump (BCP) studies, extrapolated here to 5–10 sites needed to gain predictive understanding).**	**10**

3.4	Monitoring carbon and ecological shifts	Development of autonomous methods for assessment of BCP; research needs to measure effectiveness and impacts at scale	Low New methods, especially optical to complement existing geochemical sensors and platforms and molecular tools to monitor ecological shifts	Low Any method to measure C flow and ecological shifts will have multiple uses for science and the public	10 New technologies are quick to prototype but expensive to bring to market at reliability and scale useful for CDR	10
3.5	Experimental planning and extrapolation to global scales	Full Earth system models with realistic BCP and Fe cycling, including particle cycling	Low Modeling needed to design experiments and predict impacts at local scale and in the far field do not have direct environmental impact and assist planning of more acceptable field research	Low if considering only modeling, though public acceptance of CDR is still needed, and models will be needed to assess possible impacts	5 Early for planning and later for impact assessments	10
3.6	Research on the social and economic factors and governance	Is the current London Convention and London Protocol sufficient for regulation of research and for eventual deployment?	N/A	Starting from a point of low or modest public acceptance of OIF	2	10
3.7	Document best code of conduct for research and eventual deployment	Open-data systems and peer review and independent C and impact assessments need to be codified	N/A	Needed for public acceptance of use of high seas for any open-ocean CDR	2	5–10 (early agreement of research conduct needed)

NOTE: Bold type identifies priorities for taking the next step to advance understanding of ocean fertilization as an ocean CDR approach.

4

Artificial Upwelling and Downwelling

4.1 OVERVIEW

The vertical movement of water in the ocean, termed upwelling and downwelling, acts to transfer heat, salt, nutrients, inorganic and organic carbon (C), and energy between the well-lit surface ocean and the dark, nutrient- and carbon dioxide (CO_2)-rich deep ocean. Since the 1950s, researchers have sought to artificially stimulate these physical transport processes to geoengineer localized regions of the ocean. For instance, wave-driven or density-driven artificial upwelling (AU) has been proposed as a means to supply growth-limiting nutrients to the upper ocean and generate increased primary production and net C sequestration. The latter outcome (increased C sequestration) would require that the biological production of carbon exceed the delivery of dissolved inorganic carbon (DIC) from the upwelled source water. Purposeful upwelling has also been proposed as a mechanism to sustain fisheries and aquaculture (Williamson and Turley, 2012), to generate energy (Isaacs et al., 1976), to provide a source of cold water for seawater-based air-conditioning (Hernández-Romero et al., 2019), and even to prevent the formation or severity of typhoons (Kirke, 2003). In contrast, the purposeful downward transfer of less-dense, oxygen-rich surface water, has been suggested as a mechanism to counteract eutrophication and hypoxia in coastal regions by ventilating oxygen-poor water masses (Stigebrandt et al., 2015; Feng et al., 2020). Both mechanisms may, under certain circumstances, act to enhance the oceanic sequestration of atmospheric CO_2.

According to criteria described in Chapter 1, the committee's assessment of the potential for AU and artificial downwelling (AD), as a CO_2 removal (CDR) approach is discussed in Sections 4.2–4.5 and summarized in Section 4.6. The research needed to fill gaps in understanding of AU and AD, as an approach to durably removing atmospheric CO_2, is discussed and summarized in Section 4.7.

4.2 KNOWLEDGE BASE

Despite it being known for decades that AU could be achieved without an external C-based energy source (Stommel, 1956; Isaacs et al., 1976), to date all oceanic tests of AU detailed in the

FIGURE 4.1 Examples of wave pumps. SOURCE: (A) White et al. (2010); © American Meteorological Society. Used with permission. (B) Isaacs and Schmitt (1980).

peer-reviewed literature have been relatively small in scale with deployments typically less than a week and impacting an area no larger than tens of kilometers (Huppert and Turner, 1981; Liu and Jin, 1995; Ouchi et al., 2005; White et al., 2010; Maruyama et al., 2011; Pan et al., 2016; Fan et al., 2020). Enhanced oceanic C sequestration has never been documented in any sea trials to our knowledge. Pan et al. (2016) summarized the various types of AU mechanisms, spanning wave-driven pumps, electrical pumps, salt fountains, air-bubble pumps, and air-lift pumps (Table 4.1 and Figure 4.1). In all cases, if net C sequestration were to be achieved, the elemental composition of the source water would need to be carefully considered (e.g., Karl and Letelier, 2008) given that the proximate limiting nutrient varies across the global ocean as does the stoichiometry of potentially limiting elements and the concentration of DIC and alkalinity with depth. The energy utilization and pumping efficiency is also an important consideration and varies between pumping mechanisms (Table 4.1). In short, as opposed to ocean iron fertilization (OIF; Chapter 3), there exists no proof of concept that AU could act to sequester carbon below the ocean pycnocline.

In an open-ocean test in 1989–1990, Ouchi et al. (2005) report successful delivery of deep, nutrient-rich water from 220 meters, which promptly downwelled below the euphotic zone due to the high density of the upwelled seawater. In 2008, White et al. (2010) conducted sea trials of a commercially available wave pump with the aim of stimulating a two-phased phytoplankton bloom (e.g., north of the Hawaiian Islands in the North Pacific subtropical gyre). Sensors on the wave pump documented delivery of deep cold water to the surface for a period of ~17 hours after which catastrophic failure of the pump materials occurred. More successful tests of AU have been completed in semi-enclosed bodies of water such as Sagami Bay in Japan where an electrical pumping system moored to the seafloor was operated for ~2 years. This system delivered deep water from 200- to 20-meter depth and generated enhanced concentrations of picophytoplankton and nanophytoplankton (Masuda et al., 2010); C fluxes were not measured. While operational, the main limitations of this approach were reported to be high energy costs as well as construction and maintenance expenses (Pan et al., 2016). Additional successful trials were held in the western Norwegian Fjord in 2004–2005 using air-bubble pumps and uplift pumps (Aure et al., 2007; McClimans et al., 2010), generating flow of nutrient-rich deep water at a rate of 2 m^3/s, which led to a tripling of the concentration of the phytoplankton pigment chlorophyll-*a* in a 10 km^2 plume; again no measurements of C flux were made.

And so, in principle and at least in semi-enclosed water bodies where the impact of wind and waves are lessened, AU is a viable means of fertilizing the ocean with growth-limiting nutrients. In practice, there have been no proof-of-concept sea trials to confirm or deny the potential for net C sequestration, nor has any technical design proven to be sufficiently robust for long-term deployments in the open ocean. Were this technology to be utilized for CDR, pilot studies would first need to be conducted in the open ocean. Such studies would need to measure C fluxes as well as evaluate compensating flows that may lead to decreased thermocline stability and changes in surface temperature patterns that could lead to changes in atmospheric circulation and weather.

In contrast, there have been a wide range of modeling studies aiming to assess the efficacy of AU either as a means of ocean CDR or in support of aquaculture. By and large, these models suggest that AU would be an ineffective means for large-scale C sequestration (Dutreuil et al., 2009; Yool et al., 2009; Oschlies et al., 2010b; Keller et al., 2014; Pan et al., 2015) and would require a persistent and effective deployment of millions of functional pumps across the global ocean (Yool et al., 2009). For example, Oschlies et al. (2010b) use an Earth system climate model to simulate the impact of a network of "ocean pipes" of varying lengths and generating variable upwelling velocities. These authors conclude that even under "the most optimistic assumptions" AU would lead to sequestration of atmospheric CO_2 at a rate of about 0.9 petagrams (Pg) C/yr (most of this on land due to the slowdown of respiration in terrestrial soils under a cooler atmosphere), generally below the targets needed to mitigate anthropogenic emissions (>1 Pg C/yr).

Moreover, models indicate that upwelling of deep water may lead to cooling of the lower atmosphere, reduced precipitation, promotion of ocean acidification in certain regions, and even enhancement of terrestrial C storage (Keller et al., 2014). Although not yet considered in Earth system models, electrical pumps and air-bubble pumps have also shown promise as upwelling tools (Pan et al., 2016); however, these methods come with added energy costs and so are less likely to be effective in CDR.

Beyond direct ocean CDR, AU has been proposed as a means to support local aquaculture from fisheries to seaweed farms, which may have the cascading effect of enhancing ocean C sequestration (see Chapter 5). In a recent sea trial, Fan et al. (2020) demonstrated that a rigid AU system supplied sufficient nutrient-rich water to enhance seaweed growth in an enclosed bay. While this is but one trial, coupling of AU and large-scale seaweed farming could feasibly provide a means to offset upwelled CO_2. In contrast, models developed by Williamson et al. (2009) determined that nutrient levels injected via AU would be maintained at ~0.1 percent of their source concentrations, far below what would be needed to sustain ocean aquaculture. Hence, the nutrient supply rate and nutrient stoichiometry would need to be closely evaluated relative to aquaculture needs in order to determine potential efficacy.

AD, the engineered downward generation of vertical currents, has also been suggested as a means to enhance sequestration of dissolved and particulate organic carbon. Downwelling of surface water has never been tested in the field as a means of CDR; however, regional tests have been conducted to assess mitigation of hypoxia via downwelling (Stigebrandt and Liljebladh, 2011; Stigebrandt et al., 2015). Although the technology is promising, the efficacy and the biogeochemical consequences are less certain. AD could also be directly coupled with AU as a means to pump recalcitrant dissolved organic carbon to depth or prevent outgassing. As discussed above, upwelling of cold CO_2-rich water into the surface ocean comes with the risk of net outgassing if community production does not exceed the C inputs. Downwelling of recently upwelled water masses may reduce this risk, but again this is a wholly untested scenario. Modeling studies have indicated that modification of ocean downwelling to enhance the ocean's solubility pump is "highly unlikely to

TABLE 4.1 Summary of Types of Artificial Upwelling and Their Advantages and Disadvantages

Type	Description	Advantages	Disadvantages
Wave-pump (Verhinskiy, 1987; Liu and Jin, 1995; White et al., 2010)	Extracts energy from the surface gravity waves to draw DOW	(1) Test in north of Oahu, Hawaii in June 2008; and (2) Self-powered	Pump fails after < 2 h
Electrical pump (Ouchi et al., 2005; Mizumukai et al., 2008; Masuda et al., 2011)	Uses a high power electrical pump to draw DOW	(1) Operated in Sagami Bay from 2003; (2) Robust technology and longevity structure; and (3) Large amount of uplifted DOW	Low efficiency and extremely high cost
Perpetual salt fountain (Tsubaki et al., 2007; Maruyama et al., 2011)	Uses salinity and temperature differences between layers of the DOW and the euphotic to draw DOW	(1) Test in the Mariana area of the tropical Pacific Ocean in 2002; (2) Higher Chl was detected around the pipes; and (3) Self-powered	Low amount of uplifted DOW to support an ocean farming project
Brackish water uplift pump (Aure et al., 2007; McClimans, 2008; McClimans et al., 2010)	Pumps down low density brackish water to uplift DOW of the same depth	(1) Test in a western Norwegian fjord from May to September in 2004 and 2005; (2) Enhancing and adjusting the nutrient concentration and the N/P ratio; and (3) Chl tripled, diatom biomass increased in a large extent within an influence area of 10 km^2	Lower efficiency compared to air-bubble and air-lift pump and limited applied region
Air-bubble pump (McClimans et al., 2010; Handa et al., 2013)	Pumps air through a horizontal pipe to uplift the DOW to a certain depth	(1) Tested in inner part of Arnafjord in September 2002; (2) High efficiency with an DOW to air supply of > 88;[a] and (3) Expected biological and biogeochemical responses of sea trials	Limited uplifting DOW depth
Air-lift pump (Liang and Peng, 2005)	Injects compressed gas in the pipe to uplift DOW from deeper depths	High efficiency with an DOW uplift to air supply approximately 100 m^3/min DOW	No sea trial data to date

[a] An air supply of 1 m^3/min could uplift > 88 m^3/min.
SOURCE: Pan et al., 2016.

ever be a competitive method of sequestering carbon in the deep ocean" due to impracticalities and costs (Zhou and Flynn, 2005).

Although not directly related to ocean CDR, AU has also been proposed as a means of providing energy and cooling and hence reducing terrestrial C emissions (e.g., ocean thermal energy conversion [OTEC]; Kim et al., 2021). Since the 1970s, it has been proposed that industrial-scale OTEC plants could pump high volumes of deep cold seawater to power turbines and generate electricity in an ecologically and economically sound manner to coastal and island communities.[1] Demonstration-scale plants (production on the order of hundreds of kilowatts) have even been established in the Hawaiian Islands[2] to provide power and desalinated water. Generation of power was one of the first proposed applications of AU (Isaacs et al., 1976), and certainly does have implications for C emissions, even if unrelated to oceanic CDR. Upwelling of deep cold water has also been proposed as a means of mitigating thermal stress in coral ecosystems (Sawall et al., 2020). Simulated AU experiments off Bermuda show early indications that controlled upwelling could abate coral bleaching during heat stress events potentially allowing corals to adapt to rising temperatures more steadily (Sawall et al., 2020). These examples are only meant to illustrate that while open-ocean applications of AU are unproven as a means of CDR, the principle of simulated upwelling has many applications.

In summary, AU and AD are proven means to transport water against concentration gradients without the need for C-based energy sources; small-scale pilot experiments conducted over the past few decades have shown that these technologies do work largely as expected, but the technology is not proven to be robust in the open ocean over timescales needed for CDR (longer than months). The larger challenge is not functionality, it is scaling, verification of CDR, and monitoring of ecological and biogeochemical responses to the magnitude, duration, and rate of perturbations. Undoubtedly, AU technologies are valuable as means to study ecosystem responses to nutrient disturbances (Masuda et al., 2010; Williamson et al., 2012); however, significant advances in technological readiness as well as durability and development of monitoring programs need to be made before this can be a reliable means of CDR. These issues are explored in more detail below. As done before, we also make comparisons to the other ocean CDR methods in this report and note when the same consequences would result from other less-well-studied ocean CDR methods.

4.3 EFFICACY

As noted above, there are no existing sea trials that have assessed C fluxes as a result of AU or AD. As a result, we simply do not have enough information about the long-term operation and efficacy of either AU or AD systems, and hence there has been a reliance on model simulations to assess potential efficacy. Nearly all model simulations suggest that large-scale deployment of ocean pipes would be a costly and ineffective means of ocean CDR with large uncertainty as to whether net C drawdown is attainable (Lenton and Vaughan, 2009; Yool et al., 2009; Keller et al., 2014). Model simulations of global-scale AU predict a range of perturbations to air–sea C flux, spanning <0 Pg C/yr (net outgassing, Dutreuil et al., 2009; Yool et al., 2009) to up to 3.6 Pg C/yr (Keller et al., 2014). In general, this is because most carbon exported from the surface is remineralized in the mesopelagic, and ocean upwelling would ultimately return this to the surface in excess of any enhancements in export production.

Yet, ocean physics and biogeochemistry are often more complex and nuanced than can be captured in even the most sophisticated models. There is a real need for a pilot-scale test of these technologies to assess the potential for AU and AD as either a component of CDR tools or as a

[1] See, for example, see https://www.makai.com/ocean-thermal-energy-conversion/.

[2] See http://nelha.hawaii.gov/.

means to sustain small-scale aquaculture or generate energy. Such pilot-scale tests would be where pumps can be operated autonomously for long periods of time with data transparency and international scientific cooperation, and the environmental risks and benefits can be more accurately weighted. If these were to be conducted, a determination of efficacy would be very similar to OIF in terms of cost and necessary components; however, the carbon potentially sequestered would need to be weighed relative to the upwelled carbon. Specifically, for AU to effectively sequester carbon from the atmosphere, we need to consider not only the C export ratios, that is, the ratio of inorganic carbon upwelled relative to the carbon sinking out of the upper ocean, but also the extent to which sinking organic carbon is attenuated with depth. As nutrients are delivered to presumably nutrient-limited surface waters, there is an expectation that there would be predictable changes in the phytoplankton community structure, with a shift to larger, faster-sinking cells (Karl and Letelier, 2008) that may increase export efficiency. Whether or not this would be the case is debatable because the nutrient stoichiometry of upwelled water as well as any concomitant changes in the mixed layer will also influence any shifts in community structure. The stoichiometry of limiting nutrient to CO_2 will also vary with depth and region. For this reason, the outcome of models upwelling water from a uniform depth across basins, as that discussed by Dutreuil et al. (2009), may not reflect the real potential of upwelling water masses from depth where we find the maximum deficit of CO_2 relative to other nutrients (see Karl and Letelier, 2008). The lack of spatial uniformity in the stoichiometry of growth-limiting elements relative to CO_2 is that the potential net C sequestration efficiency of a free-drifting upwelling pump may change over time. Moreover, the persistence or intermittency of upwelling will influence community structure and resultant changes in primary and export production.

Beyond the biogeochemistry, several additional research and development needs should be addressed before efficacy of these technologies for CDR can be determined. For example, siting analyses that evaluate potential wave energy needed for upwelling relative to the magnitude of nutrient injections and the physiological status of surface populations need to be conducted to determine ideal locations for pilot studies (Figure 4.2). This exercise is likely to reveal trade-offs between accessibility of coastal sites for maintenance and monitoring versus the relative permanence of C storage in other areas such as the highly stratified oligotrophic gyres. Lastly, it is important to note that the limited field studies have largely been constrained to coastal regimes with limited operational periods (shorter than weeks) and relatively shallow source waters and low upwelling rates ($<0.1\ m^3/s$), whereas model simulations have explored C sequestration potential using temporally and spatially extensive deployments with much deeper source waters and higher rates of upwelling (Figure 4.3). **In effect, there is a gap between the technological readiness of AU and projected sequestration potential. Modeled flow rates far exceed those achieved in sea trials to date and so even the highly variable C export predictions that have been made should be considered unrealistic until AU technology could feasibly be deployed at demonstration scale.**

Additionality

Downstream impacts of AU should also be further addressed because model simulations indicate the potential for ocean acidification, hypoxia, and changes in plankton successional patterns far afield from the site of upwelling. For example, Keller et al. (2014) estimated that AU would reduce ocean pH by up to 0.15 units below what is expected with the "business as usual" trajectory for ocean acidification as a result of anthropogenic CO_2 emissions. Additional potential consequences of large-scale AU are similar to OIF and could include production of greenhouse gases such as nitrous oxide, methane, or dimethyl sulfide, disturbance of benthic ecosystems, or mid-water oxygen depletion (Williamson and Turley, 2012). There is the potential for these ecosystem alterations to be felt far afield from the fertilization sites in both space and time, which would surely present

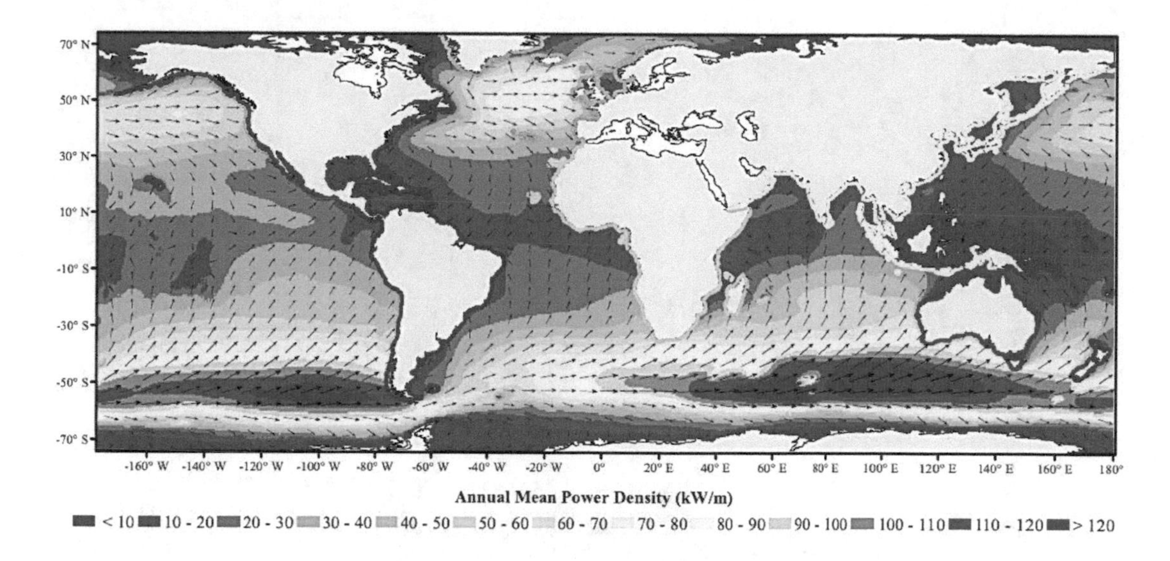

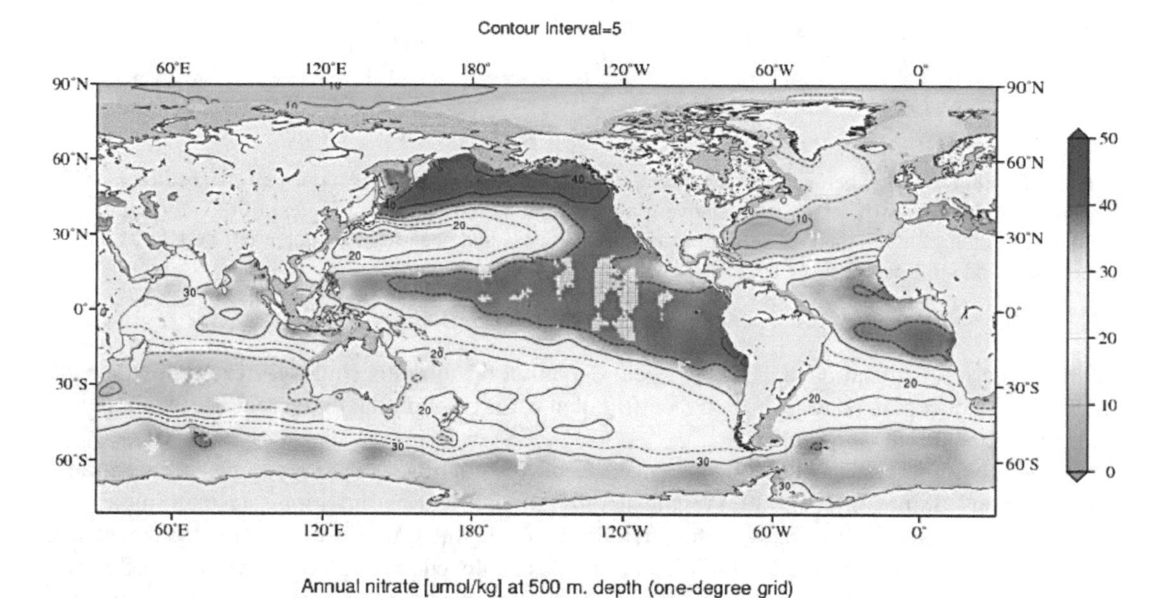

FIGURE 4.2 Annual mean wave power density and direction (top) relative to the climatological mean WOA18 nitrate at 500-meter depth (bottom). SOURCE: (top) Gunn and Stock-Williams, 2012; (bottom) World Ocean Atlas 2018, National Centers for Environmental Information, NOAA.

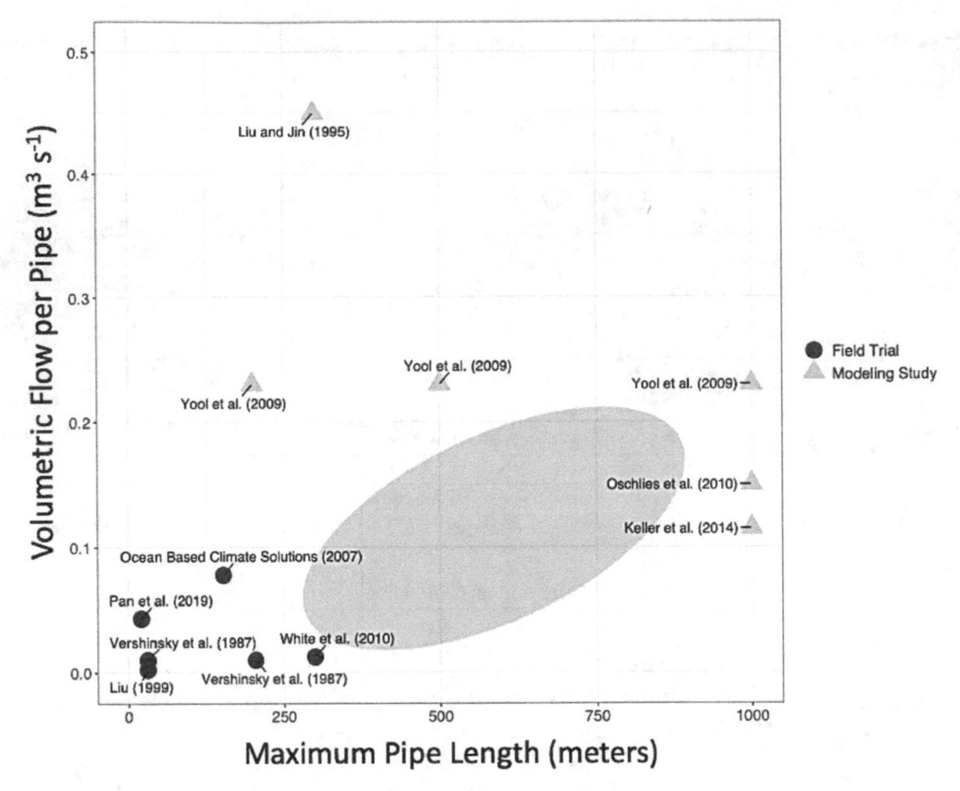

FIGURE 4.3 Disparity between the maximum pipe length and volumetric flow rate (~upwelling velocity) used in field trials (black circles) relative to model simulations (yellow triangles) of the efficacy of AU. The pink area highlights the gap between technical readiness and sequestration potential. SOURCE: David Koweek. Used with permission.

a challenge for verification of efficacy. Local depletion of nutrients could, for example, alter the productivity of regions downstream and shift global patterns of productivity, remineralization, and export (Aumont and Bopp, 2006).

The persistence and scale of AU also need further evaluation. Should pumping be intermittent or persistent? Baumann et al. (2021) indicates that the nutrient perturbation rates can impact the nutrient quality and sinking rates of the phytodetritus generated, which would affect permanence. Does AU need to be on the scale of millions of pumps deployed in the global ocean to be effective, as suggested by models? If so, this would have consequences for international shipping, fishing, and other unrelated activities at sea. Would there be any geochemical consequences if upwelling was stopped at some point? To this last point, prior models suggest termination of large-scale AU could result in a rapid net increase in global temperatures to levels even higher than in a world that had never engaged in AU (Oschlies et al., 2010b). Would AU lead to changes in global precipitation patterns or weather? Ricke et al. (2021) indeed suggest that variations in sea-surface temperature could lead to alteration of rainfall and drought. The full answers to all of these questions will surely require a carefully planned series of pilot or demonstration studies as well as incorporation of resultant data into Earth system models and evaluation of near-field and far-field biogeochemical effects.

Permanence

A commonly used time period to assess permanence of C sequestration in the ocean is ~100 years. For AU, this requires that exported organic matter exceeds the inorganic carbon delivered from source waters and that a significant fraction of particles are efficiently transferred below the winter mixing layer. If these criteria are not met, then permanence will not be achievable. Recent model results suggest that this ~100-year benchmark will not be widely achievable via ocean CDR; using an inverse ocean circulation model, Siegel et al. (2021) conclude that enhancement of ocean C export (e.g., via OIF or AU) will have a shorter-term influence on atmospheric CO_2 levels because ocean circulation and mixing will transport ~70 percent of the sequestered carbon back to the surface ocean within 50 years. This finding underscores the clear fact that the ocean's response to the magnitude, duration, and rate of perturbations will vary widely across ocean basins and with time. It also suggests that ocean CDR approaches seeking to stimulate the biological pump may be a short-term solution to buy time but unlikely to support permanent (≥100 years) sequestration. This has implications for the representativeness and wider applicability of oceanic experiments, which are often necessarily limited in space and time.

For downwelling, it will be necessary to determine how to subduct buoyant organic matter and ensure that it does not return to the surface. Limited trials have proven the capability to upwell and downwell seawater against gradients in density, nutrients, and C resources. Yet again, with no "farm scale" tests of AU or AD, we cannot assess permanence of C storage. In lieu, we must currently rely on mechanistic model simulations of ocean biogeochemistry. Dutreuil et al. (2009) simulated deployment of 200-meter-deep pipes throughout the global ocean at a spatial resolution of 20° longitude and ~ 10° latitude operating at vertical velocities of 0.1 m/s and found that while biological productivity and export were enhanced as might be expected, particularly in the equatorial Pacific, the air–sea CO_2 flux declined significantly at the sites of AU as a result of the mixing of respired CO_2 into the surface mixed layer where it could exchange with the atmosphere. Dutreuil et al. (2009) conclude that "overall, our analyses demonstrate that the enhancement of biological productivity is never enough to compensate for the additional supply of DIC to surface waters" (see Figure 4.4). Using a different model structure, a biogeochemical model coupled to the Ocean Circulation and Climate Advanced Modeling (OCCAM) physical model with embedded ocean AU pipes of either 200 meters, 500 meters, or 1,000 meters, Yool et al. (2009) came to a similar conclusion. Over a 10-year simulation, strong regional heterogeneity was observed, and changes in the air–sea flux as a result of AU were found to be both positive and negative across the global ocean.

Reversibility is also an important consideration as the intentional upwelling of CO_2-rich deep water or downwelling of O_2-rich surface waters has the potential to significantly alter ocean ecosystems from the epipelagic to the benthic. It may be difficult or impossible to distinguish far-field changes in ecosystem structure and function from changes due to decadal shifts (e.g., El Niño-Southern or Pacific Decadal Oscillation–driven shifts in production patterns). More research is needed to determine the C sequestration potential of AU and AD and whether permanence is achievable, what ocean sites would best achieve permanence, and what the environmental risks and co-benefits would be.

Monitoring and Verification

An array of devices have been proposed to facilitate either AU or AD, including wave-powered systems, airlift and bubble pumps, and salt fountains (Pan et al., 2016). In a handful of limited sea trials (described above), these technologies have proven capable of vertically pumping seawater, even if the durability and longevity of the devices tested have not been sufficient for large-scale trials (e.g., White et al., 2010) and no tests to date have ever evaluated potential changes in C

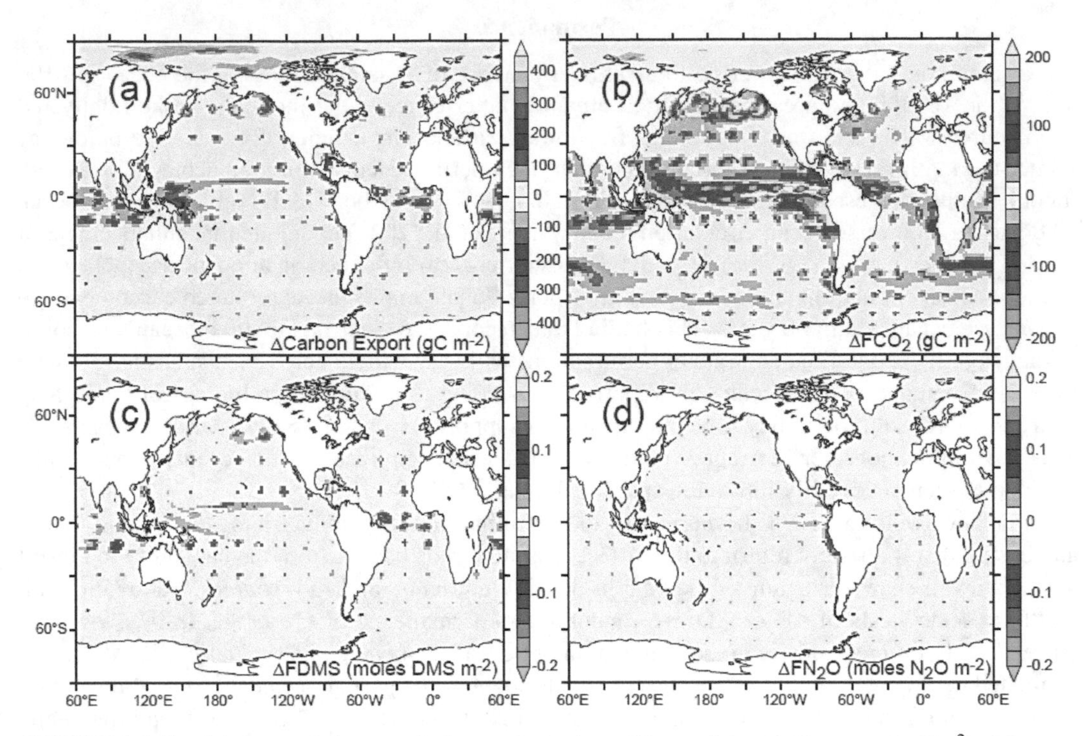

FIGURE 4.4 Spatial maps of the cumulative anomaly (over 20 years) in (a) C export (g C/m^2), (b) ocean CO_2 uptake (FCO_2, g C/m^2), ocean-to-atmosphere dimethyl sulfide flux (FDMS, moles/m^2), and (d) ocean-to-atmosphere N_2O flux (FN_2O, moles N_2O/m^2). SOURCE: Dutreuil et al., 2009, Licensed by Creative Commons CC BY 3.0.

sequestration at any temporal or spatial scale in situ. In lieu, prior tests have evaluated the "efficacy" of pumps as CDR tools based on whether surface chlorophyll was enhanced or whether water masses with the temperature and salinity signatures of the source water were detected. This is, of course, insufficient evidence of efficacy; verification cannot focus on evidence of enhanced growth of phytoplankton in the surface ocean because the fate of that material could be remineralized in the upper ocean. Rather changes in the sinking flux of particulate organic carbon into the deep ocean (below 1,000 meters) must be determined to assess efficacy and permanence of sequestered C pools. In addition to measuring efficacy for C sequestration, such pilot-scale studies would also need observations of ecosystem changes that might lead to co-benefits, such as enhanced fisheries, or negative impacts related to harmful algal blooms or alteration of food webs that are considered undesirable. Large-scale trials of AU or AD would need to employ proven technologies such as sediment traps, gliders, and instrumented profiling floats to verify C sequestration as a result of AU and AD. Verification cannot rely on proxies for production such as chlorophyll concentrations in the surface ocean.

4.4 SCALABILITY

Geographic and Temporal

The location and spacing of open-ocean pumps required for AU to serve as effective tools for CDR would first and foremost need to consider the deep-water nutrient stoichiometry (the ratio of macronutrients to one another) and nutrient-to-metabolic DIC stoichiometry, both of which vary globally and with depth. Karl and Letelier (2008) predicted that in the North Pacific Subtropical Gyre (NPSG), C export could only exceed the upwelled DIC if a two-stage bloom were triggered, with the first stage characterized by growth of phytoplankton on the upwelled nitrate and the second stage following nitrate depletion and supported by nitrogen (N_2)-fixing cyanobacteria capable of consuming the residual phosphate (~0.05–0.5 mol/L depending on source-water horizon) in the absence of nitrate and leading to net C sequestration. Examination of the elemental stoichiometry of nitrate:phosphate:DIC in this region indicated that upwelling of water from 300 to 350 meters would be ideal to promote a two-stage bloom, optimally in summer months when diazotrophs were most abundant. White et al. (2010) sought to test this hypothesis using wave-powered pumps, and despite brief upwelling at rates of 45 m^3/h, materials failures prevented a further assessment of the biological response. So although the Karl and Letelier (2008) hypothesis remains untested in the NPSG, it is still critical for open-ocean tests to evaluate the potential production and export relative to the magnitude of upwelled DIC and the nutrient stoichiometry needed for growth, which also varies globally with changes in community composition. Singh et al. (2021) show that N_2 fixation rates in long-term mesocosms (55 days) can respond positively to enhanced pulses of upwelling after a lag of several days and rapid depletion of nitrate, supporting the hypothesis of expected successional patterns in plankton communities that may be key to net CDR. In extrapolating bloom patterns resulting from AU, the rate of perturbation as well as the initial community structure in any impacted biome will also need to be considered, as suggested by Karl and Letelier (2008).

Using an ocean model framework, Dutreuil et al. (2009) found that AU would be particularly ineffective in iron (Fe)-limited regions such as the equatorial Pacific and Southern Ocean where the addition of limiting macronutrients leads to communities with an increased iron demand, resulting in a weak "fertilization effect." While no regions were strong C sinks in response to simulated AU in that work, the subarctic Pacific was considered promising because it is a region where the upwelled water contains sufficient alkalinity to compensate for the additional supply of DIC to surface waters. In contrast, Yool et al. (2009) found the strongest effect of simulated AU in centers of subtropical gyres, albeit their model results also predict considerable spatiotemporal variability. If AU were confined to the tropics and C fixation efficiency was 2.2 percent relative to air–sea uptake, Yool et al. (2009) calculated that between 189 million and 776 million pumps would need to be deployed to increase ocean C sequestration by 1 Pg C/yr above current rates. Deployments on this scale would need to consider how a network of pipes would compete with (or complement) other needs such as transatlantic shipping routes and fishing activity. In summary, siting analyses will need to identify optimal nutrient, light, and wave conditions for growth and C sequestration potential as well as potential conflicts with other marine industries.

The temporal scaling of AU deployments also needs to be considered relative to the timescale of biological responses. Should pumping be continuous or pulsed? Are there regions where efficacy is seasonally dependent, for example, in the NPSG where a diazotrophic response may be central to net C export (Karl and Letelier, 2008)? What will be the potential lifetime or durability of a network of upwelling pumps? Can upwelling be conducted in a manner that prevents rapid downwelling and subduction of nutrient-rich water below the mixed layer or euphotic zone? Are there termination effects such as pressure effects or circulation compensations that might lead to rapid warming and maybe overheating? All of these questions should be considered in a research agenda and coordinated with high-resolution models that can account for complex interactions of ocean physics, chemistry, and biology as well as downstream impacts. These issues are explored further in the research agenda.

Impact Potential

As discussed in prior sections, model simulations of large-scale AU deployments diverge significantly in the sign, magnitude, and regionality of efficacy and potential permanence of CDR. In general, model simulations of the impact are consistent with the conclusion by Fennel (2008) that "controlled upwelling is unlikely to scale up and serve as a climate stabilization wedge as defined by Pacala and Socolow (2004), i.e., it would not sequester 1 Pg C yr^{-1} over 30 yr." Moreover, the sea trials needed to assess C export potential and downstream impacts have not been conducted, and further research and development are needed to achieve technical readiness. Large-scale deployments are expected to impact weather and climate, may have implications for ocean heat and oxygen content depending on the spatial footprint and upwelling rates or frequency, and have downstream impacts such as "nutrient robbing" and impacts to biodiversity and ecosystem function. On a more practical note, there are potential conflicts with global shipping routes and fishing efforts, again depending on the scale of deployment. None of these potential impacts has high certainty without "farm-scale" or "deployment scale" trials. Such trials should be coupled to feasibility studies to address upscaling potential of the technology as well as allow for adaptive governance of the research before further investments regarding CDR potential are made. The necessary components of a research agenda addressing CDR and AU/AD are described below in Section 4.7.

4.5 VIABILITY AND BARRIERS

Environmental Impact

Because of mass conservation, any up- or downwelling has to be balanced by down- or upwelling of waters elsewhere. The spatial patterns and controls of the balancing counterflow is still unclear, but a net effect of the induced vertical translocation of water parcels is a reduction in density stratification, similar to that of enhanced vertical mixing, essentially mixing heat downward against the mean stratification of the upper ocean. Changing the density structure of the ocean is expected to change ocean circulation on scales larger than the Rossby radius, that is, a few tens of kilometers. Theory and model studies predict an enhanced overturning circulation under enhanced vertical mixing and reduced stratification. Viewed globally, this will likely reduce the vertical gradient of DIC and, if nothing else changes, lead to a net outgassing of CO_2 from the deep ocean to the atmosphere. Not surprising, natural upwelling regions in the ocean are regions of elevated partial pressure of CO_2 in the surface waters and a general outgassing of CO_2 from the ocean to the atmosphere. Upwelling regions are also strongly influenced by the phase of natural climate oscillations such as El Niño/La Niña, which lead to decadal variability in C fluxes (Bonino et al., 2019).

If AU can stimulate upper-ocean biological production and subsequent export of organic carbon to the ocean interior, enhanced oxygen consumption and production of respiratory CO_2 as well as non-CO_2 greenhouse gases in regions underlying areas of enhanced productivity are to be expected. These have been identified in model simulations (Keller et al., 2014) and are similar to effects induced by other marine CDR methods that aim to enhance marine biological productivity. Even though biological production is not intended in the AD concept, it may well be stimulated by the fertilization effect of the compensating upward return flow.

Ecological impacts of AU or AD are a net cooling of surface waters and a net warming of subsurface waters. Model simulations of large-scale massive deployment (millions of pumps throughout the global ocean) suggest that subsurface waters may warm by a few degrees for a century-long deployment of AU (Oschlies et al., 2010b; Keller et al., 2014). In theory, subsurface warming would enhance microbial and geochemical remineralization rates and thus decrease net C flux to the deep ocean (Cavan et al., 2019). This will affect metabolic rates of the mesopelagic region with likely shifts in remineralization profiles and associated vertical C fluxes.

While AU and AD will essentially lead to enhanced storage of heat in the subsurface ocean, part of this heat will be released back to the surface ocean, and hence to the atmosphere, upon termination of these CDR methods. In simulations with an Earth system model of intermediate complexity, this was found to generate higher global mean air temperatures after termination of AU than in a control experiment in which AU was never applied (Oschlies et al., 2010b).

Additional impacts that should be considered include pollution of the oceans if pump materials were to fail; depending on materials, this could introduce significant plastic, metal, and/or concrete pollution. If pumping were highly effective, there might also be practical impacts such as navigational hazards or ecological impacts such as biofouling and transport of invasive species, changes in light penetration, spectral quality of light penetration, and changes in surface heating.

Co-benefits

Several co-benefits may be realized with effective AU, including potential stimulation of the climate-cooling gas dimethyl sulfide (Taucher et al., 2017), localized lowering of surface ocean temperatures to prevent coral bleaching, and support of fisheries or aquaculture efforts (Kirke, 2003). The latter topic (see Figure 4.5) is an area of active research spanning mesoscale-based studies of food web changes to support fisheries (Taucher et al., 2017; also see https://ocean-artup.eu/vision) to offshore seaweed farms as part of a portfolio of solutions for nutrient remediation in, and C removal from, our oceans (ARPA-e, 2021b). These programs are nascent and not yet at the stage of evaluating technological readiness but should be critical to assessing the feasibility, effectiveness, associated risks, and possible side effects of AU as well as the potential co-benefits. Note also that if aquaculture yields are to be fairly considered a co-benefit, then the C budget of fisheries should also be considered. For example, Mariani et al. (2020) estimate that fossil fuel consumption and hence CO_2 emissions by fisheries are on the order of ~0.01 Gt CO_2/yr (~2.5 Mt C/yr) over the period of 1950–2014. This does not include blue carbon extractions (see Chapter 6). Also see Chapter 3 for additional discussion on the feasibility of marine aquaculture as a means for ocean CDR.

Cost

Model simulations of the efficacy of AU indicates that CDR with the potential of on the order of several Pg C per year would very likely require substantial expansion from small coastal pilot studies to sustained operations in deeper, offshore (>3 nautical miles from the coast) ocean habitats. To do so will require that scientists and engineers work together to develop upwelling prototypes capable of sustaining significant upwelling velocities (see Figure 4.3) with materials designed to resist biofouling and remain operational in a variety of sea states. Costs to do so will include materials for the pumps themselves, deployment costs, costs for development and maintenance of offshore monitoring and verification programs, any energy needed to power the pumps (e.g., ocean thermal energy conversion plants; Avery and Wu, 1994; Matsuda et al., 1998), as well as any costs for removal of pumps at the end of their life cycle or after sustained damage and any necessary maintenance. While there are a number of design specifications for AU devices with various dimensions and flow velocities (see Pan et al., 2016), Kirke (2003) estimated that a moored 500-m-deep, 12.9-m-diameter, wave-driven inertial pump made of carbon steel would cost $4.68 million in 2003 dollars plus the costs of floats, anchors, and mooring cables for an overall cost somewhere below $10 million; this value would be less if construction material were concrete or fabric, although flow rates and durability would be compromised from the author's perspective. Johnson and Dicicco (1983) estimated the total cost for design of salt-fountain–style pumps (thousands to tens of thou-

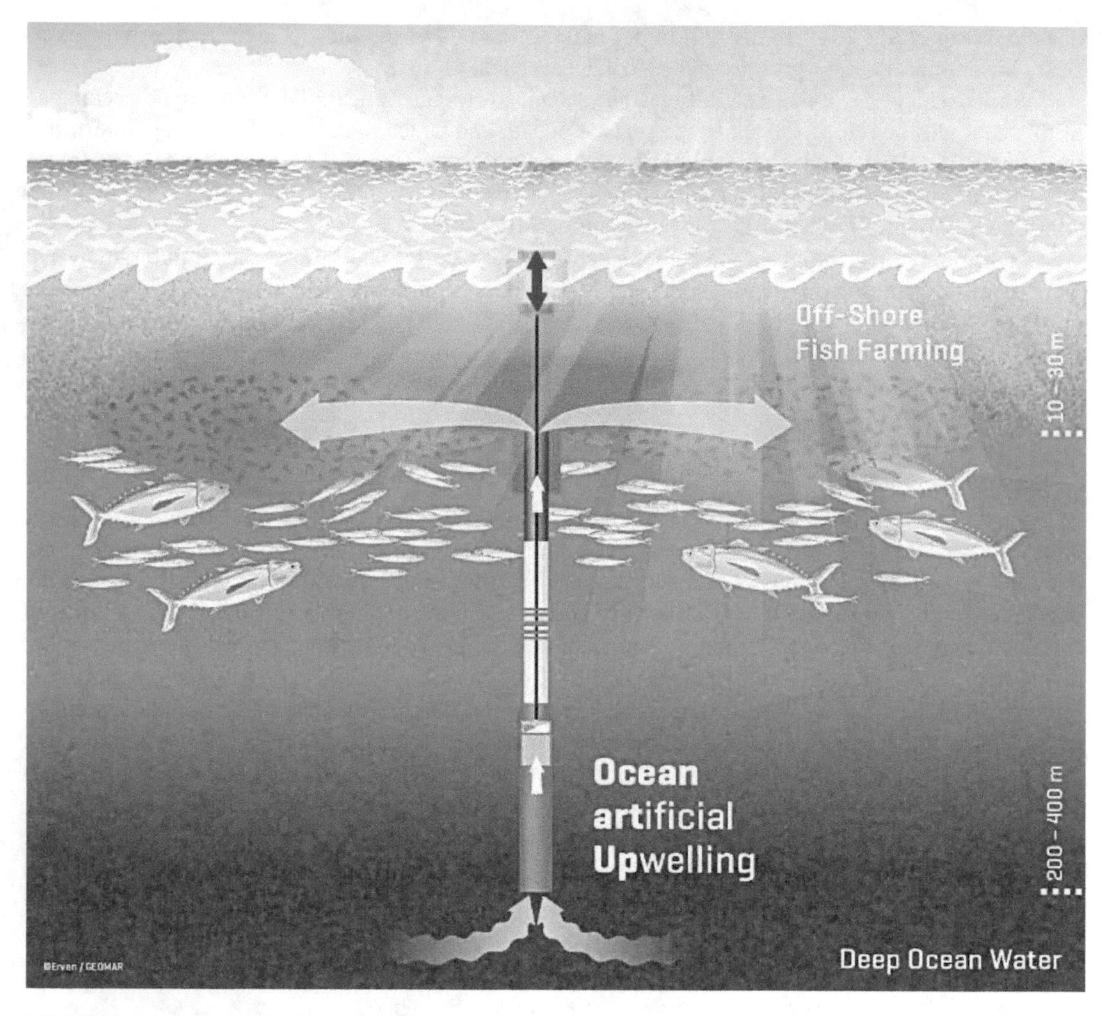

FIGURE 4.5 Depiction of the potential for AU to support offshore fish farming. SOURCE: https://ocean-artup.eu/approach.

sands of pumps) made of plastic and concrete suitable to support a 10-acre kelp farm in an optimal location (the Gulf of Mexico) and suboptimal location (the Pacific Ocean off Chile) to range from \$24.2M to \$139M in 1983 dollars with estimates scaling on pump dimensions and hence materials needs (Figure 4.6). These costs do not include installation or maintenance costs, or costs to measure C sequestration effectiveness or ecological impacts. Given the inflation rate between 1983 and 2020 of 167.02 percent (InflationTool, 2021), these costs would scale to ~\$40M at a minimum, which might be considered as a lower bound for a demonstration-scale AU trial. On the low end of potential costs, a present-day estimate from a company called Ocean-Based Climate Solutions, Inc. suggests that fabrication, assembly, shipping, and ocean operations would cost ~\$60,000 per 500-meter tube, which they estimate could sequester 250 t CO_2/yr; scaling to even 0.1 Gt CO_2/yr and, neglecting costs of verification, would then require millions of pumps and hence tens of millions in costs. For comparison, the mesocosm-based research on ecological impacts of AU was funded at €2.5M in 2017 by the European Research Council (Ocean artUp, 2021), and the Advanced Research Projects Agency-Energy (ARPA-e) MARINER program in the United States has funded ~\$50M of research "to develop the tools to enable the United States to become a global leader

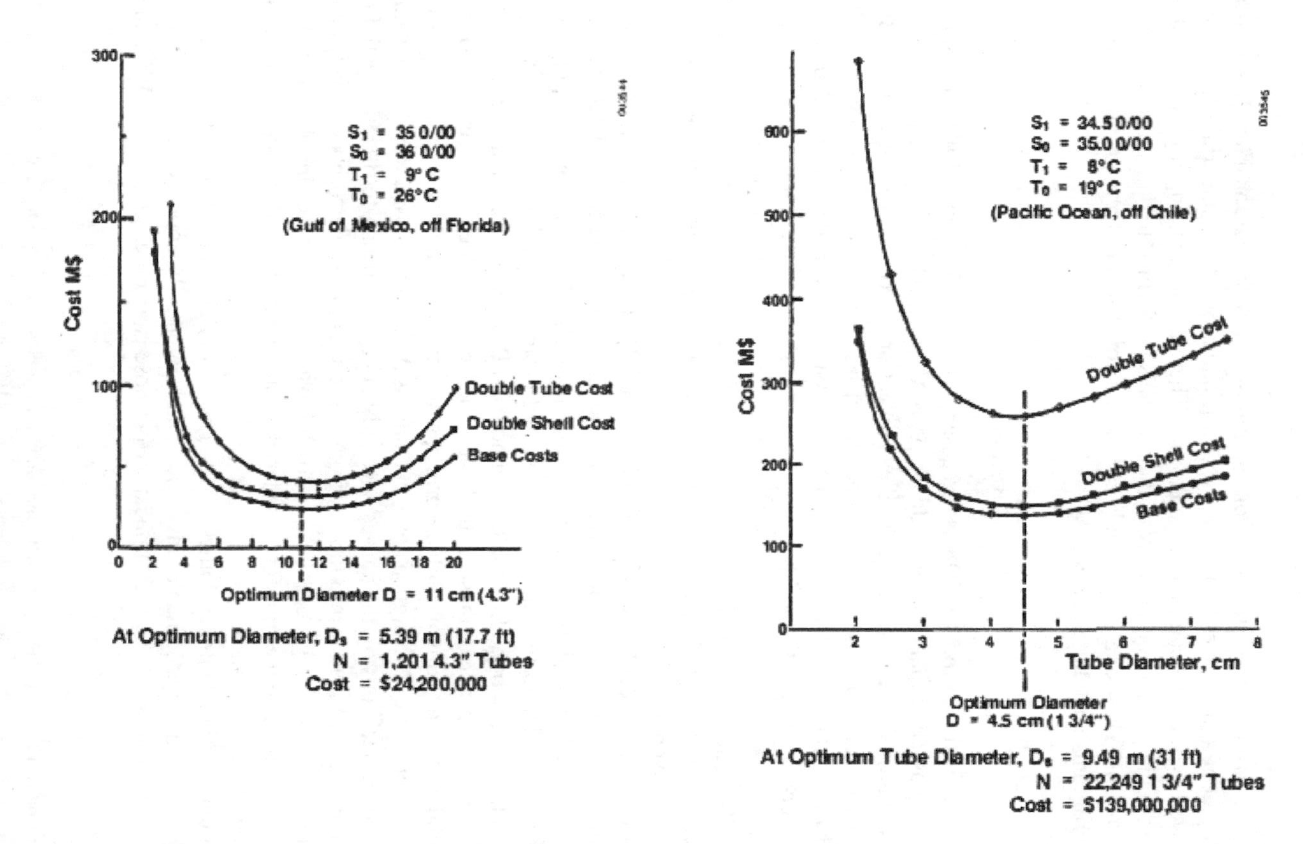

FIGURE 4.6 Calculations for salt-fountain pump networks where *N* is the number of pumps required to sustain a 10-acre kelp farm in (left) the Gulf of Mexico and (right) off the coast of Chile in the Pacific. SOURCE: Johnson and Decicco (1983).

in the production of marine biomass" which includes an assessment of AU technology (ARPA-e, 2021a). Given these limited cost comparisons, there would clearly need to be significant increases in research investments in AU to support even medium-scale deployments (hundreds of pumps).

Energy

The energy costs for AU would scale on the pump design being considered as well as any costs for deployment and recovery, monitoring, and production of the materials and supplies for the pumps as well as for verification and monitoring. There are no existing life-cycle analyses for a demonstration-scale project nor have any of the sea trials performed to date considered an energy budget for operations. These analyses would need to be a component of any proposed research and development programs aimed at CDR via AU.

Governance

The legal framework for ocean CDR is discussed in Chapter 2. Many of the international and domestic laws discussed in that chapter could apply to AU and AD.

At the international level, the parties to the Convention on Biological Diversity (CBD) have adopted a series of decisions governing "geoengineering," the definition of which is likely to encompass AU and AD.[3] The decisions recommend that parties to the CBD and other governments avoid geoengineering activities that may affect biodiversity, except for "small scale scientific research studies . . . conducted in a controlled setting."[4] The decisions are not legally binding, however. The CBD itself arguably does not prevent countries from undertaking or authorizing AU and AD projects, provided that they comply with all applicable consultation and other requirements imposed by the Convention (Webb et al., 2021).

There is significant uncertainty as to whether AU or AD constitutes "pollution" of the marine environment under the United Nations Convention on the Law of the Sea (UNCLOS) or marine "dumping" under the Convention on the Prevention of Marine Pollution by Dumping of Wastes and Other Matter (London Convention) and associated Protocol (London Protocol). In Resolution LC/LP.1 (2008) the parties to the London Convention and Protocol agreed that "ocean fertilization activities" fall within the scope of those instruments. The Resolution defines "ocean fertilization" as "any activity undertaken by humans with the principal intention of stimulating primary productivity in the oceans," which some have argued could encompass AU. However, the Resolution is nonbinding, and subsequent decisions by the parties have only applied to ocean fertilization activities that involve "the placement of matter into the sea from vessels, aircraft, platforms or other man-made structures." Some scientists (e.g., Brent et al., 2019) have argued that AU and AD are not covered because they merely involve the transfer of materials from one part of the ocean to another and not the introduction of new materials.

The treatment of AU/AD under domestic U.S. law is similarly uncertain. Table 2.3, in Chapter 2, lists key U.S. federal environmental laws that could potentially apply to AU/AD. No detailed research has been conducted on the application of those and other U.S. laws. We understand

---

[3] Report of the Conference of the Parties to the Convention on Biological Diversity on the Work of its Tenth Meeting, Decision X/33 on Biodiversity and Climate Change, Oct. 29, 2010 (hereinafter "Decision X/33"); Report of the Conference of the Parties to the Convention on Biological Diversity on the Work of its Eleventh Meeting, Decision XI/20 on Climate-related Geoengineering, Dec. 5, 2012 (hereinafter "Decision XI/20"); Report of the Conference of the Parties to the Convention on Biological Diversity on the Work of its Thirteenth Meeting, Decision XIII/4, Dec. 10, 2016 (hereinafter "Decision XIII/4").

[4] Para. 8(w), Decision X/33; Para. 1, Decision XI/20; Preamble, Decision XIII/4.

that such research is being conducted as part of an ongoing project, led by Columbia University researchers, but they had not published their findings at the time of writing.

4.6 SUMMARY OF CARBON DIOXIDE REMOVAL POTENTIAL

The criteria for assessing the potential for AU and AD as a feasible approach to ocean CDR, described in Sections 4.2–4.5, is summarized in Table 4.2.

4.7 RESEARCH AGENDA

Proof-of-concept field experiments are needed in open-ocean conditions to assess technological readiness to monitor biological responses to upwelling, determine C sequestration potential relative to upwelled macronutrients and inorganic C, and monitor or model local and downstream environmental impacts of AU and potential concomitant downwelling. There are several natural oceanic analogs that can inform our understanding of the CDR potential of upwelling, including eastern boundary currents and the Southern Ocean. Notably, these regions are characterized by large cell-sized phytoplankton communities capable of fast sinking rates, yet the upwelling of DIC tends to lead to net C outgassing (Takahashi et al., 1997; also see Chapter 5). As the technology for AU matures and field trials become practical, monitoring and verification plans need to be developed to assess the volumetric flux rate of upwelling, nutrient delivery rates, the biological response of the upper ocean, changes in the particle flux below the euphotic zone, and the air–sea CO_2 flux in both experimental and control stations. Necessary components of this flux include the following elements, summarized as a research agenda in Table 4.3.

Components

Necessary components of a complete research program include:

- Significant advances in marine engineering to develop durable pumping systems capable of sustaining long-term deployments (months to years) in a range of sea states at upwelling velocities sufficient for a sustained biological response. These systems also need to maximize mixing in the surface layer so that negatively buoyant, nutrient-rich, deep-ocean water does not immediately sink out of the euphotic zone.
- Parallel modeling efforts to estimate the feasibility of CDR potential globally and regionally given achievable upwelling velocities, elemental stoichiometry of deep-water sources, and potential biological responses. Feasibility studies would be based on model upscaling from results of technical trials and existing literature on biological response to nutrient perturbations. Mesocosm and laboratory experiments would help constrain biological responses. Efforts would help refine which regions of the global ocean would be optimal for a scaled-up research program.
- A robust monitoring plan focused on particulate and dissolved organic C export and air–sea gas exchange at the local site of upwelling as well as targeted downstream sites. This plan should span across water measurements, remote sensing, and high-resolution coupled physical/ecological models to assess large-scale and downstream impacts and the timescales and depth scales of sequestration.
- Monitoring that will assess the vertical extent of C export and any changes in remineralization length scales.

TABLE 4.2 CDR Potential of Artificial Upwelling and Downwelling

Knowledge base What is known about the system (low, mostly theoretical, few in situ experiments; medium, lab and some fieldwork, few carbon dioxide removal (CDR) publications; high, multiple in situ studies, growing body of literature)	**Low–Medium** Various technologies have been demonstrated for artificial upwelling (AU), although primarily in coastal regimes for short duration. Uncertainty is high and confidence is low for CDR efficacy due to upwelling of CO_2, which may counteract any stimulation of the biological carbon pump (BCP).
Efficacy What is the confidence level that this approach will remove atmospheric CO_2 and lead to net increase in ocean carbon storage (low, medium, high)	**Low Confidence** Upwelling of deep water also brings a source of CO_2 that can be exchanged with the atmosphere. Modeling studies generally predict that large-scale AU would not be effective for CDR.
Durability Will it remove CO_2 durably away from surface ocean and atmosphere (low, <10 years; medium, >10 years and <100 years; high, >100 years), and what is the confidence (low, medium, high)	**Low–Medium** <10–100 years As with ocean iron fertilization (OIF), dependent on the efficiency of the BCP to transport carbon to deep ocean.
Scalability What is the potential scalability at some future date with global-scale implementation (low, <0.1 Gt CO_2/yr; medium, >0.1 Gt CO_2/yr and <1.0 Gt CO_2/yr; high, >1.0 Gt CO_2/yr), and what is the confidence level (low, medium, high)	**Medium** Potential C removal >0.1 Gt CO_2/yr and <1.0 Gt CO_2/yr (low confidence) Could be coupled with aquaculture efforts. Would require pilot trials to test materials durability for open ocean and assess CDR potential. Current model predictions would require deployment of tens of millions to hundreds of millions of pumps to enhance C sequestration. (Low confidence that this large-scale deployment would lead to permanent and durable CDR).
Environmental risk Intended and unintended undesirable consequences at scale (unknown, low, medium, high), and what is the confidence level (low, medium, high)	**Medium–High** (low confidence) Similar impacts to OIF but upwelling also affects the ocean's density field and sea-surface temperature and brings likely ecological shifts due to bringing colder, inorganic carbon and nutrient-rich waters to surface.
Social considerations Encompass use conflicts, governance-readiness, opportunities for livelihoods, etc.	Potential conflicts with other uses (shipping, marine protected areas, fishing, recreation); potential for public acceptability and governance challenges (i.e., perception of dumping).
Co-benefits How significant are the co-benefits as compared to the main goal of CDR and how confident is that assessment	**Medium–High** (low confidence) May be used as a tool in coordination with localized enhancement of aquaculture and fisheries.
Cost of scale-up Estimated costs in dollars per metric ton CO_2 for future deployment at scale; does not include all of monitoring and verification costs needed for smaller deployments during R&D phases (low, <\$50/t CO_2; medium, ~\$100/t CO_2; high, >>\$150/t CO_2) and confidence in estimate (low, medium, high)	**Medium–High**. >\$100–\$150/t CO_2 (low confidence) Development of a robust monitoring program is the likely largest cost and would be of similar magnitude as OIF. Materials costs for pump assembly could be moderate for large-scale persistent deployments. Estimates for a kilometer-scale deployment are in the tens of million dollars.

TABLE 4.2 Continued

Cost and challenges of carbon accounting Relative cost and scientific challenge associated with transparent and quantifiable carbon tracking (low, medium, high)	**High** Local and additionality monitoring needed for carbon accounting similar to OIF.
Cost of environmental monitoring Need to track impacts beyond carbon cycle on marine ecosystems (low, medium, high)	**Medium** (medium–high confidence) All CDR will require monitoring for intended and unintended consequences both locally and downstream of CDR site, and these monitoring costs may be substantial fraction of overall costs during R&D and demonstration-scale field projects.
Additional resources needed Relative low, medium, high to primary costs of scale-up	**Medium–High** Materials, deployment, and potential recovery costs.

- Assessment of the CDR potential at a range of pumping frequencies including episodic versus continuous upwelling as well as any seasonal impacts on CDR potential and the optimal source-water horizons.
- A monitoring plan that would be able to estimate the additionality of sequestration—how much production and export would have occurred in natural phytoplankton communities in the absence of AU? The monitoring plan should be able to differentiate between the response to AU and the natural variability of the system, which will differ between regions (e.g., Fe-limited versus N-limited ecosystems).
- Siting studies that address potential conflicts with shipping lanes, fishing effort, and other ocean usage as well as regions where costs are minimized relative to source-water horizons (e.g., shallower pumps require reduced material costs).
- A data management plan with clear plans for data dissemination, accountability, and data transparency following the FAIR principles: findability, accessibility, interoperability, and reuse.[5]
- Complete life-cycle analyses of costs for materials, deployment, and local and downstream monitoring costs.
- Interactions with social scientists, legal experts, and economists to assess public perception, acceptability, governance, and cost feasibility as well as potential for coupling AU to macroalgae or fisheries production.

Cost and Time Frame of Research Agenda

Stable funding streams need to first be identified to build capacity and develop an active research program as well as begin the planning of complementary monitoring and modeling components necessary to evaluate environmental impacts, CDR, and additionality. Costs increase as we move from the technological readiness stages of pump development, deployment, and testing of operational needs (Can water be delivered at needed upwelling rates? Can pumps survive a range of sea states?) to large-scale deployment and monitoring of intended and unintended ecological effects. Based on prior cost estimates (i.e., Johnson and Dicicco, 1983), regional-scale networks of pumps could cost on the order of ~$40 million for technological development alone, which is

[5] See https://www.go-fair.org/fair-principles/.

within the range of expected funding for CDR development funding in the United States and the European Union (see, e.g., Burns and Suarez, 2020). Given this cost magnitude and the current lack of technological readiness, model-based feasibility studies should lead the research agenda to identify optimal siting and scaling of pump networks and CDR potential. This should be followed by expansion of technological development and small-scale proof-of-concept studies intended to show durability of materials and achievement of necessary upwelling velocities coupled to life-cycle assessments of materials and deployment costs.

Environmental Impacts of Research Agenda

Upwelling of deep nutrient-rich seawater into the surface euphotic zone brings a source of DIC in addition to potentially growth-limiting elements. The ecological consequences may include stimulation of autotrophic growth and enhanced sinking of detrital carbon. Alternatively, if the photosynthetic drawdown of DIC does not exceed the carbon introduced in the source water, outgassing of carbon may occur, which is obviously contrary to sequestration. This latter potential is the most significant "Achilles heel" of AU as a CDR strategy. Otherwise, much like OIF (Chapter 3), nutrient additions via AU may lead to shifts in plankton community structure and, potentially, alternations of productivity of higher trophic levels including commercially harvested fisheries. Any monitoring plan should also consider the possibility of harmful algal blooms, other greenhouse gas production, and hypoxia.

4.8 SUMMARY

A range of studies have conclusively demonstrated that deep nutrient-rich seawater can be delivered to the surface ocean via a number of pumping mechanisms, each of which has different energy costs, deployment modes, and pumping capabilities. A recent summary by Pan et al. (2016) describes the current state of these technologies (see Table 4.1). These limited field trials have largely been conducted in short-term (less than weeks) experiments and primarily in coastal regimes, and none have yet verified enhanced C sequestration. A research agenda aimed at testing AU as a component of a CDR portfolio would principally need to address the long-term durability and efficacy of AU technology as well as the siting for sea trials. Since AU would deliver remineralized DIC as well as potentially limiting nutritional resources, it is key to understand the nutrient-use stoichiometry of the local plankton populations relative to the stoichiometry of deep-source waters (which varies widely across the global ocean); **simply, C export flux would need to exceed the upwelled carbon.** Additionally, the input of other elements, phosphorus or iron, for example, can also govern bloom dynamics and need to be considered. For example, in the oligotrophic North Pacific Subtropical Gyre, Karl and Letelier (2008) hypothesized that upwelling of water with excess phosphorus relative to nitrogen could trigger a two-stage bloom with net C sequestration being driven by the production of diazotrophic (N_2-fixing) microorganisms that may thrive after a primary pulse of nondiazotrophic plankton. While yet untested, this hypothesis underscores the importance of understanding C:N:P stoichiometry of source waters relative to plankton growth requirements as a driver of potential C sequestration exceeding that delivered by the upwelling process. And since nutrient stoichiometry varies across ocean basins and with depth, one would expect that the ecological consequences of AU would be site and depth specific and, perhaps, time dependent.

The current state of knowledge otherwise, via model simulations, indicates that even a persistent and effective deployment of millions of functional pumps across the global ocean would not meet CDR goals for sequestration or permanence. Moreover, natural analogs where upwelling

TABLE 4.3 Research and Development Needs: Artificial Upwelling and Downwelling

#	Recommended Research	Question Answered	Environmental Impact of Research	Social Impacts of Research ($M/yr)	Estimated Research Budget	Time Frame (yr)
4.1	**Technological readiness: Limited and controlled open-ocean trials to determine durability and operability of artificial upwelling technologies**	**Can pumps be developed to withstand open-ocean conditions and sustain upwelling velocities for prolonged deployments?**	**Modest; potential for harmful algal blooms in some regions or outgassing of greenhouse gases (GHGs)**	**Modest for short term**	**5 (~100 pumps tested in various conditions)**	**5**
4.2	Feasibility studies	Given limited technological trials and achieved upwelling velocity, can these technologies be scaled up?			1 Modeling-based feasibility studies based on results of technical trials	1
4.3	Tracking C sequestration	How can we track enhanced C fluxes? Development of plan to track enhancement of biological pump should precede and parallel any field trials. Should inform composite monitoring plan.	Likely done as part of larger field experiments	See field experiments	$3M/yr New methods for tracking C from surface to depth needed	5
4.4	Modeling of C sequestration based upon achievable upwelling velocities and known stoichiometry of deep-water sources. Parallel mesocosm and laboratory experiments to assess potential biological responses to deep water of varying sources	What is the CDR potential given outcomes of sea trials and technological advancement of pumps? Given known ratios of growth-limiting nutrients and estimated biological responses, what are the optimal regions for a robust research program?			5	5

continued

TABLE 4.3 Continued

#	Recommended Research	Question Answered	Environmental Impact of Research	Social Impacts of Research ($M/yr)	Estimated Research Budget	Time Frame (yr)
4.5	Planning and implementation of demonstration-scale in situ experimentation (>1 yr, >1,000 km) in region sited based on input from modeling and preliminary experiments	What are CDR efficiencies at demonstration scale and what are the intended and unintended ecological impacts? Planning should include complete life-cycle analyses of costs for materials, deployment, and local and downstream monitoring costs.	Modest Regional impacts during upwelling period and some concerns beyond test boundaries. Observations and models needed.	Modest/High Public may view these activities as dumping or negatively as ocean geoengineering due to possible unknown ecological shifts, i.e., harmful algal blooms and co-production of other GHGs. Recent emphasis on co-benefit of enhanced fisheries yet to be verified	25 Research needs to measure all possible geochemical, physical, and ecological impacts to gauge effectiveness and impacts at scale	10
4.6	Monitoring C and ecological shifts	What are the large-scale and downstream impacts and the timescales and depth scales of sequestration? Development of autonomous and remote methods for assessment of biological carbon pump (BCP) coupled to high-resolution coupled physical/ecological models is needed and should be conducted early in coordination with planning and implementation of field trials and synthesis of those efforts.	Low New methods, especially optical to complement existing geochemical sensors and platforms and molecular tools to monitor ecological shifts	Low Any method to measure C flow and ecological shifts will have multiple uses for science and the public	10 New technologies are quick to prototype but expensive to bring to market at reliability and scale useful for CDR	10
4.7	Experimental planning and extrapolation to global scales	What is realistic BCP and elemental cycling, including particle cycling, as shown in full Earth system models?	Low Modeling needed to design experiments to predict impacts at local scale and in the far field do not have direct environmental impact and assist planning more acceptable field research	Low if considering only modeling, though public acceptance of CDR still needed and models will be needed to assess possible impacts	5 Early for planning and later for impact assessments	10

4.8	Research on the social and economic factors and governance	Is the current London Convention and London Protocol sufficient for regulation of research on the high seas? And for eventual deployment?	N/A	Starting from a point of low/modest public acceptance of ocean geoengineering	2	10
4.9	Document best "code of conduct" for research and eventual deployment	What are best practices regarding open data systems and peer review and independent C and impact assessments? These need to be codified.	N/A	Needed for public acceptance of use of the high seas for any open-ocean CDR	2	5–10 (early agreement of research conduct needed)

NOTE: Bold type identifies priorities for taking the next step to advance understanding of artificial upwelling and downwelling.

occurs are generally net sources of CO_2 to the atmosphere (Takahashi et al., 1997) versus net sinks. Even if these predictions are correct and AU proves too costly or impractical for large-scale ocean CDR, AU may prove to be a valuable tool to promote aquaculture or fisheries (assuming the extraction costs do not exceed the C sequestration potential) or simply as a research tool to better understand the biological responses of microbial communities to nutrient perturbations. Pilot studies are principally needed to address the CDR potential of AU and the ecological consequences of these activities. Only targeted, regulated, and transparent field studies can help to minimize current uncertainties and determine if this strategy could be an effective component of an ocean CDR portfolio.

5

Seaweed Cultivation

5.1 OVERVIEW

Large-scale seaweed cultivation and its purposeful sequestration is a potential ocean-based strategy for reducing atmospheric carbon dioxide (CO_2) levels. Large-scale farming of seaweed would incorporate dissolved CO_2 from the upper ocean into tissue that then can be sequestered at depth either by pumping biomass to depth or by its sinking through the water column. As many seaweeds grow, they release large amounts of dissolved organic carbon (DOC) into the upper ocean. Some fraction of that is thought to be recalcitrant to microbial activity over long timescales, enabling an additional pathway for seaweed cultivation to sequester carbon. Macrophyte biomass farms could be created on large scales in environments where there is sufficient solar illumination and available nutrients. Much progress has been made in the past decade in developing both commercial seaweed farms for human consumption and animal feed as well as pilot studies for the development of large-scale farms for biofuel production. The goal here is to assess the present state of knowledge and address what research and investments are needed to make *purposeful* seaweed cultivation and sequestration CO_2 removal (CDR) worthy.

According to the criteria described in Chapter 1, the committee's assessment of the potential for seaweed cultivation as a CDR approach is discussed in Sections 5.2–5.4 and summarized in Section 5.5. The research needed to fill gaps in understanding of seaweed cultivation and sequestration, as an approach to durably removing atmospheric CO_2, is discussed in Section 5.6.

5.2 KNOWLEDGE BASE

Studies on natural macrophyte-dominated ecosystems and in aquaculture facilities have provided much of the information needed to assess the CDR potential of seaweed cultivation. These include determinations of biomass density, carbon content, rates of net primary production (NPP), nutrient ratios, seasonality of growth, farming techniques that lead to higher yields, and release of DOC during production among others (e.g., Wheeler and North, 1981; Reed et al., 2008, 2015; Stewart et al., 2009; Rassweiler et al., 2018; Azevedo et al., 2019; Bak et al., 2020; Forbord et al.,

2020; Matsson et al., 2021). These data provide the information needed to assess the scalability of seaweed cultivation as a CDR strategy (see Section 5.3).

Considerably less is known about the fates of macrophyte carbon in natural ecosystems and how this carbon contributes to long-term carbon sequestration. Several studies have attempted to quantify the role of natural macrophyte ecosystems in global carbon sequestration (e.g., Smith, 1981; Chung et al., 2010; Wilmers et al., 2012; Krause-Jensen and Duarte, 2016; Watanabe et al., 2020). In a particularly influential contribution, Krause-Jensen and Duarte (2016) concluded that natural macroalgal ecosystems could make substantive contributions to global ocean carbon sequestration primarily through the export of plant biomass to depth and the seafloor and the production of recalcitrant DOC. Recalcitrant DOC is that fraction of the DOC pool that is resistant to rapid microbial degradation (e.g., Hansell, 2013). Via a synthesis of a broad range of previously published results over a wide range of taxa (Figure 5.1), Krause-Jensen and Duarte (2016) suggest that macroalgal ecosystems could at most sequester ~0.17 petagrams (Pg) C/yr or ~0.6 Gt CO_2/yr globally with a wide range of uncertainty based upon the assumptions applied (roughly 0.06 to 0.27 Pg C/yr).[1] They find that the export of recalcitrant DOC to below the mixed layer is the dominant contribution to their potential sequestration budget (roughly 70 percent of total sequestration).

The Krause-Jensen and Duarte (2016) paper attempts to quantify the carbon sequestration by naturally occurring macrophyte ecosystems if these ecosystems were to inhabit all regions in the world ocean where there is sufficient light on the benthos (see notes in Table 2 of Krause-Jensen and Duarte, 2016). Therefore, their estimate is the potential that natural macrophyte afforestation could

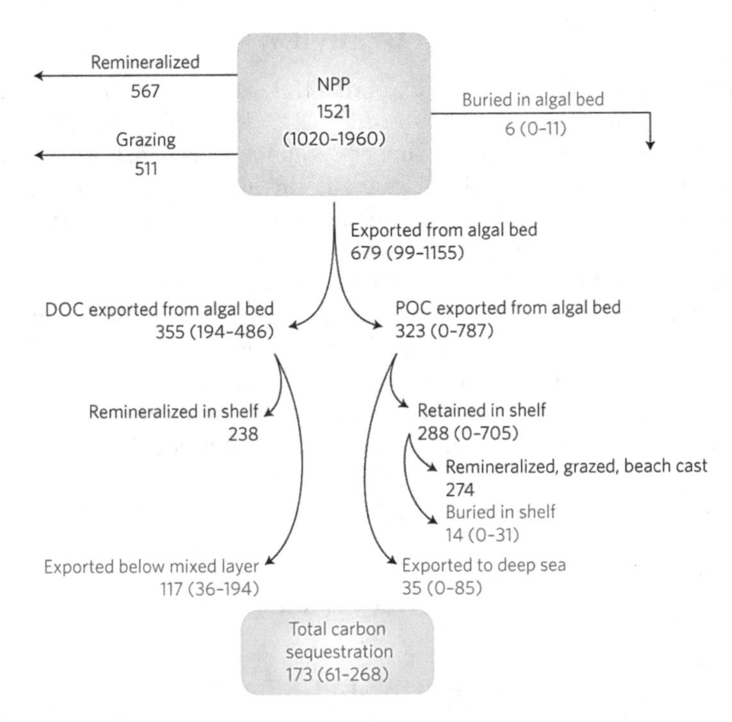

FIGURE 5.1 Synthesized pathways of carbon export and sequestration by natural macrophytes if these seaweed populations could be grown to their maximal extent (values in Pg C/yr). The blue text represents exported carbon that is thought to be sequestered for long times; black text is rapidly remineralized and exchanged with the atmosphere. Values in parentheses represent 25 percent and 75 percent quartile uncertainty levels. SOURCE: Redrawn from Krause-Jensen and Duarte (2016).

[1] 1 Pg C = 1,015 g C = 3.7 Gt CO_2.

contribute to global carbon sequestration and not the actual amount that these ecosystems presently sequester. Here, the focus is on the large-scale cultivation of seaweed biomass and its purposeful sequestration in durable ocean reservoirs. Hence, an independent assessment of its carbon budget is made in Section 5.3. Afforestation of seaweed ecosystems and their potential as a CDR strategy are discussed in Chapter 6.

Purposeful macrophyte cultivation has recently become a particularly popular CDR strategy due to its potential to scale to CDR-relevant amounts with a wide range of enticing co-benefits (e.g., GESAMP, 2019; Gattuso et al., 2021; oceanvisions.org and similar websites). There are also important barriers to its viability as an effective CDR strategy that will be discussed below. One useful conceptualization of macrophyte cultivation as an ocean CDR strategy would be considering large-scale farming of seaweeds to assimilate CO_2 from the surface ocean and then purposefully convey this fixed carbon to deep oceanic reservoirs that will remain out of contact with the atmosphere over some relevant planning time horizon, say 100 years. The air–sea CO_2 equilibrium timescales relative to surface water residence times also need to be considered, as is the case with all marine CDR approaches considered here (see Section 1.3).

The elements of purposeful macrophyte cultivation as an ocean CDR strategy are depicted in Figure 5.2. Seaweed farming converts dissolved CO_2 to seaweed biomass that is slowly replaced by air–sea fluxes from the atmosphere (e.g., GESAMP, 2019; Gattuso et al., 2021; Sala et al., 2021). The large-scale cultivation will also produce DOC during growth (Reed et al., 2008; Paine et al., 2021), some of which, it has been suggested, will be recalcitrant on decadal timescales (Krause-Jensen and Duarte, 2016). Both pathways may contribute to long-term carbon sequestration. In many regions of the world's oceans, nutrients are naturally depleted in the upper layers of the water column but can be found just beneath the euphotic zone. Thus, it is likely that artificial upwelling devices will be needed to supply required nutrient concentrations in some settings, which brings with its application other consequences, such as upwelling of enriched dissolved CO_2 deep waters and other possible ecological impacts (see Chapter 4).

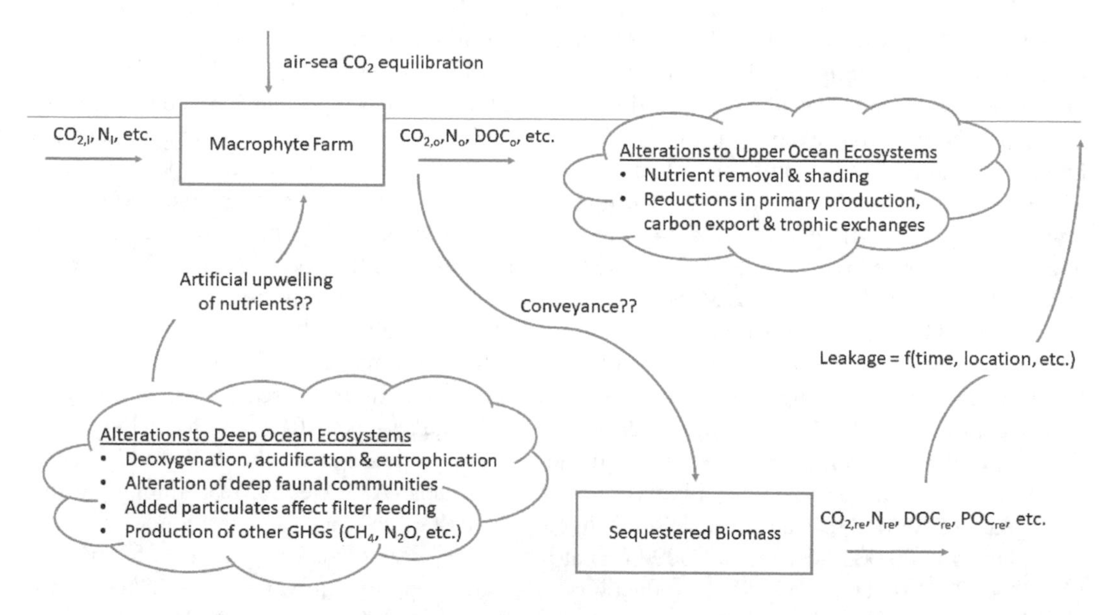

FIGURE 5.2 Elements of purposeful macrophyte cultivation as a CDR strategy. Shown in blue are the CDR elements encapsulating the large-scale farming, sequestering at depth, and issues associated with determining its efficiency (cf., need for conveyance to bring biomass to depth or artificial upwelling to supply the necessary nutrients, or leakage of remineralized CO_2 back to the sea surface). In green are the expected alterations to the upper-ocean and deep-ocean ecosystems associated with purposeful macrophyte cultivation as a CDR tool.

The harvested biomass needs to be conveyed by some means deep enough in the water column that most of the carbon biomass, when remineralized back to CO_2, will remain out of contact with the atmosphere. The leakage of sequestered material back to the sea surface will be driven by ocean circulation and mixing processes and hence will be a function of depth, location, and planning timeline (Siegel et al., 2021a). Further, the sequestered biomass will likely need to be injected to depth rapidly to ensure that little carbon is lost near the ocean surface, potentially requiring a conveyance device. The injected biomass will decompose back to CO_2 and potentially recalcitrant DOC due to food web, particularly heterotrophic microbial, processes (Figure 5.2).

The cultivation and purposeful injections of seaweed biomass will likely affect ocean ecosystems both in the euphotic zone where the biomass is grown and at depth. In the near-surface ocean, these effects include reducing ambient nutrient levels and available light. Subsequently, that will likely reduce phytoplankton primary production rates, decrease carbon export from the surface ocean, and may affect trophic exchanges of energy that support fisheries and marine mammal populations. At depth, these perturbations may alter the natural balances of organic matter decomposition and remineralization, reducing oxygen concentrations (deoxygenation) and increasing subsurface CO_2 and nutrient levels, leading to increases in acidification and eutrophication of these mesopelagic ecosystems. However, the strength of these impacts will depend upon the amount of CDR performed and the efforts made to displace their influences. There may also be several direct societal impacts of large-scale farming, including hazards to navigation and co-benefits such as reducing excess nutrients from fish and shellfish aquaculture facilities. Furthermore, the farming of macrophytes for products (e.g., food, biofuels, animal feed, etc.) may contribute to long-term carbon sequestration, possibly due to the production and release of recalcitrant DOC. The sequestration timescales of the fixed carbon in macrophyte biomass products are likely to be short (≤10 years at best) compared with the goal of CDR (≥100 years).

Much of the knowledge base for large-scale seaweed cultivation comes from experiences in large-scale seaweed farms for animal and human food as well as a potential carbon source for biofuels (Camus and Buschmann, 2017; Bak et al., 2018; Camus et al., 2018; Azevedo et al., 2019; ARPA-e, 2021a; Navarrete et al., 2021). Macroalgal farming and the products these farms create have become a multibillion dollar industry, with Asia being the main supplier (FAO, 2016). On a global basis, seaweed aquaculture production exceeds 30 million (wet) Mt and $5.6B on a monetary basis, and these totals are rapidly growing (FAO, 2016). Currently, farmed macroalgal uses include human and animal food, fertilizers, other products, and a feedstock for biofuel production (e.g., Milledge et al., 2014; ARPA-e, 2021a). The majority of algal biomass comes from a relatively small number of species (Milledge et al., 2014). Much focus has been placed on improving yields and the quality of the farmed macrophyte crops while lowering costs (e.g., Bak et al., 2018, 2020; Azevedo et al., 2019; Forbord et al., 2020; Matsson et al., 2021). Macrophyte farms are typically conducted on suspended longline ropes with embedded sporophytes moored in shallow (≤100 meters) coastal or estuarine waters (Peteiro and Freire, 2013; Camus and Buschmann, 2017; Camus et al., 2018). Existing farms are a few up to a few thousand hectares in size. Recent research supported by the U.S. Department of Energy Advanced Research Project Agency-e's (ARPA-e's) MARINER program is aimed at creating prototypes for scalable farmed systems that can be deployed in deeper waters for the purpose of growing biomass stocks for biofuels (ARPA-e, 2021a). Among the specific goals of MARINER are to develop technologies for farming macrophyte biomass that can be scaled to ≥100,000 hectares (≥1,000 km^2) at production costs of ≤$80/dry metric ton biomass (ARPA-e, 2021a). The MARINER production cost goal is roughly equivalent to ~$75/metric ton CO_2 in macrophyte biomass (assumes C content is 30 percent of the dry weight). Innovative farm designs even include systems that cycle vertically on a daily time course to optimize the acquisition of light energy during the day and subsurface nutrients at night (Navarrete et al., 2021). Important elements for successful macrophyte cultivation include selecting the cultivars to be farmed based upon environmental conditions and local macrophyte strains; culturing spores to embryonic sporophytes and efficiently attaching these sporophytes on longline ropes; installing longline ropes in the field; permitting, installing, and maintaining farm facilities; monitoring the crop status as well as the environmental conditions; and developing an understanding of the effects of the environment on the crops and the crops on the environment, etc. (Camus et al., 2018; Bak et al., 2020; Bell et al., 2020; Visch et al., 2020; Matsson et al., 2021).

The knowledge base for the long-term fates of farmed macrophyte organic matter is far less known. It is important that the farmed macrophyte carbon be removed from the surface ocean so that the assimilated carbon will remain out of contact with the atmosphere on sequestration timescales (decades to centuries). The long-term incorporation of vast amounts of macrophyte carbon in ocean sediments either by purposeful placement on the seafloor or by sinking seaweed biomass from the surface seems highly unlikely. For example, Bernardino et al. (2010) studied the degradation of 100-kg bales of giant kelp placed on the seafloor at 1,670-meter depth in the Santa Cruz Basin in the North Pacific Ocean using remotely operated vehicles. They found significant changes after 6 months in macrofaunal community abundances and diversity near (within 1 meter) to the kelp bales, but very little discernible changes in sediment total organic carbon concentrations. Similar rapid decomposition rates of kelp biomass on the benthos (roughly 5 percent biomass per day) were found in shallow continental shelf depths (80–350 meters) off Carmel Canyon, California (Harrold et al., 1998) and at the bottom (1,300 meters) of the Santa Catalina Basin (Smith, 1983). Hence the water column seems the most likely reservoir for injected macrophyte carbon. Over centennial timescales appropriate for ocean CDR, this will be in the form of respired CO_2.

A recent analysis of purposeful CO_2 sequestration timescales using an ocean circulation inverse model showed that deeper discharge locations sequester injected CO_2 for much longer than shallower ones and median sequestration times are typically decades to centuries, and approach 1,000 years in the deep North Pacific (Siegel et al., 2021a). Further, large differences in sequestration times occur both within and between major ocean basins. The Pacific and Indian basins generally have longer sequestration times than the Atlantic Ocean and Southern Ocean. Assessments made of the injected CO_2 retained over a 100-year time horizon illustrate that most of the injected carbon will still be in the ocean at injection depths greater than 1,000 meters, with several geographic exceptions such as the western North Atlantic (see further descriptions of this work below). Ocean circulation and mixing thus place important constraints on the timescales over which injected CO_2 remains in the water column.

Methodologies for the conveyance of macrophyte biomass to depth are just in their infancy. If the pneumatocysts are forced to burst (either by pressure due to being injected to depth or the tissue masticated), it is thought that the macrophyte biomass will sink until it reaches the seafloor (Krause-Jensen and Duarte, 2016; see also https://www.runningtide.com). Engineered solutions for pumping macrophyte biomass to depth via vertical tubes have also been developed, and a system for removing nuisance mats of *Sargassum* from the surface ocean has been prototyped recently (Gray et al., 2021). Technologies need to be developed to verify the delivery of organic carbon to depth with minimal losses.

Last, very little is known about the fate of DOC produced by macrophyte populations as they grow. Field work on natural giant kelp forests suggests that ~14 percent of macroalgal NPP is released as DOC of all labilities (Reed et al., 2015), the proportion of that DOC that is recalcitrant to degradation is largely unknown. Krause-Jensen and Duarte (2016) simply assume that the partitioning of macrophyte DOC into labile and recalcitrant fractions follows the ratio of global DOC export from the upper ocean to the global NPP, which results in an estimate for the production of recalcitrant DOC that is roughly 8 percent of macrophyte NPP. The validity of that assumption remains a major knowledge gap. A recent study of temperate *Sargassum* beds showed that 56 percent to 78 percent of the released DOC collected in these ecosystems was resistant to decomposition after

150 days (Watanabe et al., 2020). However, the timescales relevant for assessing CDR strategies are decadal to centennial, and more research in this area is clearly needed.

5.3 EFFICACY AND SCALABILITY

We can investigate the impact potential and scalability of purposeful macrophyte cultivation using a simple scaling analysis. A reasonably successful CDR goal would be to grow and sequester enough macrophyte biomass to remove 0.1 Gt CO_2/yr (0.027 Pg C/yr = Seq_{goal}) from the upper ocean over a time horizon of more than 100 years. Clearly at these levels of CDR, seaweed cultivation is envisioned as contributing to a portfolio of terrestrial- and ocean-based CDR approaches. To assess the potential of purposeful macrophyte cultivation as a scalable CDR tool, we will estimate the areal size of a farm required to grow that amount over that time horizon and use this formulation to discuss the scalability of purposeful macrophyte cultivation to climate control–relevant scales.

Following the synthesis of Krause-Jensen and Duarte (2016), we consider the two primary pathways linking farmed macrophytes and carbon sequestration—the purposeful injection of particulate macrophyte carbon to depth and the release and eventual sequestration of recalcitrant DOC from the growing macrophyte farms, or

$$Seq_{goal} = 0.1 \text{ Gt } CO_2/\text{yr} = 0.027 \text{ Pg C/yr} = Seq_{Bio} + Seq_{DOC} \quad (5.1)$$

The C budget for natural macrophyte ecosystems created by Krause-Jensen and Duarte (2016) suggests that the Seq_{DOC} should be 8 percent of global macrophyte NPP (f_{DOC}; see Figure 5.1). Assuming this holds for a farmed system, Seq_{DOC} should be

$$Seq_{DOC} = f_{DOC} * NPP_{Farm} \quad (5.2)$$

where NPP_{Farm} is the NPP of the farm. Some fraction (f_{loss}) of the farmed biomass will be lost due to herbivory, frond senescence, storms, and inefficiencies in the conveyance of biomass to depth, etc. We will assume that the value of f_{loss} is 20 percent, but it is recognized that this value is, at present, poorly constrained. Thus, the accumulation of macrophyte C biomass of the farm per year (Seq_{Bio}) will be equal to

$$Seq_{Bio} = (1 - f_{loss}) * NPP_{Farm} \quad (5.3)$$

Thus, the amount of macrophyte NPP required to grow up each year to reach the 0.1 Gt CO_2/yr goal would be

$$NPP_{Farm} = Seq_{goal}/(1 + f_{DOC} - f_{loss}) = 0.033 \text{ Pg C/yr} \quad (5.4)$$

Contrasting the natural example above (Figure 5.1), the sequestration of macrophyte carbon from farm systems will be dominated by the direct injection of C biomass to depth as the entire crop (minus losses in the upper ocean) will be sequestered. Thus, the release of recalcitrant DOC on growth should be a small part of the total carbon sequestration budget (~9 percent of Seq_{goal}). This is much smaller than the Krause-Jensen and Duarte (2016) synthesis, which suggests that recalcitrant DOC release dominates (~70 percent) the total sequestration from natural macrophyte ecosystems. This difference arises because the biomass losses and remineralization in the surface ocean for CDR using seaweed cultivation and purposeful sequestration will be much smaller than in Krause-Jensen and Duarte's (2016) synthesis of a natural seaweed ecosystem. Hence, the contributions of recalcitrant DOC for CDR via seaweed cultivation will be much smaller than has been suggested for natural systems.

The areal extent of a seaweed farm capable of sequestering 0.1 Gt CO_2/yr can be determined knowing NPP_{Farm}, the number of crops per year that can get grown and sequestered (N_{Crop}), the expected biomass density in the farm (Yield) and the C density of crop ($C_{Content}$), or

$$A_{Farm} = NPP_{Farm}/(Yield * C_{Content} * N_{Crop}) \tag{5.5}$$

Assuming that the farmed system will maximize seaweed density and carbon quality, we use giant kelp observations of the maximum biomass observed from the Santa Barbara Coastal Long-Term Ecological Research data record (Rassweiler et al., 2018) or a maximum yield (Yield) equal to 1 kg DW/m^2 and 30 percent of that dry weight (DW) will be carbon by DW mass ($C_{Content}$). We also assume that one can grow and sequester on average 1.5 crops each year, which may be an optimistic assumption based on present macrophyte farms, particularly for higher-latitude sites. Together, the required area needed to sequester 0.1 Gt CO_2/yr by seaweed cultivation comes to

$$A_{Farm} = 0.033\ e(15\ g\ C/yr)/([1e(3\ g\ DW/m^2)] * [0.3\ g\ C/g\ DW] * [1.5\ crops/yr]) = 7.3e(10\ m^2) = 73{,}000\ km^2 = 7.3\ \text{million hectares} \tag{5.6}$$

Thus, the size of a farm required to grow enough biomass to sequester 0.1 Gt CO_2/yr would be a single square farm, 270 km on a side. These farms would logically be multiple farms spread out across the globe. This area is approximately equivalent to half the size of the State of Iowa or if one considers a 100-m-wide continuous belt of seaweed farm along all continents and islands, it would require 730,000 km of coastline. That is 63 percent of the global coastline. If placed along the coastline of the United States only, it would comprise a nearly 0.5-km-wide continuous belt of seaweed farm. The amount of ocean surface area required to sequester 0.1 Gt CO_2/yr demonstrates the size of the engineering and logistical tasks at hand associated with scaling seaweed-cultivation CDR solutions to climate-relevant scales.

The formulation above enables the assessment of controls on the required size of macrophyte farms needed to grow enough biomass to sequester climate-relevant amounts of CO_2 from the upper ocean. Increasing the sequestration goal (Seq_{goal}) will linearly increase the farmed area required, while increasing the yield, carbon content, and number of crops per year, or reducing the biomass losses before sequestration will similarly reduce the required farmed area. Work from ARPA-e's MARINER project[2] suggests that innovation could lead to increases in yields by up to fivefold from the 1-kg DW/m^2 base value used here. However, only slight changes in the carbon content per unit biomass would be expected. Innovations could also help increase the number of crops per year for a given farm installation by dramatically increasing growth rates via cultivar selection and careful breeding of macrophyte strains and by engineering solutions that reduce biomass losses before sequestration.

There are many requirements to farm vast amounts of macrophyte carbon biomass. Optimal growth of macrophyte biomass requires adequate nutrient concentrations and light levels (e.g., Jackson, 1977, 1987; Gerard, 1982; Zimmerman and Kremer, 1986). Achieving both will be difficult because throughout most of the world's oceans, vertical regions where there is enough solar radiation to drive photosynthetic processes (the euphotic zone) are often depleted in macronutrients, while vertical strata where adequate nutrients are available are often too deep to support growth, which is why macrophyte populations naturally inhabit nearshore habitats where both nutrients

[2] Von Kietz workshop presentation, https://www.nationalacademies.org/event/02-02-2021/a-research-strategy-for-ocean-carbon-dioxide-removal-and-sequestration-workshop-series-part-3.

and light are often adequate to support their growth. The nutrient requirements for concentrated seaweed farming will be particularly intense. A recent paper estimates that current seaweed production in China will by 2026 utilize all of the excess anthropogenic phosphorus discharged into Chinese waters (Xiao et al., 2017). To alleviate these issues, the MARINER program has made efforts to select cultivars that grow efficiently under low-nutrient conditions or employ artificial upwelling devices to supply nutrients from deeper depths (see details in Chapter 4) or implement novel mechanical means to bring the crop to depth at night (ARPA-e 2021a; Navarrete et al., 2021). Another important siting requirement is to create farming facilities that are robust to storms and protect macrophyte biomass and the farm infrastructure from storm losses. Access to nearby ports would also be essential to help reduce costs for farm operations and maintenance.

There remains the question of how the cultivated biomass should be sequestered to optimize the amount of respired CO_2 that will be retained in the ocean before outgassing to the atmosphere. Macrophyte biomass will be recycled through oceanic food webs and eventually be respired to CO_2 on decadal timescales. Thus, efforts must be made to deposit this biomass at depth where it will be assimilated by deep-sea ecosystems and respired back to CO_2, but not in immediate contact with the atmosphere. Very little will likely end up in seafloor sediments; however, recent environmental DNA analysis has shown evidence of macrophyte DNA in surface sediment (Geraldi et al., 2019). Very little is known about the timescales of degradation of macroalgal carbon or DNA in seafloor sediments.

The timescales over which CO_2 injected within the ocean interior remains sequestered from the atmosphere has recently been assessed using a model of steady-state global ocean circulation and mixing (Siegel et al., 2021a). This model shows challenges ahead for any purposeful water-column injections of CO_2 aimed at sequestering CO_2 from the atmosphere. First, there will be a wide range of sequestration times linking a discharge location with the sea surface due to the infinite number of pathways connecting them. The resulting probability distribution is highly skewed, with a large fraction of relatively young transit times and a long tail of very long transit times. Second, deeper discharge locations will sequester CO_2 longer than shallower ones, and median sequestration times are typically decades to centuries. Third, large differences in sequestration times occur both within and between the major ocean basins, with the Pacific and Indian basins generally having longer sequestration times than the Atlantic Ocean and Southern Ocean. Last, assessments made over a 100-year time horizon illustrate that most of the injected carbon will be retained for injection depths greater than 1,000 meters, with several geographic exceptions such as the western North Atlantic (Figure 5.3). Retention is nearly ensured by depositing macrophyte biomass on the seafloor at depths greater than 2,000 meters (Figure 5.3). Conveyance apparatuses are likely needed to reduce the fragments of particulate carbon and DOC that could potentially be unintentionally released into the upper ocean.

5.4 VIABILITY AND BARRIERS

Barriers to Implementation

The viability of seaweed cultivation as a CDR strategy has a range of barriers, incentives, and issues related to its implementation that need to be addressed. First, one must consider its potential environmental impacts. The sequestering of large amounts of organic matter at depth will surely have detrimental effects on the ecology of the deep sea, which is likely both the largest (by volume) and least understood biome on Earth (e.g., Martin et al., 2020). This is because the purposeful inputs of macrophyte organic matter at depth will eventually respire back to CO_2, leading to local to regional increases in deoxygenation, acidification, and eutrophication. Further anthropogenic inputs in vast amounts of particulate matter will also influence visibility and contacts among mesopelagic

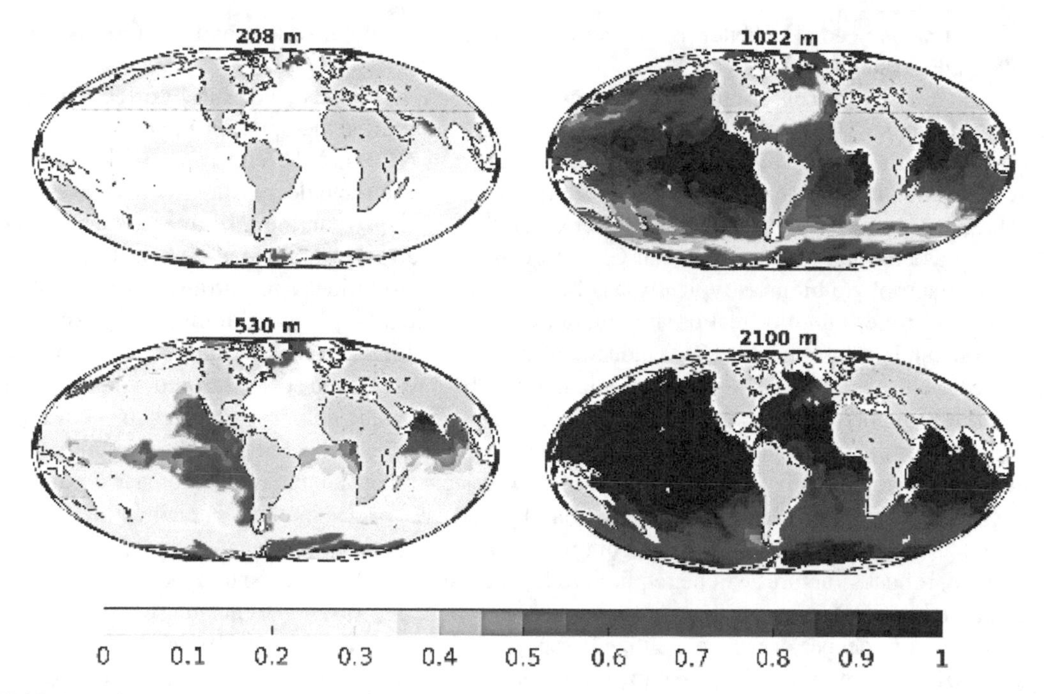

FIGURE 5.3 Maps of the fraction of CO_2 retained after 100 years for injection depths of 208, 530, 1,022, and 2,100 meters in the purposeful sequestration metrics modeling of Siegel et al. (2021b). Figure redrafted from the original journal submission. Licensed under Creative Commons CC BY 4.0.

organisms, similar to what might be expected from the improper disposal of tailings from deep-sea mining operations (e.g., Drazen et al., 2020).

The biological pump exports roughly 10 Pg C/yr from the upper ocean to depth over the entire globe (e.g., Siegel et al., 2014; DeVries and Weber, 2017; Boyd et al., 2019). If this natural export flux decays with depth following the so-called Martin curve (Martin et al., 1987; Buesseler et al., 2007, 2020), the natural flux of organic matter that arrives at 2,000 meters averaged over the entire world ocean will be ~1.2 Pg C/yr. For the above scaling analysis for sequestering 0.1 Gt CO_2/yr, the purposeful input of seaweed will increase the global delivery of organic matter by nearly 5 percent. An input of 1 Gt CO_2/yr (0.27 Pg C/yr), as suggested in the X-Prize competition,[3] will have a substantive impact on the natural inputs of organic matter delivered to a horizon of 2,000 meters (increasing the global flux of organic matter at that depth by ~25 percent). These alterations will surely alter mesopelagic and deep-sea food webs by adding foreign biomass with potentially different food qualities. The decomposition of the added biomass will lower oxygen levels and increase acidity and nutrient levels, leading to increased deoxygenation, acidification, and eutrophication. Further, the nature of the purposeful inputs will likely be highly heterogeneous in space and likely intermittent in time because it seems difficult to ensure that the inputs of organic matter will be or could be dispersed uniformly at depth, especially given that the infrastructure required to cultivate macrophyte biomass carbon to CDR-relevant scales will be localized to a few port cities. This concentration of organic matter inputs will greatly increase the local-scale environmental impacts of the purposeful additions of organic matter. Further, there remains a great deal of uncertainty in the fate of the dissolved organic matter produced as seaweed grows. Coordinated research in all

[3] See https://auto.xprize.org/prizes/elonmusk.

of these areas is needed to better assess the impacts of seaweed cultivation and sequestration as a viable CDR strategy.

Further, the growing of large-scale crops of macrophyte biomass will surely affect euphotic zone ecosystems. The farm will divert ambient nutrients that drive the upper ocean ecosystem into cropped biomass, likely reducing rates of phytoplankton NPP and thereby in turn reducing the fluxes of C export from the surface ocean into the ocean interior and decreasing the flow of energy into higher trophic levels that support fisheries and other valued marine resources. These effects may increase the need for more ocean CDR to offset these losses. This may be offset somewhat because macrophyte biomass typically has higher carbon-to-nutrient concentrations (either phosphorus or nitrogen) than typical organic matter concentrations in plankton-dominated ecosystems (e.g., Rao and Indusekhar, 1987). Understanding the ecological and environmental impacts of large-scale cultivation of seaweeds is a critical area where research is sorely needed. Another issue to resolve is the fact that seaweed cultivation will likely introduce nonnative species to ecosystems where the farming occurs because nearshore macrophyte species will need to be farmed in offshore biomes. It is also likely that cultivars will need to be selected that can maximize biomass production in low-nutrient environments. The introduction of nonnative species may have detrimental ecological impacts and legal implications that complicate the permitting processes.

Huge, robust structures will be required to farm enough biomass carbon to make seaweed cultivation a viable CDR strategy. Much work is needed to engineer robust systems that will likely need to be situated in deeper ocean waters than present-day seaweed farms, typically in shallow waters. Again, ARPA-e's MARINER program is conducting important work developing prototypes for these systems (ARPA-e, 2021a). The size of these structures suggests that there should be concerns about the risks of entanglement with whales and other air-breathing vertebrates, displacement of fishing and other ecosystem services, and the hazards to navigation created by the structures needed for seaweed cultivation. Further, efficient ship systems that minimize their carbon emissions need to be developed to manage cultivation and sequestration systems.[4]

Monitoring and Verification

Every CDR strategy will need a monitoring and verification program to prove its veracity so that the end-to-end costs, benefits, and environmental impacts can be quantified. For seaweed cultivation, these assessments will need to be conducted on both local (e.g., farm, injection location, etc.) as well as on global scales. Local-scale efficacy could be assessed via coupled observations and modeling similar to that of a standard oceanographic process study. The goal will be to conduct fieldwork in contrasting sites and farm types so that detailed process numerical models can be developed, tested, and applied to other sites. Developing systems to achieve this goal should follow the planning for major oceanographic process studies focused on the biological carbon pump (e.g., Siegel et al., 2016). This will require substantive resources. Using the EXPORTS program as a basis, a month-long assessment of carbon export pathways costed out at roughly $115M (including ship time, scientists, staffing, sample and data analyses, logistics, etc.).

Monitoring and verification on global scales will be harder. Numerical modeling, informed by observations, will be one critical component. For example, aerial optical observations, both satellite and drone based, would be very useful for mapping farm biomass and productivity for seaweed taxa that form canopies (Bell et al., 2020; Cavanaugh et al., 2021) and may be useful for assessing the displaced natural productivity of the farm. Future satellite missions, such as NASA's upcoming

[4] For an example of a recent prototype, see https://arpa-e.energy.gov/technologies/projects/autonomous-tow-vessels.

Plankton, Aerosol, Cloud and ocean Ecosystems[5] and Surface Biology and Geology[6] missions, will provide hyperspectral imagery on a variety of spatiotemporal scales that will be useful for mapping seaweed canopy biomass and productivity (Bell et al., 2015a,b, 2020). Existing global biogeochemical monitoring systems, such as the emerging Biogeochemical (BGC) Argo array,[7] could also be useful for this task by assessing levels of dissolved oxygen utilization over time, especially on local spatial scales near where the biomass has been conveyed. However, relying on the BGC Argo array to assess impacts on global scales may be problematic. For example, the global sum of present-day global dissolved oxygen utilization levels corresponds to roughly 1,500 Pg C with an uncertainty of about 200 Pg C (Carter et al., 2021). Thus, detecting even a 1-Gt CO_2 (0.27-Pg C) change in global inventories due to purposeful injections of cultivated seaweed biomass will require severe constraints on the required accuracy and precision of oxygen (O_2) measurements and the timescales required to see these changes. For example, an input of 1 Gt CO_2/yr would represent less than a 1:2,000 reduction in global mean oxygen concentrations, potentially requiring detailed analyses and many years detecting these small global-scale changes. Further separating already-committed anthropogenic changes expected (e.g., Moore et al., 2018) from the purposeful CDR changes may be very difficult to assess. Thus, numerical modeling informed by local-scale process studies may be the best way to assess the impacts of seaweed cultivation and sequestration. Note also that these models probably need to better account for higher trophic levels because these processes are generally left out of most Earth system models (e.g., Bonan and Doney, 2018).

Required Resources

The costs needed to construct, operate, and maintain the farm infrastructure will be substantial. Scaling seaweed cultivation to useful CDR levels will require immense farm structures (7.3 million hectares in extent for sequestering 0.1 Gt CO_2/yr) to be built, maintained, and operated. Large power sources will be required in this process as well to transport and maintain these facilities. To this end, offshore seaweed farms may involve combining renewable energy sources. There are two current seaweed farm pilot projects offshore Belgium and the Netherlands that are using existing offshore renewable energy farms as their foundation. In essence, the farms are colocated between turbines or solar plants. One project is colocated with a seafloor wind farm (the Norther Wind Farm), and the other is colocated with the North Sea Farmers offshore floating solar farms (United). The goal of these pilot projects is to move forward with almost complete automation of the growth and harvesting of seaweed by using some of the renewable power generated in situ. Results of these pilot projects are slated for 2022 (Durakovic, 2020).[8]

Lessons learned from these and other pilot studies will be important to quantify life-cycle cost analyses that will be needed for determining the net benefits of seaweed cultivation to CDR.

Co-benefits

There are a wide variety of co-benefits created by seaweed cultivation CDR. Building, operating, and maintaining farms will provide a huge enhancement to the blue economy. Seaweed cultivation CDR could provide jobs and livelihoods for many. The quality of these jobs matters to people: for example, in a study of attitudes about seaweed cultivation generally, interviewees in Scotland were wary of large-scale internationally owned seaweed farms. They have witnessed a pattern in

[5] See https://pace.gsfc.nasa.gov.

[6] See https://sbg.jpl.nasa.gov.

[7] See https://biogeochemical-argo.org/.

[8] See also https://www.msp-platform.eu/projects/multi-use-platforms-and-co-location-pilots-boosting-cost-effective-and-eco-friendly.

the salmon farming industry where individual operators were bought out by international owners, which was perceived to reduce benefits and jobs. They emphasized the need for social innovation that would provide decent pay for all involved and benefits for the community more broadly, and were more accepting of locally based, cooperative development (Billing et al., 2021).

Seaweed cultivation also has potential ecological co-benefits. On local scales, seaweed farming, particularly in coastal waters, could act to reduce the effects of anthropogenically driven acidification on shellfish aquaculture farms and could help reduce the effects of deoxygenation and eutrophication created by these farms (e.g., Neori et al., 2004; Xiao et al., 2017). Other co-benefits with potential greenhouse gas reductions include the additions of macrophyte biomass to animal feeds that could reduce methane emissions (e.g., Maia et al., 2016)

Governance

The legal framework for ocean CDR is discussed in Chapter 2. Many of the international and domestic laws discussed in that chapter could apply to seaweed cultivation.

With respect to international law, Webb et al. (2021) concluded that seaweed cultivation undertaken for the purpose of CDR is likely to be considered a form of "geoengineering" under the Convention on Biological Diversity (CBD). As discussed in Chapter 2, in 2010, the parties to the CBD adopted a nonbinding decision recommending that countries take a "precautionary approach" and avoid geoengineering activities that could affect biodiversity.[9] The decision provided an exemption for "small scale scientific research studies . . . conducted in a controlled setting" that "are justified by the need to gather specific scientific data and . . . subject to a thorough prior assessment."[10] However, because the decision is not legally binding and merely offers guidance on the conduct of geoengineering activities, countries could conduct or authorize other projects that do not meet the specific requirements (Webb et al., 2021).

Seaweed cultivation projects could also implicate provisions of the United Nations Convention on the Law of the Sea (UNCLOS). Parties to UNCLOS must "take all necessary measures" to prevent and control pollution resulting from, among other things, the introduction of alien species to "a particular part of the marine environment" where they "may cause significant and harmful changes."[11] Prior to conducting or authorizing seaweed cultivation projects involving the growing of nonnative species, parties to UNCLOS would need to conduct a risk assessment, consult with other potentially affected countries, and take other steps to minimize any adverse effects (Webb et al., 2021).

It is unclear whether the Convention on the Prevention of Marine Pollution by Dumping of Wastes and Other Matter (London Convention) and associated Protocol (London Protocol) would apply to seaweed cultivation projects. There is an open question as to whether the sinking of cultivated seaweed could constitute "dumping" for the purposes of the London Convention and Protocol (Webb et al., 2021). In theory, "dumping" could also occur if growth-stimulating materials were added to ocean waters to fertilize seaweed crops, and/or waste products (e.g., nets and lines) from farms were disposed of in the water (Webb et al., 2021).

---

[9] Report of the Conference of the Parties to the Convention on Biological Diversity on the Work of its Tenth Meeting, Decision X/33 on Biodiversity and Climate Change, Oct. 29, 2010 (hereinafter Decision X/33). See also Report of the Conference of the Parties to the Convention on Biological Diversity on the Work of its Eleventh Meeting, Decision XI/20 on Climate-related Geoengineering, Dec. 5, 2012 (hereinafter Decision XI/20); Report of the Conference of the Parties to the Convention on Biological Diversity on the Work of its Thirteenth Meeting, Decision XIII/4, Dec. 10, 2016 (hereinafter Decision XIII/4).

[10] Para. 8(w), Decision X/33. Affirmed in Para. 1, Decision XI/20 & Preamble, Decision XIII/4.

[11] Art. 196, United Nations Convention on the Law of the Sea, 1833 U.N.T.S. 397, Dec. 10, 1982.

Webb et al. (2021) examined the application of domestic law to seaweed cultivation projects. The applicable law will depend on precisely where and how projects are conducted. State, and in some cases local, law would apply to projects undertaken in state waters (i.e., generally within 3 nautical miles of the coast). There has been no comprehensive analysis of all applicable state and local laws. However, Webb et al., (2021) found that at least three states—Alaska, California, and Maine—have laws requiring permits or other approval for seaweed cultivation projects. Webb et al. (2021) noted that some other states have more general aquaculture permitting laws that could apply to seaweed cultivation. However, some only provide for the issuance of permits for shellfish or finfish farming and do not anticipate the permitting of seaweed cultivation. Federal permits (e.g., from the U.S. Army Corps of Engineers) may also be required for some seaweed cultivation projects in state waters (e.g., those involving the placement of structures in the water). State and federal agencies must generally consult with affected Native American tribes prior to issuing permits (Webb et al., 2021).

Federal law would apply to seaweed cultivation projects undertaken in federal waters (i.e., generally 3 to 200 nautical miles from the coast). Webb et al. (2021) reviewed the potentially applicable federal laws. Although there is no federal permitting regime for seaweed cultivation, projects that require use of the seabed (e.g., to anchor structures or lines) may require a seabed lease or other authorization (e.g., under the Outer Continental Shelf Lands Act). Permits would also be required under the Marine Protection, Research, and Sanctuaries Act (MPRSA) for projects involving the dumping of materials, including nets and lines, into ocean waters. It is uncertain whether the sinking of cultivated seaweed would constitute dumping and thus require a permit. Projects affecting other marine species or ecosystems may be subject to additional requirements, for example, under the Endangered Species Act (ESA) and Marine Mammal Protection Act (MMPA).

Domestic law would have limited application to seaweed cultivation projects outside U.S. waters (i.e., more than 200 nautical miles from shore). Where U.S. citizens or vessels are involved, the ESA and MMPA could apply. Dumping from U.S. vessels, or other vessels that were loaded in the United States, may also be subject to domestic permitting requirements under the MPRSA.

Community engagement will be required for effective governance, especially when activities affect coastal populations and livelihoods (see Chapter 2). Community engagement will also be critical in capturing potential co-benefits.

5.5 SUMMARY OF CDR POTENTIAL

The criteria for assessing the potential for seaweed cultivation as a feasible approach to ocean CDR, described in Sections 5.2–5.4, is summarized in Table 5.1.

5.6 RESEARCH AGENDA

A research agenda for assessing whether seaweed cultivation and sequestration is CDR worthy will require an assessment of the components compiled in Figure 5.2. These are

1. Improve existing technologies that enable the cost-effective, large-scale farming and harvesting of seaweed biomass;
2. Create and assess the means to convey large amounts of harvested biomass to depth in the ocean interior or to the seafloor without large losses of carbon;
3. Understand the long-term fates of seaweed carbon (i.e., both biomass at depth and DOC released during growth) and use this understanding to develop numerical models of the fates of seaweed carbon in the environment;

TABLE 5.1 CDR Potential of Seaweed Cultivation

Knowledge base What is known about the system (low, mostly theoretical, few in situ experiments; medium, lab and some fieldwork, few carbon dioxide removal (CDR) publications; high, multiple in situ studies, growing body of literature)	**Medium–High** Science of macrophyte biology and ecology is mature; many mariculture facilities are in place globally. Less is known about the fate of macrophyte organic carbon and methods for transport to deep ocean or sediments.
Efficacy What is the confidence level that this approach will remove atmospheric CO_2 and lead to net increase in ocean carbon storage (low, medium, high)	**Medium Confidence** The growth and sequestration of seaweed crops should lead to net CDR. Uncertainties about how much existing net primary production (NPP) and C export downstream would be reduced due to large-scale farming.
Durability Will it remove CO_2 durably away from surface ocean and atmosphere (low, <10 years; medium, >10 years and <100 years; high, >100 years), and what is the confidence (low, medium, high)	**Medium–High** >10–100 years Dependent on whether the sequestered biomass is conveyed to appropriate sites (e.g., deep ocean with slow return time of waters to surface ocean).
Scalability What is the potential scalability at some future date with global-scale implementation (low, <0.1 Gt CO_2/yr; medium, >0.1 Gt CO_2/yr and <1.0 Gt CO_2/yr; high, >1.0 Gt CO_2/yr), and what is the confidence level (low, medium, high)	**Medium** Potential C removal >0.1 Gt CO_2/yr and <1.0 Gt CO_2/yr (medium confidence) Farms need to be many million hectares, which creates many logistic and cost issues. Uncertainties about nutrient availability and durability of sequestration, seasonality will limit sites, etc.
Environmental risk Intended and unintended undesirable consequences at scale (unknown, low, medium, high), and what is the confidence level (low, medium, high)	**Medium–High** (low confidence) Environmental impacts are potentially detrimental especially on local scales where seaweeds are farmed (i.e., nutrient removal due to farming will reduce NPP, C export, and trophic transfers) and in the deep ocean where the biomass is sequestered (leading to increases in acidification, hypoxia, eutrophication, and organic carbon inputs). The scale and nature of these impacts are highly uncertain.
Social considerations Encompass use conflicts, governance-readiness, opportunities for livelihoods, etc.	Possibility for jobs and livelihoods in seaweed cultivation; potential conflicts with other marine uses. Downstream effects from displaced nutrients will need to be considered.
Co-benefits How significant are the co-benefits as compared to the main goal of CDR and how confident is that assessment	**Medium–High** (medium confidence) Placing cultivation facilities near fish or shellfish aquaculture facilities could help alleviate environmental damages from these activities. Bio-fuels also possible.
Cost of scale-up Estimated costs in dollars per metric ton CO_2 for future deployment at scale; does not include all of monitoring and verification costs needed for smaller deployments during R&D phases Low, <\$50/t CO_2; medium, ~\$100/t CO_2; high, >>\$150/t CO_2 and confidence in estimate (low, medium, high)	**Medium** ~\$100/t CO_2 (medium confidence) Costs should be less than \$100/t CO_2. No direct energy used to fix CO_2.

TABLE 5.1 Continued

Cost and challenges of carbon accounting Relative cost and scientific challenge associated with transparent and quantifiable carbon tracking (low, medium, high)	**Low–Medium** The amount of harvested and sequestered carbon will be known. However, an accounting of the carbon cycle impacts of the displaced nutrients will be required (additionality).
Cost of environmental monitoring Need to track impacts beyond carbon cycle on marine ecosystems (low, medium, high)	**Medium** (medium–high confidence) All CDR will require monitoring for intended and unintended consequences both locally and downstream of CDR site, and these monitoring costs may be a substantial fraction of overall costs during R&D and demonstration-scale field projects.
Additional resources needed Relative low, medium, high to primary costs of scale-up	**Medium** Farms will require large amounts of ocean (many million hectares) to achieve CDR at scale.

4. Build and test a demonstration-scale system for seaweed cultivation and sequestration CDR that in principle can be scaled up to 0.1-Gt CO_2/yr levels and deploy these systems in diverse oceanographic settings;
5. Validate and monitor the CDR performance of the demonstration-scale seaweed farming and sequestration systems on local scales;
6. Evaluate the environmental impacts of large-scale seaweed farming and sequestration systems both in the upper ocean where the farming occurs and at depth where the seaweed is transported for sequestration;
7. Understand better the legal framework required for seaweed-based CDR and the socioeconomic factors that would affect coastal communities and Indigenous groups; and
8. Document "best practices" and perform spatial planning exercises to assess the best places for conducting seaweed cultivation CDR.

The first research component should build upon the successes achieved and challenges identified by ARPA-e's MARINER program. ARPA-e has made a substantial investment in MARINER (>$30M over 3 years), and work in this area needs to continue so that development-scale facilities can be developed. Given the investments made to date, this work can be done quickly, with development-scale ($\geq$1 km^2; $\geq$100 hectare) farms in place in the next decade. At CDR scale, many million hectares of farm facilities need to be established. Hence, seaweed cultivation systems will very likely need to be engineered to operate efficiently at ocean depths of hundreds to thousands of meters. This engineering expertise exists within the oil and gas industries. Research should also include a focus on increasing the farm's C yield and reducing crop durations via the selection of appropriate phenotypes and the deployment of specific apparatuses (e.g., AU, vertically profiling farms, etc.) that will enhance nutrient concentrations in the farm. Improvements in the abilities to monitor and model crop growth as a function of environmental conditions are also needed. Further, efficient ocean transportation systems that minimize their C emissions and appropriate harvesting techniques need to be developed so that the CDR gained is not lost due to transport and to protect those organisms that forage or inhabit the farms. Last, permitting has been a challenge in developing and testing pilot-scale farms in the MARINER program and will likely be a challenge for further research. Addressing permitting challenges is essential to enable researchers to develop and test new farming and harvesting technologies.

Harvested biomass needs to be conveyed to depth without losing C biomass in the surface ocean and deposited in durable oceanic reservoirs. There has been little research conducted in this area. Demonstration-scale engineering studies need to be conducted, illustrating that the harvested biomass can be conveyed mechanically to a durable reservoir in the ocean with minimal losses, either in the water column or on the seafloor. Freely sinking biomass has also been suggested, and the decomposition rates of freely sinking biomass need to be fully quantified to assess whether this conveyance strategy would be successful. Research budgets in this area should be comparatively modest.

Assessments of the long-term fates of seaweed biomass and its by-products (cf., both biomass at depth and released DOC during growth) are required to develop predictive models. This could be accomplished via a set of biomass and dissolved organic matter decomposition experiments conducted in situ, in mesocosms, or in laboratory settings. The results of this need to be synthesized into a numerical model of seaweed fates. Challenges are to conduct these experiments on timescales that are relevant to seaweed cultivation CDR—years to decades. This work also needs to lead to predictive numerical models where farm, harvest, and sequestration scenarios can be explored.

Demonstration-scale systems for seaweed cultivation and sequestration aimed to be scaled to 0.1-Gt CO_2/yr levels need to be developed and tested. This work will build from the first two components in this research plan and logically are supported after clear paths emerge from these two elements. This will answer the question whether seaweed cultivation and sequestration is a viable CDR strategy. The demonstration-scale system will also be useful for validating its CDR performance as well as assessing its environmental impacts (see below). Many of the same environmental and social impacts of the research for farming apply to the implementation of a demonstration-scale system. These goals would be most readily achieved if these systems were deployed in diverse ocean settings.

The CDR performance of the demonstration-scale seaweed farming and sequestration system will need to be monitored on local spatiotemporal scales. This will require both process oceanographic field studies (similar to the recent NASA EXPORTS campaign; Siegel et al., 2021b) and selected sensor arrays embedded in the farm and conveyance infrastructure. These data will address the system performance of a demonstration-scale system. Regional-scale numerical modeling of ocean circulation and mixing coupled with ecological and biogeochemical modules will also be a big part of the validation and monitoring of these systems, which in turn will require data from the process studies and system models on macrophyte fates and influences of the farms and harvesting on the environment (discussed previously).

One would also need to evaluate the environmental impacts of large-scale seaweed farming and sequestration both in the upper ocean where the farming occurs and at depth in the water column or the seafloor where the seaweed is conveyed for sequestration. This work would need to include the downstream impacts of displaced nutrients on ecosystems and ecosystem services. Fieldwork would also be needed to achieve the previous task (validating and monitoring CDR performance). Hence, portions of these tasks could be done simultaneously. Additional measurements would be needed to understand the effects of seaweed sequestration on the macrofaunal communities in the water column and on the seafloor.

Research on the social/economic factors and governance for seaweed cultivation CDR is also required. The legal framework for seaweed-based CDR is murky, and many questions remain unanswered. Further, many socioeconomic factors affect coastal communities and Indigenous groups that need to be considered. Additionally, an understanding of public perceptions and whether there is a social license to conduct this work is required.

Last, the above components need to be synthesized into a "best practices" manual, and spatial planning exercises need to be performed to assess the best places for conducting seaweed cultivation

CDR. This synthesis of the emerging state of knowledge on seaweed cultivation CDR needs to be completed before one should consider implementing these technologies at scale.

5.6 SUMMARY

In summary, seaweed cultivation and sequestration could be a compelling ocean CDR strategy (see Table 5.2). There is a good understanding of the underlying biology, ecology, and biogeochemistry of macrophytes and their cultivation, although many advances are needed to grow seaweed biomass to meet CDR requirements. In principle, it should work (i.e., reduce atmospheric CO_2), but there is a large degree of uncertainty about how much productivity and C export it would displace from planktonic ecosystems and the durability of the sequestered carbon if not properly conveyed to an appropriate site. Scaling to CDR-worthy levels (≥0.1 Gt CO_2/yr) will be difficult due to the large amount of farming area required. However, much has been learned already in the MARINER program, and there should be recognition of the many marine engineering accomplishments made by the global oil and gas industries to date. The costs should be less than $100/metric ton CO_2; assuming that the MARINER's cost target for growing macrophyte biomass is met ($75/metric ton CO_2), the other costs (e.g., conveyance, monitoring, etc.) should be smaller. Research needs to be continued to help ensure that this cost target is achieved. The energy expenditures should be small relative to some other CDR strategies because solar energy can fix CO_2 into organic matter. On the other hand, several potentially detrimental environmental factors exist where farming occurs and where the biomass is sequestered. The scale of these impacts is highly uncertain at this time. There are both positive and negative social impacts from CDR via seaweed cultivation and sequestration. If conducted at scale, it will enhance the blue economy, which will benefit both coastal communities and many marine industries. There may also be several co-benefits from placing farms adjacent to other uses (e.g., fish farming, etc.), which may help mitigate some environmental damages conducted by aquaculture facilities. On the negative side, the vast farms represent hazards to navigation, and they may displace fishing and other uses via the placing of farms or the reduction in planktonic productivity and trophic exchanges due to the large-scale cultivation of seaweed biomass.

TABLE 5.2 Research and Development Needs: Seaweed Cultivation and Sequestration

No.	Recommended Research	Question Answered	Environmental Impact of Research	Social Impacts of Research	Estimated Research Budget ($M/yr)	Time Frame (years)
5.1	**Technologies for efficient large-scale farming and harvesting of seaweed biomass**	**Can we build efficient demonstration-scale farms to grow and harvest seaweed biomass that have the potential to be CDR worthy?**	**Moderate on local scales. Demonstration-scale farms will affect local ecosystems, reducing ambient nutrients, net primary productivity, C export, etc. Farming of nonnative cultivars would be an environmental risk. Farms could also reduce the effects of acidification in the upper ocean.**	**Concerns with permitting due to macrofaunal entanglements, introduction of nonnative cultivars, hazards to navigation, displacing fishing effort, etc.**	**15 based upon MARINER funding as a starting point**	**10**
5.2	**Engineering studies focused on conveying harvested biomass to a durable oceanic reservoir with minimal losses of carbon**	**How do we convey large amounts of seaweed biomass to depth or seafloor with minimal losses?**	**Minimal for engineering testing**	**Minimal for engineering testing**	**2**	**10**
5.3	**Assessment of long-term fates of seaweed biomass and by-products**	**Can we predict the long-term fates of seaweed carbon?**	**Minimal**	**Minimal**	**5**	**5**
5.4	Implement and deploy a demonstration-scale seaweed cultivation and sequestration system	Can a system be built that will scale to CDR-worthy scales?	Moderate on local scales, affecting upper ocean and conveyance depths	Moderate—Concerns with permitting due to macrofaunal entanglements, introduction of nonnative cultivars, hazards to navigation, displacing fishing effort, etc.	10	10 starting 5 years from now
5.5	Validate and monitor the CDR performance of a demonstration-scale seaweed cultivation and sequestration system	Is seaweed cultivation and sequestration a viable CDR strategy and can we monitor its performance?	Minimal	Minimal	5	10 starting 5 years from now (done in concert with Task 5.6)

5.6	**Evaluate the environmental impacts of large-scale seaweed farming and sequestration**	**What are the environmental impacts of seaweed sequestration?**	**Minimal**	**Minimal**	**4**	**10**
5.7	Research on the socioeconomic factors and governance	Would seaweed cultivation CDR affect communities and stakeholders?	N/A	N/A	1	2
5.8	Document "best practices" and perform spatial planning exercises	How should seaweed cultivation be conducted and where?	N/A	N/A	1	2

NOTE: Bold type identifies priorities for taking the next step to advance understanding of seaweed cultivation as an ocean CDR approach.

6

Recovery of Marine Ecosystems

6.1 OVERVIEW

In recent years, there has been increased attention to nature- or ecosystem-based solutions in confronting climate change. In 2005, enhanced forest conservation (reducing emissions from deforestation and forest degradation) was promoted as a global initiative for greenhouse gas removal in developing countries (Angelsen et al., 2009). The United Nations (UN) started a Trillion Tree Campaign in 2007 to rehabilitate degraded land. In marine systems, much of the focus has been on coastal ecosystems, including mangroves, salt marshes, and seagrasses, perhaps because they can be managed in ways similar to those for terrestrial systems (NASEM, 2019). These coastal blue carbon systems largely comprise vascular plants, and several research efforts are already under way to study these systems and their potential for long-term C storage.

To date, much of the discussion around ecosystem-based solutions has focused on terrestrial and urban context, with risk reduction, such as the buffering ability of salt marshes and mangroves, on a local level (Solan et al., 2020). Benthic, pelagic, and offshore ocean systems are an emerging area of interest, yet there is a considerable amount of uncertainty about the potential for these marine systems to store and sequester carbon. Unlike terrestrial systems, a large fraction of the biomass in marine systems generally comprises animals, with consumers, including microbes and protists, far outweighing producers (Figure 6.1). Many of these organisms, especially the largest predators, are at risk from human activities such as fishing, shipping, pollution, introduction of invasive species, and habitat destruction. Although the fraction of biomass carbon in the oceans is relatively small compared to that of terrestrial systems, the role of animals in biogeochemical cycles and ecosystem structure has garnered attention, especially in light of the widespread impact of human activities on the state of the oceans and the co-benefits of ecosystem recovery approaches. At the same time, studies of kelp and other macroalgae such as *Sargassum*, benthic algae, and phytoplankton, have increased our understanding of their potential role in the carbon cycle.

The interconnected nature of climate change and biodiversity loss means that resolving either issue requires consideration of the other (Pörtner et al., 2021). In this chapter, we use the term "ecosystem-based" solutions to avoid concerns about the framing of the terms "natural," "nature-based

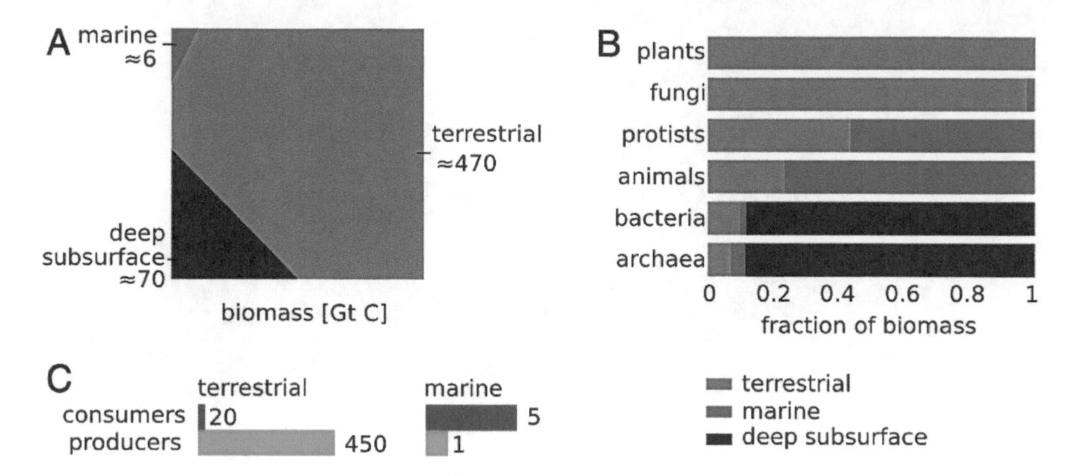

FIGURE 6.1 Although marine biomass is smaller than terrestrial biomass (A), animals have a higher fraction of biomass in the ocean than on land (B). Plants are mainly terrestrial, animals are mainly marine, and bacteria and archaea are dominant in the deep subsurface. Unlike terrestrial systems, there is greater consumer than producer biomass in marine systems (though estimates for marine macroalgae can span several orders of magnitude); bars represent gigatons of carbon (C). SOURCE: Bar-On et al., 2018. Licensed by Creative Commons CC BY 4.0.

solutions," or NBS (Bellamy and Osaka, 2019). As in terrestrial systems, extensive monocultures of macroalgae, for example, could come at the expense of marine biodiversity. Yet there have been some important policy guidelines that have emerged from the NBS literature:

1. Ecosystem-based solutions are not a replacement for the rapid phase-out of fossil fuels and must not postpone critical action to decarbonize our economies.
2. They encompass the protection, restoration, and management of a wide array of natural and semi-natural ecosystems on land and in the sea.
3. They are created, employed, managed, and monitored in collaboration with Indigenous peoples and local communities, respecting local rights and generating local benefits.
4. Ecosystem-based solutions potentially support or enhance biodiversity from the level of the gene up to the ecosystem.[1]

Here, we focus on species and ecosystems in their native ranges, many of which have been depleted or altered. We do not discuss blue carbon efforts, which are often framed as coastal systems dominated by vascular plants and have been examined in earlier reports (NASEM, 2019). Efforts to enhance ecosystem functions, such as artificial upwellings and iron fertilization, are discussed elsewhere in this report and are not considered ecosystem-based in this context, though, of course, they both involve ecological responses to human interventions.

The protection and restoration of marine ecosystems and the recovery of fishes, whales, and other animals have the potential to aid in carbon dioxide (CO_2) removal (CDR) and sequestration in the oceans, though considerable uncertainties remain, including the role of marine animals in nutrient cycles, the global areal extent of macroalgal and other ecosystems, and the fraction of carbon that is stored or sequestered during many of these ecological processes. See Section 1.4 for discussion of the ocean carbon cycle and the biological pump. In this chapter, we focus on manage-

[1] See https://nbsguidelines.info/.

ment actions that could result in ocean CDR, using a framework that compares natural, preindustrial processes to present-day fluxes, the difference being anthropogenic perturbation (see, e.g., Regnier et al., 2013). Rather than focus on present-day fluxes, the emphasis is on the restoration of natural processes and the impact of these activities on the carbon cycle.

Despite the uncertainty of these estimates, ecosystem-based solutions focused on conservation and recovery are likely to be an attractive part of CDR approaches, in part because they offer low-regret solutions, with many perceived co-benefits and the potential for global governability (Gattuso et al., 2021). The United Nations, for example, has declared 2021 to 2030 as the UN Decade on Ecosystem Restoration.[2] Critical to this approach will be taking a whole-ecosystem perspective, going beyond individual populations or species and considering the changes and responses on various timescales that will be necessary to understand the role of CDR in restoration and conservation.

According to criteria described in Chapter 1, the committee's assessment of the potential for ecosystem recovery, as a CDR approach, is discussed in Sections 6.2–6.5 and summarized in Section 6.6. The research needed to fill gaps in understanding of ecosystem recovery, as an approach to durably removing atmospheric CO_2, is discussed and summarized in Section 6.7.

6.2 KNOWLEDGE BASE

The ocean is a sink for about 25 percent to 30 percent of the atmospheric CO_2 emitted by human activities, approximately 2.5 to 2.6 petagrams (Pg) C/yr (Gruber et al., 2019; Watson et al., 2020). For this carbon to be sequestered at the scale of multiple decades, it must be transferred through food webs, enter sediments, or sink below the surface layer to the deep ocean, generally more than 1,000 meters (Passow and Carlson, 2012; Martin et al., 2021).

The biological carbon pump, which results from photosynthetically produced organic matter being transported from the surface layer to depth, is responsible for a large portion of the observed vertical gradient of natural carbon in the ocean (see Figure 1.8). The complex pathways of the biological carbon pump are driven by numerous biophysical and chemical interactions, including phytoplankton productivity, zooplankton grazing, marine vertebrate interactions, oceanic mixing and turbulence, advection, and the sinking of particles and aggregates (Ducklow et al., 2001; Nayak and Twardowski, 2020; Martin et al., 2021). Where possible, we have included estimates for C sequestration below the permanent pycnocline, or deep ocean, though in some cases estimates are only available to depths of 100 meters or so.

The preindustrial ocean carbon cycle is thought to have been in approximate steady state, with the biological uptake and downward export of surface CO_2 largely balanced by the return of respired CO_2 from the deep sea by the physical circulation. Despite the large role of marine systems in the carbon cycle, our understanding of changes or perturbation in ocean CO_2 storage from past degradation is not well constrained, and our knowledge of the effects on ocean C storage by restoring and protecting marine organisms, ecological functions, and ecosystems is still emerging. For too long, the disciplines of population biology (and, as a subset, fisheries management), ecosystem ecology, and biological oceanography have proceeded on separate paths. A first step toward integrating these approaches would be to examine the ecological and biogeochemical baselines of the preindustrial ocean, including animals, macroalgae, phytoplankton, microbes, and their functions, while at the same time exploring how these baselines have shifted in the present ocean. Uniting these disciplines is even more complex in a changing ocean, with shifting circulation and residence times and the associated problems of using traditional metrics to infer impacts of marine biology on C storage (Koeve et al., 2020). One of the benefits of examining the CDR potential of

[2] See https://www.decadeonrestoration.org/.

oceans is that it could spark greater collaboration between these fields and break down barriers to interdisciplinary studies.

The ocean's ability to retain and remove carbon from the atmosphere is well established, and there is an extensive international effort to document the physicochemical uptake of anthropogenic carbon into the global ocean (e.g., Talley et al., 2016a; Gruber et al., 2019). In the *Second State of the Carbon Cycle Report*, Hayes et al. (2018) estimated that coastal waters of North America take up 0.16 Pg C/yr (Figure 6.2; Fennel et al.. 2018). In Table 6.1, we list a few of the studies that we are aware of that have attempted to quantify the carbon stored in marine biomass, in deadfall carbon, or through ecological functions such as trophic cascades and nutrient subsidies. Many of these cases refer to present-day biological fluxes and are not directly comparable to estimates of net ocean uptake of anthropogenic carbon.

For the purposes of this report, we focus on the role of restoring marine organisms and ecosystems as a method of removing and sequestering anthropogenic CO_2 in the oceans. For many of these systems, there have been few efforts to quantify the C impact of anthropogenic change over the past decades or centuries, though this is changing for fisheries and whales. Mariani et al. (2020), for example, note that fisheries have reduced the C sequestration potential of large fish, preventing the sequestration of about 22 Mt C through blue carbon extraction. This estimate does not include nutrient transfer and other potential indirect effects, which deserve further study. Restoration, of course, will not result in any net gains in C storage and sequestration if existing species, habitats, and ecological functions are not maintained. Conservation, in the form of protecting and preserving marine ecosystems, will also need to occur to make any gains in CDR.

The collective impact of anthropogenic changes, including rising CO_2 levels, changes in nutrient inputs, and ecosystem alteration, on C processing and exchanges along ocean margins is complex and difficult to quantify (Regnier et al., 2013; Fennel et al., 2018). In this report, we largely focus on the potential role of conservation policy and recovery on the carbon cycle and biological pump, but clearly changes that occur because of climate change and other human impacts can have large effects on the flux of carbon in the oceans. Steffen et al. (2015) reported that the risk of transgressing planetary boundaries, such as biosphere integrity, could destabilize the current state of Earth systems. There are also potential trade-offs between fisheries and food security (e.g., Sustainable Development Goal 14) and efforts to promote additional CDR by restoring ocean ecosystems to something that might resemble a pristine state.

Although ecological processes have been well studied in terrestrial and coastal systems, the role of large marine animals and other organisms in the carbon cycle is still relatively new. One compelling question is whether animals and macroalgae have functional impacts that are disproportionate to their biomass, much as mat-forming mosses and lichens play important regulatory roles in the water cycle and inhibit microbial decomposition in some land ecosystems (Smith et al., 2015). Covering roughly 3 percent of Earth's land, moss-dominated peatlands store nearly 33 percent of all global terrestrial carbon, about 540 Gt C (Turetsky, 2003; Yu et al., 2011).

Although marine ecosystems have been proposed as a climate solution, there is a fair amount of uncertainty surrounding our understanding of the ocean's ability to store and sequester additional organic carbon, both in scale and in permanence. Given that ocean recovery is likely to be among the most publicly acceptable CDR approaches in the ocean, an approach with co-benefits and low regrets (Gattuso et al., 2021), it is likely to continue as a fruitful area of study. Yet how does it scale relative to other CDR schemes? Below we discuss a few of the mechanisms and systems that have been proposed as ecosystem-based solutions or areas that could merit further study.

TABLE 6.1 Selected Studies of Carbon Stored or Sequestered in Existing or Restored Ecosystems

Mechanism	System	Region	C Stored or Sequestered	Reference
Macroalgae flux to detritus	Macroalgal–sediment systems	UK	0.70 Tg/yr	Queirós et al., 2019
Coralline algae growth and deposits	C storage by coralline algae and the beds they create	Global	1.6 × Gt C/yr	van der Heijden and Kamenos, 2015
Potential C sequestered by global macroalgae	Export of C to the deep sea and burial in coastal sediments	Global	173 Tg C/yr	Krause-Jensen and Duarte, 2016
Global ocean C export	Biological pump, export of biogenic particles from surface waters to stratified interior	Global	~6 Pg C/yr	Siegel et al., 2014
Diel vertical migration of fish and other metazoans	Global C fluxes and sequestration by fish and metazoans due to respiration, fecal pellets, and deadfalls	Global	5.2 Pg C/yr	Pinti et al., 2021
Gelatinous zooplankton C flows	Sinking of jelly-falls and fecal matter below 100 meters	Global	1.6–5.2 Pg C/yr	Luo et al., 2020
Fish-based contributions to ocean C flux	Passive and active flux from fish feces and migration	Global	1.5 ± 1.2 Pg C/yr	Saba et al., 2021
Recovery of marine vertebrates and increase of wild biomass	Role of vertebrates in the ocean C cycle, including living biomass, deep-sea carcasses, and nutrient cycling	Global	No total provided	Martin et al., 2021
Baleen whale biomass and carcasses	Storage of C in living whales and sequestration of C in deep-sea whale carcasses	Global	0.0089 Gt C/yr, with potential for significant increase in restored populations	Pershing et al., 2010
Sea otter and kelp forest trophic cascade	Increase in C storage by kelp forests via sea otter suppression of herbivorous sea urchins	Northwest North America	4.4–8.7 Tg C total stored	Wilmers et al., 2012

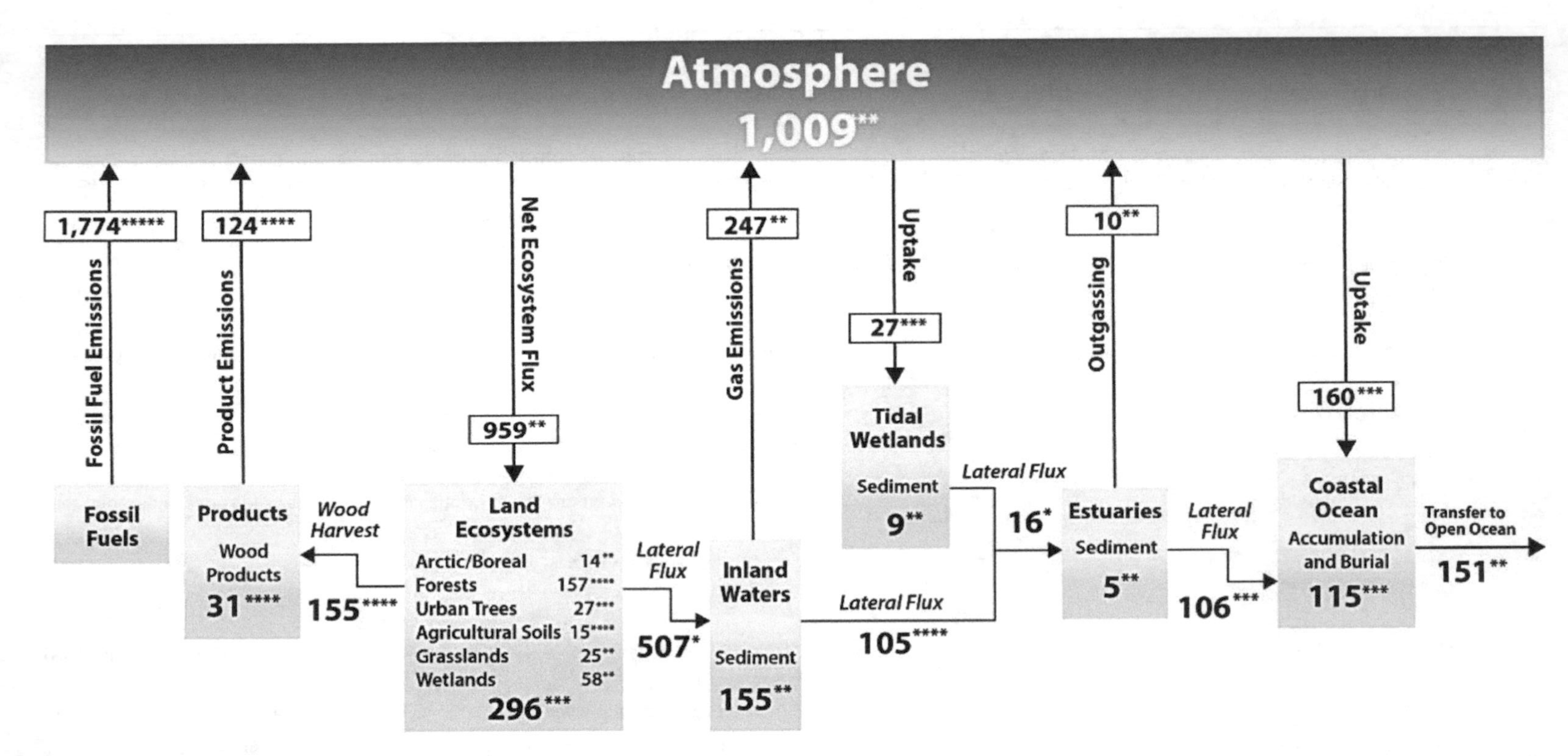

FIGURE 6.2 Major components of the North American carbon cycle. North American coastal oceans, excluding tidal wetlands and estuaries, have a net uptake of 160 Tg C/yr. SOURCE: Hayes et al., 2018.

Macroalgae

Macroalgae form extensive and highly productive benthic marine habitats. Yet macroalgae afforestation has largely been overlooked as a natural source of sequestration in blue carbon schemes, in part because of challenges in documenting the amount of carbon that is sequestered during particulate organic carbon (POC) export (see Figure 5.1). The flowering plants that make up mangrove, seagrass, and salt marsh ecosystems have rhizomes and roots that can retain and store carbon in the soil for centuries. In contrast, macroalgae tend to be shorter lived, with all of their biomass growing above the seafloor: kelp typically have holdfasts attached to rocky surfaces, and planktonic seaweeds such as *Sargassum* typically float near the surface. Despite these differences, the carbon that macroalgal systems supply to sediment stocks in angiosperm habitats has been included in earlier blue carbon assessments, so they have already been recognized as playing a role in CDR (Krause-Jensen et al., 2018). In this section, we discuss the protection and restoration of native seaweeds. Seaweed cultivation and sequestration CDR is discussed in Chapter 5. Given that our understanding of seaweed aquaculture is related to natural macroalgal systems, see the knowledge base section of that chapter for further discussion of macroalgae and CDR.

Among the largest and most widely distributed macroalgae are the brown algae known as kelps and rockweeds (orders Laminariales, Tilopteridales, Desmarestiales, and Fucales), growing in dense populations or forests (Duffy et al., 2019). Although they cover about 28 percent of the world's coastlines, kelp forests are declining faster than coral reefs and tropical forests (Feehan et al., 2021). Approximately 82 percent of kelp productivity becomes detritus (Krumhansl and Scheibling, 2012), and much of the carbon in macroalgae is assumed to return to its inorganic form in the water column through herbivory. Sinking speed of detritus depends on tissue type, size, and whether it has been consumed by herbivores (Wernberg and Filbee-Dexter, 2018). Export depths and sinking rates can vary by several orders of magnitude. Kelp forests in high-energy environments have a high potential for long-distance export, especially in areas with steep grades to depth (Wernberg and Filbee-Dexter, 2018).

In their study of the role of macroalgae in C sequestration, Krause-Jensen and Duarte (2016) estimated that about 173 Tg C/yr (or 11 percent of net primary production) could potentially be sequestered in marine sediments and deep-sea waters. They further suggest that roughly 70 percent of the sequestered carbon is due to the export of recalcitrant dissolved organic carbon (DOC) to the deep sea. The scale of this sequestration is comparable to all other coastal blue carbon habitats combined (Duarte et al., 2013). Note that Krause-Jensen and Duarte (2016) calculated the amount of carbon that could be sequestered if all ocean regions with enough light on the seafloor grew macrophytes. Hence, these estimates do not necessarily reflect the long-term sequestration by present-day macrophyte populations. Future research could refine these calculations for macroalgae afforestation sequestration in different oceans and ecoregions.

Queirós et al. (2019) estimated that the magnitude of detritus uptake within the deep-sea food web and sediments varies seasonally in the English Channel, with an average net sedimentary organic macroalgal C sequestration of 8.75 g C/m^2, about 4–5 percent of estimates for mangroves, salt marshes, and seagrass beds. The average net sequestration per year of POC in sediments of the English Channel is 58.74 g C/m^2, about 26–37 percent of blue carbon habitats (Figure 6.3). In a study of the Great Atlantic Sargassum Belt, Bach et al. (2021) found that biogeochemical feedbacks, nutrient reallocation, and calcification could greatly reduce the CDR efficacy of floating seaweed, though it could play an important role in increased ocean albedo.

Proposed restoration techniques include afforestation through seeding of macroalgae; the reversal of trophic cascades through the protection of sea otters and other predators; and exclusion methods such as flexible fencing to reduce sea urchin densities (Sharma et al., 2021).

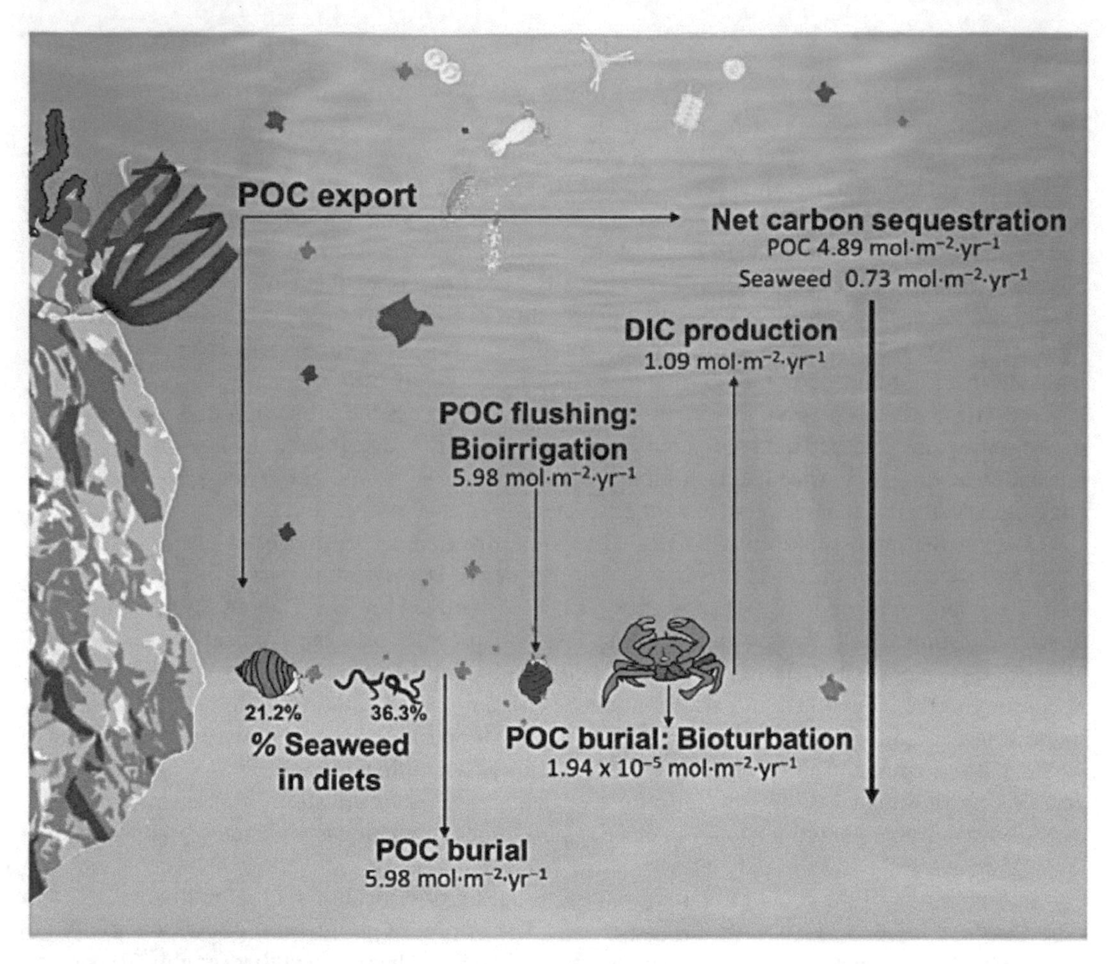

FIGURE 6.3 Macroalgal sediment system in the English Channel, including seasonally averaged carbon fluxes and contributions to faunal diets. Note: DIC = dissolved inorganic carbon; POC = particulate organic carbon. SOURCE: Queirós et al., 2019.

Benthic Algae

Coralline or nonfleshy benthic algae have a worldwide distribution, comprising systems that can be thousands of years old (van der Heijden and Kamenos, 2015; Riosmena-Rodríguez et al., 2017). They have a large organic C stock (Figure 6.4) with the potential to store and sequester carbon. Yet relatively little is known about the role of calcified algae in C burial, including the balance between burial and the biological processes that release carbon (van der Heijden and Kamenos, 2015; Mao et al., 2020). Calcifying algal systems could represent a substantial, global-scale C repository with millennial longevity, though they are sensitive to climate variability and disturbance from trawling and other activities. The calcification process also reduces seawater alkalinity and increases the partial pressure of CO_2, reducing the CDR efficacy of calcifying algae.

Animals and the Carbon Cycle

It has been suggested that conserving and restoring large marine vertebrates and other animals could provide an ecologically sound alternative to more intensive ocean CDR schemes such as iron

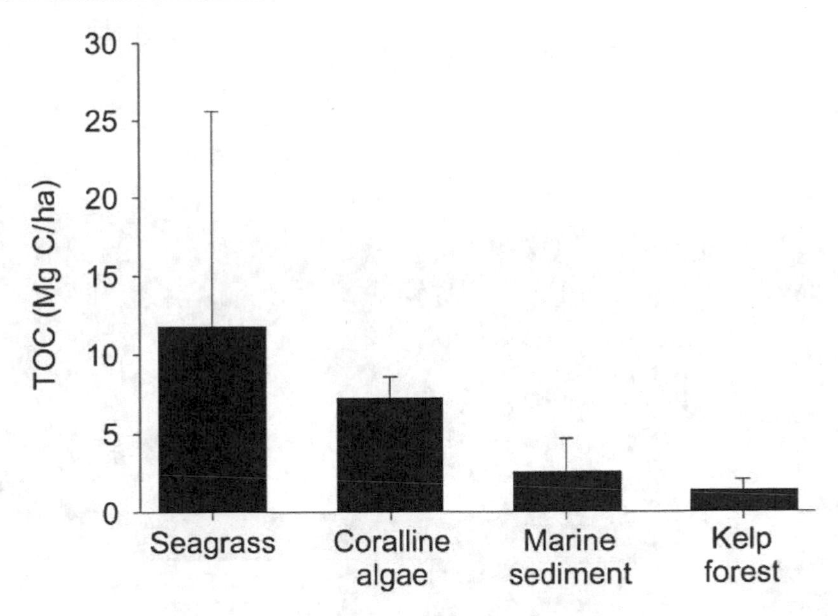

FIGURE 6.4 Sediment organic carbon stock (total organic carbon Mg C/ha) in systems classified as blue carbon repositories. SOURCE: Mao et al., 2020.

fertilization (e.g., Pershing et al., 2010). There are several ways that animals can influence the ocean carbon cycle (Figure 6.5): through the carbon stored in animal biomass; in carcasses and excretion, especially when they are exported to the deep sea; through nutrient subsidies and fertilization; and through trophic interactions. We briefly discuss these pathways in the sections below.

Animal Biomass

Carbon can be stored or sequestered in the living biomass of marine populations. Since marine animals can self-perpetuate through reproduction, they can continually generate new biomass. The carbon stored in recovering populations can be considered sequestered, as Martin et al. (2021) noted, "perhaps infinitely." The rebuilding of fish stocks after decades of overfishing, and the increase of deadfall carbon, will reactivate a natural C pump driven by large marine fish (Mariani et al., 2020). Many of these fishing activities, especially on the high seas, would not be economically viable without subsidies. Bianchi et al. (2021) estimate that the pre-exploitation biomass was 3.3 ± 0.5 Gt, cycling roughly 2 percent of global primary production and producing 10 percent of surface biological export. The ecological impact of recovery can enhance or reduce the amount of carbon stored in ecosystems, depending on its ecological function and whether a species is an herbivore or predator (Schmitz et al., 2018).

Bar-On et al. (2018) estimated that marine animals comprise approximately 1.4 Gt C (Figure 6.1). This includes whales and other marine mammals, fishes, and other vertebrates; zooplankton; benthic species; and other invertebrates. (It does not, however, include echinoderms, one of the most diverse groups of marine animals, and is thus likely an underestimate.) Although we could not find total estimates for marine mammals, marine arthropods comprise about 1.0 Gt C, and fish (freshwater and marine) comprise about 0.7 Gt C. Pershing et al. (2010) estimated that rebuilding populations of eight whale species would store and sequester 8.7 Mt C in living biomass. Although these numbers are relatively small compared to terrestrial plant-based carbon, many animal populations in the ocean have been overexploited or reduced because of indirect human activities such as

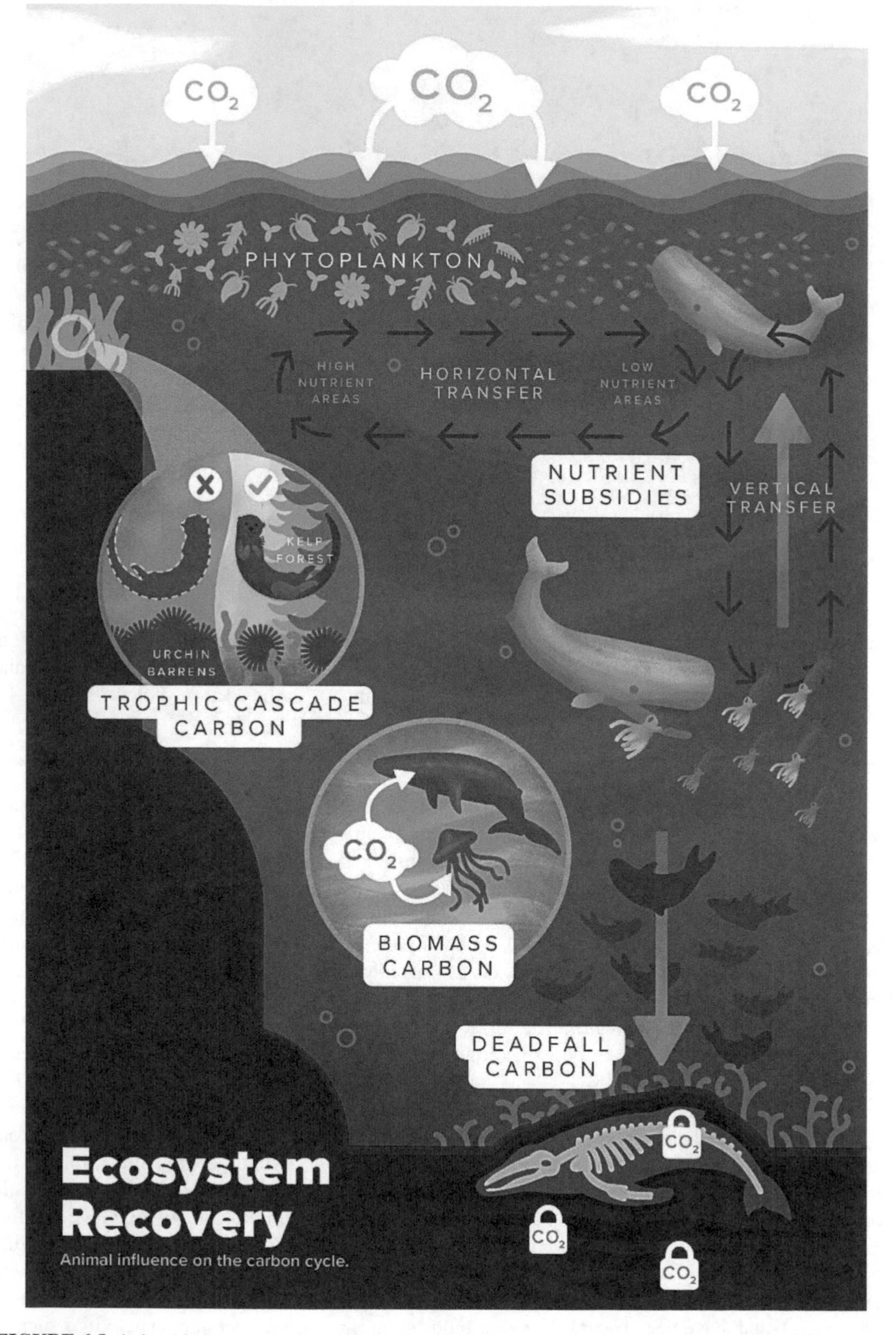

FIGURE 6.5 Animal influence on the marine carbon cycle.

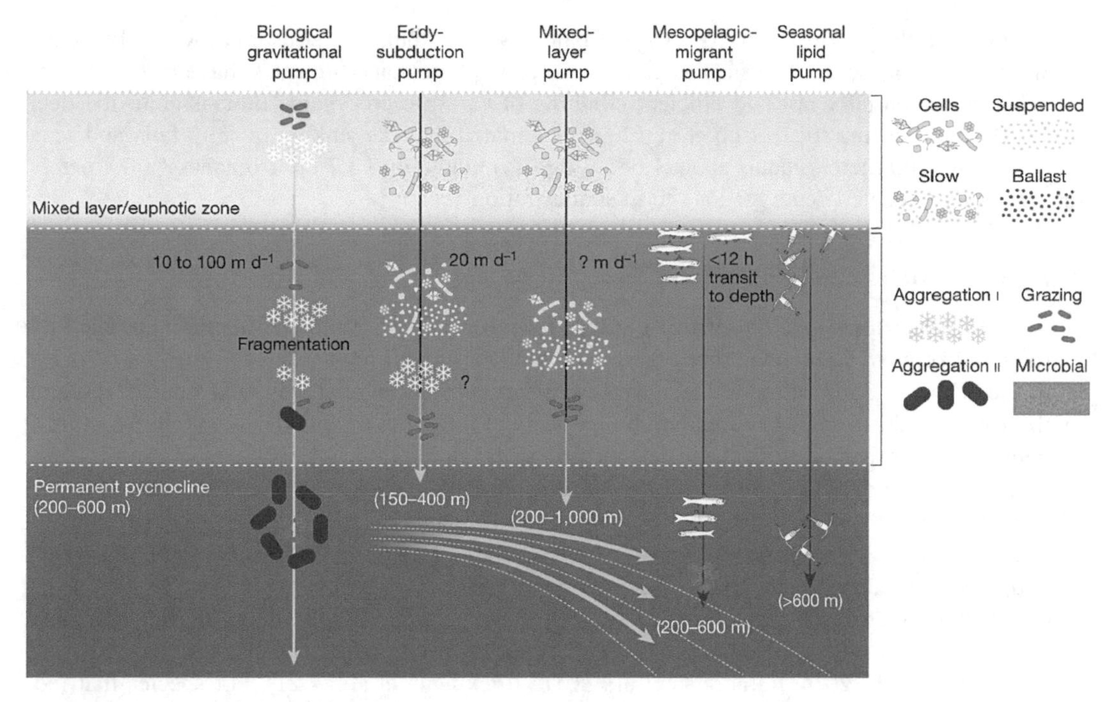

FIGURE 6.6 Particle pumps, including the mesopelagic migrant pump. SOURCE: Boyd et al., 2019.

pollution and habitat destruction. The recovery of historic numbers of marine animals presents an opportunity in terms of biomass, deadfall carbon (see below), and the restoration of zoogeochemical pathways in the carbon cycle (Schmitz et al., 2018).

There have been a few attempts to calculate these values for vertebrates, for example, an assessment of oceanic C for the United Arab Emirates for fishes, marine mammals, sea turtles, and seabirds (Pearson, 2019). But a global assessment of the biomass and geochemistry of marine animals in the present, past, and future under various recovery scenarios has not been completed.

Deadfall and Excreted Carbon

The primary movement of carbon from surface waters to the stratified interior is through the biological pump, which involves the sinking of biogenic particles that range from marine snow to metazoans, such as fish and zooplankton, and their feces (e.g., Falkowski et al., 1998). Active vertical migration also has a large impact on the C export flux of the biological pump, with organisms moving large distances between the surface and depths, potentially enhancing C export (Archibald et al., 2019). The carbon in carcasses and excreted material becomes available to food webs or bacteria.

Boyd et al. (2019) estimated that the mesopelagic migrant pump could export more than 2 Pg C/yr. The concept for this pump is based on long-established observations of diurnal migration in zooplankton and fishes, the movement of the animals beneath the permanent pycnocline, and the release of fecal pellets at depth, with rapid sinking rates (Figure 6.6). In another study, Saba et al. (2021) calculated that fish contribute 16 percent ±13 percent to the global downward flux of carbon, equivalent to 1.5 ± 1.2 Gt C/yr. Whale falls, or cetacean carcasses that sink to the deep seafloor can sequester carbon for hundreds to thousands of years. Pershing et al. (2010) estimated that although relatively small in present numbers, especially when compared to the global C sink, whale falls would remove 1.9 x 10^5 t C/yr if baleen whale stocks were restored. Recently, gelatinous zooplankton, or jellyfish such as cnidarians, ctenophores, and tunicates, have received some attention as potentially fast and efficient conduits of carbon-rich organic biomass into the deep ocean (Lebrato et al., 2019). Luo et al. (2020) estimated that the sinking of jelly-falls and fecal matter below 100 meters equals about 1.6–5.2 Pg C/yr, and 0.6 to 3.2 Pg C reaches 1,000 meters, the generally accepted depth for long-term sequestration.

Nutrient Subsidies and Fertilization

Ocean fertilization—adding nutrients (usually iron) to the upper ocean to stimulate plankton blooms and remove CO_2—is perhaps the most well-known and highly debated method of ocean CDR (see Chapter 3). Whereas artificial iron fertilization has been relatively well studied, research on the role of marine animals in nutrient fluxes (natural fertilization) and its potential impact on C storage and sequestration is for the most part still emerging.

Relief of iron deficiency in areas of the sub-Antarctic Southern Ocean can enhance C export into the deep ocean and sediment (Pollard et al., 2009), and natural supplies of iron appear to result in much higher export efficiency than artificially added iron (see Chapter 3). Such findings on natural supplies of iron, along with recent research on the high iron content in the feces of whales, salps, and other marine animals, have prompted several efforts to quantify the role of marine animals in the iron cycle and other biogeochemical cycles, such as nitrogen and phosphorus (Lavery et al., 2010; Roman et al., 2014; Ratnarajah et al., 2018; Böckmann et al., 2021). For species that feed at the surface, such as many baleen whales, the recycling of iron and other nutrients likely aids in the retention of autochthonous, or recycled, iron, but does not necessarily support new production (e.g., Lavery et al., 2014).

Animals such as large fishes, sea turtles, marine mammals, and seabirds might be specifically essential for the biogeochemical cycle of several elements because they are widespread, highly mobile, and globally distributed (Wing et al., 2014; Tavares et al., 2019). Sperm whales, for example, feed on deep-living prey and defecate at the surface, transporting allochthonous iron into the euphotic zone and raising the nutrient standing stock of surface waters (Lavery et al., 2010). Lavery et al. assumed that about 75 percent of the total iron defecated by sperm whales persists in the photic zone, with iron in the beaks of cephalopod prey sinking. This nutrient contribution can enhance new production, increasing net uptake of CO_2 from the atmosphere and C export to the deep ocean, though the amount of carbon that is stored, sequestered, or released back into the surface waters remains an area of active debate (for a review of marine vertebrates in the carbon cycle, see Martin et al., 2021; Roman et al., 2021).

Many baleen whales are capital breeders, storing resources in highly productive high-latitude areas, for annual migration to lower-latitude and often oligotrophic systems where they breed. They can transport nutrients between these systems via urine, placentas, carcasses, and sloughed skin (Roman et al., 2014). These can be considered external, or allochthonous, nutrient inputs that could ultimately lead to new production and C sequestration, because nutrients are released at locations thousands of kilometers from their origin, often in areas that are nutrient limited. (Areas that are not nutrient limited may not have a large impact on the global total.) Nitrogen released by the modern-day population of Southern Ocean blue whales on their calving grounds, for example, is estimated to result in the storage and sequestration of 5.1 x 10^2 t C/yr; yet before whaling greatly reduced this species, estimates would have been closer to 1.4 x 10^5 t C/yr (Table 6.2; Roman et al., 2014).

Despite the attention given to marine mammals, it is likely that the far more abundant krill, salps, and other zooplankton play an important role in the biogeochemical cycles. Vertical movement by marine organisms has the potential to increase ocean mixing across pycnoclines, though these contributions might be small when compared to internal gravity waves (Kunze, 2019). A recent study found that fecal pellet material from salps released more bioavailable iron to Southern Ocean phytoplankton than krill pellets (Böckmann et al., 2021). One estimate suggests that about two-thirds of these salp pellets stay near the surface (Iversen et al., 2017). Phytoplankton communities took up to five times more iron from the salps' fecal pellets than from krill feces. The variation in iron availability in different types of fecal pellets could influence biogeochemical cycles and C fixation of the Southern Ocean, potentially representing a negative feedback to climate change as salp populations increase with reduced sea-ice coverage (Böckmann et al., 2021). Such regime shifts are linked to changes in sea-ice coverage and are not considered part of an ecosystem recovery approach to CDR.

Trophic Interactions

Several studies have examined the food web dynamics of animals in the carbon cycle. Planktivorous fish and invertebrates can convert carbon into longer-lived biomass and rapidly sinking fecal pellets, increasing C export (Martin et al., 2021). At the same time, grazing animals such as dugongs and green turtles can decrease seagrass meadow structure and biomass, decreasing the quantity fated to the detrital pool (Scott et al., 2021).

Predators also play a role in these systems. A few studies have examined the role of predators in enhancing C sequestration, including sharks (predation risk) and sea otters (consumptive trophic cascade). The risk of predation by sharks limits herbivore abundance to zones close to reefs, resulting in reduced grazing, increased algal biomass, and improved sedimentary C stocks and sequestration away from reef refuges (Atwood et al., 2018). In a classic example, sea otters are considered keystone predators in the North Pacific, consuming sea urchins and other benthic species, which can have a positive impact on kelp and C storage and sequestration. Wilmers et al. (2012) estimated that sea otter predation could increase the storage of 0.015–0.043 Gt C/yr and the sequestration of 0.0013–0.023 Gt C/yr by restoring kelp forests. Gregr et al. (2020) predicted a net benefit of CAD$2.2M for the C sequestered by sea otters.

In vegetated coastal systems, Atwood et al. (2015) noted that there was sufficient evidence to suggest that intact predator populations can be vital to preserving or increasing carbon reserves stored in coastal or marine ecosystems. As such, policy and management need to be improved to reflect these realities. Beyond these coastal systems, there is a need for a deeper exploration of the role in trophic interactions on macroalgae as well as in pelagic and offshore benthic ecosystems.

Ecosystem-Based Solutions

Ecosystem-based solutions, also known as nature-based solutions, have the potential to protect natural habitats and restore modified ecosystems while providing other societal benefits for humans, such as health and ecosystem services, and protecting other species (Solan et al., 2020). Ecosystem-based solutions to climate change rely largely on traditional conservation efforts to protect and restore marine ecosystems and native species. Roman et al. (2015), Duarte et al. (2020), and others have recently reported on the ongoing recovery of many large marine animals, including humpback whales, sea turtles, and elephant seals, after decades of overharvesting. A variety of measures, including reduction of direct and indirect take, expansion of protected areas, and improved enforcement have helped many of these species, and dedicated efforts have the potential to restore the diversity of marine life.

Although the overriding theme of this chapter is CDR in the ocean, it would be counterproductive to ignore the role of marine conservation efforts to protect current stores and flows in the carbon cycle. Lovelock et al. (2018) noted that two steps are necessary for C sequestration in coastal systems such as mangroves, a process that likely holds true for pelagic and benthic systems: conservation is necessary to avoid additional CO_2 emissions from further habitat loss, and restoration is used to restore C pools and, during their buildup, act as CO_2 sinks. Similar steps are in play in protecting benthic and pelagic systems: to avoid leakage, current C removal processes must be protected, while others are restored. Benthic algae and sediments are at risk from bottom trawling, for example. Leaving sediments undisturbed from bottom trawling can protect stored carbon and enhance anoxic remineralization, which in turn generates alkalinity and reduces partial pressure of CO2 (pCO_2). At the same time, new algal systems can be restored.

One of the big questions for any ecosystem-based research agenda is how much additional carbon is expected to be removed as a result of habitat restoration and population recovery. Answering this will require tests of concepts for the CDR potential of restoring benthic and pelagic systems. Oreska et al. (2020) showed that restoring seagrass meadows in Virginia removed 9,600 tons of CO_2 from the atmosphere over 15 years, but it also heightened methane (CH_4) and nitrous oxide (N_2O) production, releasing 950 tons of CO_2 equivalents. The meadow now offsets 0.42 t CO_2 equivalents/ha per year. Similar efforts to examine the removal and release of CO_2 and other greenhouse gases in different systems will be necessary to estimate the sequestration rates of restored marine ecosystems. In some cases, new methods for greenhouse gas accounting will likely be needed to calculate the net benefit of marine systems that include macroalgae, phytoplankton, metazoans, and microbes.

Marine Protected Areas and Habitat Restoration

The protection of marine ecosystems has typically followed two paths: (1) management of individual species or groups of species, such as the Marine Mammal Protection Act in the United States and protections under the International Union for Conservation of Nature (IUCN) and the International Whaling Commission, and (2) protection of habitats and communities through marine protected and conservation areas and other types of restrictions on habitat use. Developing effective marine protected areas (MPAs) for biodiversity and CDR requires identifying vulnerable habitats and restoring areas that will be the most effective inside and outside of protected areas. There is a rich literature on the value of spillover effects, networks, and dispersion capability for MPAs, but the focus on carbon is relatively new.

Roberts et al. (2017) discussed the ability of protected areas to help ecosystems and people adapt to the impacts of climate change by helping to mitigate sea-level rise, intensification of storms, shifts in species distribution, and other expected changes. Protected areas can also play direct and indirect roles in the many processes discussed in this chapter: promoting the storage of carbon in biomass and carcasses, protecting the seafloor from trawling and mining, protecting and restoring food webs and reversing human-induced trophic cascades, and reducing the pressures of overharvesting.

The push to protect more of the oceans was adopted by the IUCN World Conservation Congress in 2016, calling for 30 percent protection by 2030. This effort and similar ones proposed by the Biden administration in the United States and elsewhere provide an opportunity for researchers to examine the role of benthic and pelagic species in CDR. As populations and ecosystems change as a result of protections, expected increases in C sequestration can be measured using established and emerging technologies. To be effective, and to understand the role of protected areas in CDR, MPAs will need resources for monitoring and enforcement (see *Monitoring and Verification* below).

In addition to MPAs and habitat restoration, restrictions on destructive benthic and deep-sea activities may be necessary to retain carbon already captured in the seafloor. Marine sediments store approximately 2,322 Pg C in the top 1 meter, nearly twice that of terrestrial soils (Atwood et al., 2020). Seventy-nine percent of the global marine C stock is in abyss and basin zones that have largely been undisturbed. Areas that have been extensively modified have likely released a fair amount of carbon already. Only about 2 percent of C stocks is located in fully protected areas, where it is safe from legal disturbance of the seafloor (Atwood et al., 2020). Recognizing the need to protect deep-ocean ecosystems and biodiversity, the IUCN recently called for a moratorium on seabed mining.[3] The restriction or prohibition of seabed mining and bottom trawling will likely help store and retain old carbon and sequester new carbon in the seafloor.

Macroalgae Protection and Restoration

Although there are no global estimates for loss of macroalgal habitats, one study showed declines in kelp forests in 38 percent of ecoregions, with other areas showing increases, indicating that there is high geographic variability (Krumhansl et al., 2016). Pessarrodona et al. (2021) showed a loss of forest-forming seaweeds and rise of turf algae across four continents, resulting in the miniaturization of underwater habitat structure. Protection and restoration can avoid global losses in many of these ecoregions. Some losses of kelp, however, can be attributed to increasing water temperatures. In such cases, restoration and protection might not help retain kelp forests, and other methods such as assisted migration of kelp at the leading edge of their range might be necessary to increase kelp forests. Several other algae species, such as the commercially important *Gelidium*, a red seaweed that is the source of laboratory agar, have been overharvested and are subject to trade restrictions and regulations (Callaway, 2015).

There are several approaches to restoring degraded seaweed habitats and protecting existing ones, with a focus on CDR. Each of the three items below focuses on CDR, but they all are expected to enhance biodiversity and some ecosystem services, such as tourism and finfish fisheries.

Habitat Protection

Protecting areas with extant macroalgae and restoring historic areas is the first step to enhancing algal growth and CDR. Measures can include MPAs, reduced nutrient loading, reduced disturbance of the seafloor, and bans on extraction of stones and rocks from sensitive areas.

Protection of Carbon Sinks in Sediments Beyond the Habitats

Carbon export and sequestration from macroalgae typically depends on POC export to the seafloor. Trawling and seafloor mining can release this carbon back to the surface, so protecting these areas from human disturbance is essential in the CDR process.

Restoration of Macroalgae

In addition to protection, restoration of disturbed macroalgal systems is essential to the CDR approach. Methods can include the control of herbivores, such as sea urchins (see Reversing Trophic Cascades and Restoring Food Webs below) and the seeding of macroalgae, including techniques such as "green gravel," small rocks seeded with kelp (Fredriksen et al., 2020).

Japan probably has the most experience in kelp restoration (Figure 6.7; Duarte et al., 2020). In the United States, kelp restoration is typically linked to the return of predators, such as sea otters (Wilmers et al., 2012). The U.S. seaweed and kelp industry is small compared to production in East Asia, but there is rising interest in the market for human consumption (Janasie and Nichols, 2018). A new partnership in British Columbia among the Haida Nation, Parks Canada, and Fisheries and Oceans Canada, academic institutions, nongovernmental organizations (NGOs), and commercial

[3] See https://www.iucncongress2020.org/motion/069.

FIGURE 6.7 Rocky bottom after kelp loss (left) and kelp beds (*Saccharina japonica*) after restoration in Hokkaido, Japan. SOURCE: Duarte et al., 2020; photo credit, Nippon Steel Corporation.

fishers is focused on two coastal restoration areas for recovery of kelp and sea otters (Goldman, 2019).

Fish and Fisheries Management

The discussion of food systems in relation to CO_2 emissions has typically focused on terrestrial agriculture and land-use changes (e.g., Crippa et al., 2021). In contrast, fisheries management has generally looked at the expected response of fish stocks to global change and related uncertainties (e.g., Howard et al., 2013). The potential global fisheries catches, for example, are expected to decrease by about 3 million tons for each increase per degree Celsius of global surface warming, with a high number of species turnover (Cheung et al., 2016).

This discussion has recently begun to change. The majority of commercially targeted fish populations are predatory, and the removal of biomass and alteration of the ecological structure and function of the ocean, even under regimes that are considered sustainable such as maximum sustainable yield, are likely to have impacts on the ocean carbon cycle and storage (Yodzis, 2001; Spiers et al., 2016; Stafford, 2019). As such, strategies intended to increase harvest, by keeping populations at 50 percent of carrying capacity or lower, will likely be ineffective as a C strategy. In addition to the direct take from fisheries and habitat destruction, indirect take can affect whales and other marine mammals. Currently, about 11 percent of commercially captured fish is discarded (Gilman et al., 2020), and marine mammal bycatch endangers and limits the recovery of many populations (Read et al., 2006).

Several new studies have estimated the potential role of rebuilding fish populations in the ocean carbon cycle. Marine fisheries have released at least 0.2 Gt of carbon into the atmosphere since 1950, with greater than 43 percent of the carbon released by high seas fisheries coming from areas that would be economically unprofitable without subsidies (Mariani et al., 2020). These calculations include the amount of carbon that was extracted rather than sequestered by dead fish sinking to the deep sea (21.8 Mt C), the amount of carbon extracted from the ocean as living biomass (37.5 Mt C), and the carbon emitted from burning fossil fuels and processing fish on land (202.8 Mt C). Bianchi et al. (2021) estimated that fish cycled roughly 2 percent of global primary production (9.4 ± 1.6 Gt/yr) and produced 10 percent of surface biological export before industrial fishing. Sala et al. (2021) estimated that bottom trawling and dredging release 0.4 Pg C/yr. This rather large number needs to be reconciled with ocean inorganic CO_2 budget estimates from Gruber et al. (2019) and other researchers. By expanding the value of fish to include greenhouse gas mitigation and CDR, in the form of fishery industrial emissions, biomass, deadfall carbon, and trophic cascade carbon, it is possible that there will be a shift in values that can help restore wild fish populations, while also maintaining protein supplied by seafood.

Restoring Populations of Large Marine Organisms

Given their potential to store and sequester carbon, conserving large species, and the largest individuals within a species, should be a top priority (Pershing et al., 2010). Historical records, genetic studies, and food web models all suggest that apex predators, large fishes, and marine mammals were more abundant in the past (Ferretti et al., 2010; Magera et al., 2013; Christensen et al., 2014). Unlike terrestrial systems, marine ecosystems should be top-heavy because of complex food webs and top-down effects, with more biomass stored in large predators than in lower trophic levels; the current bottom-heavy trophic structure could be a result of widespread human defaunation of the ocean (Woodson et al., 2020).

Marine organisms, including predatory fish and marine mammals, are especially vulnerable to human activities in the ocean. In addition to direct threats from commercial fishing, marine mammals and other vertebrates are also at risk from ship strikes, ocean noise, plastics, and other types of marine pollution (e.g., Rockwood et al., 2017; López Martínez et al., 2021). The reduction of these risks provides an opportunity for restoration and experimentation. The restoration of large marine organisms, through fisheries management and mitigation, MPAs, pollution reduction, and other protective policies, can result in increased biomass, deadfall carbon, and nutrient transfer and subsidies.

Reversing Trophic Cascades and Restoring Food Webs

Animals perform a complex set of trophic and nontrophic interactions that can cascade through food webs and affect carbon processes (e.g., Schmitz et al., 2018). The conservation and recovery of marine keystone species have been proposed as one way of restoring and enhancing C sequestration in kelp forests and other habitats. The repair of trophic interactions between sea otters and sea urchins through otter recovery, for example, could lead to a 1,200 percent increase in kelp biomass carbon within the otter's range in the North Pacific (Wilmers et al., 2012). The loss of predators such as sharks can have a large effect on C burial in coastal ecosystems, potentially reducing burial by up to 90 percent in some habitats (Atwood et al., 2015). The C stock of macroalgae growing in high-risk areas for herbivorous fish that are away from coral reefs is 24 percent higher than in areas close to reef refuges (Atwood et al., 2018).

The restoration of intact food webs and the reversal of human-induced trophic cascades can enhance or reduce C storage (Estes et al., 2011; Schmitz et al., 2018). The recovery of herbivorous species, while good for biodiversity conservation, might not increase carbon sequestration. The return of the green turtle to the Lakshadweep Archipelago in the Indian Ocean, for example, resulted in overgrazing of seagrass, reduced in seagrass fish diversity, biomass, and abundance, and major declines in sediment-stored carbon (Gangal et al., 2021).

6.3 EFFICACY

The protection of marine ecosystems—including protection, relocation, restoration, and the reduction of pollution and overexploitation—is feasible and durable (Gattuso et al., 2018). Yet there have been no comprehensive attempts to estimate the CDR potential of restoring entire marine ecosystems and only a few efforts to quantify the global impact of restoring and protecting taxa or functional groups: for example, for macroalgae (173 Tg C/yr; Krause-Jensen and Duarte, 2016) or restricting certain destructive practices such as bottom trawling (0.4 Pg C/yr; Sala et al., 2021). In contrast to terrestrial and coastal systems, with a large literature on natural climate solutions, there has been little quantitative research on the impact of management tools, such as protected areas and harvest restrictions, on C fluxes in marine ecosystems.

Perhaps the most important step toward including ecosystem restoration in a CDR scheme is measuring the net greenhouse gas benefits for a given project. This would likely involve measuring the carbon that is stored in biomass, sequestered as deadfall or fecal carbon, or enhanced and sequestered via processes such as nutrient fertilization. The production of CH_4 or N_2O, or the release of CO_2, through respiration should also be accounted for.

Additionality

Restoring marine ecosystems offers two clear and overlapping benefits: reducing biodiversity loss and restoring the role of marine organisms in the carbon cycle. Reduced harvest, restored habitat, and larger no-take zones will almost certainly sequester more atmospheric carbon, though the scale of many of these reductions remains debated and is an important area for new research. There trade-offs between commercial fishing, food security, and CDR will need to be considered in management plans. One challenge in quantifying additionality in the protection of marine fauna is that changes in protection in one area can result in increased fishing in others, and many fish and marine mammals are migratory, with their ranges shifting as temperatures change and prey populations move (Cheung et al., 2016). The monitoring of many migratory species will need to go beyond political boundaries.

Durability

One of the advantages of ecosystem restoration is that marine organisms can self-perpetuate through reproduction, generating new biomass: carbon can be stored in stable populations "perhaps infinitely" (Martin et al., 2021). As biomass continues to stay relatively stable, more carbon would be stored through death and deposition, resulting in a long-term C drawdown. This process would happen naturally if humans were not continually altering biogeochemical cycles. Unfortunately, these cycles are expected to change as a result of rising temperatures, ocean acidification, and ocean deoxygenation, with the C capture and storage potential of the oceans, along with productivity and biomass, likely to decline (Gruber, 2011). Changes in surface nutrients and phytoplankton and zooplankton biomass could also reduce total biomass and body size of fish and other vertebrates (Boyce et al., 2010; Britten et al., 2016).

One of the greatest risks to the CDR potential of marine ecosystems, and an existential risk to humans and many other animals, is the current extinction crisis combined with an increased atmospheric concentration of CO_2. The fossil record shows major responses in the oceans during global C perturbations and mass extinctions. In benthic systems, responses include reduced diversity, decreased burrow size and bioturbation intensity, and extinction of trophic groups (Bianchi et al., 2021). Models developed by Tulloch et al. (2019) show predicted declines and local extinctions of baleen whales and krill in the Southern Ocean by 2100 as a result of climate change. In the case of widespread ecosystem shifts, there is a potential for rapid changes that could result in the reversal of CDR gains.

There is some evidence, however, that diverse and abundant fish communities are more resilient in the face of warming waters and other climate shifts (Sumaila and Tai, 2020). MPAs could help build resilience to climate change for coral reefs and other ecosystems (e.g., Chung et al., 2019). In addition to the effects of increased greenhouse gases, management and protection regimes can be subject to changes in policy, which would affect the permanence of C removal.

TABLE 6.2 Estimated Amount of Carbon Fixed, Stored, or Sequestered in Pre-exploitation and Modern Cetacean Populations

Mechanism	Species	Region	Pre-exploitation C Estimate (*N*)	Modern C Estimate (*N*, year)	C Fixed, Stored, or Sequestered	Reference
Living biomass carbon	Pantropical spotted dolphin (*Stenella attenuata*)	Eastern Tropical Pacific	5.9×10^4 t C (3.6×10^6)	1.4×10^4 t C (8.6×10^5, 2006)	Stored	Martin et al., 2016
Living biomass carbon	Spinner dolphin (*Stenella longirostris*)	Eastern Tropical Pacific	2.4×10^4 t C (1.8×10^6)	1.4×10^4 t C (1.1×10^6, 2006)	Stored	Martin et al., 2016
Living biomass carbon	8 baleen whale taxa[a]	Global	2.0×10^7 t C (2.6×10^6)	3.1×10^6 t C (8.8×10^5, 2001)	Stored	Pershing et al., 2010
Deadfall carbon	8 baleen whale taxa[a]	Global	1.9×10^5 t C/yr (2.6×10^6)	2.9×10^4 t C/yr (8.8×10^5, 2001)	Sequestered	Pershing et al., 2010
Whale pump	Sperm whale (*Physeter macrocephalus*)	Southern Ocean	2.4×10^6 t C/yr (1.2×10^4)	4×10^5 t C/yr (1.2×10^3, 2001)	Fixed	Lavery et al., 2010
Whale pump	Blue whale (*Balaenoptera musculus*)	Southern Ocean	1.3×10^8 t C/yr (2.4×10^5)	2.8×10^6 t C/yr (5.2×10^3, 2012)	Stored	Lavery et al., 2014
Great whale conveyor belt	Blue whale	Southern Ocean	1.4×10^5 t C/yr (3.4×10^5)	5.1×10^2 t C/yr (4.7×10^3, 2001)	Stored and sequestered	Roman et al., 2014

NOTES: Mechanisms include storage in living biomass, sequestration in whale carcasses sinking to the deep sea, and nutrient transport (vertical pump and horizontal conveyor belt) via excretion and defecation of nitrogen, phosphorus, and iron. C values are gross and do not account for the amount of carbon respired by cetaceans. Exploitation refers to bycatch for dolphins and industrial whaling for all other cetaceans.

[a]*Balaenoptera acutorostrata, B. bonaerensis, B. borealis, B. brydei, B. musculus, B. physalus, Balaena mysticetus, Eschrichtius robustus, Eubalaena spp., Megaptera novaeangliae.*

SOURCE: Courtesy of Heidi Pearson, University of Alaska Southeast, and Roman et al., 2021.

Monitoring and Verification

Monitoring and verification are essential to protection efforts, especially when they have particular goals such as CDR and biodiversity protection. One concern about CDR is that an extensive monitoring and evaluation system would require large infrastructure for a relatively small return. Should there be a minimum requirement to invest in monitoring and evaluation? The current ocean uptake of anthropogenic carbon from physicochemical process is approximately 2.5 Pg C/yr; if the CDR approaches we discuss can increase uptake on a petagram scale, monitoring programs such as the Global Ocean Hydrographic Investigations Program, or GO-SHIP, should be able to detect these changes.

All of the processes we discuss will require documenting that CDR is occurring. This process will require going beyond indirect calculation to demonstrate enhanced sequestration at sink sites (Krause-Jensen et al., 2018). On land, the Verified Carbon Standard is the most commonly used standard. It includes several requirements: the emissions reduction or removal must be real, measurable, permanent, unique, and additional. Whether these measures, designed for terrestrial systems, will apply in marine C schemes remains to be seen (Krause-Jensen et al., 2018). Mesoscale nitrogen and iron enrichment experiments could test if marine mammal and seabird feces effectively modify phytoplankton processes, enhancing diatom biomass and increasing atmospheric CO_2 drawdown (Boyd et al., 2000; Pollard et al., 2009).

Environmental Monitoring of Marine Protected Areas

Solutions for Monitoring MPA Ecosystem Health

Good monitoring and data collection are essential if MPAs are to deliver their intended biodiversity and CDR outcomes. Monitoring ecosystem health and environmental stressors in offshore and remote MPAs is particularly challenging because of limited access for data collection. Custom solutions can be developed to fit individual geographic settings, environmental conditions, and conservation priorities. Satellite remote sensing approaches, for example, can be used to monitor the evolution and environmental conditions of new MPAs (Kachelreiss et al., 2014).

Offshore and Remote MPAs

Deep-ocean and seamount MPAs In the Northeast Pacific off Canada, there are deep-ocean cabled sensor networks that have provided real-time access to the Endeavour Hydrothermal Vents, which was Canada's first MPA, established in 2009. Data products from this site were developed to support management of the Endeavour MPA (Juniper et al., 2019). Monitoring was geographically extended in 2018 to support surveys of offshore Pacific seamount MPAs using remotely operated vehicles, autonomous sensor platforms, multibeam mapping, and vessel traffic data products (virtual automatic identification system, or AIS, fences). Although cable-based sensor networks are limited in their geographic extent, existing systems should be used to deliver monitoring data products. All of the remote technologies and data products could be made available for long-term, site-specific ecosystem health monitoring of deep-ocean and seamount MPAs.

Regional-scale monitoring Management of other deep-ocean ecosystems requires monitoring of regional-scale threats, such as marine heat waves and ocean deoxygenation, in addition to site-specific monitoring. Regionally deployed flotillas of new-generation Argo profiling floats to monitor ocean temperature, dissolved oxygen, and other seawater properties could be used for delivering management data products. Ships of opportunity could be coordinated and engaged to ensure that the floats remain in the regions of interest (e.g., Smith et al., 2019a). The ships themselves could also provide data and data products. These autonomous-sensor platforms profile between the sea surface and ocean depths every few days, measuring water properties and relaying data via satellite. Data and data products from decades-long "Line" surveys in regions (e.g., Line P in the Northeast Pacific) provide a historic framework for understanding long-term ocean change in areas where these surveys take place.

Marine mammal activity in remote protected areas Cabled and autonomous mooring networks can deliver marine mammal monitoring solutions, using underwater hydrophone systems that record whale sounds and ship noise, and AIS data that track vessel movements with a focus on data products that inform long-term management by reporting on monthly and annual levels of marine mammal and large vessel activity in MPAs. In the case of real-time operational use, these data can be used for compliance and enforcement. In areas of known marine mammal activity, virtual AIS fences could be implemented to identify ship traffic and inform ships' captains to reduce vessel strikes.

Ecosystem Health Solutions: Coastal Zone MPA Monitoring

Coastal zone MPA monitoring could be delivered in the United States through the existing regionally based Ocean Observing Systems (IOOS Strategic Plan 2018–2022; IOOS, 2018). The Strategic Plan already includes delivery of the benefit to "more effectively protect and restore healthy coastal ecosystems," and this could be expanded to include observations and measurements of ocean CDR efficacy and environmental impacts. In areas not covered by IOOS, community-based monitoring programs could be implemented. These approaches are especially important in Indigenous communities (see, e.g., Kaiser et al., 2019).

Ecosystem Health Solutions: MPA Data Products

State of the ocean data products The collection of long time-series data by offshore and coastal observing systems provides critical information for evaluating local ocean health trends and their relationship to regional ocean change. Some permanent observatory installations host core sensors for monitoring seawater properties and noise levels. These properties and their relationships to ecosystem health include

- temperature—informs on long-term warming trends and marine heat waves;
- salinity (salt concentration)—detects anomalies in freshwater input to coastal waters;
- pressure (depth)—informs on sea-level rise, tsunami and storm surge events, tidal cycles; and
- dissolved oxygen concentration—key to detecting hypoxia events and long-term oxygen depletion that can have deleterious impacts on marine ecosystems.

Daily noise index data product Noise generated by surface vessels is well known to be harmful to marine mammals, and there is growing evidence of negative impacts on fish and invertebrates. To capture ship-source signature noise profiles, multihydrophone arrays are needed to ensure that the data meet ANSI standards. Once source signatures are captured, these data can be combined in real time to model the noise in heavily trafficked areas (Zhang et al., 2020) and potentially could be used to reduce noise through voluntary or mandatory measures. Long-term fixed hydrophone networks can provide a benchmark to assess increases or decreases in noise (Thomson and Barclay, 2020).

Biodiversity and stock assessments Information on trends in species diversity and fish abundances is often a key metric for monitoring conservation objectives in MPAs and plays an important role in understanding the carbon dynamics of these systems. Imagery from fixed observatory underwater cameras and regular ship-based video transect surveys could be used to develop data products on species diversity and abundance to inform MPA management and CDR research. Manual, machine vision, and crowdsourcing tools are several approaches that extract biological information from video and photo archives to reveal trends in species numbers (Matabos et al., 2017).

Solutions for Compliance and Enforcement Monitoring

Tools that deliver summary and real-time data products of vessel and marine mammal activity in MPAs, using data from vessel AIS and passive acoustic monitoring (hydrophone) data from cabled and autonomous listening stations, are important tools to support monitoring for compliance and enforcement. Hydrophone systems are expensive to invoke in all regions, but research is under way to use satellite data for detection of some marine mammals that could fill large areas (Platonov et al., 2013). Summary data products will permit monthly review of vessel and marine mammal activity in remote and coastal MPAs, for compliance monitoring. Real-time vessel tracking from AIS data supports enforcement by providing alerts of vessels crossing a virtual fence. For those vessels not emitting AIS signals, real-time detection is possible using directional arrays of hydrophones, over-the-horizon radar, and satellite-based detectors.

6.4 SCALABILITY

It is well accepted that oceanic biological processes are an important component in the global carbon cycle (Table 6.1). But perhaps the biggest question surrounding ecosystem protection and recovery regarding carbon removal is whether such conservation efforts will scale to the level of other proposed C removal efforts in the oceans and on land. To date, many of the studies on marine carbon have been conducted on relatively small geographic and temporal scales, although there have been several efforts to examine global C budgets for macroalgae, benthic algae, marine sedimentary carbon, the mesopelagic pump, and whale and fish carbon. The opportunistic nature of several of these studies and a reliance on intricate food web models compounded by a scarcity of observations to constrain models on a global scale can result in large uncertainties (Bianchi et al., 2021). Below, we briefly discuss questions of geographic and temporal scalability.

Geographic Scalability

The restoration of marine ecosystems could play an important role in the removal of carbon from the atmosphere, but to our knowledge, there have been no models addressing scalability on an ocean-basin or ecosystem scale that cut across species and processes. There is a need to develop an ecosystem-level approach to understanding the role of organisms and their ecosystem functioning in benthic, pelagic, tropical, temperate, and polar systems. New efforts to protect and restore the oceans, such as expanding MPAs and improving fisheries management, provide an ideal opportunity to study changes in C sequestration between degraded and restored ecosystems and between depleted and abundant marine species.

Marine ecosystems are diverse, and distributions and productivity are patchy. Continental shelves, for example, have been argued to contribute disproportionately to the oceanic uptake of CO_2 (Fennel et al., 2018). There is a need to identify hot spots for C storage and sequestration. Locating connected macroalgal–sediment systems, for example, will be an important step toward including them in management actions aimed at carbon removal (Queirós et al., 2019). In the case of nutrient cycling by large vertebrates and other animals, models have been used to assess scale and the potential impact on the global carbon cycle, often focusing on the Southern Ocean or other high-nutrient, low-chlorophyll (HNLC) regions (e.g., Martin et al., 2021; see also the scalability section in Chapter 5). As with ocean iron fertilization (OIF), the upwelling of the nutrients and respired carbon trapped at depth in the Southern Ocean could offset a substantial fraction of the naturally mediated downward flux of POC (Jin et al., 2008).

Even if the scale of some of these systems is relatively small, there could be an opportunity for island nations with limited land cover but large ocean Exclusive Economic Zones (EEZs) to offset their carbon emissions by protection efforts that explicitly value marine organisms and ecosystems in their waters. Similarly, the positive consequences of population, species, and habitat recovery and restoration could be substantial for ocean resilience and biodiversity on a global scale.

Temporal Scalability

Marine ecosystems have the potential to store carbon for long periods of time. Yet given current challenges, such as overfishing, pollutants, and climate change, what is the likelihood that they can be restored on a policy-relevant timescale? Duarte et al. (2020) suggest that substantial recovery of the abundance, function, and structure of marine life could be achieved by 2050 if major pressures are reduced. Response to protection efforts will vary by taxa and ecosystem, and some recovery efforts, such as the protection of coral reefs and endangered cetaceans such as North Atlantic and North Pacific right whales, will require substantial policy changes and sharp reduction in greenhouse gases.

Just as there are hot spots of C sequestration, there are likely hot moments, with sequestration rates varying seasonally and annually. As C emissions approach net zero, a diverse portfolio of marine conservation efforts may have the potential to contribute to global CDR approaches. (See the section on Durability for further discussion of temporal aspects of C sequestration under current and future scenarios.)

6.5 VIABILITY AND BARRIERS

Many people have strong emotional connections to oceans (Spence et al., 2018). These connections can result in greater resistance to engineered solutions and greater attraction to ecosystem-based solutions that endorse healthy and productive oceans while protecting marine wilderness (Cox et al., 2020). There is some support among the environmental NGO community for CDR efforts that involve ecosystem recovery on land and in the ocean. Social science research has indicated that ecosystem-based C removal efforts have greater public support than engineered approaches (Wolske et al., 2019). There is real value in conservation, yet the question of scale of ecosystem protection—how much carbon can be stored and removed by conservation and restoration efforts—is an area of active debate, one that is ripe for research.

Widespread efforts to protect and restore ecosystems could face barriers, because marine conservation can come into conflict with other uses, such as commercial fisheries, shipping, marine renewable energy, and mining. The designation of MPAs, for example, has faced opposition from fisheries and other industries. Conversely, the designation of large MPAs could reduce fossil fuel extraction, but also impede the expansion of renewables such as offshore wind. If the goal of establishing an MPA includes CDR, then managers should be open and honest about the objectives, uncertainties, and need for monitoring. Carbon benefits are just one of many ecosystem services that MPAs can provide, and they might need to be balanced against other objectives (Howard et al., 2017).

Research on MPAs has documented how MPA development and stricter fisheries management can impact local ocean-dependent communities and island nations. MPAs that do not incorporate local interests and priorities can cause impacts on food security, displacement, and economic strain. Although MPAs can enhance food security, impacts vary among social groups (Mascia et al., 2010), and strategies must take this into account. Local perspectives are critical to understanding social impacts (Gollan and Barclay, 2020), and participatory approaches are one way to take small-scale fisheries into account (Kockel et al., 2020).

Any attempt to use ecosystem protection and recovery for C removal would likely benefit from the four guidelines proposed by a large community of researchers and conservation practitioners based in the UK for nature-based solutions to climate change: (1) ecosystem-based C removal must not be an alternative to the rapid phaseout of fossil fuels and must not delay decarbonization, (2) it must include a diversity of landscapes and in this case marine ecosystems, (3) it must be employed with the full engagement and consent of local communities and Indigenous peoples and promote adaptive capacity, and (4) it must be designed to provide quantifiable benefits for biodiversity (Seddon et al., 2021).

Environmental Impact

Despite the many uncertainties surrounding ecological restoration as a CDR tool, a clear advantage is that the environmental impact is almost certain to be positive, at least from the perspective of environmental stewardship and ecosystem integrity. There is also high economic value in marine stewardship (Roman et al., 2018). The environmental impact of any research agenda, deployment of monitoring equipment or research cruises, is likely to be small compared to C offsets or other more active CDR approaches.

In establishing a research agenda, it is important to acknowledge that we are not certain that there will be major C benefits, but we need to understand mechanisms and cycles at play. Such an approach could attract strong public support.

Co-benefits

Unlike some of the other ocean-based approaches considered in this report, there are considerable co-benefits in restoring ocean ecosystems, including biodiversity conservation and the restoration of many ecological functions and ecosystem services damaged by human activities such as pollution and overfishing. The recovery of kelp forests, for example, has C, biodiversity, and fisheries benefits (Duarte et al., 2020). The restoration of many species and habitats would also likely be beneficial for tourism and recreational and subsistence fishing; by encouraging the support of thriving marine life and habitats it is even possible that such restoration could mitigate the climate impacts of recreation.

In addition to any C benefits, reversing trends of biodiversity loss also have important nonuse values such as existence and spiritual values. The loss of whales and whale-fall habitats as a result of commercial whaling, for example, likely caused the endangerment and extinction of species dependent on these lipid-rich carcasses in the deep sea (Smith et al., 2019b). In addition to protecting the whales themselves, the recovery of whale populations could have C benefits and help protect and restore deep-sea communities. In some cases, these co-benefits might be the primary advantage of ecosystem restoration, and the uncertainties surrounding the geographic and temporal scale of CDR suggest that research into the long-term viability of many ecosystem-based approaches is essential.

Cost

Relative to other interventions, ecosystem-based solutions have the potential to be cost-effective, providing many benefits beyond CDR (Dasgupta, 2021). Nonetheless, nature-based solutions have been undercapitalized, and this lack of finance has been one of the barriers to the implementation and monitoring of such projects (Seddon et al., 2020). Similar to coastal blue carbon initiatives, the costs of implementing ecosystem recovery efforts for CDR depend on project size, intervention type, and monitoring of C removal (NASEM, 2019). If the effort is initiated as part of broader policies such as the Marine Mammal Protection Act, 30x30,[4] or international treaties, only the costs of measuring and monitoring CDR need to be considered, but these policy actions come with their own costs. Although many of these costs are likely to be borne by government as part of efforts to protect the ocean, there is increased interest among private financial investors to support restoration and conservation projects: approximately 3 percent of impact assets, or $3.2B, are associated with conservation, and interest is growing among investors (Bass et al., 2018; Dasgupta, 2021).

Note that although direct costs are relatively low for the implementation of many ecosystem-based natural climate solutions, there can be negative economic impacts to some industries, such as oil and gas, fisheries, and mining, that provide jobs and lease payments for use.

Energy

As is true with costs, under most of the scenarios presented here, the energy required for research activities would be restricted to measuring and monitoring CDR associated with ecosystem and species recovery. These studies would likely be focused on research cruises and equipment necessary to measure marine biogeochemistry and changes in the carbon cycle.

Governance

Conservation efforts, such as eliminating overexploitation, protecting habitats, and restoring extirpated species, are generally feasible, local, and have high global governability (Gattuso et al., 2018). In addition to the co-benefits and public acceptance of ecosystem-based approaches, one of the great advantages of focusing on the protection and restoration of marine habitats and species is that the goals tend to align with those of several broad-reaching international agreements and domestic laws.

Chapter 2 provides an overview of the current state of the legal framework for ocean CDR. Many of the international and domestic laws discussed in that chapter could apply to the ecosystem recovery approaches discussed in this chapter. Several international agreements recognize the importance of protecting marine species and their habitats. Most notably:

- Parties to the United Nations Convention on the Law of the Sea (UNCLOS) have an "obligation to protect and preserve the marine environment" and must, among other things, take steps to prevent pollution and other activities that "may cause significant and harmful changes" to the marine environment.[5] UNCLOS further provides that, whereas parties have a "sovereign right to exploit their natural resources,"[6] they must adopt "proper conservation and management measures" to ensure that living resources within their EEZs

[4] See https://www.oceanunite.org/30-x-30/.

[5] Art. 192, 194, & 196, United Nations Convention on the Law of the Sea, 1833 U.N.T.S. 397, Dec. 10, 1982 (hereinafter "UNCLOS").

[6] Art. 193, UNCLOS.

are "not endangered by over-exploitation."[7] Parties must also take steps to conserve the "living resources of the high seas."[8]

- The Convention on Biological Diversity (CBD) requires parties to "[d]evelop national strategies, plans or programmes for the conservation . . . of biological diversity."[9] To that end, parties must "[e]stablish a system of protected areas or areas where special measures need to be taken to conserve biological diversity" and take steps to "restore degraded ecosystems and promote the recovery of threatened species."[10]

A new international agreement, aimed at conserving marine biodiversity in areas beyond national jurisdiction, is also being developed by the UN General Assembly.

Many U.S. laws similarly recognize the importance of protecting marine ecosystems; for example:

- The Endangered Species Act (ESA) notes that many marine, freshwater, and terrestrial species have become extinct or at risk of extinction "as a consequence of economic growth and development untempered by adequate concern and conservation" (16 U.S.C. § 1531(a)). The ESA aims to reverse this trend by establishing a program for "the conservation of . . . endangered and threatened species" and "the ecosystems upon which they depend" (16 U.S.C. § 1521(b)).
- The Marine Mammal Protection Act (MMPA) similarly recognizes that certain "marine mammals are, or may be, in danger of extinction or depletion as a result of man's activities" (16 U.S.C. § 1361(1)). The MMPA further declares that such species are an important resource that "should be protected" so as to "maintain the health and stability of marine ecosystems" (16 U.S.C. § 1361(6)). The MMPA aims to maintain stocks of marine mammals at their optimum sustainable populations, such that they continue to be "significant functioning element[s]" of their ecosystems (16 U.S.C. § 1461(2)).
- The National Marine Sanctuaries Act (NMSA) indicates that "certain areas of the marine environment" have "special national, and in some cases international, significance" and thus require special protection (16 U.S.C. § 1431(a)). The NMSA establishes a program through which such areas are identified and conserved (16 U.S.C. § 1431(b)(1)-(2)). A key goal of the program is to "maintain the natural biological communities in the national marine sanctuaries and to protect . . . natural habitats, populations, and ecological processes" (16 U.S.C. § 1431(b)(3)).

The ecosystem recovery approaches proposed in this chapter are broadly consistent with the above goals. They are also in line with the goal set by President Biden in January 2021 of conserving at least 30 percent of U.S. land and waters by 2030.[11] A preliminary report on pathways for achieving the goal was published by the Departments of Agriculture, Commerce, and Interior and the Council on Environmental Quality (National Climate Task Force, 2021). The report emphasizes the importance of restoring ocean ecosystems, while also recognizing the "need to fight climate change with the natural solutions that . . . the oceans provide" (National Climate Task Force, 2021, p. 6). The conservation efforts discussed in this chapter are able to do both.

[7] Art. 61, UNCLOS.

[8] Art. 117, UNCLOS.

[9] Art. 6(a), Convention on Biological Diversity, June 5, 1992, 1760 U.N.T.S. 79, June 5, 1992 (hereinafter CBD).

[10] Art. 8, CBD.

[11] Executive Order 14008 of January 27, 2021: Tackling the Climate Crisis at Home and Abroad, 86 Fed. Reg. 7619, 7627 (Feb. 1, 2021).

6.6 SUMMARY OF CARBON DIOXIDE REMOVAL POTENTIAL

The criteria for assessing the potential for ecosystem recover as a feasible approach to ocean CDR, described in Sections 6.2–6.5, is summarized in Table 6.3. The assessment presented is considered conservative and may change with future research; for example, future research may indicate that upper limits are higher for both scalability and durability.

6.7 RESEARCH AGENDA

We developed a research agenda (Table 6.4) that examines the potential role of protecting and restoring marine ecosystems in reducing additional C emissions and enhancing CDR. This agenda includes basic scientific research on the C removal potential and permanence of different organisms, ecosystems, and processes, an examination of the expected outcomes from different policy tools, and the socioeconomic and governance aspects of managing marine ecosystems and organisms for C removal. The framework for these studies combines experimental research, demonstration sites, and adaptive management.

For each of these systems, there is a need for research to help understand historical baselines in the context of the carbon cycle, past habitat degradation and organism losses due to human activities, and the potential changes to the characteristics of these systems as a result of climate perturbation and ocean acidification. It would be helpful to estimate the current and historical contribution of different species across the entire size spectrum in terms of annual C flux. Although there have been some efforts to estimate the CDR potential of marine recovery, such as changes in fisheries management and kelp afforestation, to our knowledge, there have been no broad solicitations for proposals at the federal level to address this nascent field.

Research will be required to follow the changes in C storage due to CDR as a result of management interventions such as the restoration of macroalgal habitats; the management of fish, whale, and other animal populations; and the establishment of MPAs. How much carbon is stored if ecosystems are restored? How much carbon is stored indirectly? These questions require more study. Given the somewhat piecemeal approach to studies of the coupling of oceanic ecosystems and carbon to date, there is a strong need for an ecosystems approach that combines both population dynamics, nutrient ecology and subsidies, and anthropogenic impacts. The framework developed for this study should be an integral part of long-term ecosystem-based management for the oceans, including CDR, fisheries, and other human uses.

This research agenda will likely require the development or use of current observing systems that can detect a change in C sequestration over a relatively short period of time (e.g., 10 years). A modeling effort will almost certainly be necessary to understand the C impacts of improved habitat and recovered populations over a measurable area. Indirect evidence from remineralized carbon and nutrients or associated oxygen consumption at the seafloor could help understand the role of animals and other marine organisms in these processes. For all of the approaches below, it will be essential to quantify the balance between C storage, burial, and release. It will also be important to show changes in C accumulation over time, both in the short term and the expected retention over decades and even centuries.

These efforts will require interdisciplinary work: the long separation of ecosystem ecology, population biology, and oceanography must end and go broader, expanding into economics, governance, and social science. Although each of the approaches below is not detail, it is expected that a full rollout would require several steps, including basic research, development, demonstration, deployment, and monitoring and verification (see NASEM, 2019). Key to the development of any place-based project is understanding the location and areal extent of the species or habitats considered below. More work is needed to develop remote sensing approaches to estimate potential areas

TABLE 6.3 CDR Potential of Ecosystem Restoration Efforts

Knowledge base What is known about the system (low, mostly theoretical, few in situ experiments; medium, lab and some fieldwork, few carbon dioxide removal (CDR) publications; high, multiple in situ studies, growing body of literature)	**Low–Medium** There is abundant evidence that marine ecosystems can uptake large amounts of carbon and that anthropogenic impacts are widespread, but quantifying the collective impact of these changes and the CDR benefits of reversing them is complex and difficult.
Efficacy What is the confidence level that this approach will remove atmospheric CO_2 and lead to net increase in ocean carbon storage (low, medium, high)	**Low–Medium Confidence** Given the diversity of approaches and ecosystems, CDR efficacy is likely to vary considerably. Kelp forest restoration, marine protected areas, fisheries management, and restoring marine vertebrate carbon are promising tools.
Durability Will it remove CO_2 durably away from surface ocean and atmosphere (low, <10 years; medium, >10 years and <100 years; high, >100 years), and what is the confidence (low, medium, high)	**Medium** 10–100 years The durability of ecosystem recovery ranges from biomass in macroalgae to deep-sea whale falls expected to last >100 years.
Scalability What is the potential scalability at some future date with global-scale implementation (low, <0.1 Gt CO_2/yr; medium, >0.1 Gt CO_2/yr and <1.0 Gt CO_2/yr; high, >1.0 Gt CO_2/yr), and what is the confidence level (low, medium, high)	**Low–Medium** Potential C removal <0.1–1.0 Gt CO_2/yr (low–medium confidence) Given the widespread degradation of much of the coastal ocean, there are plenty of opportunities to restore ecosystems and depleted species. However, ecosystems and trophic interactions are complex and changing, and research will be necessary to explore upper limits.
Environmental risk Intended and unintended undesirable consequences at scale (unknown, low, medium, high), and what is the confidence level (low, medium, high)	**Low** (medium–high confidence) Environmental impacts would be generally viewed as positive. Restoration efforts are intended to provide measurable benefits to biodiversity across a diversity of marine ecosystems and taxa.
Social considerations Encompass use conflicts, governance-readiness, opportunities for livelihoods, etc.	Trade-offs in marine uses to enhance ecosystem protection and recovery. Social and governance challenges may be less significant than with other approaches.
Co-benefits How significant are the co-benefits as compared to the main goal of CDR and how confident is that assessment	**High** (medium–high confidence) Enhanced biodiversity conservation and the restoration of many ecological functions and ecosystem services damaged by human activities. Existence, spiritual, and other non-use values. Potential to enhance marine stewardship and tourism.
Cost of scale-up Estimated costs in dollars per Mt CO_2 for future deployment at scale; does not include all of monitoring and verification costs needed for smaller deployments during R&D phases (low, <$50/t CO_2; medium, ~$100/t CO_2; high, >>$150/t CO_2) and confidence in estimate (low, medium, high)	**Low** <$50/t CO_2 (medium confidence) Varies, but direct costs would largely be for management and opportunity costs for restricting uses of marine species and the environment. No direct energy used.
Cost and challenges of carbon accounting Relative cost and scientific challenge associated with transparent and quantifiable carbon tracking (low, medium, high)	**High** Monitoring net effect on C sequestration is challenging.

TABLE 6.3 Continued

Cost of environmental monitoring Need to track impacts beyond carbon cycle on marine ecosystems (low, medium, high)	**Medium** (medium–high confidence) All CDR will require monitoring for intended and unintended consequences both locally and downstream of CDR site, and these monitoring costs may be substantial fraction of overall costs during R&D and demonstration-scale field projects. This cost of monitoring for ecosystem recovery may be lower.
Additional resources needed Relative low, medium, high to primary costs of scale-up	**Low** Most recovery efforts will likely require few materials and little energy, though enforcement could be an issue. Active restoration of kelp and other ecosystems would require more resources.

for restoration and protection for macroalgae and benthic algae, as well as marine animal carbon, with a deeper understanding of historic biomass and expectations for populations in a future altered by climate change and habitat modifications.

In addition to determining the C sequestration potential of ecosystem restoration, this work is expected to make important contributions to marine ecology. Uniting researchers in ecosystem ecology, biological oceanography, population biology, and social science—disciplines that have long operated on separate paths—to examine marine CDR can provide co-benefits by coordinating research, analyzing and synthesizing information, and drafting and implementing management plans.

Habitat Protection and Restoration

The current push to increase MPAs provides a unique opportunity to examine the impacts of marine conservation and restoration on CDR. Policies such as 30 x 30, calling for 30 percent protection of the oceans in less than a decade, could provide research platforms to measure changes in C uptake and sequestration in relation to macroalgae, marine animal carbon, benthic algae, and sedimentary carbon. By protecting multiple sites, there will be opportunities to compare ecosystem functions across protected areas as well as inside and outside of protected areas. These could serve as experimental treatments that can be leveraged to test ideas about ocean CDR linked to ecosystems. As with coastal blue carbon systems, assessing the C stocks of offshore MPAs, and their changes over time, will require mapping, sample collection and analysis, monitoring C movement into and out of the system, and determining emissions avoided because of management activities (Howard et al., 2017). One potentially effective approach to understanding the CDR potential of restored ecosystems is to compare the present-day C fluxes in the ocean to preindustrial C transfer. By quantifying the human perturbations of C flux in the ocean, we can estimate the potential role of a restored ocean in the carbon cycle (see, e.g., Regnier et al., 2013).

Historically, much of the ocean economy has been based on commercial extraction, such as hunting fisheries, whaling, and guano, though this has been changing in recent years (e.g., Roman et al., 2018). A shift away from the overharvest of marine resources in the twentieth century is likely to result in changes in the carbon cycle: by reduction in the extraction of fish biomass, fuel consumption, and habitat disturbance and through changes in trophic structure. There have been many calls to halt the loss of biodiversity as fast as possible and to reach net positive results through restoration and regeneration of marine and terrestrial ecosystems (e.g., naturepositive.org). We strongly recommend that C accounting and research be a central aspect of the monitoring of MPAs and restoration efforts in the coming decades, including the examination of multiple systems with geographic, ecological, and taxonomic differences and varying levels of anthropogenic pressure. There is a need to develop consistent methodologies to conduct economic analyses and examine trade-offs between C benefits and other ecosystem services (e.g., Boumans et al., 2015).

Community engagement is essential here. Social science research can help us understand the institutions, policies, and cultural practices that lead to community support and engagement in marine ecosystem recovery. It can also help determine why efforts sometimes fail and what can make them successful. A governance component could examine how the legal framework for MPAs could be structured to increase C benefits. Since almost 50 percent of sedimentary C stocks reside outside of EEZs, international cooperation will be required to prevent remineralization of C stored beneath the high seas (Atwood et al., 2020). Ultimately, a transdisciplinary research program that includes questions of governance, oceanographic, conservation, and ecological research activities, and stakeholder engagement is likely to result in the greatest success.

Macroalgae

Research is needed to address numerous ecological, social, and biogeochemical questions in relation to the protection and restoration of macroalgal communities. It is essential to quantify the balance between C burial and release. What is the permanence of this buried material, and how vulnerable is it to rising temperatures and elevated climate change disturbances? Measurements of detritus production by whole macroalgal communities are needed to estimate how much released macroalgal particulate organic carbon is likely to become sequestered in deep, coastal sediment or other habitats (Queirós et al., 2019). Sequestration rates could be small if much of the POC is consumed and respired back to the surface ocean or if it settles on the seabed along the continental shelf. What is the natural mechanism for transporting POC away from the coast and then downward into the deep ocean? Stable isotopes and eDNA could be employed more widely to quantify and trace the contribution of macroalgae to deep-sea sediments, both in the past and present.

There is also a need for improved estimates of the global area and production of macroalgae at the level of major functional groups (Krause-Jensen et al., 2018). There has been some progress on this front using high-resolution satellite imagery, such as efforts to map the distribution and persistence of giant kelp (*Macrocystis pyrifera*) forests across a 35-year time series (Arafeh-Dalmau et al., 2021). Such techniques could be applied to monitor kelp protection and restoration efforts. Modeling frameworks would also be essential to understand the spatial distribution of macroalgal biomass and its contribution to climate budgets and C sequestration (van Son et al., 2020). There is also a need to better understand the impacts of elevated temperatures and other climate impacts on macroalgal distribution and C storage.

Natural and social science research on the actions that could be used for expanding natural macroalgae is needed, including studying the costs and benefits of deployment and if the actions are effective. Kelp restoration is becoming a prominent management intervention, but it can be quite expensive (more than $6,000 per hectare; Eger et al., 2020). Studying restoration success at appropriate spatial scales will be essential to encourage institutions to support kelp restoration.

Benthic Communities

There is a need to extend our understanding of the contributions that benthic communities make in relation to greenhouse gases, C sequestration and storage, and other climate change–related measures (Solan et al., 2020). Although there has been a substantial amount of work on shallow and intertidal benthic systems, such as mangroves and seagrasses, offshore and deep-sea benthic communities have largely been overlooked. Even shallow-water benthic communities, such as oyster reefs and other bivalve populations, could benefit from increased scrutiny: calcium carbonate ($CaCO_3$) shell production decreases alkalinity and can act as a source for CO_2, yet shellfish reefs contain significant pools of carbon and facilitate atmospheric C drawdown via filtration and biodeposition (Fodrie et al., 2017; Solan et al., 2020). Understanding the balance between these processes is essential to quantify the C dynamics of benthic communities.

A research agenda that explores benthic protection and restoration would expand our understanding of these and surrounding areas. Benthic algae rarely occur in isolation: in designing a research agenda, it is important to move beyond individual species to consider ecosystem-level responses. Approaches should measure the balance of C release and burial, or net community calcification versus photosynthesis, at the ecosystem level.

What is the extent of C removal at the ecosystem level as a result of human activities such as trawling and recreation? How can these activities be mitigated? An interdisciplinary approach that examines how these changes could affect key stakeholders is more likely to succeed from a carbon, conservation, and societal perspective.

Marine Animals and Carbon Dioxide Removal

Given the many uncertainties surrounding CDR and marine animals, quantifying the direct and indirect roles of marine fauna in the carbon cycle is a priority for further research (Stafford et al., 2021). Synthesis studies to identify the potential significance of marine vertebrates in regional and ocean basin scales is essential, to see whether their functional impacts are greater than the relatively small biomass (compared to phyto- and zooplankton) and compare them to other carbon cycle components (Martin et al., 2021).

Of perhaps equal value is examining the role of fisheries and other human activities, such as shipping, on marine populations and how the mitigation of some of the resulting environmental impacts could affect CDR. Essential in these studies is understanding the approaches necessary to ensure that reducing an activity such as bottom trawling in one area does not increase trawling in another, resulting in no net gain for carbon. An international and multistakeholder approach will likely be necessary here. A governance component of this research could examine how the legal framework for protecting marine mammals, seabirds, fish, and other large animals could be structured to increase C benefits.

A recent report and workshop at the International Whaling Commission identified data gaps and research needs on the role of whales in the carbon cycle (Roman et al., 2021). Although several studies have examined the ecological role of cetaceans, understanding these mechanisms, especially regarding nutrient transport and cycling, in the context of CDR is still in its infancy. Empirical investigations and additional study sites are needed to enhance, support, or refute hypotheses regarding the role of cetaceans in the carbon cycle. Research on the vulnerability of different species and populations to rising sea surface temperatures and reduced sea ice will help determine which species and populations, along with their CDR potential, are likely to be affected by climate change in the next century (e.g., van Weelden et al., 2021).

Marine Biomass and Deadfall Carbon

The CDR potential of pelagic communities has not been as well examined as coastal and nearshore benthic systems. The development stage of research for the restoration of marine animals involves understanding historic numbers and ranges, but also projecting changes with altered habitats and climates. As biomass and population numbers change, it will be essential to look for signals, via changes in organic and inorganic C stocks, nutrients, or associated oxygen consumption, in the water column and at the seafloor, on both local and larger scales. In addition to changes prompted by conservation, there is a need for further analyses of shifts in species habitat and the potential influence on food webs and the carbon cycle. Understanding the impacts of the expansion of salps into former krill ranges in the Antarctic and the potential biogeochemical changes, for example, will help track the role of animals in the carbon cycle now and in the future ocean (Böckmann et al., 2021).

Among cetaceans, generalist feeders such as humpback whales appear to be thriving under protection regimes. Other species such as the North Atlantic right whale have not responded to restrictions on direct harvest and remain endangered. There is an urgent need to examine past population sizes, current trends in marine vertebrate populations, and expectations for species and ecological processes under climate change scenarios. With data on life history and population structure, it is possible to quantify the amount of carbon stored in living biomass and sequestered in deadfalls. The most robust quantification currently available was based on global population estimates from 2001 (Pershing et al., 2010). This analysis could be updated with new population estimates and better understanding of pre-exploitation and future numbers.

Nutrient Cycling

The role of iron and nitrogen in the ocean carbon cycle has been well established, yet much of the focus on nutrient fertilization has been on open-ocean iron addition experiments and a few macronutrient (nitrogen and phosphorus) studies or schemes (see Chapter 3). There is a tendency to favor more natural approaches over engineered or artificial approaches; this preference could result in a dilemma between approaches with high C sequestration potential but low levels of acceptability and approaches with possibly low sequestration potential but high levels of acceptability (Bertram and Merk, 2020). At present, we do not know the sequestration potential of restoring historic populations of marine animals such as whales and seabirds or of shifting regimes under climate change, such as jellification, including salps.

There is a strong need for empirical data to test the role of nutrient cycling in the ocean carbon cycle. Although several models have been developed to look at the role of fecal nitrogen and iron in primary productivity (e.g., Lavery et al., 2010; Roman et al., 2014; Ratnarajah et al., 2016), it is unknown how closely these results approximate field conditions. Further quantification of cetacean diving behavior is needed to determine what proportion of foraging dives occur above or below the mixed layer and thus contribute to autochthonous or allochthonous nutrient cycling. It is also important to determine how available fecal plume nutrients are to phytoplankton and the impact of the microbial loop on nutrient uptake (Ratnarajah et al., 2018). Finally, it is critical to estimate what proportion of the carbon fixed as a result of whale fecal fertilization sinks below the mixed layer and becomes stored or sequestered (Roman et al., 2021). A spatial understanding of sequestration timescales is important here. For many areas of the ocean, it is likely that upper-ocean ecosystem productivity will be transported back to the surface within 50 years, and thus will have only a short-term influence on atmospheric CO_2 (Siegel et al., 2021a).

Predator and Prey Dynamics

The loss of apex predators in ocean ecosystems has resulted in changes in marine food webs, ecosystems, and productivity. The restoration of these and other large animals has the potential to increase C sequestration (e.g., the trophic interactions between sea otters, sea urchins, and kelp in the North Pacific) or decrease C storage (green sea turtles feeding on seagrasses in the Indian Ocean). The examination of changes in C dynamics as these predator–prey dynamics are restored offers a unique opportunity for marine ecologists, population biologists, oceanographers, and social scientists to work together. They can help develop the tools necessary to prompt these changes and examine the shifts in biomass, ecological functions, ecosystem services, and C sequestration over time.

6.8 SUMMARY

Efforts to protect and restore marine ecosystems cannot be considered in isolation from the challenge of reducing the production of greenhouse gases. These goals of restoring habitats and recovering overharvested species are ambitious but necessary and could help reduce atmospheric carbon in coming decades, achieving other co-benefits along the way (Hoegh-Guldberg et al., 2019). This chapter identifies some of the approaches for marine ecosystem restoration that could contribute and the processes by which this restoration contributes to C removal, going beyond the current baseline of sequestration.

Most of the signatories of the Paris Agreement have committed to nature-based solutions in their climate change programs (Seddon et al., 2019). A whole ecosystem perspective, which includes abundant species and protected ecosystems as well as those in need of restoration, will be essential to these approaches, along with better understanding of the ecological functions provided by marine species. Leveraging naturally occurring C fixation, storage, and sequestration interactions in the oceans and on land, can be a strategy with lower risk and cost than many geoengineering solutions (Griscom et al., 2017; Martin et al., 2021).

Whether ecosystem restoration is scalable is a matter of perspective and timing. Although the CDR potential of ecosystem recovery in the oceans is relatively small in comparison to present greenhouse gas emissions, this is perhaps more a reflection of the magnitude of current global emissions than a diminished role in the seas. As emission rates come down, ocean CDR could rise in importance. In anticipation of that transition, ecosystem-based approaches merit discreet attention and funding now.

The challenge for ecosystem-based CDR does not center around functionality—clearly animals, plants, and microbes all play an important role in the carbon cycle of the ocean—rather it is a question of scaling, verification, and concerns about reversibility. Restoring and protecting the ocean will likely remove a substantial amount of carbon from the atmosphere (though even approximate estimates remain challenging at this point) with relatively low costs for CDR verification above those already allocated for conservation efforts. As C emissions approach net zero, a diverse portfolio of marine conservation efforts could make a substantial contribution to global CDR approaches.

TABLE 6.4 Research and Development Needs: Ecosystem Recovery

#	Recommended Research	Gap Filled	Environmental Impact of Research	Estimated Cost of Research ($M/yr)	Time Frame (years)
6.1	**Restoration ecology and carbon**	**Estimate the change in C storage between natural and present-day marine ecosystems**	**Low**	**8**	**5**
6.2	**Marine protected areas (MPA): Do ecosystem-level protection and restoration scale for marine CDR?**	**Estimate the ability of ocean conservation and MPA protection to enhance the storage and sequestration of carbon per year until 2050**	**Low**	**8**	**10**
6.3	**Macroalgae: Carbon measurements, global range, and levers of protection**	**Improve our understanding of the fate of macroalgal carbon, the range of different species and habitats, and the socioeconomic levers and costs of restoring kelp and other macroalgal habitats**	**Low**	**5**	**10**
6.4	Benthic communities: disturbance and restoration	Improve our understanding of the impacts of human disturbance on benthic communities and the potential rate of change under different protection scenarios	Low	5	5
6.5	**Marine animals and CDR**	**Carbon interactions and outcomes have been estimated for only a few marine species. Efforts will improve our understanding of the direct and indirect impacts of marine animals on CDR, including biomass, deadfall carbon, nutrient transfer, and trophic cascades**	**Low**	**5**	**10**
6.6	Animal nutrient-cycling	Test of models of movement of iron (vertical) and nitrogen (horizontal) as mediated by whales and other air-breathing vertebrates	Low	5	5
6.7	Commercial fisheries and marine carbon	Improve our understanding of fisheries emissions, fish populations and ecological function, and impacts on sedimentary carbon	Low	5	5
6.8	Community engagement with marine ecosystem recovery	Understand what institutions, policies, and cultural practices lead to community support and engagement in marine ecosystem recovery; why efforts might fail; and what can make them successful	Low	2	5

NOTE: Bold type identifies priorities for taking the next step to advance understanding of ecosystem recovery as a CDR approach.

7

Ocean Alkalinity Enhancement

7.1 OVERVIEW

Current concern about the accelerated rate of carbon dioxide (CO_2) diffusion from the atmosphere into the surface ocean has prompted the marine scientific community to explore ocean CO_2 removal (CDR) approaches. Land-based CDR methods such as afforestation or bioenergy with carbon capture and storage have received much attention recently. However, meeting climate mitigation targets with land-based CDR alone will be extremely difficult, if not impossible, because the ocean governs the atmospheric CO_2 concentration and acts as the natural thermostat of Earth, simply because the ocean contains more than 50 times as much carbon as the atmosphere (Sarmiento and Gruber, 2002). One proposed ocean-based CDR technique is ocean alkalinity[1] enhancement (OAE) (Figure 7.1), also termed enhanced weathering (EW), proposed by Kheshgi (1995). This approach is broadly inspired by Earth's modulation of alkalinity on geological timescales. Adding alkalinity via natural or enhanced weathering is counteracted by the precipitation of carbonate, which reduces alkalinity and, in today's ocean, is driven almost entirely by calcifying organisms. For example, on geologic timescales, the dissolution of alkaline silicate minerals plays a major role in restoring ocean chemistry via addition of alkalinity to the ocean and conversion of CO_2 into other dissolved inorganic carbon (DIC) species (Archer et al., 2009). To date, most attention has been paid to terrestrial EW applications (Köhler et al., 2010; Schuiling and Tickell, 2010; Hartmann et al., 2013), with potential co-benefits in addition to CDR including stabilization of soil pH, addition of micronutrients, and crop fertilization (e.g., Manning, 2010). When applied to the ocean, EW of minerals is achieved by adding large amounts of pulverized silicate or carbonate rock or their dissolution products, which adds alkalinity to the surface ocean and thereby "locks" CO_2 into other forms of DIC, which is expected to promote atmospheric CO_2 influx into the ocean. Specifically, following alkalinity addition, CO_2 is converted into bicarbonate ions (HCO_3^-) and carbonate ions (CO_3^{2-}), and these chemical changes lead to a rise in pH (Kheshgi, 1995; Gore et al., 2019). There-

[1] Alkalinity can be defined as the number of moles of hydrogen ion equivalent to the excess of proton acceptors over proton donors in seawater (Dickson, 1981).

fore, this approach has the potential to not only remove atmospheric CO_2 but also counteract ocean acidification and thus contribute to the restoration of ecosystems threatened by it.

OAE involves the dissolution of large amounts of naturally occurring silicate (Schuiling and Krijgsman, 2006; Köhler et al., 2010), carbonate minerals (Kheshgi, 1995; Rau and Caldeira, 1999; Caldeira and Rau, 2000; Harvey, 2008; Rau, 2011), and mineral derivatives or other alkaline materials, such as some industrial waste products (Figure 7.1). Dissolution may either occur in the ocean, following processing, grinding, and dispersal of mineral, or it might be achieved in chemical reactors on land or on board ships (Figure 7.1). Another OAE approach is the generation of sodium hydroxide (NaOH) electrochemically (electrodialytically or electrolytically) resulting in the drawdown of atmospheric CO_2 into bicarbonate (Rau, 2008).

Natural increases in ocean alkalinity via rock weathering can lead to the removal of at least 1.5 moles of atmospheric CO_2 for every mole of dissolved magnesium (Mg)- or calcium (Ca)-based minerals (e.g., wollastonite, olivine, and anorthite) and 0.5 mole for carbonate minerals (e.g., calcite and dolomite) (see Dissolution of Naturally Occurring Minerals below). These processes are responsible for a net removal of 0.5 billion t CO_2/yr from the atmosphere (Renforth and Henderson, 2017).

Based on modeling studies, adding large amounts of alkalinity globally could be an effective CDR method (Ilyina et al., 2013; Keller et al., 2014) that has the potential to promote CO_2 uptake from the atmosphere while mitigating ocean acidification. Some of the proposed minerals are abundant in Earth's crust, but the costs, logistics, and environmental footprints of mining, industrial transformation (e.g., the conversion of limestone into quicklime via calcination), pulverizing, and transporting minerals require careful consideration when designing OAE deployments in open ocean and coastal locations (Figure 7.1). Of particular concern are the local environmental impacts and the CO_2 release associated with a considerable increase in mining, industrial transformation, and transportation activities required for large-scale deployments of minerals (see Sections 7.3 and 7.4). Like other CDR approaches and negative-emissions technologies, OAE must be examined carefully after experimental validation to determine its potential benefits and environmental and social side effects (both intended and unintended), as well as assessment of the geographic boundaries for its deployment. The vast majority of results on OAE impacts on chemical and biological processes come from modeling studies, and there is therefore an urgent need to generate empirical data from laboratory and mesocosm studies, and to explore these questions in small field trials. This chapter brings together the state of knowledge on OAE and explores the feasibility potential and anticipated limitations of this CDR approach.

According to criteria described in Chapter 1, the committee's assessment of the potential for OAE as a CDR approach is discussed in Sections 7.2–7.7 and summarized in Section 7.8. The research needed to fill gaps in understanding of OAE as an approach to durably removing atmospheric CO_2 is discussed and summarized in Section 7.9. Although OAE and electrochemical approaches are explored in separate chapters, there are intrinsic commonalities in the methods.

7.2 KNOWLEDGE BASE

The ocean absorbs up to ~30 percent of the CO_2 that is released to the atmosphere through physicochemical processes and, as atmospheric CO_2 increases, equivalent levels of CO_2 are measurable in the surface ocean within timescales of months to years, causing ocean acidification. Briefly, as CO_2 diffuses in seawater, it combines with H_2O to form carbonic acid (H_2CO_3), which dissociates into HCO_3^- and protons (H^+), causing ocean acidification. A proportion of the excess protons combine with CO_3^{2-} to form more HCO_3^-. The concentration of carbonate in the ocean is governed by the global alkalinity cycle (Broecker and Peng, 1982). Essentially, the production of calcium carbonate ($CaCO_3$) in the surface ocean exceeds the supply of alkalinity from rivers. To

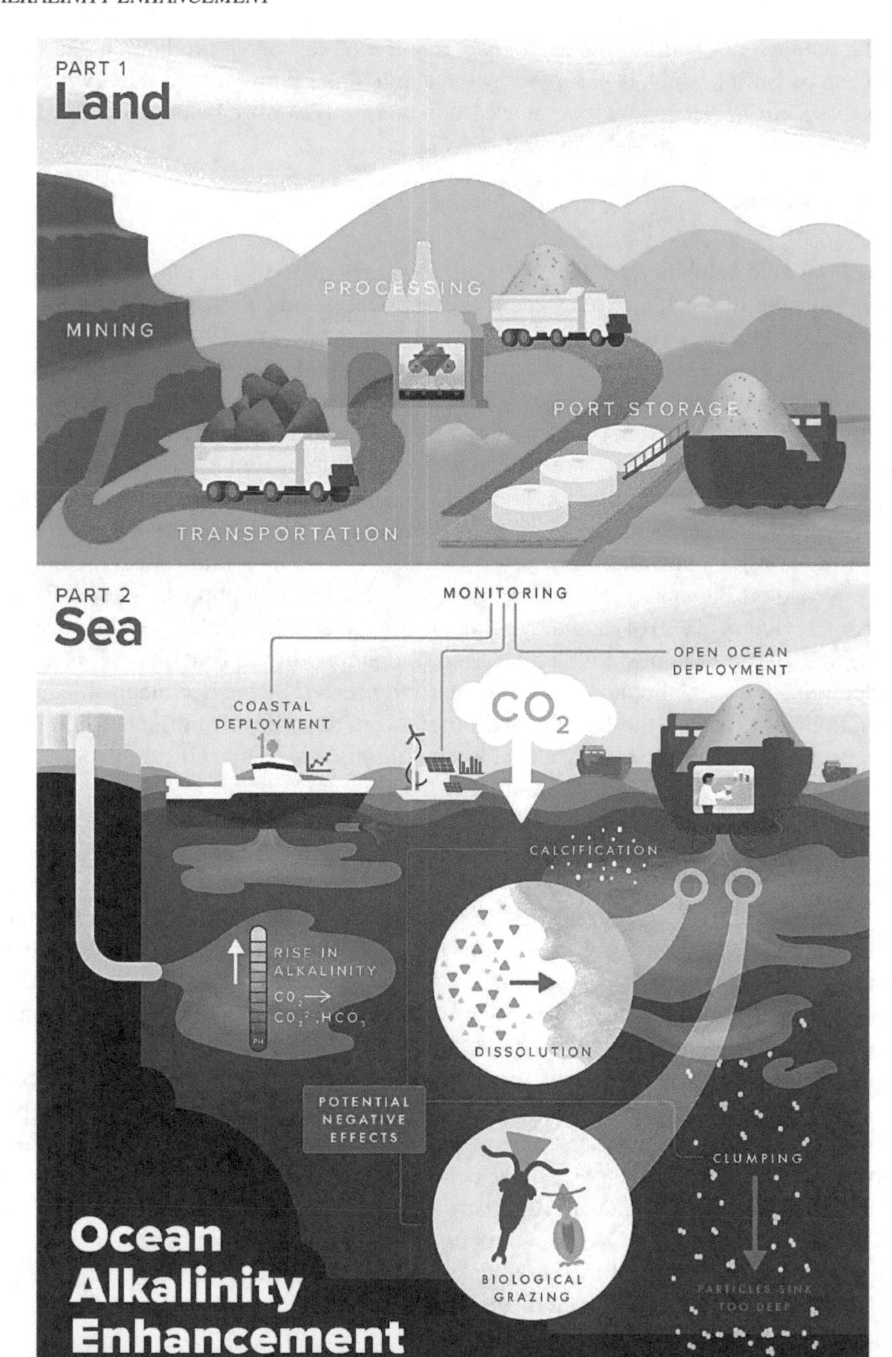

FIGURE 7.1 Steps and potential impacts of ocean alkalinity enhancement in coastal and open-ocean environments. (Part 1) Mining and transportation and required infrastructure at ports for storage of alkaline products. (Part 2) Addition of minerals or alkaline seawater (via prior dissolution of minerals or addition of NaOH generated electrochemically) to coastal waters through pipelines and ships, and to open-ocean waters through ships; monitoring of the carbonate system through autonomous measurements on buoys and flow through systems aboard ships; potential impacts of alkalinity addition.

maintain the balance of alkalinity fluxes, only a fraction of carbonate production can be removed from the ocean by burial, which is governed by the depth of the saturation horizon (where precipitation equals dissolution, dictated by inorganic carbon thermodynamics[2]), defined as the depth where the saturation state of the carbonate mineral is 1 (7.1):

$$\Omega = [Ca^{2+}]\,[CO_3^{2-}]/K_{sp} \tag{7.1}$$

Where K_{sp} is the solubility product constant. Ω values of 1 suggest a fluid in thermodynamic equilibrium with the mineral; $\Omega < 1$ at greater ocean depths indicates undersaturation, and based on inorganic carbon thermodynamics, mineral dissolution should exceed precipitation. On the other hand, $\Omega > 1$ at shallower depths indicates oversaturation, and mineral precipitation is promoted with respect to dissolution. On timescales longer than millennia, carbonate compensation adjusts the alkalinity of the global ocean, counteracting any changes in atmospheric CO_2.

Carbonate chemistry conditions are critical to the functioning and ecological fitness of calcareous organisms, and this is exemplified in two water basins of contrasting carbonate properties: the Black Sea, with carbonate chemistry conditions comparable to those in the open ocean, and the Baltic Sea, which harbors low-alkalinity waters (Müller et al., 2016). Interestingly, while the Black Sea displays seasonal blooms of the calcareous plankton coccolithophores (Iglesias-Rodriguez et al., 2002; Kopelevich et al., 2014) and the sediments are covered by thick layers of calcareous ooze (Hay, 1988), the Baltic Sea does not harbor coccolithophores. In the Black Sea, it is also possible that the selection for coccolithophores might be partially forced by the decline in silicate following the construction of a dam on the Danube and other rivers that deliver nutrients to the basin. This was reported as a likely factor shifting an ecosystem dominated by the silicate-dependent diatoms and dinoflagellates into a coccolithophore-dominated system (Mihnea, 1997).

Only a small fraction of the total DIC pool in seawater is in the form of CO_2 gas that can exchange with the atmosphere. Adding alkalinity to seawater can be beneficial because it shifts the partitioning of inorganic species, lowering the concentration of CO_2 gas in seawater. Indeed, increasing alkalinity shifts CO_2 into other DIC species such as HCO_3^- and CO_3^{2-}, the main forms of seawater alkalinity, consequently, decreasing the partial pressure of CO_2 (pCO_2). If this process occurs at the ocean surface, air–sea flux of CO_2 will be enhanced, resulting in net oceanic CO_2 uptake. Globally, atmosphere–ocean upper mixed layer equilibration occurs on timescales of months to a year, although many factors (e.g., mixing and wind speed) will affect the exchange rate and there could be localized effects that accelerate this exchange. However, while alkalinity addition consumes CO_2 from the ocean surface, replenishment via air–sea flux can take place very slowly, possibly on timescales of years, depending on the oceanographic properties of the location where deployments take place (Harvey, 2008; Feng et al., 2017; Bach et al., 2019). Therefore OAE can possibly lead to severe carbonate chemistry perturbations until equilibration with air takes place. In the Mediterranean Sea, OAE simulations applying a constant and steadily rising discharge relative to the surface pH trend of the baseline scenario attaining comparable quantities of annual discharge by the end of the alkalinization period suggest nearly doubling the CO_2 uptake after 30 years (Butenschön et al., 2021).

The implications are twofold—the method has the potential to promote atmospheric CO_2 uptake by the ocean and mitigate acidification and its negative biological and long-term biogeo-

[2] In some $CaCO_3$-producing organisms, some organisms have been found to maintain calcification in undersaturated waters ($\Omega < 1$). In addition to mechanisms that promote low levels of H+ such as proton pumps and physical proximity of processes that remove H+, such as photosynthesis in calcifying algae (Bergstrom et al., 2019), the extent to which the shell or skeleton is covered by an organic layer can provide protection against dissolution (Ries et al., 2009). This is a common feature in most calcifiers; for example, the epicuticle in crustacea and echinoids, periostracum in mollusks, ectoderm in corals, and utricles in calcifying algae that cover the biomineral provide some protection against dissolution (Ries et al., 2009).

chemical consequences. For example, an increase in CO_3^{2-} concentrations increases the calcium carbonate saturation state of seawater, which is central to promoting and maintaining precipitation of calcium carbonate in organisms that produce shells, exoskeletons, and plates of carbonate such as some planktonic organisms, corals, and shellfish. OAE has potential benefits over other CDR schemes although empirical data are necessary to determine the effectiveness, risks, and side effects. Advantages include "permanent" CO_2 sequestration on timescales of millenia or longer in the absence of processes that remove the added alkalinity; not requiring long-term storage of large quantities of CO_2; and, in addition to its application as a CDR method, possible lessening of some of the effects of ocean acidification. OAE can also represent a source of nutrients (see Hartmann et al., 2013) including silicate and some beneficial (e.g., iron [Fe] and magnesium) micronutrients, although the potentially toxic effect of metals such as nickel (Ni) (Montserrat et al., 2017), leached from olivine, is of concern. The main unknown is how OAE deployments would alter the biogeochemical cycling of elements on local and planetary scales and the repercussions of these alterations on marine ecosystems given the permanence of these chemical changes.

Learning from the Past

The development of paleo proxies has enabled paleoreconstruction of seawater pH and the carbonate system (e.g., Foster, 2008; Wei et al., 2021); however, understanding the paleorecord of seawater alkalinity is challenging because alkalinity is poorly constrained. The balance between mineral weathering, which increases alkalinity, and the largely biotic precipitation of carbonate, which decreases alkalinity, is the primary mechanism governing seawater alkalinity. The silicate rock weathering hypothesis postulates that over timescales of a million to hundreds of million years, there is a negative feedback between the tectonic uplift of mountains and the amount of CO_2 in Earth's atmosphere. The mechanism driving this process is the increasing tectonic uplift of mountains that intensifies erosion on steep slopes, thus enhancing the rate of chemical weathering, which consumes atmospheric CO_2, thus cooling Earth. Ocean drilling science programs aim to test this hypothesis to improve forecasts of how this negative feedback that regulates atmospheric CO_2 can be used to predict whether the Earth system can regulate the rapid and vast amounts of anthropogenic carbon and over specific timescales (Koppers and Coggon, 2020).

Two periods of apparent contrasting alkalinity—the Pleistocene and the Paleocene–Eocene—are discussed by Renforth and Henderson (2017) to illustrate past forcings leading to shifts in alkalinity. During the last glacial period, in the Pleistocene (20 kya ago), high alkalinity is expected to have increased oceanic CDR and thus decreased atmospheric pCO_2 (a significant cause of glacial cooling). A combination of factors including lower glacial sea level leading to less continental shelf area for carbonate reef formation, thus decreasing coastal $CaCO_3$, and probable erosion and dissolution of old reefs ($CaCO_3 + CO_2 \rightarrow Ca^{2+} + CO_3^{2-}$), would increase CO_3^{2-} and global ocean alkalinity (Opdyke and Walker, 1992). A physical process in the deep ocean driven by alterations in ocean circulation in the Atlantic resulting in slow-moving deep waters comprising significantly more remineralized CO_2 would reduce deep-water pH, thus shoaling the lysocline, consequently adding alkalinity by dissolution of $CaCO_3$ (Boyle, 1988; Sigman et al., 2010; Renforth and Henderson, 2017).

The minor and trace element contents (ratioed to calcium) of biogenic carbonates have been used for paleoclimate reconstructions and, specifically, the uranium/calcium (U/Ca) ratio of biogenic carbonates has been used as a proxy for seawater CO_3^{2-}, particularly for the deep ocean where temperature changes are modest. Experimental calibrations indicate that U/Ca in corals and foraminifera are negatively correlated with seawater CO_3^{2-} or pH (Russell et al., 2004; Anagnostou et al., 2011; Inoue et al., 2011; Raitzsch et al., 2011; Keul et al., 2013; Raddatz et al., 2014; Allen et al., 2016). However, the correlation between U/Ca and seawater CO_3^{2-} is largely an outcome of alterations in the rates of the alkalinity pump, but can be confounded by the calcification strategy used by individual calcifiers (Chen et al., 2021). At present, it appears that a multiproxy approach combining stable isotopes and trace element/Ca ratios can only provide partial understanding of the biological mechanisms to be able to use these tracers for paleoceanographic reconstructions.

Dissolution of Naturally Occurring Rocks and Minerals

The overall approach proposed with OAE is to accelerate the slow geological process of mineral dissolution in seawater by increasing the reactive surface area of the minerals through pulverizing the alkaline rock into small particles. The hope with this approach is to lock CO_2 away from the atmosphere via conversion into proton acceptors (mainly HCO_3^- and CO_3^{2-}) over proton donors (Wolf-Gladrow et al., 2007) in a semipermanent fashion. Three main methods, in addition to electrochemical approaches (reviewed in Chapter 8), are currently being considered to enhance alkalinity in seawater: the dissolution of naturally occurring silicate-based minerals such as olivine, accelerated limestone weathering (Rau and Caldeira, 1999), and dissolution of calcium carbonate derivatives—quicklime (calcium oxide, CaO) or portlandite (calcium hydroxide, $Ca(OH)_2$)—in the surface ocean. Dissolution of other minerals such as chalk, calcite, or dolomite also affect seawater pH and total alkalinity on timescales of months. The impact of mineral addition is specific to the mineral itself and technology. The theoretical maximum potential of a rock or mineral (E_{pot}, kg CO_2/t_{rock}) to convert CO_2 into bicarbonate can be estimated by equation 7.2 (see Renforth, 2019):

$$E_{pot} = \frac{M_{CO2}}{100}\left(\alpha\frac{CaO}{M_{CaO}} + \beta\frac{MgO}{M_{MgO}} + \varepsilon\frac{Na_2O}{M_{Na2O}} + \theta\frac{K_2O}{M_{K2O}} + \gamma\frac{SO_3}{M_{SO3}} + \delta\frac{P_2O_5}{M_{CaO}}\right)\cdot 10^3\cdot\eta \qquad (7.2)$$

where CaO, MgO, SO_3, P_2O_5, Na_2O, and K_2O are the elemental concentrations of calcium, magnesium, sulfur (S), phosphorus (P), sodium (Na), and potassium (K), expressed as oxides, M_x is the molecular mass of those oxides. Coefficients α, β, γ, δ, ε, and θ consider the relative contribution of each oxide (α, β, ε, and θ are equal to 1, γ is equal to -1, and δ is equal to -2 for the pH range of the ocean); η is molar ratio of CO_2 to divalent cation sequestered during EW (equivalent to 1.4–1.7 for the ocean's temperature, pressure, and salinity ranges). E_{pot} values for common rock types is presented in Figure 7.2. Given the size of the rock resource, there may be scalable potential in the extremes of the distribution.

Silicate Minerals and Rocks (Olivine, Basalt)

Alkaline, silicate rock dissolution is an important process that consumes excess CO_2 and restores seawater chemistry on geologic timescales (Archer et al., 2009). This has inspired the idea of speeding up dissolution by crushing rock and pulverizing into small particles to enhance dissolution. The acceleration of silicate weathering generates stable dissolved calcium and magnesium bicarbonates accompanied by the release of metals.

Olivine-rich rocks are abundant and vary in composition and are among the most abundant minerals—forsterite and fayalite (Fe_2SiO_4). The most abundant olivine-rich rocks contain 30–50 percent iron (Deer et al., 2013). Forsterite is among the most desirable mineral constituents due to its high reactivity and relative high concentration in cations (e.g., Oelkers et al., 2018). The weathering of olivine in seawater can be described by

$$(Mg, Fe)_2\,SiO_4 + 4CO_2 + 4H_2O \rightarrow 2(Mg^{2+}/Fe^{2+}) + 4HCO_3^- + H_4SiO_4 \qquad (7.3)$$

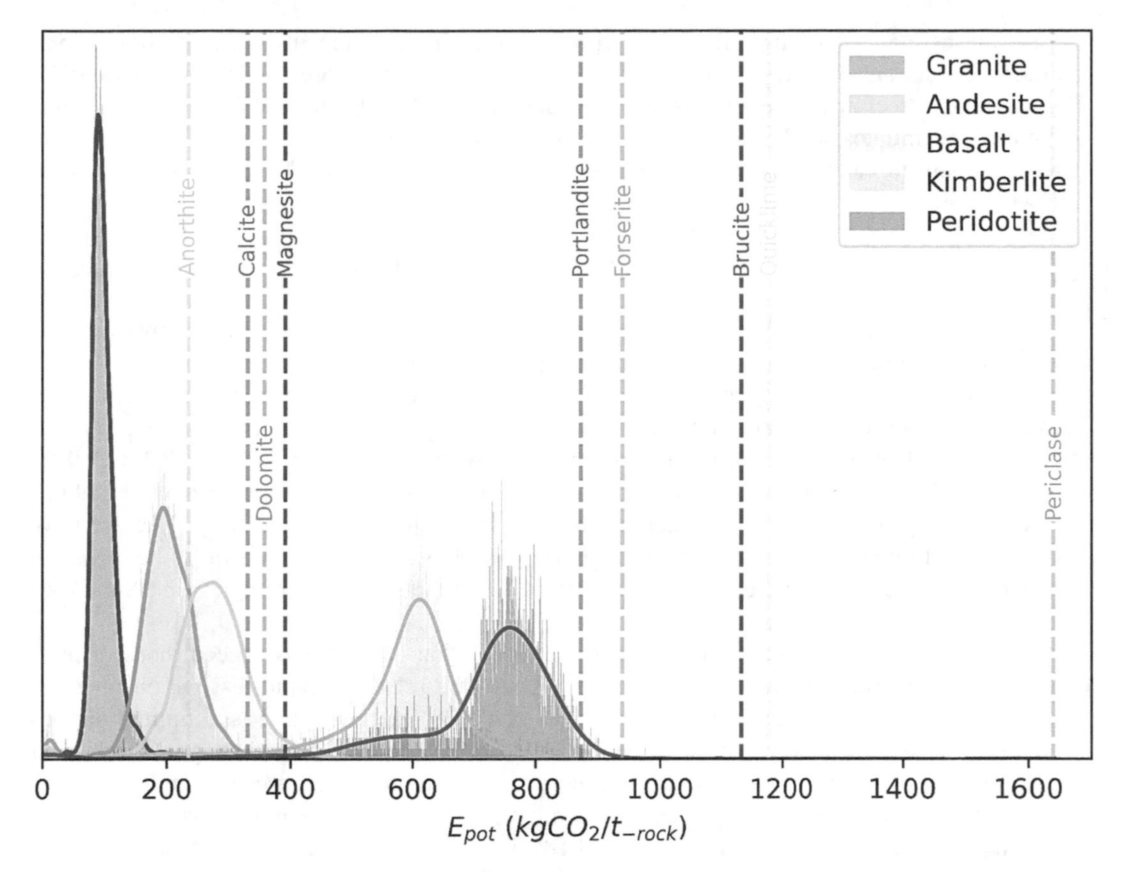

FIGURE 7.2 Enhanced weathering potential of common rock types derived from oxide data from GEOROCK database (http://georoc.mpch-mainz.gwdg.de/georoc/Start.asp). Histogram frequency (vertical axis) bars were generated using 1,000 bins across the horizontal axis scale to generate frequency of carbon capture potential for each rock type. Given that the total size of the samples varied (n = 10,839, 8,747, 10,337, 5,431, and 3,226, for granite, andesite, basalt, kimberlite, and peridotite, respectively) frequency was normalized to each total such that the vertical axis represents proportional frequency. The lines represent a Gaussian kernel density estimation fit of the histogram. The potentials for the pure minerals magnesite ($MgCO_3$), calcite ($CaCO_3$), dolomite [$CaMg(CO_3)_2$], forsterite (Mg_2SiO_4), anorthite ($CaAl_2Si_2O_8$), quicklime (CaO), portlandite [$Ca(OH)_2$], and brucite [$Mg(OH)_2$] were calculated by expressing their Mg and Ca composition as oxides and using equation 2. The efficiency factor n (equation 2) was assumed to be 1.5 for silicates and 0.75 for carbonates (Renforth and Henderson, 2017).

In the case of pure fayalite (equation 7.4) and forsterite (equations 7.5–7.7), the dissociation of olivine leads to the following products (see Griffioen, 2017, for details):

$$Fe_2SiO_4 + 0.5O_2 + 5H_2O \rightarrow 2Fe(OH)_3 + H_4SiO_4 \quad (7.4)$$

$$Mg_2SiO_4 + 2CO_2 + 2H_2O \rightarrow 2MgCO_3 + H_4SiO_4 \quad (7.5)$$

$$Mg_2SiO_4 + 2CO_2 + 2Ca^{2+} + 2H_2O \rightarrow 2Mg^{2+} + H_4SiO_4 + 2CaCO_3 \quad (7.6)$$

$$3Mg_2SiO_4 + 8CO_2 + 7^{1/2}H_2O \rightarrow 1/2Mg_4Si_6O_{15}(OH)_2 \bullet 6H_2O + 4Mg^{2+} + 8HCO_3^- \quad (7.7)$$

Equations 7.5 and 7.6 describe weathering reactions in environments where carbonate precipitation occurs and have a 1:2 molar ratio of olivine to CO_2 compared to 1:4 in equation 7.3. The secondary precipitates are important to consider because the efficiency of CDR declines when secondary precipitates are developed. For instance, the rise of pH in response to the above reactions can result in calcite precipitation and the concomitant release of CO_2 in the ocean via the following reaction:

$$Ca^{2+} + 2HCO_3^- \rightarrow CaCO_3 + CO_2 + H_2O \quad (7.8)$$

The release of CO_2 by this reaction diminishes the efficiency of OAE via olivine by roughly a factor of two compared to that without secondary precipitation.

The effect of microbes appears to be unclear and likely minor, but some studies indicate decreasing dissolution rates. Possibly, this is because of the presence of mineral coatings, and the effect microbes have on the oxidation and/or complexation of metals (Seiffert et al., 2014; Oelkers et al., 2018). Numerous studies have illustrated that thin layers of amorphous silica can establish on the olivine surface, slowing its dissolution (e.g., Wang and Giammar, 2013). The selective colonization of minerals from experiments exploring aqueous environments seems to be governed by the need of specific microbes to harvest nutrients from the dissolving olivine (e.g., Shirokova et al., 2012).

Although it is widely accepted that forsterite dissolution rates always exceed those of most other silicate minerals, this is a misconception because the relatively rapid dissolution rates are reported to occur under acidic conditions (Oelkers et al., 2018). Indeed there are limited options to enhance the rates of forsterite dissolution for EW, and the only processes that appear to significantly augment forsterite dissolution rates are increasing forsterite surface area (through grinding), decreasing seawater pH, and the presence of some organic ligands at neutral to basic conditions (Wogelius and Walther, 1991; Oelkers et al., 2018). In an analysis of the cost of grinding, Hangx and Spiers (2009) estimated over 60 kilowatt-hour (kWh) of energy would be required and >30 kg CO_2 emitted when finely ground olivine is produced to consume 1 ton of CO_2. The effect of the aqueous organic species in an aqueous solution on dissolution rates can be a result directly from the interaction of the organic species with the bonds that need to be broken to dissolve the mineral (Olsen and Rimstidt, 2008), or indirectly by changing the pH of the water (see Oelkers et al., 2018). It appears that because of very low dissolution rates per unit surface area, silicate minerals in general need to be pulverized to ≤1 μm to dissolve on relevant timescales as elevated ocean pH slows dissolution rates (GESAMP, 2019). The combination of secondary mineral precipitation and the limited routes to enhance dissolution rates of olivine are challenging issues in using EW of olivine as a CDR approach.

Limestone

Harvey (2008) proposed the idea of adding powdered $CaCO_3$ mineral (limestone) to seawater as a CDR protocol. Similar to olivine, $CaCO_3$ dissolution kinetics in seawater is expected to be slow for supersaturated conditions ($\Omega > 1$) and the release of alkalinity on short timescales only takes place under $CaCO_3$ undersaturation, and therefore this is an inefficient method to apply to the surface ocean. Limestone dissolution has been proposed as a potentially desirable method to apply to CO_2-rich waste gas streams where seawater and elevated CO_2 and limestone could facilitate the conversion of $CaCO_3$ into bicarbonate:

$$CaCO_3 + CO_{2(gas)} + H_2O \rightarrow Ca^{2+} + 2HCO_3^- \quad (7.9)$$

The dissolution rates of whole-shell biogenic carbonates increase with decreasing $CaCO_3$ saturation state, decreasing temperature, and fluctuate predictably with respect to the relative solubility of the calcifiers' polymorph mineralogy (high-Mg calcite [mol% Mg > 4] ≥ aragonite > low-Mg calcite [mol% Mg < 4]); the implications being that ocean alkalinity and temperature will continue to increase dissolution of biogenic carbonate shells and skeletons (Ries et al., 2016). The role of physics on the fate of the particles is exemplified in an experiment releasing limestone particles smaller than 10 μm in surface waters, where horizontal eddy diffusion appeared to be the major factor affecting the dispersion of chalk to concentrations below detection limit (Balch et al., 2009).

Carbonate Mineral Derivatives: Quicklime and Portlandite

Given the poor solubility of $CaCO_3$ in surface ocean seawater, Kheshgi (1995) investigated producing an alternative, quicklime (calcium CaO). Although the addition of 1 mole of CaO to seawater removes 2 moles of CO_2 (equation 7.8), the use of CaO is a less efficient CDR method because CO_2 is released in the production of quicklime by the thermal breakdown of limestone (equation 7.10).

$$CaCO_3 \rightarrow CaO + CO_2 \quad (7.10)$$

$$CaO + 2CO_2 + H_2O \rightarrow Ca^{2+} + 2HCO_3^- \quad (7.11)$$

A by-product of CaO that is generated through slaker technology is portlandite (calcium hydroxide $Ca(OH)_2$; equation 7.12).

$$CaO + H_2O \rightarrow Ca(OH)_2 \quad (7.12)$$

For average open-ocean pH, calcium hydroxide can react with CO_2 to form calcium bicarbonate:

$$Ca(OH)_2 + 2CO_2 \rightarrow Ca(HCO_3)_2 \quad (7.13)$$

Similarly to CaO, while 2 moles of CO_2 can be removed per mole of $Ca(OH)_2$, there is only a net removal of 1 mole of CO_2 because the generation of $Ca(OH)_2$ releases 1 mole of CO_2 under acidic conditions and through electrolysis:

$$CaCO_3 + 2H_2O + \text{direct current electricity} \rightarrow 0.5O_2 + H_2 + Ca(OH)_2 + CO_2 \quad (7.14)$$

7.3 EFFICACY

Dissolution Kinetics

The dissolution kinetics of minerals have been studied through laboratory experimentation for decades. However, the two main approaches to the study of chemical weathering in the field (geochemical) and laboratory (individual mineral dissolution rates) rarely produce the same results (Brantley, 2008). There are also conflicting results regarding the effect of CO_2 and carbonate ions on dissolution rates (Wogelius and Walther, 1991; Golubev et al., 2005). Nonstoichiometry and incomplete dissolution have also been reported, perhaps due to dissolution of impurities, precipitation of secondary minerals, or preferential leaching of elements from the mineral surface (Brantley, 2008). These discrepancies between field and lab observations, and the large uncertainties regarding which factors govern mineral dissolution (White and Brantley, 1995) highlight the urgent need for field experimentation.

There are many choices of independent variables and expressions describing the dissolution rates of minerals as well as spatial and temporal variations in mineral properties and external influences such as diffusion and advection. Numerical simulations capable of handling this complexity would be required to address the observed variability. Ideally the dissolution kinetics of a mineral should reflect the underlying chemistry and physics of the reactions, although there are contrasting ideas explaining how solids dissolve such that there is no single rate equation that can be used to model natural and technological processes (Rimstidt et al., 2012). Along with temperature and pH, particle size and surface area are perhaps the most important factors governing dissolution rates (Rimstidt et al., 2012). A mineral surface is composed of sites with different surface properties that constitute the reactive (White and Peterson, 1990) or the effective surface area (Aagaard and Helgeson, 1982; Helgeson et al., 1984). However, most studies use instead the total surface area (also referred to as the BET [Brunauer–Emmett–Teller] surface area) or the geometric surface (Liittge and Arvidson, 2008), which consider all the surface sites uniformly reactive (Grandstaff, 1978; White and Peterson, 1990). However, aggregation could lead to overestimation of the available surface area (when BET values are applied).

Fate of Mineral Particles in the Surface Ocean

A concern with the deployment of particles as one OAE approach (adding alkalinized seawater following mineral dissolution is also an option) is the fate of the particles spread on the ocean surface. Possible scenarios that are likely to result in increased export of particles to depth include removal of mineral grains by grazers and conditions leading to particle aggregation during deployment of the pulverized mineral (Figure 7.1). Therefore, timing of mineral deployment is expected to be hugely important. For example, the amount and type of organic matter associated with biological processes, such as the presence of high concentrations of phytoplankton and their associated mucus, can impact particle aggregation and increase the size of aggregates (see Hamm, 2002) promoting export to depths below the mixed layer.

The effect of particle size on dissolution rates has been investigated with a focus on olivine (Figure 7.3). It appears that the most challenging aspect is the need to grind the mineral to ensure CDR in a timely manner and dissolution in the surface ocean. Calculations by Hangx and Spiers (2009) suggested that grained olivine between 2 and 6 μm is required to achieve significant steady-state CO_2 uptake rates within a few decades. Grain sizes around 1 μm appear to be appropriate to facilitate dissolution before sinking out of the surface mixed layer (Köhler et al., 2013). Interestingly, for particles smaller than1 μm, Brownian motion appears to be the main mechanism driving aggregation (Köhler et al., 2013). Indeed a study by Bressac et al. (2012) using mesocosms where a Saharan dust deposition event was simulated revealed partial aggregation resulting in high sinking velocities, orders of magnitude greater than calculations based on modeled data (Köhler et al., 2013). Given the mineral characteristics of the dust—40 percent quartz, 30 percent calcite, and 25 percent clays (in addition to micro- and macronutrients), and given that the finest dust fraction is <20 μm, this type of deposition event could be considered a natural analog for mineral deployments.

7.4 SCALABILITY

Durability

Like other ocean CDR approaches, OAE is only effective if CO_2 can be "locked" away from the atmosphere for long time periods. The permanence of an ocean CDR approach has been defined

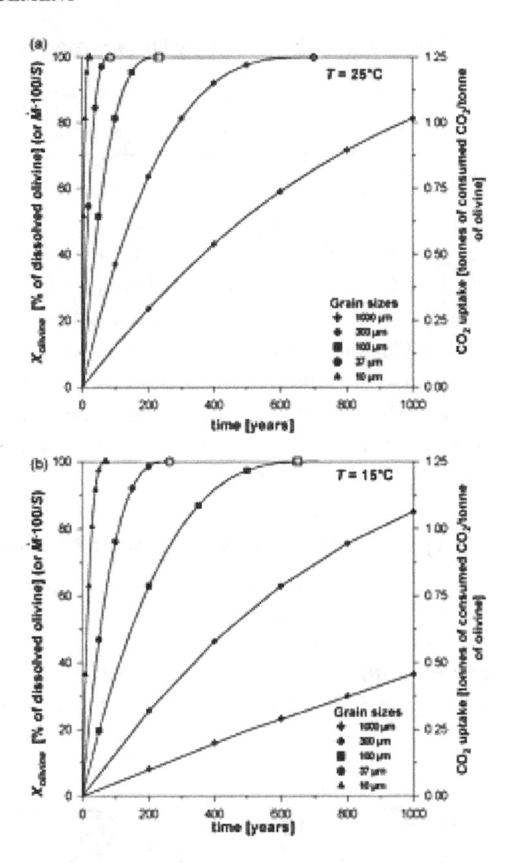

FIGURE 7.3 Percentage of granular olivine dissolution ($X_{olivine}$) (left *y* axis) and CO_2 uptake per ton of dissolved olivine (right *y* axis) as a function of time (see Hangx and Spiers [2009], for details). Open symbols indicate reaction times to reach steady-state reaction, at constant olivine supply rate. The amount of CO_2 removed per ton of olivine was calculated assuming reaction 2, which results in the formation of soluble Mg bicarbonate salts and CO_2:olivine uptake ratio of 1.25 t CO_2 per ton olivine. (a) Reaction time series at 25°C, assuming that the dissolution rate of olivine is 1.58×10^{-10} mol/m^2/s. (b) Reaction progress at 15°C, with the olivine dissolution rate being 5.19×10^{-11} mol/m^2/s (calculated using the dissolution rate at 25°C and an activation energy of 79.5 kJ/mol, following Wogelius and Walther (1992). SOURCE: Hangx and Spiers, 2009.

as one that leads to sustained CO_2 sequestration over centennial timescales. Although this has not been extended to OAE (see Siegel et al., 2021a), OAE can be considered as permanent when the conversion of CO_2 into other inorganic carbon molecules (HCO_3^- and CO_3^{2-}) remains stable for periods of time exceeding a century. The extent of permanence of OAE is critically dependent on the properties of the water column, including: (1) those that govern dissolution kinetics; (2) the abiotic (or biotic) precipitation of carbonate (an undesirable side effect because it leads to the release of CO_2); (3) the amount and type of organic matter that leads to aggregation of mineral particles, for example, the adsorption of dissolved organic matter onto dust particles or the effect of colloidal-size particles and gels on particle aggregation (Alldredge and Silver, 1988; Engel, 2000; Armstrong et al., 2001; Francois et al., 2002; Klaas and Archer, 2002; van der Jagt et al., 2018); (4) the possible removal of particles by grazers (Banse, 1994); and (5) ocean physical processes such as advection and subduction (Itsweire et al., 1993), which dilute the local chemistry impacts from the site of deployment.

Location and Frequency of Deployments

Critical issues determine the ideal frequency and location of deployments. In terms of the efficacy in removing carbon from the atmosphere, it has been suggested that only when deployments are conducted continuously is atmospheric CO_2 successfully reduced, although these reductions are modest because atmospheric CO_2 from human sources continues to increase (Keller et al., 2014). Locations for mineral deployment (see Figure 7.4) have been evaluated in the open ocean (Köhler et al., 2013) and coastal zones (Hangx and Spiers, 2009). The main challenge is that mineral particles must be ground to a very small size to avoid the rapid sinking of particles out of the euphotic zone before being dissolved, to ensure rapid CO_2 uptake at the surface (Meysman and Montserrat, 2017). It has been suggested that $CaCO_3$ powder could be added to the surface layer in regions where the depth of the boundary between supersaturated and unsaturated water (saturation horizon) is relatively shallow (250–500 meters) and where upwelling velocity is high (Harvey, 2008). The much deeper saturation horizons (depths where Ω_{cal} and $\Omega_{arag} = 1$) in the North Atlantic compared with those in the North Pacific and North Indian oceans (Jiang et al., 2015) could be important considerations when deciding deployments.

Decisions on the scale of deployments will be determined by a range of factors (see Caserini et al., 2021), including

- accessibility of sources of sufficient mineral;
- need for substantial upscaling of the current magnitude of mining and production of derivatives (and the associated environmental and other effects);

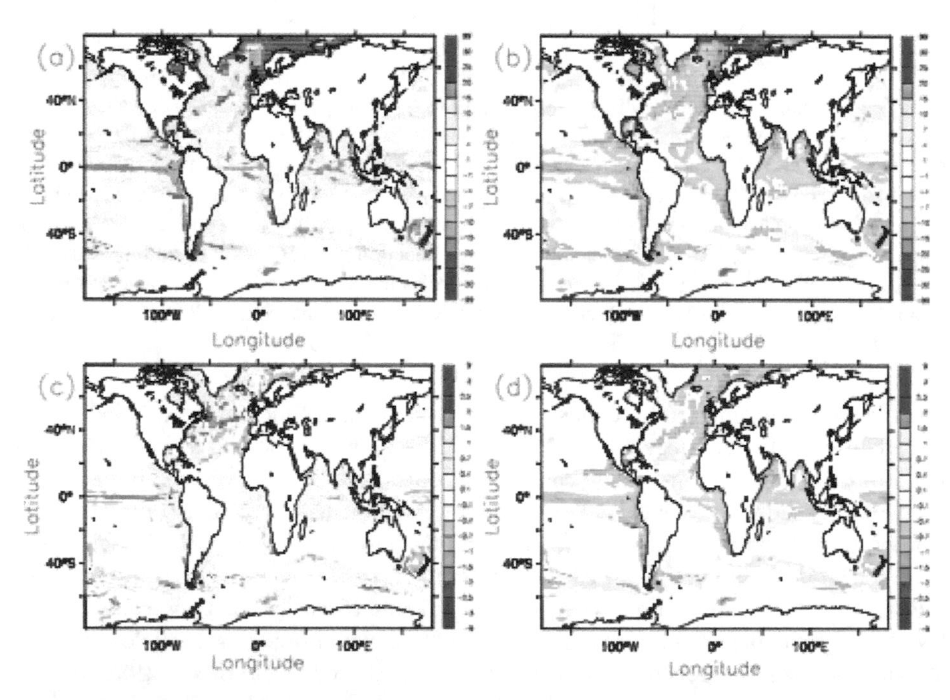

FIGURE 7.4 Simulated anomalies (g C/m^2 per year) from model runs representing scenarios when olivine is added with respect to a control without olivine (see Köhler et al., 2013). Values represent annual averages in 2009. (a) Diatom net primary production; (b) nondiatom net primary production; (c) export production of organic carbon at 87-meter depth. (d) $CaCO_3$ export at 87-meter depth. SOURCE: Köhler et al., 2013. Licensed by Creative Commons CC BY 3.0.

- economics of rock mining, crushing, and transportation;
- accessibility and adaptation of existing ships;
- implementation of loading facilities in ports;
- international and national laws or regulations;
- restrictions that could limit mining, processing, transportation, and/or discharge;
- carbon removal efficiency; and
- human elements, that is, persons' and governments' agreements to responsible intervention.

Simulation studies on the behavior of mineral particles following their release in the wake of a ship using different methodologies and fluid dynamic modeling approaches suggest that key parameters governing mineral dissolution in the upper layers of the water include particle radius and density, diffusion potential, volume discharge rate, vessel speed, waterline length, time after disposal, flow velocity at the injection point, and diameter of the circular area of discharge, with other parameters such as geometry and porosity also expected to be important (Caserini et al., 2021).

The criteria for the choice of location of deployments remain to be determined. For coastal regions, ecosystems particularly susceptible to ocean acidification and upwelling zones could perhaps be good initial candidates for trials. For open-ocean deployments, in addition to new dedicated vessels, the use of existing vessels with modifications to transport and deploy mineral, when possible, would be financially, logistically, and energetically desirable. Long routes would be well suited for low discharge rates, and new designs would optimize mineral storage and deployment (Caserini et al., 2021). An important issue with mineral deployment is how to prevent loss of mineral to depth and the secondary precipitation of minerals (e.g., $CaCO_3$ or $Ca(OH)_2$ when deploying quicklime) (see Caserini et al., 2021).

If deployments were to be conducted on a large scale, atmospheric pCO_2 removal might cause the release of CO_2 from ocean regions that are not subject to addition of $CaCO_3$, consequently raising the pH and supersaturation in these regions (see Harvey, 2008). The addition of limestone powder in small particles would potentially increase albedo and thus induce a cooling effect while decreasing penetration of solar radiation into the mixed layer, reducing the strength of the biological carbon pump (Balch et al., 1996; Harvey, 2008).

An assessment of the potential for $Ca(OH)_2$ discharge at a global scale estimates between 1.7 Gt/yr (load at departure) and 4.0 Gt/yr (one or two intermediate reloads for bulk carriers and container ships, respectively) (Caserini et al., 2021) (Table 7.1). As an example, the global potential of CDR from quicklime discharge by existing cargo and container ships while also carrying freight is 1.5–3.3 Gt CO_2/yr. However, there are significant uncertainties about the feasibility of using existing commercial vessels and what model would be applied to address the trade-offs to substitute a fraction of the profit-making transportation for mineral deployment. While additional infrastructure would be needed at ports to load vessels, and vessels would need to be adapted so that they could unload while steaming, a key limitation of using spare capacity in existing freight is that the locations for alkalinity deployments would be constrained to sea lanes. Alternatively, there is an argument for the use of a dedicated fleet of ships for mineral addition. The addition of a billion tons of mineral per year will likely require ~1,000 dedicated vessels (Harvey, 2008; Renforth et al., 2013; Caserini et al., 2021), a marginal, but substantial, increase on the 10,000 that are already operating. An increase in the fleet of vessels could cause environmental challenges including increase in ambient ocean noise levels (Kaplan and Solomon, 2016), potential transfer and establishment of nonindigenous species (Muirhead et al., 2015), and changes in concentrations and radiative forcing of short-lived atmospheric pollutants due to shipping emissions (Dalsøren et al., 2013).

TABLE 7.1 Example of an Assessment of Potential Global $Ca(OH)_2$ Deployment from Bulk Carriers and Container Ships

		Bulk carriers		Container ships	
		Single load	Multiple load	Single load	Multiple load
Average tonnage	dwt	7500		45000	
% dead weight cargo capacity (DWCC)	%	85%		85%	
% DWCC usable for $Ca(OH)_2$	%	15%		15%	
% tonnage used for transporting $Ca(OH)_2$	%	13%		13%	
$Ca(OH)_2$ load on board	t	9563		5738	
Number of vessels per category	-	9286		4855	
Distance traveled of the maritime route	km	6300		8900	
Average cruising speed	km/h	21.4		27.4	
Average at sea days per annum	d/year	181		218	
Number of intermediate stops (with $Ca(OH)_2$ reload)	n°/year	0	1	0	2
Average distance traveled per typical sea-leg	km	6,300	3,150	8,900	2,967
Navigation duration of the route	d	12.3	6.1	13.5	4.5
Average discharge rate along the route	kg/s	9.0	18.0	4.9	14.7
Number of exploitable yearly travels per vessel	n°/year	15	30	16	48
Total discharged $Ca(OH)_2$ per vessel	Mt/year	0.14	0.29	0.09	0.28
Total discharged $Ca(OH)_2$ by vessel category	Gt/year	1.3	2.7	0.4	1.3

SOURCE: Caserini et al., 2020. Licensed by Creative Commons CC BY 4.0.

Transportation

All OAE approaches require the extraction, processing, and transport of rock. If OAE deploys at a scale to contribute a meaningful quantity of U.S. removal requirements (e.g., 100–500 Mt CO_2/yr), then dedicated rock extraction on the same order of magnitude (100–1,000 Mt/yr) would be required. The United States currently produces ~1.4 Gt/yr of crushed rock (USGS, 2021), 68 percent of which is limestone, shell, or dolomite, 7 percent is basic igneous rock, and the balance is made up of silica-rich igneous or sedimentary rock (granite or sandstone), metamorphic rock, or unidentified geology. Between 1990 and 2007, production capacity increased by nearly 800 Mt/yr to 1.8 Gt/yr (Figure 7.5). While annual fluctuations in production are likely driven by short-term variability in demand, a 10-year rolling average is possibly more representative of longer-term constraints for capacity increase. However, sustaining a 3 percent increase over 30 years would be sufficient to create an additional 2-Gt/yr extraction capacity.

The deployment of alkalinity would necessitate the creation of local and regional management and monitoring structures as well as significant investment in adaptation and coordination of existing mining, industrial processing, and transportation schemes. A recent analysis of transportation based on the International Maritime Organization for bulk carriers and container ships concluded that the maximum slaked lime potential discharge from all active vessels worldwide is between 1.7 and 4.0 Gt/yr (Caserini et al., 2021). Because some of these operations are presently powered by fossil fuels, measures such as minimizing transportation by conducting operations on a local scale when possible will need to be implemented to maximize the effectiveness of OAE. In addition to the effectiveness of atmospheric CDR, global deployments would require augmentation of the current fleet and a significant proportion of new dedicated ships, which would undoubtedly cause impacts due to enhanced maritime traffic, pollution, and noise. However, this would require the expansion of lime production (from ~360 Mt/yr to >500 Mt/yr by 2100) to a scale comparable to the global cement industry (4.5 Gt/yr), with technologies that prevent or reduce the emission of the process CO_2 (e.g., Hanak et al., 2017) such as electrochemical splitting of water, salt, or mineral powered by nonfossil electricity (e.g., House et al., 2007; Rau et al., 2018).

7.5 ENVIRONMENTAL AND SOCIAL IMPACTS

A requirement for OAE is the expansion of mining operations (see Figure 7.1, Part 1). It is estimated that the additional mining effort needed to remove at least 1 Gt of CO_2/yr is equivalent to the global cement industry, which currently extracts ~7 Gt of rock per year (Renforth and Henderson, 2017). Adverse environmental impacts of mining include ground vibration from blasting, noise pollution, poor soil and air quality as a result of dust, low quality and quantity of surface water and groundwater, air and water pollution, both on and off the mine site, increase in truck traffic transporting mineral, sedimentation and erosion, and land subsidence. All these factors have direct impacts on wildlife habitat, forestland and recreational land, human habitat, physical, mental, and social wellness, food security, and cultural and aesthetic resources (Sengupta, 2021). However, it has been argued that materials produced from the iron and steel industry could play an important role in meeting our climate targets. Specifically, the use of industrial metal slag, rich in alkaline compounds and Ca and Mg ions, has the potential to enhance CO_2 capture (Gartner et al., 2020; Li et al., 2020; Chukwuma et al., 2021). The presence of mineral carbonate phases such as calcite, magnesite, ankerite, and kutnohorite in the material demonstrates carbonation reaction has occurred after slag formation and therefore there is still carbonation potential to be realized (Chukwuma et al., 2021).

Despite increasing efforts to minimize the negative impacts of mining on local communities, an expansion of the mining industry raises concerns not only about the environmental impact but also the local social landscape with implications for sustainability. Along with the promise of wealth and jobs, the expansion of mining can also bring negative social effects. Some examples include demographic changes resulting in a structural and functional transformation of the social environment including shifts in gender balance, increase in nonresident workforces, pressures on infrastructure, appropriation of land from the local communities, housing and services, social inequality, poor child development and education outcomes, pressures on families and relationships, drug and alcohol abuse, decline in community subsistence and lifestyles, and impacts on Indigenous communities (see Petrova and Marinova, 2013; Candeias et al., 2019; Sincovich et al., 2018). Significant impacts on workers' health include cancer, respiratory diseases, injuries, long-term exposure to chemical agents, and ergonomic issues (Candeias et al., 2018). In addition to mining, the release of mineral particles into seawater, particularly those smaller than 10 μm,[3] which can penetrate deep into the respiratory system, could be a concern for human health, and therefore deployment locations must be carefully selected (Daly and Zannetti, 2007).

---

[3] See https://www.epa.gov/pm-pollution/health-and-environmental-effects-particulate-matter-pm; https://www.epa.gov/pm-pollution/particulate-matter-pm-basics.

The deployment of alkalinity as pulverized rock or alkaline seawater where mineral dissolution has previously been conducted has multiple environmental impacts (see Figure 7.1, Part 2). These impacts will be more pronounced at the point of deployment, and the magnitude of these impacts will be critically dependent on mixing regimes affecting the rate of dispersal of added mineral or alkalinized water. In addition to rising pH, the deployment of alkalinity could raise the saturation state (Ω) of $CaCO_3$ well above present-day levels near the injection site. Although saturation is strongly correlated with temperature, spontaneous precipitation of calcite or aragonite does not occur in seawater, even in the warmest tropical regions. Therefore, addition of alkalinity to cold waters should not lead to precipitation, at least if levels are kept at or below the saturation state of the current warmest ocean waters (see Henderson et al., 2008).

Calcifying organisms are sensitive to changes in carbonate chemistry parameters including Ω, CO_3^{2-}, and HCO_3^-. However, the impact of ocean alkalinization (which increases Ω, pH, and levels of CO_3^{2-} and HCO_3^- and decreases CO_2) is largely unknown because most of the literature has focused on environmental responses associated with ocean acidification (decreasing Ω, CO_3^{2-}, and pH and increasing CO_2 and HCO_3^-). Biological processes can alter expected responses to shifts in Ω that can make interpretation of biotic responses challenging. One of these processes is the formation of organic films covering the mineral, a mechanism that has been described in mussels, coccolithophores, and corals (Ries et al., 2009; Tunnicliffe et al., 2009; Ho, 2013) and that appears to protect carbonate from dissolution. The biological effects of significant increases in both pH and carbonate and bicarbonate ion concentrations brought about by OAE methods remain unknown.

An increase in Ω has the potential to increase abiotic or biologically mediated calcification at the ocean surface. While the precipitation of $CaCO_3$ leads to the release of CO_2, an increase in $CaCO_3$ has the potential to increase ballast (Klaas and Archer, 2002) of particulate organic carbon (made up of fixed CO_2 via photosynthesis). Therefore, the hypothetical enhancement of calcification using some OAE protocols could have a positive or negative feedback on carbon removal, depending on the balance between the enhanced ballast feedback and the release of CO_2 feedback as a result of $CaCO_3$ precipitation (Riebesell et al., 2009; Bach et al., 2019). These hypotheses need to be tested experimentally.

Many technical, environmental, logistical, social, and ethical questions remain unanswered. If OAE was deployed at large scale, for ocean pH to be maintained above 8.0, on the order of 2–10 $\times 10^{14}$ mol alkalinity/yr for cumulative emissions of 1,500–5,000 Pg C, respectively, would be required (Paquay and Zeebe, 2013). However, a research agenda should include driving models with realistic deployment scenarios. In the case of quicklime or slaked lime dissolution, the amount of CO_2 emitted from the mining, transformation of limestone into quicklime, and transport operations needs to be considered in the calculations. For example, just the transformation of limestone into quicklime releases ~1.4 mol CO_2/mol limestone (Kheshgi, 1995) including ~0.4 mol resulting from the burning of coal during the mining operations of limestone plus 1 mol resulting from the transformation of limestone into quicklime (Kheshgi,1995). Therefore, producing large amounts of quicklime without carbon capture and storage would cause undesirable additional CO_2 release (see Pacquay and Zeebe, 2013). Some possible scenarios need to be considered; for example, it is possible that abiotic precipitation of $CaCO_3$ takes place following addition of CaO and so this scenario would result in additional CO_2 release. Nevertheless, regionally deployed OAE has the potential to help with recovery of ecosystems impacted by ocean acidification; for example, it could be effective in protecting coral reefs against ocean acidification (Feng et al., 2016), a major stressor to corals in addition to warming and eutrophication.

Biotic Impacts of Alkalinity Addition

While the biological and ecological repercussions of ocean acidification have been extensively explored, the opposite process of alkalinization and how it impacts marine life remains largely unknown, and thus there is an urgent need to generate empirical data on the impacts of OAE on marine biota. It is likely that the effects of alkalinity enhancement of terrestrial systems, for example, through agricultural liming, are already affecting the biota of some coastal and estuarine systems (see Guo et al., 2015; Müller et al., 2016), but these impacts remain unknown. It has been proposed that olivine dissolution will likely benefit silicifiers (diatoms) and possibly N_2 fixers (cyanobacteria) and the release of additional Si, Fe, and Ni could increase productivity, promoting a "green ocean" (see Bach et al., 2019). There are, however, concerns about the potential negative impact of Ni and other metals leached during mineral dissolution on marine organisms and how micronutrient addition will affect the health and structure of marine communities. Indeed Ni toxicity could alter the success of marine organisms and their processes, although there are examples of Ni-limited cyanobacteria that could potentially benefit from Ni supply (Sakamoto and Bryant, 2001; Ho, 2013; Blewett and Wood, 2015b; Montserrat et al., 2017).

In contrast with the green ocean, a "white ocean" has been proposed as the scenario where calcifiers would profit from the addition of $CaCO_3$ derivatives (Bach et al., 2019). In reality, experimental work needs to address these hypotheses and evaluate the consequences of mineral dissolution products. At present, assessing feasibility is hindered due to the complexities of navigating the legal framework to obtain permits for conducting exploratory research in the form of small-scale field deployments in parallel with monitoring and verification. Social acceptance and costs are important considerations.

The release of metals to seawater and the effect on marine organisms is a serious concern. While the impurities in limestone are likely small and negligible in $CaCO_3$ derivatives, the possibly toxic elements contained in some silicate rocks may alter the magnitude of primary production and the community structure of primary producers. An issue that has not yet been explored is the potential for bioaccumulation and biomagnification in the food chain, with potentially harmful impacts on food security. Another concern with adding powdered mineral on the ocean surface is that grazers (e.g., zooplankton, larvae of fish, echinoderms, etc.) might not discriminate between prey and mineral particles, with potentially damaging impacts. Ingestion of these particles could potentially enhance particle sinking via aggregation into feces, increasing sinking out of the euphotic zone and slowing dissolution (see Harvey, 2008).

Optical impacts of mineral addition are an important consideration when designing mineral deployments. For example, it is known that "white waters" caused by the $CaCO_3$ plates produced during coccolithophore blooms can have important ecological consequences including decreased visibility, which affects visual foragers such as seabirds and fish (Baduini et al., 2001) making it difficult to see prey at the surface and possibly excluding some prey from these white waters (Baduini et al., 2001; Lovvorn et al., 2001; Eisner et al., 2005). In an experiment releasing uniform $CaCO_3$ particles into the surface mixed layer, the optical signature for kilometer-sized surface patches of mineral particles was short-lived, largely due to horizontal eddy diffusion (Balch et al., 2009). Mesocosm experiments following the fate of Saharan dust (an analog of pulverized mineral addition to seawater) indicated rapid settling velocities, nonlinear export of particulate matter, and the formations of organic-mineral aggregates (Bressac et al., 2012). It is therefore expected that the optical signature associated with mineral deployment will be attenuated rapidly, within timescales of days (Balch et al., 2009).

Experimentation must address optimal particle size, dissolution kinetics in various water bodies, and impacts of trace elements and metals leached into seawater as the mineral dissolves. In the case of olivine, the addition of pulverized mineral in surface ocean waters (see Köhler et al.,

2010, 2013) has two main potential issues—the slow dissolution kinetics and the leaching of metals that might have unique impacts on different marine organisms. It has been proposed that olivine grain sizes of ~10 μm require on the order of 1–20 years (Hangx and Spiers, 2009; Schuiling and de Boer, 2011), while grain sizes on the order of 1 μm would sink slowly enough to guarantee a nearly complete dissolution within the mixed layer of the open ocean (Köhler et al., 2013), and the energy consumption for grinding might reduce the carbon sequestration efficiency by ~30 percent.

For olivine, the biological and ecological repercussions of an increase in nickel from its naturally occurring trace concentrations in seawater (0.002–0.16 μmol/kg; see IPCS, 1991; Burton et al., 1994) need to be evaluated. Ni toxicity is a concern for some marine organisms and processes including spawning in mysiid shrimps, DNA damage and metabolic and cytotoxic effects in the blue mussel *Mytilus edulis*, and disruption of ionic balance in the green crab *Carcinus maenas* (Millward et al., 2012; Blewett and Wood, 2015a; Blewett et al., 2015). However, some cyanobacteria representatives might potentially benefit from Ni additions. For example, in the cyanobacterium *Synechococcus*, nitrogen assimilation can be limited by low Ni availability because urease appears to be limited by natural levels of nickel (Sakamoto and Bryant, 2001). In *Trichodesmium*, nickel appears to elevate superoxide dismutase (SOD) activities and nitrogen fixation rates, suggesting SOD protection of nitrogenase from reactive oxygen species inhibition of N_2 fixation during photosynthesis (Ho, 2013).

Large-scale dispersion of mineral could also affect elemental ratios. Specifically, Ca and Mg ions and other trace metals associated with minerals could affect biogeochemistry and organisms' health. For example, increasing levels of Mg leaching from the mineral could result in increases in the Mg:Ca ratios that could affect the stability of carbonate. Indeed, calcite with a higher Mg content has been shown to be less stable in aqueous solutions (Bischoff et al., 1987) and high-Mg calcite is more susceptible to dissolution than aragonite. The amounts of Si, Ca, Mg, Fe, and Ni added alongside alkalinity strongly depend on the impurities of rocks used for OAE (Bach et al., 2019). Metal interference might also be possible; for example, high Ca^{2+} and Mg^{2+} concentrations may interfere with the uptake of nutrients by microorganisms, and this is exemplified in the freshwater cyanobacterium *Microcystis aeruginosa*, in which Ca and Mg levels can influence Fe uptake (Fujii et al., 2015). Fe and Ni enrichment from olivine deployment could promote primary production by N_2-fixers until other nutrients (e.g., P) become limiting, leading to a more productive "green(er) ocean," as depicted by Bach et al. (2019).

Corals have also been studied in relation to Fe availability, and although responses to Fe enrichment appear to be strongly dependent on local environmental conditions, particularly the availability of other nutrients (Rädecker et al., 2017), field observations suggest a negative impact of Fe enrichment or Fe toxicity at high concentrations. For example, Fe leaching from shipwrecks has been connected to the abundance of invasive *Corallimorpharia* and benthic fleshy algae (Schroeder et al., 2008; Work et al., 2008; Kelly et al., 2014). Additionally, while modest increases in Fe can promote growth, excess Fe concentrations were found to disrupt the coral–algal symbiosis through toxicity (Harland and Brown, 1989).

Impacts of Biota on Weathering

In addition to the impacts of weathering on marine biota, organisms can accelerate mineral dissolution and act as catalysts for alkalinization. The "benthic weathering engine" was described by Meysman and Montserrat (2017) as a mechanism of enhanced olivine weathering in marine sediments, fueled by microorganisms and invertebrate fauna in a low-pH environment. Indeed, the decomposition of organic matter by complex microbial consortia in the seabed releases CO_2 and organic acids into the pore solution, resulting in acidification and thus promoting olivine weathering and the dissolution of carbonates, a process termed "metabolic dissolution" (Rao et al., 2012).

In these environments, long, filamentous microbes called "cable bacteria" appear to be responsible for acidification (down to pH 5) of the top few centimeters of the sediment (Pfeffer et al., 2012), which promotes dissolution of acid-sensitive minerals (Riisgaard-Petersen et al., 2012; Meysman et al., 2015). In addition to microbes, macrofauna within the sediment can enhance olivine weathering through the process of bioturbation (Meysman et al., 2006). The transit of mineral particles through the gut of benthic fauna such as lugworms under high enzymatic activity and low pH, combined with mechanical abrasion during ingestion and digestion, appear to increase silicate mineral dissolution rates (Mayer et al., 1997; Needham et al., 2004; Worden et al., 2006). A perhaps less-studied process is the natural source of alkalinity via anaerobic degradation of organic matter. For example, the generation of alkalinity in mangrove sediments takes place via sulfate reduction, $CaCO_3$ dissolution, denitrification, and ammonification (Krumins et al., 2013). At global scale, alkalinity production in sediments could make a contribution of as much as ~15 percent of the CO_2 drawdown from the atmosphere in shelf and marginal seas (Thomas et al., 2009; Hu and Cai, 2011).

7.6 MONITORING AND VERIFICATION

The monitoring of alterations in ocean conditions on the spatiotemporal scales required to constrain carbon uptake and ecosystem impacts remains a challenge. The main mechanisms to monitor ocean carbonate chemistry are through monthly or seasonal time-series stations at single locations (Bates et al., 2014), discrete measurements on recurrent cruise transects every decade (see Gruber et al., 2007; Talley et al., 2016b), and underway measurements of pCO_2 from research vessels and ships of opportunity (Bakker et al., 2016). Existing programs have been instrumental in providing data to quantify air–sea CO_2 fluxes and the anthropogenic carbon inventories of the global ocean (Takahashi et al., 2009; Sabine and Tanhua, 2010; Bushinsky et al., 2019). There are, however, limitations for capturing trends and spatiotemporal dynamics with a degree of certainty because ship-based observations typically have poor seasonal resolution driven by logistical issues (e.g., remote locations and harsh weather conditions preventing operation at sea, disproportion of data associated with summer months, etc.) (Bushinsky et al., 2019). The development of sensors has addressed this issue, and examples include the commercially available SeaFET pH sensor using Honeywell Durafet technology (Martz et al., 2010; 2015), the recently developed ocean robot uncrewed surface vehicle that operates autonomously and provides hourly CO_2 flux estimates (Sutton et al., 2019), and the Moored Autonomous DIC (MADIC) system available for field deployments (Fassbender et al., 2015).

There is technological readiness to conduct all the measurements necessary to assess CDR potential. At least two of four parameters are required to fully describe the marine carbonate system (Millero, 2007). The choice of parameters (pH, pCO_2, DIC, and total alkalinity) depends on the specific process of interest. For example, anthropogenic carbon inventory determinations typically use DIC and total alkalinity, while pCO_2 is required to study the direction of air–sea CO_2 fluxes and total alkalinity can be used to infer calcification and dissolution processes through the water column (Wanninkhof, 1992; Sabine et al., 2004). Protocols and best practice are already well established, and sensor technology is available in three out of the four parameters of the carbonate system with the exception of alkalinity sensors for which a suite and platform combination does not exist yet on a commercially available level. Current autonomous observational capabilities include surface moorings, fixed observatories, profiling floats, and emerging technologies including autonomous surface vehicles (gliders, autonomous underwater vehicles), which involves repackaging of systems rather than development of new sensors, and the use of wind and waves to power autonomous surface vehicles (see Bushinsky et al., 2019).

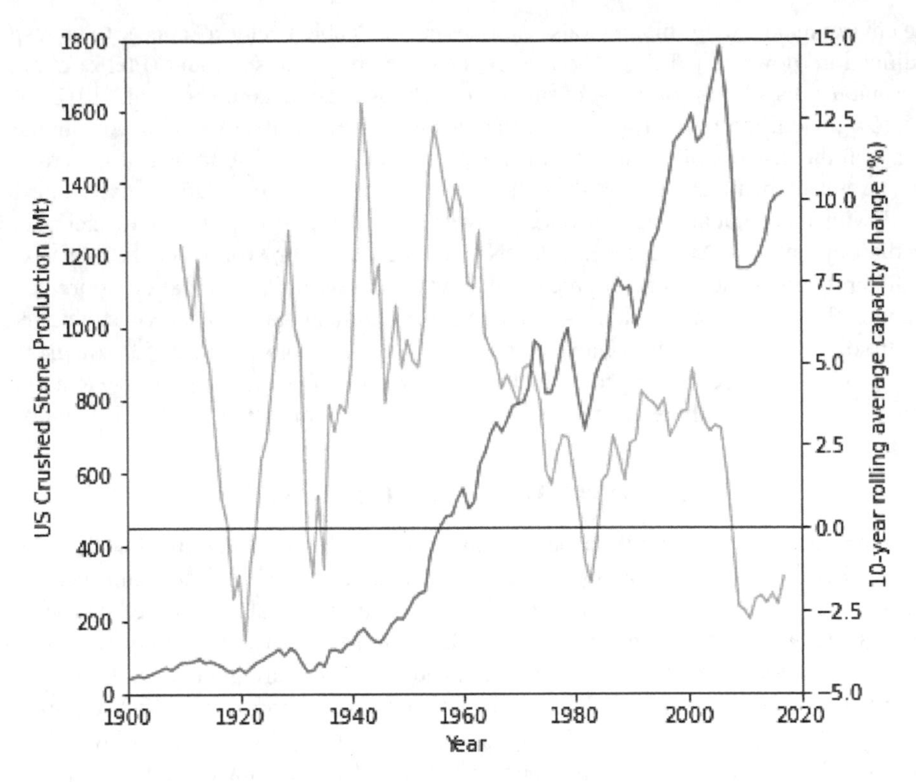

FIGURE 7.5 Annual production of crushed stone in the United States (blue) and the 10-year rolling average change in production in the United States (red). SOURCE: USGS, mineral commodity summaries.

7.7 VIABILITY AND BARRIERS

Co-benefits

The main co-benefits of OAE are the potential mitigation of ocean acidification, which would have a positive impact on many organisms, particularly the $CaCO_3$-producing community, and the potential for fertilization via the addition of metals such as iron. Assessing the impact of OAE deployments on calcification and the ecological fitness of calcifying organisms and other functional groups requires transitioning from laboratory experimentation and mesocosm-based trials to field deployments to determine the complexity of factors affecting the inorganic carbon chemistry. For example, the balance between ballasting of particulate organic carbon (enhanced by aggregation) and the contribution of calcification and its impact on CDR must be tested in the field under various seasonal and biological conditions (see Section 7.3). Similarly, the physical effect of particles (e.g., potential ingestion by grazers) and chemical impact of raised alkalinity on the physiology of organisms, which could alter important biogeochemical processes (e.g., calcification and silicification), must be explored in the field. For example, it has been suggested that a shift in phytoplankton functional group composition favoring diatoms could occur if silica-rich olivine is used to enhance alkalinity while coccolithophores might be promoted when carbonate-based minerals are used (Köhler et al., 2013; Bach et al., 2019). Additionally, specific regional sensitivities to OAE, for example, the Arctic Ocean and tropical oceans might become hot spots for biogeochemical changes following OAE (see González and Ilyina, 2016).

Energy

Proposals for OAE use a range of supply chains, technology options, and energy sources. All proposals involve the extraction, processing, transport, and dissolution of rocks or minerals and thus the energy costs of OAE include mining, transformation, grinding, transportation on land and at sea, monitoring and verification, and building of specialized vessels and pipes (see Figure 7.1). However, to date, a complete life-cycle analysis has not been conducted, and there are no empirical data for a scaled experiment or field trials with calculated energy budgets. This type of analysis is required and must be an element of OAE research schemes.

Limited information on energy consumption includes mineral extraction and grinding. Operations typically use diesel for on-site transport, and either on-site power generation when sites are small and remote (diesel generators, renewables) or power-grid integration. The energy demands to extract a rock and process it to centimeter to millimeter particle sizes is relatively small and routine practice (requiring ~5–20 kWh of electricity per ton CO_2 (Renforth, 2012). However, three times more energy is required to grind particles to diameters of ~10 μm (Hangx and Spiers, 2009). Freight energy requirements and associated emissions depend on the method of transport and distance traveled (Table 7.2). However, transport distances of tens to hundreds of kilometers could still be feasible for OAE schemes (Renforth, 2012; Moosdorf et al., 2014).

Processes that facilitate accelerated mineral dissolution typically have the largest energy demand within a supply chain. All OAE approaches require the extraction and processing of minerals. While it is relatively inexpensive to extract and crush rock (tens of kilowatt hours per ton for size reduction to millimeters), fine grinding requirements depend on mineral dissolution rate (Kelemen et al., 2020). For instance, EW may require on the order of 150–900 kWh/t to create sufficiently micrometer-size (10 μm) particles (Renforth, 2012). Ocean liming (the creation of CaO or $Ca(OH)_2$ for addition to the ocean) requires ~5 GJ/t of process heat and 150–300 kWh of electricity for the calcination of limestone, the distribution of lime into the ocean, and the geological storage of the produced gas (Kheshgi, 1995; Renforth et al., 2013). Electrochemical approaches may require thousands of kilowatt hours per ton of CO_2 (examined in greater detail in Chapter 8). Movement of materials by large ocean-going vessels is relatively less carbon intensive than other methods of transport (Table 7.2). Renforth et al. (2013) estimate that approximately 100 MJ/t of bunker fuel would be required for the distribution of lime into the ocean (which includes both onboard material handling systems and fuel for steaming) from a 300k-deadweight tonnage vessel.

From the range of OAE approaches, only ocean liming has received formalized technoeconomic assessment. Renforth et al. (2013) estimate the cost of ocean liming to be ~$120/t CO_2 for oxy-fuel flash calcination of limestone, but this may be reduced to ~$70/t CO_2 using other systems or dolomite as the mineral feedstock. Caserini et al. (2019) consider the potential integration of ocean liming with biomass energy and hydrogen production. The sale of hydrogen and greater car-

TABLE 7.2 Freight Energy Demands and Associated Emissions from Extracted and Processed Minerals

Transport Method	Gaseous CO_2 Emissions, t/km	Energy, MJ/(t/km)	Operating Expense, $/(t/km)
Heavy goods vehicle by road	62	1.3	0.07
Rail freight	22	0.2	0.04–0.05
Inland waterways	31		
Large ship distribution	7	0.2	0.001

SOURCE: Adapted from Renforth (2012).

bon removal suggest costs could be as low as \$64/t CO_2. Preliminary calculations estimate that the operational costs of accelerated weathering of limestone may be only on the order of tens of dollars per ton of CO_2 (Langer et al., 2009), and that carbonate addition to upwelling regions may be on the same order of magnitude (Harvey, 2008). Costs associated with electrochemical approaches are considered in Chapter 8.

Governance and Social Dimensions

The legal framework for ocean CDR is discussed in Chapter 2. Many of the international and domestic laws discussed in that chapter could apply to OAE.

With respect to the application of international law, researchers (e.g., Webb et al., 2021) have concluded that OAE would constitute "geoengineering" for the purposes of Decisions X/33, XI/20, and XIII/4, adopted by the parties to the Convention on Biological Diversity.[4] The decisions recommend that parties to the Convention and other governments avoid geoengineering activities that may affect biodiversity, except for "small scale scientific research studies . . . conducted in a controlled setting."[5] The decisions are not legally binding, however. Additionally, researchers (e.g., Brent et al., 2019; Webb, 2020; Webb et al., 2021) have found that OAE may be considered a form of marine "pollution" under the United Nations Convention on the Law of the Sea and marine "dumping" under the London Convention and Protocol.

As discussed in Chapter 2, in 2013, the parties to the London Protocol adopted an amendment governing "marine geoengineering," the definition of which is likely to encompass OAE.[6] The 2013 amendment establishes a framework under which parties may approve certain marine geoengineering activities.[7] However, at the time of writing, the framework applied only to activities relating to ocean fertilization. Researchers (e.g., Brent et al., 2019; Webb et al., 2021) have noted that the framework could be expanded to apply to OAE in the future. However, unless and until that occurs, OAE activities will be subject to the general requirements of the London Convention and Protocol. Both instruments require marine dumping to be permitted by the country under whose jurisdiction it occurs and impose restrictions on when permits can be issued. Webb et al. (2021) concluded that OAE projects could likely be permitted by parties to the London Convention, but not by parties to the London Protocol.

The United States is a party to the London Convention and has implemented it domestically through the Marine Protection, Research, and Sanctuaries Act (MPRSA, 33 U.S.C. § 1401 et seq.). MPRSA applies to the "dumping" of "material" in the oceans (33 U.S.C. § 1411). The statutory definition of "dumping" has been interpreted as encompassing the discharge of ground rock for the purposes of OAE (e.g., Webb et al., 2021). Under MPRSA, OAE projects must be permitted by the Environmental Protection Agency (EPA) if the discharge occurs within the U.S. territorial sea, or if the rock is transported from the United States or on a U.S.-registered vessel (regardless of where the discharge occurs) (33 U.S.C. §§ 1411-1413). EPA must comply with various consultation and other procedural requirements before issuing permits (see Chapter 2).

[4] Report of the Conference of the Parties to the Convention on Biological Diversity on the Work of its Tenth Meeting, Decision X/33 on Biodiversity and Climate Change, Oct. 29, 2010 (hereinafter Decision X/33); Report of the Conference of the Parties to the Convention on Biological Diversity on the Work of its Eleventh Meeting, Decision XI/20 on Climate-related Geoengineering, Dec. 5, 2012 (hereinafter Decision XI/20); Report of the Conference of the Parties to the Convention on Biological Diversity on the Work of its Thirteenth Meeting, Decision XIII/4, Dec. 10, 2016 (hereinafter "Decision XIII/4").

[5] Para 8(w), Decision X/33; Para 1, Decision XI/20; Preamble, Decision XIII/4.

[6] Resolution LP.4(8), Amendment to the 1996 Protocol to the Convention on the Prevention of Marine Pollution by Dumping of Wastes and Other Matter, 1972 to Regulate Marine Geoengineering, Oct. 18, 2013.

[7] Annex 4, Resolution LP.4(8).

Previous research (e.g., Webb, 2020; Webb et al., 2021) has examined other domestic laws that could apply to OAE, as well as related onshore activities, such as mining and rock grinding. The domestic laws applicable to onshore activities are not discussed here.

7.8 SUMMARY OF CARBON DIOXIDE REMOVAL POTENTIAL

The criteria for assessing the potential for OAE as a feasible approach to ocean CDR, described in Sections 7.2–7.7, is summarized in Table 7.3 and Table 7.4.

7.9 RESEARCH AGENDA

Field trials are urgently needed in both coastal and open-ocean waters to monitor mineral dissolution kinetics, the dynamics of the DIC system, biogeochemical and biological impacts, and carbon sequestration potential, and to assess technological readiness and determine environmental and societal impacts of OAE both in marine systems and on land.

Research consortia and philanthropic endeavors in coastal waters such as the nonprofit Project Vesta that proposes the EW of ground olivine on beaches to increase coastal carbon capture or the planning of pilot studies by the European Union-funded OceanNETs consortium, which includes offshore mesocosm experiments to assess the ecological impacts of OAE, are already under way. The location and characteristics of the offshore deployments remain open questions, although sites where upwelling velocities are high are likely desirable to prevent or delay export of mineral particles to depth. The importance of timing of deployment needs to be addressed in the field because seasonal properties of seawater, including those driven by biological events, alterations in community composition, and conditions leading to changes in dissolved organic matter could have major impacts on particle aggregation and export.

As knowledge of the chemistry and technology for deploying alkalinity matures, monitoring and verification plans need to be implemented to assess site- and temporal-specific dissolution kinetics, rates of alkalinization, biological responses, the fate of particles and biogeochemical impacts through the water column, and the air–sea flux of CO_2 through observations using manipulations in mesocosms and the field. The necessary elements of research and development are highlighted in Table 7.5.

Specific elements of a research program include

- **Development of an empirical framework for OAE.** Laboratory, mesocosm, and field experimentation in close collaboration with the engineering field to assess the technical feasibility and readiness level of OAE approaches (including the development of pilot-scale facilities). These should include assessments of short- versus long-term deployments under contrasting oceanographic conditions including locations of varying upwelling velocities, different DIC system properties.
- **Assessment of the relationship between approach and environmental impact.** The environmental impact of OAE is closely related to the methods or technologies that promote alkalinity changes (e.g., the choice of rock or mineral and the mechanisms by which it is dissolved). An experimental program that considers the environmental response to OAE will be most effective if they were constrained by what might be practical.
- **Technology development and assessment.** Few of the methods for OAE have been developed beyond bench scale. As such, the costs and feasibility of the approaches remain speculative, which can be substantially improved with a research program designed to accelerate technology development and demonstration of pilot-scale facilities. This would

have maximum impact if it were integrated with rigorous technoeconomic and life-cycle assessment.

- **Assessment of the relationship between point of addition and global impact.** All OAE approaches increase alkalinity at the point of addition, and mixing distributes that change across the surface, and, eventually, deep ocean. The impacts are likely to be largest around the point of addition, the magnitude of which will be controlled both by the rate of OAE and hydrodynamics. Research is needed to assess the dispersion or evolution of alkalinity, CO_2 gas exchange, and fate of dissolving particles at locations of OAE, and how these attenuate into regional and global waters.
- **Development of a strong monitoring program.** This would be focused on particulate and DIC at the surface and through the water column at the site of deployment and adjacent waters. This would require a parallel development of protocols for transparent and verifiable deployment, monitoring, and carbon accounting schemes.
- **Research on governance of OAE research.** Research into the legal framework of ocean CDR research and exploration of the legal framework relevant to obtaining permits to conduct field research.
- **Development of strong educational programs.** This would be achieved through transparent data repository management efforts and engagement with schools and the public and private sectors.
- **Development of life-cycle analyses.** This would include costs such as expansion of mining, materials, transportation, environmental impact assessment, mineral deployment, and monitoring.
- **Development of strong collaborations.** This would include those with social scientists, economists, and governance experts to gain knowledge on public perception, acceptability, and costs.

7.10 SUMMARY

Approaches to increase ocean alkalinity have been proposed since the mid-1990s but remain at a relatively early stage of development. OAE attempts to mimic natural weathering processes by adding crushed minerals either directly to the ocean, to coastal environments, or to terrestrial environments. Given that the surface ocean is supersaturated with common carbonate minerals, they cannot be directly added to the ocean. To facilitate mineral dissolution, the reaction of carbonate minerals with elevated CO_2 has been suggested, initially for reducing fossil fuel emissions, but potentially for CDR by coupling with direct air capture or biomass energy carbon capture and storage. Others have proposed converting carbonate minerals into more reactive forms (e.g., lime or hydrated lime) for addition to the ocean. Finally, electrochemical approaches may be used to create basic and acidic solutions at each electrode, the former could be used for OAE while the latter would need to be neutralized through reaction with silicate minerals.

The two key mechanisms by which OAE could impact the environment are (1) elevated alkalinity that is "unequilibrated," that is, high-pH, low-CO_2 concentrations that may be more acute around the point of addition; and (2) the addition of other biologically active elements (iron or silica, as in ocean fertilization, or nickel, chromium, or other trace metals). Research is needed to assess the ecological response to OAE. Although much can still be learned from laboratory-based and "contained" (mesocosm) experiments, including exploring the impacts of OAE on the physiology and functionality of organisms and communities, implementation pathways to responsible deployment will require field trials. Such trials will be essential to assess how euphotic and benthic biogeochemical processes are affected, the response from complex communities, the indirect effects

TABLE 7.3 CDR Potential of Ocean Alkalinity Enhancement

Knowledge base What is known about the system (low, mostly theoretical, few in situ experiments; medium, lab and some fieldwork, few carbon dioxide removal (CDR) publications; high, multiple in situ studies, growing body of literature)	**Low–Medium** Seawater–CO_2 system and alkalinity thermodynamics are well understood. Need for empirical data on alkalinity enhancement; currently, knowledge is based on modeling work. Uncertainty is high for CDR possible impacts.
Efficacy What is the confidence level that this approach will remove atmospheric CO_2 and lead to net increase in ocean carbon storage (low, medium, high)	**High Confidence** Need to conduct field deployments to assess CDR, alterations of ocean chemistry (carbon but also metals), how organic matter can impact aggregation, etc.
Durability Will it remove CO_2 durably away from surface ocean and atmosphere (low, <10 years; medium, >10 years and <100 years; high, >100 years), and what is the confidence (low, medium, high)	**Medium–High** >100 years Processes for removing added alkalinity from seawater generally quite slow; durability not dependent simply on return time of waters with excess CO_2 to ocean surface.
Scalability What is the potential scalability at some future date with global-scale implementation (low, <0.1 Gt CO_2/yr; medium, >0.1 Gt CO_2/yr and <1.0 Gt CO_2/yr; high, >1.0 Gt CO_2/yr), and what is the confidence level (low, medium, high)	**Medium–High** Potential C removal >0.1–1.0 Gt CO_2/yr (medium confidence) Potential for sequestering >1 Gt CO_2/yr if applied globally. High uncertainty coming from potential aggregation and export to depth of added minerals and unintended chemical impacts of alkalinity addition.
Environmental risk Intended and unintended undesirable consequences at scale (unknown, low, medium, high), and what is the confidence level (low, medium, high)	**Medium** (low confidence) Possible toxic effect of nickel and other leachates of olivine on biota, bio-optical impacts, removal of particles by grazers, unknown responses to increased alkalinity on functional diversity and community composition. Effects also from expanded mining activities (on land) on local pollution, CO_2 emissions.
Social considerations Encompass use conflicts, governance-readiness, opportunities for livelihoods, etc.	Expansion of mining production, with public health and economic implications; general public's potential for public acceptability and governance challenges (e.g., if perceived as "dumping").
Co-benefits How significant are the co-benefits as compared to the main goal of CDR and how confident is that assessment	**Medium** (low confidence) Mitigation of ocean acidification; positive impact on fisheries.
Cost of scale-up Estimated costs in dollars per metric ton CO_2 for future deployment at scale; does not include all of monitoring and verification costs needed for smaller deployments during R&D phases (low, <\$50/t CO_2; medium, ~\$100/t CO_2; high, >>\$150/t CO_2) and confidence in estimate (low, medium, high)	**Medium–High** >\$100–\$150/t CO_2 (low–medium confidence) Cost estimates range between tens of dollars and \$160/t CO_2. Need for expansion of mining, transportation, and ocean transport fleet.
Cost and challenges of carbon accounting Relative cost and scientific challenge associated with transparent and quantifiable carbon tracking (low, medium, high)	**Low–Medium** Accounting more difficult for addition of minerals and non-equilibrated addition of alkalinity, than equilibrated addition.

continued

TABLE 7.3 Continued

Cost of environmental monitoring Need to track impacts beyond carbon cycle on marine ecosystems (low, medium, high)	**Medium** (medium–high confidence) All CDR will require monitoring for intended and unintended consequences both locally and downstream of CDR site, and these monitoring costs may be substantial fraction of overall costs during R&D and demonstration-scale field projects.
Additional resources needed Relative low, medium, high to primary costs of scale-up	**Medium–High** Adaptation and likely expansion of existing fleet for deployment; infrastructure for storage at ports. Infrastructure support for expansion of mineral extraction, processing, transportation, and deployment.

TABLE 7.4 Costs and Energy Needs for Ocean Alkalinity Addition

OAE Method, Material	Cost (\$/t CO_2)	Energy
Land, silicate rock	50–150	100–1,000 kWh/t
Coast, silicate rock	No data	~100–1,000 kWh/t
Ocean, silicate rock	No data	100–1,000 kWh/t
Ocean liming	70–130	5 GJ/t and 150–300 kWh
Accelerated weathering of limestone	10–40 (Opex)	No data
Ocean, carbonate addition to upwelling	20 (Opex)	~100 MJ/t

of OAE, and optimal environments and treatment methods. Coupling this research to a rigorous monitoring program will be essential in accounting for CDR and the environmental impact of the experiments, but would help scope the requirements of monitoring of scaled-up OAE.

The environmental impact of OAE would be closely related to the methods or technologies that promote alkalinity changes (e.g., the choice of rock or mineral and the mechanisms by which it is dissolved). An experimental program that considers the environmental response to OAE will be most effective if it were constrained by what might be practical. Research and development is required to explore and improve the technical feasibility and readiness level of OAE approaches (including the development of pilot-scale facilities). Research on social and governance considerations associated with contained and pilot-scale experiments and deployment (if any) is also required.

TABLE 7.5 Research and Development Needs: Ocean Alkalinity Enhancement

#	Research Needs	Gap Filled	Environmental Impact of Research	Estimated Research Budget ($M/yr)	Timeframe (years)
7.1	Research and development to explore and improve the technical feasibility and readiness level of OAE approaches (including the development of pilot-scale facilities)	Can we develop optimized approaches to enhance alkalinity in seawater? What are the conditions leading to undesirable effects such as aggregation and reverse weathering?	Low impact of pilot facilities. Initially, most of the work will be lab and mesocosm based.	10	5
7.2	**Laboratory and mesocosm experiments to explore impacts on physiology and functionality of organisms and communities**	**What are the physiological effects of rising alkalinity on marine biota?** **What are the effects of OAE on community structure?**	**Negligible impacts**	**10**	**5**
7.3	**Field experiments**	**What are the optimal sites to deploy alkalinity?** **How are biogeochemical processes affected in the euphotic zone as at depth?** **How do complex communities respond to OAE?** **What are conditions to avoid (e.g., possibly high dissolved organic matter) and those desirable (upwelling regions or estuaries?) to conduct deployments?**	**Modest impact for experiments conducted on a small spatial and temporal scale**	**15**	**5–10**

continued

TABLE 7.5 Continued

#	Research Needs	Gap Filled	Environmental Impact of Research	Estimated Research Budget ($M/yr)	Timeframe (years)
7.4	Research into the development of appropriate monitoring and accounting schemes, covering CDR potential and possible side effects.	Are ships of opportunity (coastal and open ocean) conducting measurements of the carbonate chemistry (through continuous systems) appropriate vehicles to detect change in alkalinity and carbonate chemistry? To what extent do they address CDR potential? In the mid to long term (>5 years), what are the most appropriate locations and depths to deploy alkalinity sensors?	Modest impact because most of the research and technology supporting monitoring of the dissolved inorganic carbon system and biogeochemical impacts (e.g., calcification) are well established.	10	5–10

NOTE: Bold type identifies priorities for taking the next step to advance understanding of ocean alkalinity enhancement as an ocean CDR approach.

8

Electrochemical Engineering Approaches

8.1 OVERVIEW

Electrochemistry considers chemical reactions that result in the production or consumption of electricity. Technologies use electrochemistry to take measurements of chemical systems, use chemical reactions to generate electricity, or use electricity to drive chemical reactions. Several approaches have been proposed that use electricity to promote or to drive reactions that ultimately result in carbon dioxide (CO_2) removal from the atmosphere (House et al., 2007; Rau, 2008; de Lannoy et al., 2012; Datta et al., 2013; Rau et al., 2013; Eisaman et al., 2018; Zhao et al., 2020; La Plante et al., 2021; Oloye et al., 2021).

Direct CO_2 Removal

A range of approaches have been proposed to extract CO_2 from seawater, analogous to methods that remove it directly from the atmosphere (Willauer et al., 2011; Eisaman, 2020). Acid approaches exploit acidic conditions created around the anode to shift the equilibrium of the carbonate system (see Box 8.1) toward a greater concentration of aqueous CO_2, which is evolved/degassed from the solution and collected for permanent storage (de Lannoy et al., 2012; Datta et al., 2013). Basic approaches exploit high-pH conditions created around the cathode to shift the equilibrium of the carbonate system toward a greater concentration of bicarbonate and/or carbonate ions (La Plante et al., 2021). This creates conditions in which carbonate precipitation can occur and promotes an increase in aqueous CO_2, which may be evolved and collected similarly to the acid approach (Rau, 2008; de Lannoy et al., 2012). Dissolved inorganic carbon is removed and collected as both solid carbonate residues and as evolved CO_2 gas. Basic approaches that force the precipitation of solid carbonate without restoring alkalinity do not result in the net removal of CO_2 from the atmosphere, although they reduce the concentration of CO_2 that is dissolved in solution. In both approaches, the base and acid streams are recombined and discharged back into the ocean.

Alkalinity Creation

It is possible to use electrochemical processes to increase the alkalinity of seawater, and/or to force the precipitation of solid alkaline materials (i.e., hydroxide minerals). In the former process, reactions at the cathode increase the alkalinity of the surrounding solution, which can then be discharged into the ocean. In the latter, the reaction cell is separated by an ion-selective membrane, a physical barrier, or internal hydrodynamics, and seawater or brine is introduced into the cathode compartment (Rau et al., 2013; Zhao et al., 2020; La Plante et al., 2021; Oloye et al., 2021). The potential difference across the cell promotes the migration of ions (usually sodium) across the barrier/cell, and ultimately promotes the formation of solid metal hydroxide residues (e.g., brucite or portlandite) that could be added to the ocean to increase its alkalinity.

These two broad categories of ocean CDR electrochemical engineering are, in some cases, not mutually exclusive, and could be deployed as a hybrid approach that both extracts CO_2 from seawater, in the form of a gas, or as mineral carbonates, while increasing the alkalinity of the effluent solutions. In this chapter we do not refer to applications for direct CO_2 reduction (e.g., to produce syngas, fuels, or chemicals), but specifically focus on approaches for removing or absorbing CO_2 from/into seawater.

According to criteria described in Chapter 1, the committee's assessment of the potential for electrochemical processes, as an approach to CDR, is discussed in Sections 8.2–8.5 and summarized in Section 8.6. The research needed to fill gaps in understanding how electrochemical processes can be used as a mechanism for durably removing atmospheric CO_2 is discussed and summarized in Section 8.7.

8.2 KNOWLEDGE BASE

Electrochemical Engineering

The relationship between electricity and chemistry has been explored since the 18th century, with anatomists demonstrating movement of muscles with the application of electricity (Galvani, 1841). This was followed by a spate of new discoveries in the early 19th century including the production of hydrogen and oxygen through electrolysis (Grove, 1839), electroplating, and a chemical-based description of a galvanic cell. The first fuel cell was demonstrated in the 1830s (Grove, 1839), and the Hall-Heroult process for producing aluminum metal from bauxite ore was invented in the 1880s (Charles, 1889).

In processes that use electricity to drive reactions, a potential difference is applied to a liquid using electrodes. Positively charged cations in the solution migrate to the cathode (negatively charged electrode), which promotes reduction reactions, and the negatively charged anions migrate to the anode (positively charged electrode), which promotes oxidation. Typically, these reactions are undertaken in a "cell," and the electrodes are separated by a permeable separator. Figure 8.1 shows a simplified schematic of a diaphragm cell used in the production of chlorine and sodium hydroxide, chlor-alkali.

The two largest industrial users of electrochemistry are in the production of aluminum, in which 50 million tonnes (Mt) are produced annually (USGS, 2021) and chlor-alkali (sodium hydroxide and chlorine), in which 72 and 65 Mt are produced annually, respectively (Egenhofer et al., 2014; Renforth, 2019). Innovation in these industries over the 20th century was largely concerned with improving energy efficiencies, reducing cell costs, and improving the purity of the final product.

BOX 8.1
Fundamentals of Electrochemical Engineering

The electrical charge (q, coulombs) transferred is proportional to the amount of electrons (n, moles) involved in the reaction. Work (ΔG, joules/mol) is done when charge moves through a potential difference (E, volts), such that

$$\Delta G = qE = -nFE \tag{8.1}$$

Where F is the Faraday constant (9.649 × 10^4, C/mol of electrons). Electrons liberated through oxidation reactions flow from the anode to the cathode for incorporation in reduction reactions. ΔG is normalized to the mass of solid product and typically given in joules per mole. The work done is proportional to the potential difference applied across the electrolysis cell, which is related to the difference in thermodynamic potentials at each electrode ($E^C - E^A$), and the electric overpotentials at each electrode (η_C, η_A) and within the cell/circuit.

$$E = E_e^C - E_e^A - |\eta_C| - |\eta_A| - |\eta_{cell}| \tag{8.2}$$

For example, the chlor-alkali process uses a reaction cell containing aqueous sodium chloride (NaCl). The anode and cathode are separated by a semipermeable membrane that allows for the migration of Na ions. Chloride ions are oxidized at the anode to produce chlorine gas (equation 8.3). At the cathode, water is reduced to hydrogen gas and hydroxide ions (equation 8.4). The overall reaction results in the formation of sodium hydroxide (NaOH, equation 8.5)

$$2Cl^-_{(aq)} \rightarrow + Cl_{2\,(g)} + 2e^- \tag{8.3}$$

$$2Na^+_{(aq)} + 2e^- + 2H_2O_{(l)} \rightarrow 2NaOH + H_{2\,(g)} \tag{8.4}$$

$$2NaCl_{(aq)} + 2H_2O_{(l)} \rightarrow Cl_{2\,(g)} + H_{2\,(g)} + 2NaOH \tag{8.5}$$

The Gibbs free energy of equation 8.5 is 483.24 kJ/mol, suggesting that a minimum potential difference of 2.5 V is required to drive the forward reaction. Typically, chlor-alkali systems require 3–4 V due to overpotentials at the electrodes, overpotentials within the cell, or inefficiencies due to reverse reactions or by-product formation.

Design Choices in Electrochemical Engineering

A range of design choices may be available in the construction and the operation of electrochemical reaction systems. These choices are made so that overpotentials at the electrodes or within the cell are minimized, while maximizing current and yield efficiency either when using electricity as an input, or in the case of multiple energy inputs such as electrons and photons/phonons (Nørskov et al., 2004). An exhaustive discussion of these technology options for ocean-based CDR is beyond the scope of this report, but a few possibilities are articulated below (Brinkmann et al., 2014; Paidar et al., 2016; Yan et al., 2020).

Choice of Electrolyte

The electrolyte serves the dual purpose of conducting electric current between electrodes and as a reactant in the chemical processes. The ohmic overpotential (η_{ohmic}, mV) created by the elec-

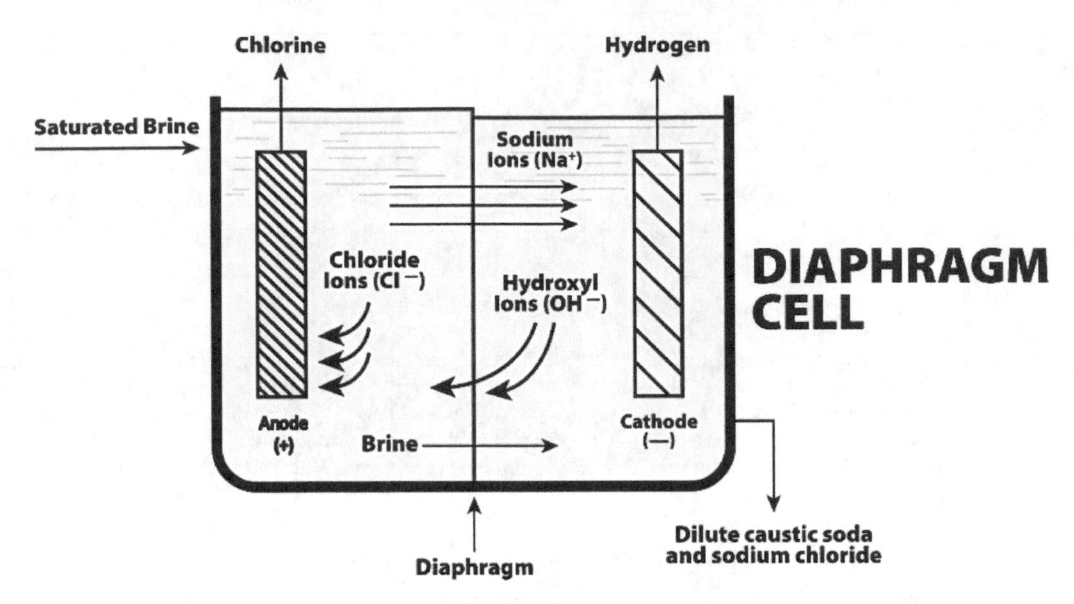

FIGURE 8.1 A simplified schematic of a chlor-alkali electrolysis system consisting of an anode and cathode separated by a diaphragm. SOURCE: Adapted from Kroschwitz, 1991; Brinkmann et al., 2014. Licensed under Creative Commons CC BY 4.0.

trolyte is inversely proportional to its conductivity (equation 8.6; Uhlig and Revie, 1985; O'Brien et al., 2005):

$$\eta_{\text{ohmic}} = \frac{10^5}{k} \cdot i \cdot d \tag{8.6}$$

where, k is the specific conductivity (μS/cm), d is the spacing between electrodes (cm), and i is the current density (A/m^2). A less conductive electrolyte will result in greater resistive losses for a given geometry and current density of an electrolysis cell. The composition of the electrolyte also influences resistive losses resulting from mass transport of ions across boundary layers at the electrodes and the cell divider, such that a limiting current (i_{lim}) is imposed on the system (equation 8.7; O'Brien et al., 2005):

$$i_{\text{lim}} = \frac{[X] \cdot D_X \cdot F \cdot n}{\delta} \tag{8.7}$$

where, $[X]$ denotes the concentration of either the cation or anion in bulk solution (mol/kg), δ is the diffusion layer thickness (m), and D_x is the diffusion coefficient of the anion or cation X (m^2/s). Finally, the reaction of an electrolyte may create unwanted reaction products (e.g., hypochlorite ions in the chlor-alkali process) or reverse reactions may occur (e.g., the production of water in alkaline electrolysis). These undesirable reactions consume some of the electrical energy of the process, resulting in an inefficiency (Bennett, 1980). The ratio of the theoretical electric current needed for creating the desired product to the total electric current used by the system is the Faraday or current efficiency (Bard and Faulkner, 2001). With appropriate process engineering, current efficiencies of >95 percent have been reported in the chlor-alkali industry (Brinkmann et al., 2014).

Table 8.1 shows the possible ohmic overpotential and limiting current density for a range of mineral-saturated electrolytes. For instance, use of an NaCl-saturated brine, a common feedstock in chlor-alkali, results in low ohmic overpotential (<40 mV) and high limiting current densities

(>100 kA/m^2). While it may be desirable to exclude chlorine (Cl) from the electrolyte for the CO_2 removal (CDR) processes described below, an electrolyte based only on calcium (Ca)/magnesium (Mg) carbonates or hydroxides would result in high ohmic overpotential and low current densities. For comparison, current densities of 1–6 kA/m^2 are typical in chlor-alkali (Brinkmann et al., 2014).

Choice of Cell Divider

Early electrolysis experiments used electrolysis cells in which the cathode or anode were not separated (undivided cells; Konishi et al., 1983), as is commercial best practice for aluminum production (Haupin, 1983) or electrochemical wastewater treatment (Chen, 2004). However, to increase selectivity, limit unwanted reactions, and increase product purity (Paidar et al., 2016), the two halves of the electrolysis cell are often separated (or "divided") by either a diaphragm or an ion-permeable membrane, the former allowing both ion and solvent transport, while the latter allows only ions. Diaphragm materials that are porous, thin, and chemically stable in the electrolyte and cell operating conditions are preferred, with asbestos being used extensively in the past, and polymer membranes finding use in more recent times (Strathmann, 2004).

Choice of Electrode Material

Electrode materials (i.e., herein referring predominantly to the active/electrocatalyst surface, and not the current collector) have been developed and used to improve the performance and to reduce the cost of electrolysis cells. The lowest exchange current densities and the highest catalytic properties for the lowest material cost are desirable. Nickel-based alloys are used extensively in alkaline water electrolysis (for both electrodes) and for the cathode in chlor-alkali (the anode is often composed of ruthenium-titanium oxides; Brinkmann et al., 2014). Challenges associated with disposal of chlorine gas created during the electrolysis of seawater have promoted the development of oxygen-selective electrodes (e.g., Izumiya et al., 1997; Petrykin et al., 2010; Vos et al., 2018, 2019; Okada et al., 2020). Chlorine evolution at the anode involves fewer electrons and intermediary products compared with oxygen, and thus has a kinetic advantage such that chlorine is typically the only detected gas (Vos and Koper, 2018). Reduction in chlorine production could be essential in the development of some of the systems considered below, although oxygen evolution could accelerate electrode degradation (Kolotyrkin et al., 1988) and will also need to be carefully managed in systems that also evolve hydrogen.

Choice of Cell Configuration

In electrodialysis the electrodes are separated by an ion-selective membrane, with the intention of separating the acid and base streams (Figure 8.2). It is possible to separate the electrodes with additional membranes that are selective for either anions or cations (bipolar membrane electrodialysis). A bipolar membrane configuration is intended to promote water dissociation into hydrogen ions (H^+) and hydroxide (OH^-) (rather than ion transport in monopolar membranes) (Pärnamäe et al., 2021), producing separate streams of acid and base.

Description of Proposed Systems

A range of systems have been proposed for CDR or ocean alkalinity enhancement (OAE), summarized in Table 8.2. For simplicity we represent the products of electrolysis as acid and base. Both CDR processes "'swing" alkalinity of seawater to extract the largest amount of dissolved inorganic

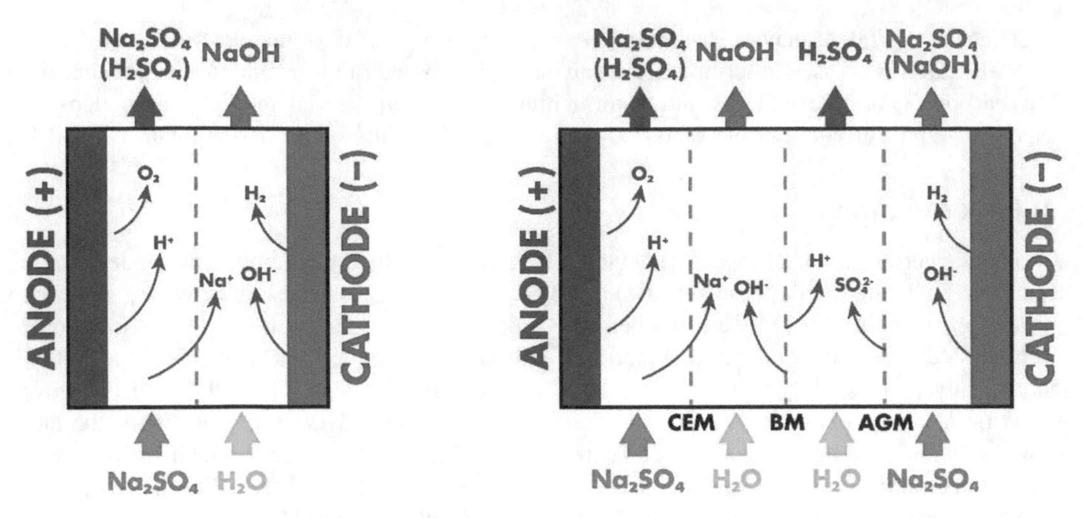

FIGURE 8.2 Simplified schematic of monopolar electrodialysis (left) and bipolar electrodialysis (right). NOTE: CEM = cation exchange membrane, BM = bipolar membrane, and AEM = anion exchange membrane.

carbon—whether as a gas or as mineral carbonate. However, there is no net impact on alkalinity when acid and base streams are recombined.

Several systems have been proposed that do not use electrochemistry, but either remove CO_2 from seawater or increase its alkalinity. Willauer et al. (2009) considers a system that filters bicarbonate, and Willauer et al. (2011) investigated CO_2 extraction using ion exchange resins. These experiments were conducted on solutions with elevated partial pressure of carbon dioxide (pCO_2). Given that they do not modify alkalinity, the water requirements at ambient CO_2 pressures may be prohibitive (see discussion in Section 8.3). Davies (2015) investigated the thermal decomposition of desalination rejected brine. Similar to seawater electrolysis, the products are solid magnesium hydroxide ($Mg[OH]_2$) and liquid hydrochloric acid (HCl), although the thermal energy requirements possibly limit wide application

Acid Process

CO_2 Removal

This approach (Figure 8.3) uses electrolysis to create streams of acid (HCl, at the anode) and base (NaOH, at the cathode). When the acidic anolyte is mixed with CO_2-"rich" seawater or brine, the acidification shifts the equilibrium of the carbonate system toward CO_2, which is vented and can be subsequently pressurized and geologically stored. The acidified CO_2-lean seawater is mixed with the base liquid stream and then released to the ocean where it absorbs CO_2 from the atmosphere to an extent described by its alkalinity and temperature.

In the system explored by Eisaman et al. (2012) and de Lannoy et al. (2018), a small proportion of influent seawater is treated or demineralized using nanofiltration, mineral precipitation, CO_2 desorption, and resin towers (e.g., as typical of a system used in the chlor-alkali process to limit electrode fouling) before being passed into a three-compartment bipolar membrane electrodialysis system. The acid generated in electrolysis (0.4–2 mol/kg) is used to acidify 160–800 times the volume of seawater. Prior to acidification, nitrogen (N_2) and oxygen (O_2) are extracted at 30-mbar pressure, such that pure CO_2 gas is extracted at 30–80 mbar pressure (i.e., by vacuum stripping).

TABLE 8.1 Comparison of Electrochemical Properties of a Range of Mineral-Based Electrolytes

Electrolyte (equilibrium mineral phase)	Saturated Cation Conc. ($\log_{10}$ mol/kg)	Saturated Anion Conc. ($\log_{10}$ mol/kg)	Anion	Ionic Strength (mol/kg)	Conductivity (μS/cm)	Electrolyte Ohmic Overpotential[a] (mV)	Mass Transfer Limit Current Density—Cation[b] (kA/m^2)	Mass Transfer Limit Current Density—Anion[b] (kA/m^2)
NaOH[c]	0.80	0.69	OH^-	4.86	319,321	6	81	246
KOH[c]	0.65	0.54	OH^-	3.50	230,963	8	85	177
$Ca(OH)_2$ (portlandite)	−1.82	−1.56	OH^-	0.04	2,731	659	0.2	1.4
$Mg(OH)_2$ (brucite)	−4.10	−3.80	OH^-	0.0002	17	>100,000	<0.01	<0.01
NaCl (halite)	1.12	1.12	Cl^-	6.88	450,675	4	170	261
$CaCl_2$ (antarcticite)	0.93	1.23	Cl^-	8.38	548,029	3	129	664
$MgCl_2$ (bischofite)	0.93	1.23	Cl^-	11.85	773,002	2	116	664
Na_2CO_3 (Na_2CO_3)	1.00	0.70	$NaCO_3^-$	7.25	474,900	4	127	53
K_2CO_3 ($K_2CO_3 \cdot 1.5H_2O$)	1.10	0.80	CO_3^{2-}	18.85	1,224,502	1	237	111
$CaCO_3$ (calcite)	−3.92	−4.09	HCO_3^-	0.0004	26	>50,000	<0.01	<0.01
$MgCO_3$ (magnesite)	−3.74	−3.95	HCO_3^-	0.0006	41	>40,000	<0.01	<0.01
Na_2SO_4 (thenardite)	0.68	0.38	SO_4^{2-}	3.84	253,241	7	61	49
$CaSO_4$ (gypsum)	−1.80	0.20	SO_4^{2-}	0.05	3,127	576	0.2	32
$MgSO_4$ (epsomite)	0.87	0.87	SO_4^{2-}	3.38	223,119	8	101	150
Seawater	–	–		0.64	43,119	42	–	–

NOTES: Solution saturation concentrations were calculated using PHREEQC v3 (Parkhust and Appelo, 1999) and the LLNL.dat database. Mineral phases (in parenthesis) were brought into equilibrium with deionized water at 25°C and 1 bar.

[a] Assuming - 5mm electrode spacing and a current density of 6 kA/ m^2 across 3-m^2 cell divider.

[b] Assuming a diffusion layer thickness of 0.01 mm.

[c] Considers a solution of 25% w/w.

TABLE 8.2 Summary of Ocean-Based Electrochemical Approaches for CDR from the Atmosphere

Direct CDR—acid process	The acid stream from the anode in the electrochemical cell is used to decrease the pH of seawater to evolve CO_2. The base stream from the cathode is then mixed with the decarbonized seawater to capture additional CO_2 from the air, resulting in a continuous closed cycle where CO_2 is effectively removed from the air via seawater.
Direct CDR—base process	The base stream from the cathode in the electrochemical cell is used to precipitate carbonate from seawater with or without the evolution of CO_2.
Ocean alkalinity enhancement (OAE)—seawater/brine electrolysis	Using seawater as the electrolyte, sodium hydroxide (NaOH) is concentrated at the cathode (which is added to the ocean to increase alkalinity; see Chapter 7). The acid stream from the anode is neutralized through reaction with silicate rocks.
OAE—water electrolysis	Similar to seawater/brine electrolysis but using an alternative non-NaCl-based electrolyte (e.g., based on Ca or Mg).
OAE—salt recirculation	Similar to seawater/brine electrolysis but the salt form, through reaction with silicates, is recycled back into the catholyte.
Hybrid approaches	A combination of electrochemical approaches that results in an increase in ocean alkalinity and the removal of CO_2 from seawater; e.g., as a gas, or in mineral carbonates.

Base Process

In this approach the base liquid produced by electrochemistry is mixed with seawater/brine to force the precipitation of carbonate minerals. Like the acid process, carbonate precipitation shifts the equilibrium of the carbonate system toward CO_2, which if not suitably compensated (e.g., by addition of base) would require venting, pressurization, and storage. The basic seawater is mixed with the acid stream from the electrodialysis unit before being returned to the ocean. Addition of base to seawater increases pH and potentially results in the oversaturation of a range of minerals. Relatively smaller additions of NaOH would oversaturate carbonate minerals, leading to their precipitation. But, if the addition rate of NaOH was greater than the carbonate precipitation rate, then it is possible that a higher pH could also lead to the precipitation of calcium or magnesium hydroxide minerals, which would result in a net reduction in alkalinity of seawater if they were not redissolved (Figure 8.4). De Lannoy et al. (2018) suggest that maintaining a pH between 9.3 and 9.6 would prevent this. Figure 8.5 shows the effect of the progressive addition of NaOH to seawater at different rates (slow and fast) and the resulting saturation indices for brucite ($Mg(OH)_2$), a hydrated magnesium carbonate mineral hydromagnesite ($Mg_5(CO_3)_4(OH)_2{\bullet}4H_2O$), and strontianite ($SrCO_3$) for solutions equilibrated with calcite (Figure 8.5a, i.e., by slow addition of NaOH to allow calcite to precipitate) and those unequilibrated to a mineral phase (Figure 8.5b, i.e., by rapid addition of NaOH).

Production of Alkalinity by Seawater Electrolysis

A system similar to that used in the chlor-alkali process could also be used for the production of solid or liquid bases for ocean alkalinity enhancement (Chapter 7). In these systems, seawater or brine are introduced into the anolyte/anode. A potential difference between cathode and anode encourages the migration of sodium across the cell divider such that an acid solution is produced at the anode and a basic solution at the cathode. In the system espoused by House et al. (2007) chlorine gas produced

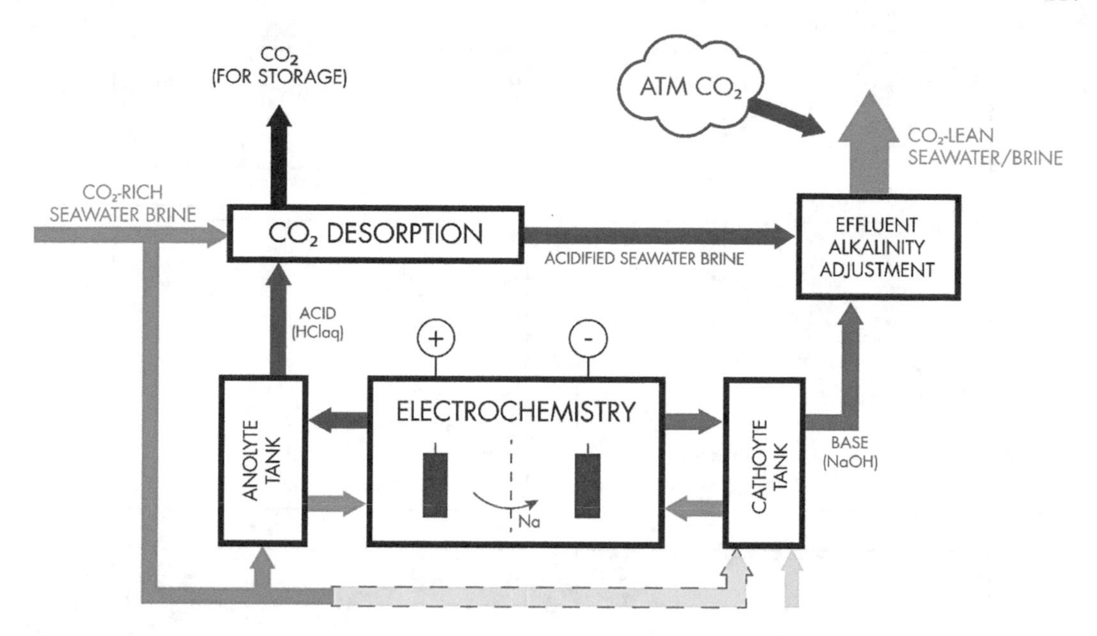

FIGURE 8.3 Simplified schematic of the acid process to remove CO_2 from seawater. Systems for precipitating sodium hydroxide (NaOH) are used in the chlor-alkali process, but it is also possible to sweep the catholyte tank with seawater or fresh water if a dissolved base is required (light green arrows).

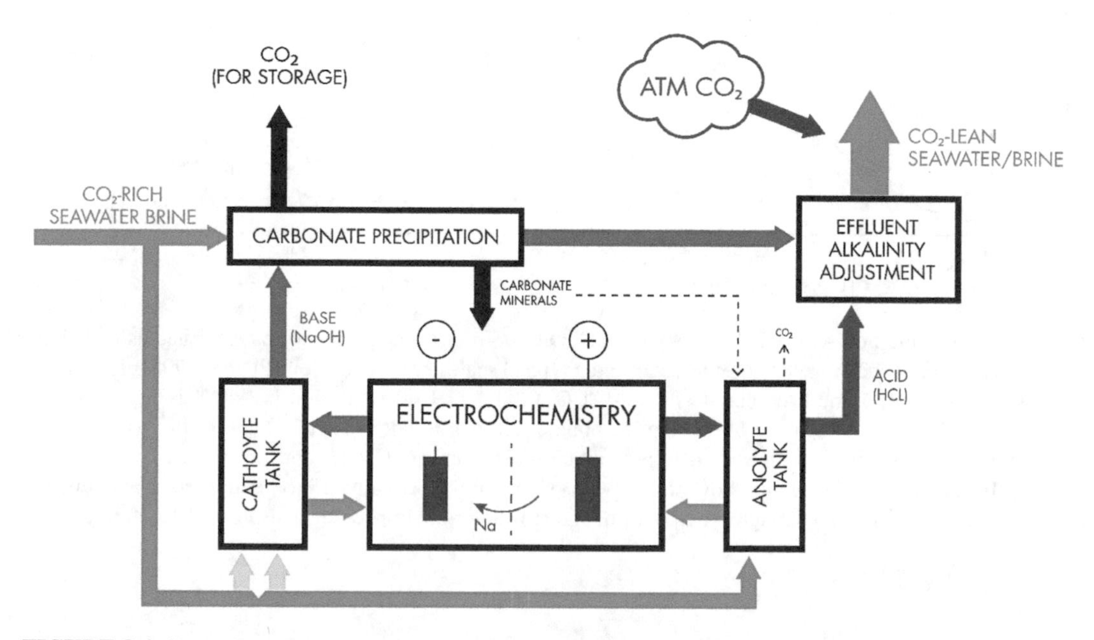

FIGURE 8.4 Simplified schematic of the base process to remove CO_2 from seawater.

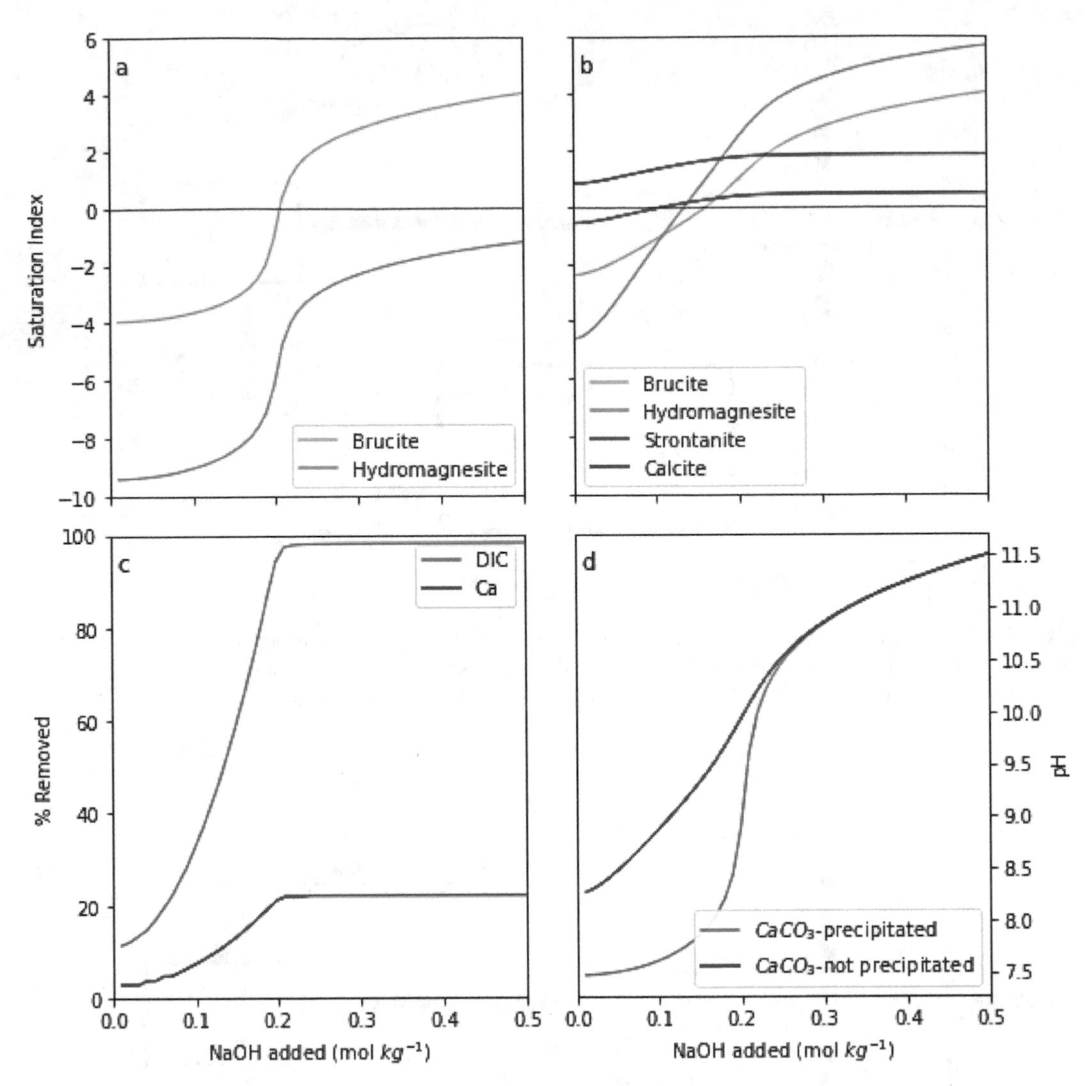

FIGURE 8.5 Impact on seawater chemistry from NaOH addition. Calculations were performed using PHREEQC v3 (Parkhurst and Appelo, 1999) at 25°C and a typical seawater composition (Pilson, 1998). The impact on brucite ($Mg(OH)_2$) and hydromagnesite ($Mg_5(CO_3)_4(OH)_2{\bullet}4H_2O$) saturation if NaOH is added (a) at a rate equal to the precipitation rate of calcium carbonate (saturation index [SI] of calcite = 0) or (b) if NaOH is rapidly added without calcium carbonate precipitation (SI of calcite $CaCO_3$ and strontianite $SrCO_3$ is also shown). In the case of calcite, (c) shows how much of the original dissolved inorganic carbon and calcium in the solution are removed through precipitation, and (d) shows the impact on seawater pH for both cases.

at the anode and hydrogen gas at the cathode are combined in a fuel cell to produce hydrochloric acid. The acid is then neutralized through the reaction with a silicate mineral to produce a salt and silica (equation 8.8), which will need to be disposed of, or could potentially be valorized in other industries. The base is discharged into the ocean (as a solid or a liquid) to increase ocean alkalinity (Figure 8.6). Davies et al. (2018) presents a system that is coupled to the reject brines from reverse osmosis, where the alkalinity is removed as a solid $MgOH_2$ base. This system is simulated by Tan et al. (2019), who suggest that reverse osmosis water production could be carbon negative, providing an appropriate carbon incentive (\$10–\$100/t CO_2) and/or revenue stream from HCl production if implemented and low-carbon electricity is used.

$$4HCl_{(aq)} + Mg_2SiO_{4(s)} \rightarrow 2MgCl_{2(s/aq)} + SiO_{2(s)} + H_2O_{(l)} \quad (8.8)$$

Production of Alkalinity by Water Electrolysis

A simplified version of a system proposed by Rau (2008) and Rau et al. (2013) is shown in Figure 8.7. In this system a basic or ultrabasic silicate mineral is reacted directly with the anolyte to produce a solution rich in cations (e.g., Ca and Mg), silicon, and other elements in the rock. This solution is passed to the anode chamber of an electrolysis cell, and the cations selectively migrate to the cathode across the cell divider. The base concentrated solution can be passed to a system for concentration and precipitation of Ca or $MgOH_2$ for subsequent addition to the ocean, or the elevated alkalinity solution could be added directly to the ocean. A solution containing only Ca^{2+} or Mg^{2+} charge balanced by OH^- ions will likely result in prohibitive ohmic losses in the electrolyte or mass transport losses at the electrode (Table 8.1). As such, the presence of a highly soluble salt (e.g., SO_4^{2-} used by Rau et al., 2013) is essential.

Production of Alkalinity by Salt Recirculation

This involves a system that promotes the migration of anions (e.g., Cl^-, SO_4^{2-}) across the electrolysis cell divider to form an acidic solution for neutralization through reaction with basic or ultrabasic (Ca, Mg) silicate minerals (hypothesized by Rau et al., 2013). The solid or aqueous

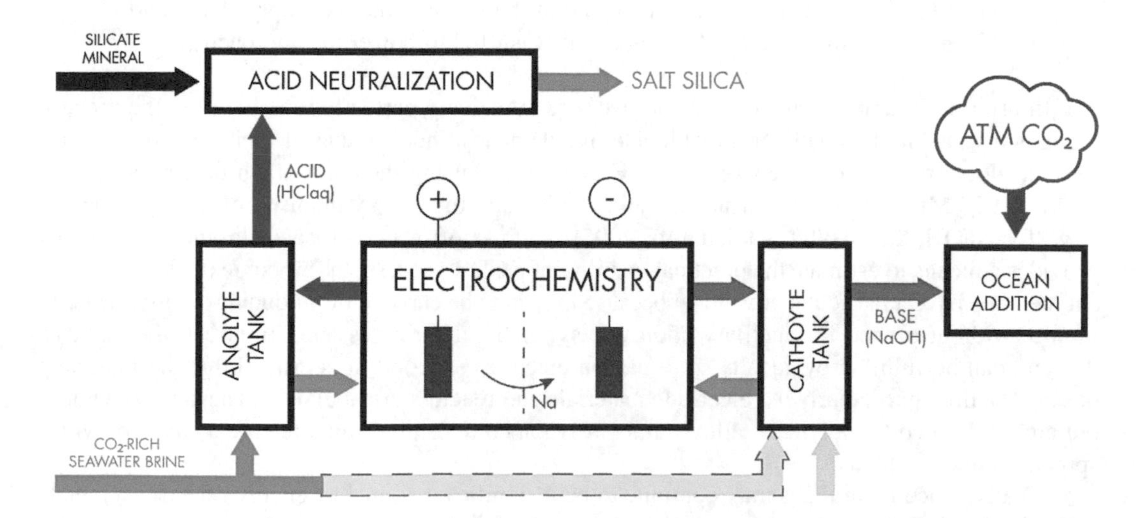

FIGURE 8.6 Simplified schematic of seawater electrolysis for ocean alkalinity enhancement.

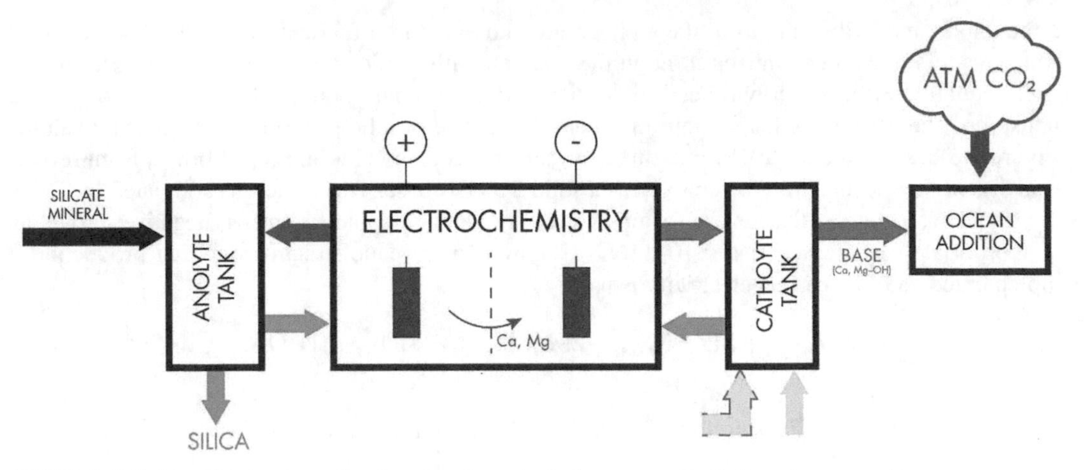

FIGURE 8.7 Simplified schematic of water electrolysis for ocean alkalinity enhancement.

salt is then added to the catholyte, resulting in an accumulation of Ca or Mg, and an increase in alkalinity (Figure 8.8).

8.3 EFFICACY

Achievable Scale

When considering electrochemical methods for ocean-based carbon removal, it is important to contextualize the scale that such processes could achieve if they were to be successful. As such, important considerations around achievable scale are based around (see Chapter 1 for discussion on scale:

1. The need/consumable demand for stoichiometric additives (e.g., acids, bases, salts),
2. The materials of construction of electrochemical systems (e.g., alloys),
3. The net energy intensity of the life cycle of the process and its supply chain, and
4. The grid emissions factor of electricity that is used to power the approach.

In brief, first, unless reagents and additives are effectively perfectly reusable (i.e., the process features a vanishingly small consumable additive demand), independent of their cost, on account of their global production levels (e.g., NaOH is produced using the chlor-alkali process at a level of around 70 Mt/yr (World Chlorine Council, 2017)), and the energy intensity of their production (e.g., for NaOH, 2.5 MWh/t (Chlistunoff, 2005)), the use of or need for stoichiometric additives provides a means to estimate the practical viability of a CDR process. In this context, electrochemical methods have a particular advantage because they may be capable of producing acidity/basicity, in situ, without a need for ancillary additives, synthetic electrolytes, etc.; in effect, the additive demand can be fulfilled by, ideally, zero-carbon electrons. Second, it is critical that the materials of construction, particularly the electrode materials and reactors are abundant. These may include, but are not limited to, polymers, alloys, and or carbon composites, which feature a basis for widespread production today.

Finally, since natural gas may continue to be a significant source of energy (electricity), generally in the short to medium term, it would be attractive if carbon removal processes featured a

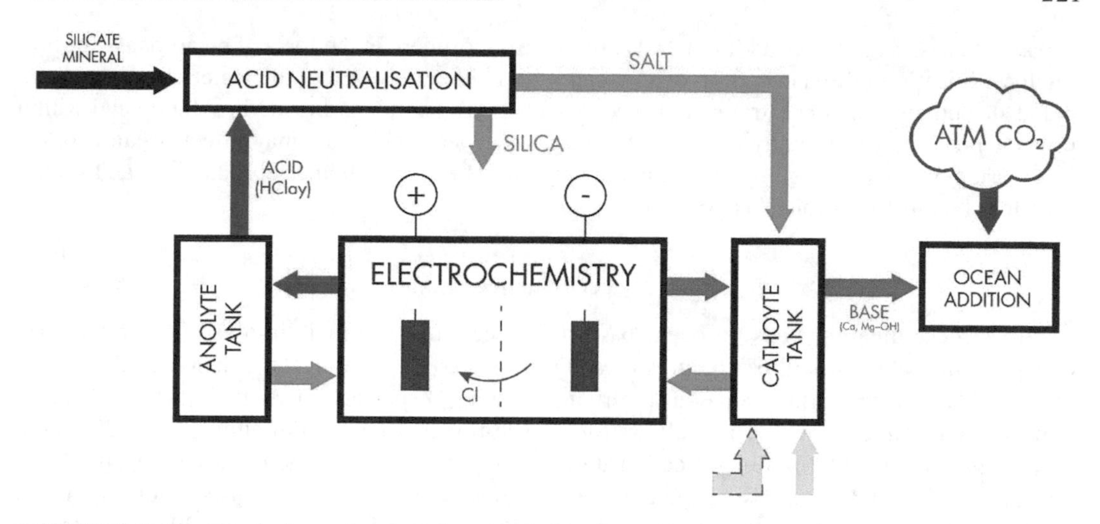

FIGURE 8.8 Simplified schematic of a salt recirculation process for ocean alkalinity enhancement.

net (i.e., uncompensated by the potential embodied energy benefit of any co-produced hydrogen) energy intensity/demand of less than 2.4 MWh/t CO_2 removed—to deliver a (net) negative carbon removal outcome.[1] Maximum (net) energy intensity is quantified by assessing the specific intensity ratio (SIR) of the energy source, that is, the reciprocal of the grid emissions factor (e.g., for natural gas combustion–based electricity generation, this is around 0.42 t CO_2 emitted per MWh of electricity produced;[2] see Section 8.3). Of course, it is necessary that, in general, the net energy intensity be substantially inferior to the maximum value to realize the highest CDR benefit for example, as ensured by use of renewable energy as far as possible, particularly during periods of excess generation capacity (i.e., when curtailment is ongoing). All that said, it is particularly desirable to develop electrochemical processes that have the potential to produce valuable co-products, for example, hydrogen, oxygen, and/or minerals (see Section 8.4). The former are particularly attractive because if these "clean fuel" co-products can partially power the electrochemical process, they in effect allow for reduced reliance on fossil fuel sources (e.g., up to 50 percent less fossil-fuel energy would be required depending on the co-produced hydrogen yield). Also, in an energy storage–constrained market, they offer a means to bridge the energy gap between time periods when renewable electricity may be in short supply (e.g., at night) while reducing the net carbon intensity of the fuel source that is used to power the CDR process (e.g., when natural gas–based electricity may be a possible energy input).

Additionality and Downstream Effects

Despite their enormous potential, electrochemical processes that couple with the world's oceans may exert unintended consequences. Such consequences include some that we can imagine today, and other nonlinear and downstream effects that may only be ascertained when electrochemical CDR processes are operated at the pilot or larger scales. In addition to general issues associated with ocean-based CDR, electrochemical CDR processes seek to (1) remove dissolved CO_2 from seawater as a means of reducing its acidity or (2) alkalinize seawater as a means of enhancing its

[1] Here the term "removal" implies end-to-end carbon abatement including capture, sequestration, and immobilization for periods generally well in excess of 10,000 years.

[2] See https://www.eia.gov.

capacity to absorb and store additional CO_2 from the atmosphere in the form of bicarbonate species. Such acidification or alkalinization of seawater will, deliberately, affect the water chemistry (e.g., pH, distributions and concentrations of species such as CO_2 and bicarbonate in solution and within mineral precipitates, turbidity, etc.). Such changes in water chemistry may affect ocean ecology and marine organisms (e.g., by changing the optimal pH for growth and reproduction) and should therefore be limited as much as possible.

Permanence

Electrochemical methods for ocean-based CDR seek to (1) extract dissolved CO_2 in the form of a gas, after which it will be compressed and sequestered in geological formations; (2) stabilize dissolved CO_2 in the form of aqueous bicarbonate (HCO_3^-) species, for example, when alkalinity enhancement of the aqueous phase allows for the additional drawdown of atmospheric CO_2; and/ or (3) immobilize CO_2 in mineral carbonates. If effectively implemented, particularly the latter approaches (2 and 3), can ensure the durable and effectively permanent sequestration of CO_2 at Earth's surface while eliminating the risk of accidental release (La Plante et al., 2021) due to, for example, fault activation in geological sequestration (Jahediesfanjani et al., 2018). Notwithstanding changes in sea-surface temperature, CO_2 stabilized in the form of bicarbonate anions is expected to remain stable in surface waters (seawater) for periods on the order of 100,000 years (Falkowski et al., 2000; Caldeira et al., 2018). CO_2 stabilized within mineral carbonates, as highlighted by the persistence of limestone deposits on Earth's surface, is expected to remain stable for periods on the order of hundreds of millions of years (Lackner, 1995).

Monitoring and Verification

Electrochemical methods, on account of being contained engineering processes, are readily amenable to assessments of mass, energy, and CO_2 balances, that is, both embodied in and as related to CDR, within the boundaries of the process. For example, processes that generate acidity or basicity can be readily instrumented to assess the extent of acid and base generation within a liquid stream (e.g., of the anolyte and catholyte), the amount of hydrogen, oxygen, and/or chlorine produced (e.g., in the case of seawater electrolysis) or CO_2 evolved (in the case of CO_2 stripping processes), the amount of solid precipitates formed (e.g., mineral carbonates and/or hydroxides), or minerals dissolved (e.g., silicate rock dissolution), and the consequent energy expenditures and seawater fluxes. Such monitoring can be carried out by analysis of the influent and effluent compositions, typically within electrochemical cells or reactors, and of the gases evolved within a closed headspace. The challenge with electrochemical processes, however, is that their CDR basis may often be indirect. For example, except in the case of CO_2 stripping or solid carbonate mineral formation, these processes often rely on the alkalinization of seawater directly or by the dissolution of minerals within it (e.g., brucite, silicate rock, etc.) to enhance its CO_2 storage capacity. The enhancement of the CO_2 storage capacity results in the additional dissolution of atmospheric CO_2 in or its absorption into seawater, which in turn results in CDR. If mineral dissolution may occur within the system boundary, the temporal dynamics and the amount of atmospheric CO_2 removal that will occur can be readily established. But, if mineral dissolution may occur slowly, and beyond the process boundaries (e.g., following discharge of the effluent into the ocean), it will be necessary to rely on an indirect basis of quantifying the CDR benefit.

8.4 SCALABILITY

Resource Availability

A large, and practically unlimited, elemental reservoir is found in the ocean for the production of acid and base creation (HCl and NaOH). If all NaCl in seawater were split, the equivalent of 26×10^6 Gt of NaOH would be produced, which could capture 30×10^6 Gt of CO_2 (equivalent to 1,000x more than the carbon in the Earth system).[3] Furthermore, terrestrial brines (e.g., produced water, desalination brines) could also provide resources for such acid and base production, although to a smaller extent. From stoichiometry and charge balance considerations, 2 mol of base (NaOH) are needed to neutralize 1 mol of CO_2, or more than 1 mol of acid is needed to dissolve 1 mol of alkaline mineral (e.g., $CaCO_3$, $MgSiO_3$, etc.). So, to effect 500 Mt/yr of CDR, an equivalent quantity of acid and base is needed. This is far larger than current scales of production ~70 Mt/yr for NaOH (and most future projections, see Figure 8.9) and ~300 Mt/yr for H_2SO_4.[4] Mined salt is a common feedstock for chlor-alkali, but as such, this requires not only the development of new facilities for acid and base production but also new infrastructure to store, transport, or dispose of acid and base materials.

The ocean offers a considerable resource for absorbing gaseous CO_2. Systems designed to extract CO_2 from seawater unavoidably create mass handling challenges. The ocean contains approximately 2 mmol/kg (88 g CO_2/m^3) of dissolved inorganic carbon; aqueous CO_2 constitutes only a small proportion of this (~10 μmol/kg at atmospheric pressure). The majority of the carbon is contained in the form of HCO_3^- and CO_3^{2-} ions. As such, without modifying alkalinity, it may only be possible to extract hundreds of milligrams of CO_2 per cubic meter of seawater. While the prevalent volumetric concentration of CO_2 in the ocean (>80 g CO_2/m^3) is greater than that in the atmosphere (<1 g CO_2/m^3), the mass concentration is less (0.09 and 0.6 gCO_2/kg in the ocean and atmosphere, respectively). A facility designed to remove 0.1 Mt CO_2/yr from seawater would need to treat 1.3 Gt seawater/yr, or about 3 million m^3/day, which is 5–10 times greater than the world's largest desalination facilities (Eke et al., 2020). In other words, treating a water flux as great as present-day global desalination would amount to around 1–2 Mt CO_2/yr removal. Treating a similar seawater volume to increase its alkalinity could remove ~90 Mt CO_2/yr. Thus, there may be advantages to processes that create a concentrated solid or liquid base, ex situ or in situ, and distribute this in the ocean.

Some of the processes described above require the extraction, processing, and dissolution of rock either to increase alkalinity or to replenish alkalinity, which is lost during carbonate precipitation. The scalability implications of rock extraction is discussed in Chapter 7. If CO_2 is extracted from seawater, there is a need to store this fluid-state CO_2, for example, either in terrestrial reservoirs or by pumping it into the very deep ocean.[5] It is likely that large volumes of CO_2 derived from either the atmosphere or captured emissions will need to be disposed of over the next century. This challenge was explored in a recent National Academies report (NASEM, 2019, Chapters 6 and 7) and elsewhere (Kelemen et al., 2019).

---

[3] Assumes the sodium concentration of seawater is 0.4 mol/L (Pilson, 1998) and the volume of the ocean is 1.3 × 109 km^3 (https://ngdc.noaa.gov/mgg/global/etopo1_ocean_volumes.html). The Earth system contains 871, 1,400, 450, 900, 700, 1,750, 37,100 (~43,000 in total) Gt C in the atmosphere, permafrost, soils, surface ocean, dissolved organic carbon in the ocean, shallow sediments, and deep ocean, respectively (Canadell et al., 2021).

[4] See https://www.essentialchemicalindustry.org/chemicals/sulfuric-acid.html.

[5] Note that the liquid CO_2 relevant to a deep-ocean disposal approach relies on the basis that liquid CO_2 is denser than liquid water.

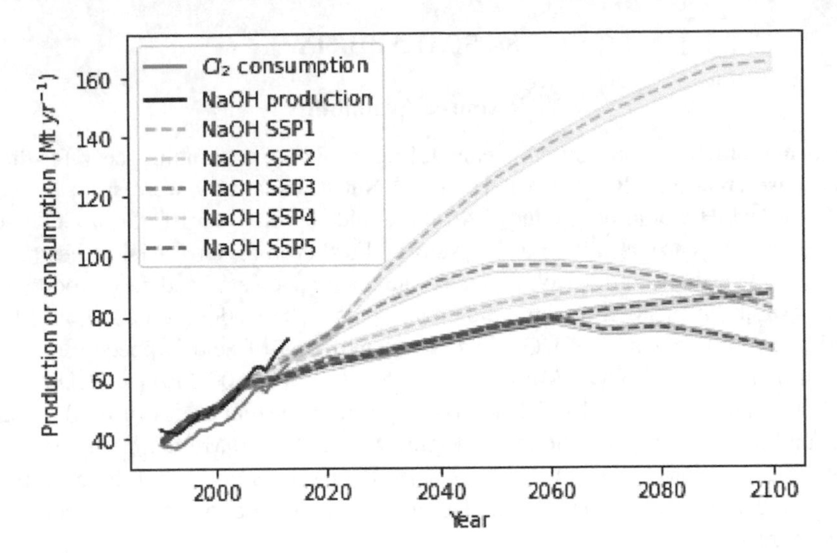

FIGURE 8.9 Historical production of NaOH and consumption of chlorine (chlorine production from Egenhofer et al., 2014; NaOH production calculated from chlorine). The estimated projections of NaOH production are based on the method in Renforth (2019) for a range of shared socioeconomic pathways (SSPs). SOURCE: Egenhofer et al. (2014), Renforth (2019).

Low-Carbon Electricity

Global annual electricity supply is currently 25,000 TWh and will possibly increase to ~35,000 by 2040 (EIA, 2021). If an electrochemical approach consumed ~2 MWh/t CO_2 removed (see below), then 1 Gt CO_2/yr would require around 2,000 TWh/yr (20 percent addition on the projected increase). The approaches described above will likely require low- or zero-carbon electricity to be feasible. While the supply and generation capacity of renewable electricity are increasing, current forecasts estimate that fossil fuels will contribute ~45 percent of electricity generation for the next three decades at least (EIA, 2021). As such, it is necessary to reduce the energy intensity of CDR approaches to below 1.5–9.5 MWh/t CO_2 (i.e., an inverse of the 105–680 kg CO_2/MWh grid emissions intensity of electricity produced in the United States;[6] Figure 8.10).

It may be possible to use excess electricity generation at times of the day in which demands of the grid are lowest, and thus the power is "cheap." Such electricity is only available for no more than ~8 hours when curtailment is in effect.[7] To operate a capital-intensive facility using such electricity implies accepting a capacity factor of 30 percent, which produces a capital inefficiency (due to lack of operation for the majority of the day), and a need to oversize the plant to meet production goals. Although such reduced utilization could be remedied, it requires access to cheap long-duration electricity storage and/or net emission technologies to produce their own fuels (e.g., hydrogen) to power operations when curtailment is not in effect or cheap renewable electricity is not available.

Geospatial and Adjacent Infrastructure Considerations

Unquestionably the need for seawater in electrochemical ocean CDR requires the siting of such processes and plants at sites that offer ready access to seawater. While the optimal attributes of sit-

[6] See https://www.epa.gov/egrid/data-explorer.

[7] See https://www.pge.com.

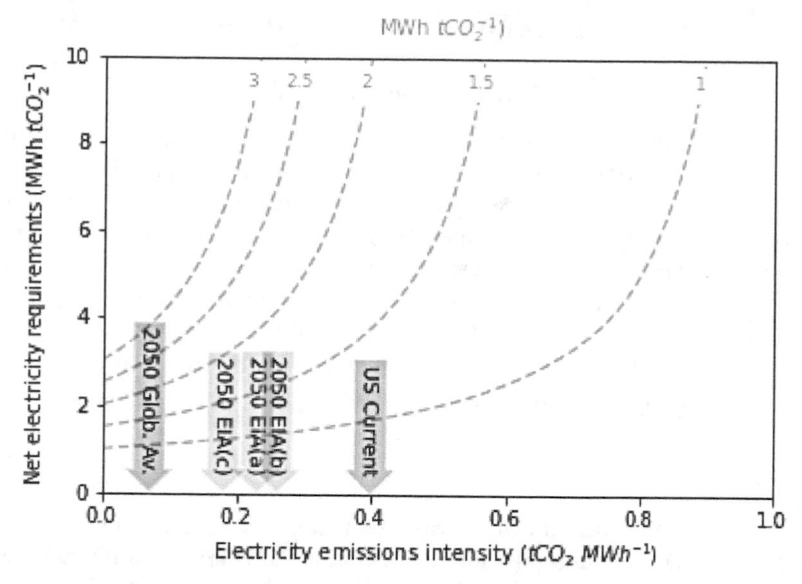

FIGURE 8.10 The relationship between electricity requirements per net tonne of CO_2 removed (*y* axis), the emissions intensity of the electricity (*x* axis). The blue contours represent the energy requirements (MWh) of the system to remove 1 ton of CO_2 from the atmosphere (see Section 8.5 for estimated energy requirements). Arrows indicate current U.S. grid average emission intensity, projected emission intensity, or the U.S. Energy Information Administration (a) AEO2020 reference scenario, (b) high-cost renewable energy scenario, and (c) low-cost renewable energy scenarios (EIA, 2021), and the 2050 global average emission intensity of electricity projected by the Intergovernmental Panel on Climate Change. SOURCE: Bruckner et al., 2014.

ing, for example, in the open ocean, along the coast, or in river deltas are not fully clear, there are adjacent infrastructure considerations that are relevant: (1) seawater access, (2) low-carbon energy supply, (3) infrastructure for supply of raw materials and disposal of waste, and (4) labor.

First, a significant cost associated with desalination plants (and other facilities that require seawater) is the infrastructure for seawater intake. Thus, colocation at sites that host water intake infrastructure (e.g., desalination plants, power plants, etc.) is valuable not only for reducing the overall capital costs of eventual commercial plants, but also conceivably for the potential to utilize operating water intake permits of existing facilities (see section on Governance below). Second, the most effective means for CDR implies the use of renewable energy. Thus, it is important to locate eventual facilities in regions that feature substantial potential for renewable energy generation including wind and solar (either photovoltaic or concentrated solar power based). Third, if CDR is accomplished in a manner that yields a fluid, or solid, it is necessary to build out infrastructure including pipelines (and compression stations) to convey the CO_2 to sequestration sites, or to develop solid waste handling and valorization facilities (e.g., for the potential use of mineral carbonates and hydroxides as construction materials). Although some of these facilities already exist (e.g., landfills for solid waste handling), given the enormous scale of CDR required, it would be necessary to expand such facilities considerably, to achieve scale relevance.

8.5 VIABILITY AND BARRIERS

Environmental Impact

A comprehensive life-cycle assessment for electrochemical ocean-based approaches has yet to be completed. The following provides an overview of potential issues, summarized in Figure 8.11, for best available membrane processes. Mercury (Hg) amalgam systems that were historically used for chlor-alkali typically yielded elevated Hg in effluent waters and surrounding sediments, accompanied by other potentially toxic metals (Arribére et al., 2003), and some early diaphragm systems that use asbestos for their cell divider report particle emissions to water and air. Many of these systems are being replaced by membrane or alternative diaphragm systems, such that their environmental impacts would not be applicable to future ocean-based electrochemical approaches. The environmental impact associated with rock extraction is discussed in Chapter 7.

Gaseous Chlorine Production

In seawater or brine electrochemical reactions (equation 8.5) as relevant to the chlor-alkali and conceptually similar processes, around 1 tonne of chlorine gas is produced for sufficient alkalinity to be generated to remove a tonne of CO_2. There are approximately 23 Mt of gaseous chlorine in the atmosphere, most of which is derived from human sources (Khalil, 1999), which impacts atmo-

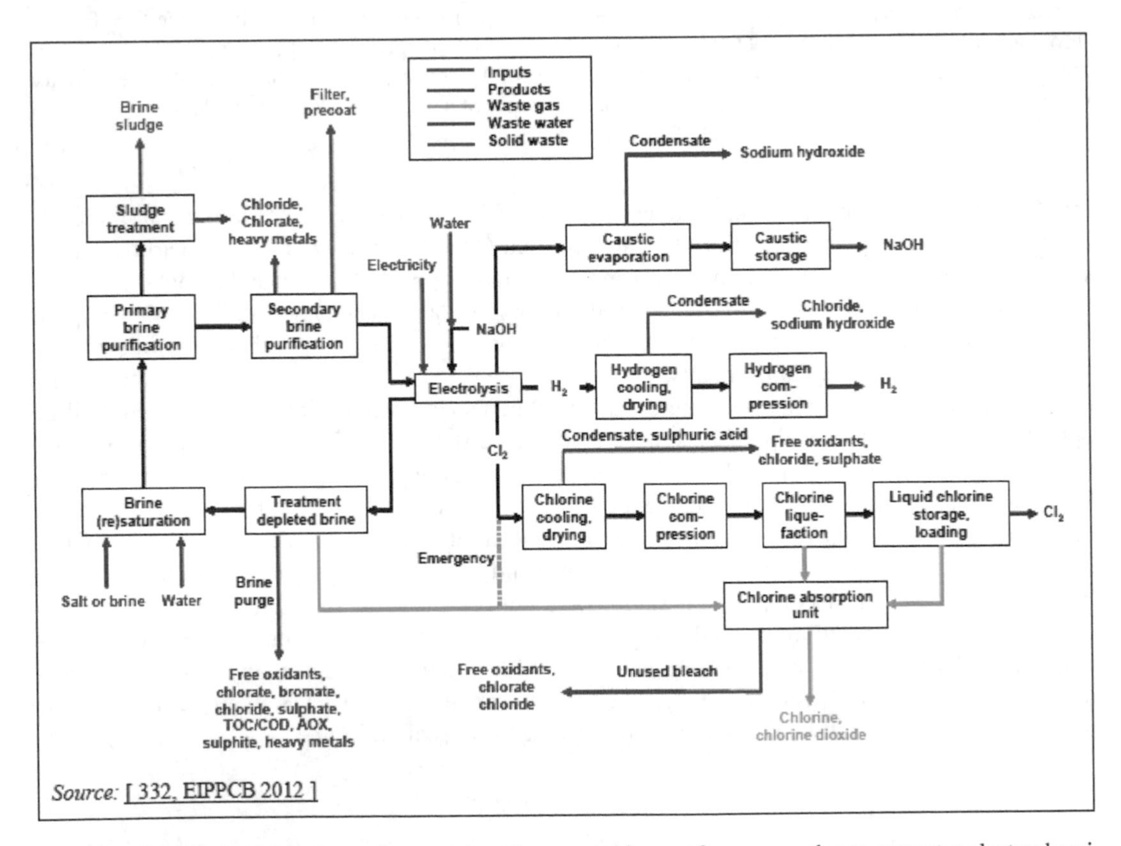

FIGURE 8.11 Potential sources of material environmental impact from a membrane seawater electrochemical/chlor-alkali process. Herein, CO_2 is considered an implicit input in the process. SOURCE: Redrawn from Brinkmann et al., 2014. Licensed under Creative Commons CC BY 4.0.

spheric N, OH, and CH_4 concentrations (Zhai et al., 2021). Given that the global chlorine market is on the order of 60–70 Mt/yr (Figure 8.9), systems operating at climate-relevant scales will likely require a highly efficient method of managing and disposing of excess chlorine. It may be possible to reduce or eliminate chlorine emissions by using selective electrodes (i.e., that promote the oxygen evolution reaction as opposed to the chlorine evolution reaction), or by reaction of generated Cl_2 with H_2 to produce HCl and its reaction with cation-rich silicates to produce salt. In addition to chlorine, CO_2 is often reported in waste gases.

Wastewater

Table 8.3 presents ranges of emissions to water for best available technology from chlor-alkai processes. Wastewaters are generated in feedstock brine purification and depleted brine treatment, as condensates from product treatment, or bleach from the chlorine gas absorption unit. It is unclear how applicable these are to the CDR or OAE systems considered in this chapter, but an industrial-scale plant will likely produce some wastewater, the generation and treatment of which is not explored in the literature.

Energy Use

The life-cycle impact of the chlor-alkali industry is dominated by energy consumption in electrolysis, which has implications for both the global warming potential impact and the impact on human health (Garcia-Herrero et al., 2017), with membrane-based processes performing favorably.

Reduced Dissolved Inorganic Carbon in Effluent Waters

Processes that remove CO_2 from seawater will, by intention, result in very low dissolved inorganic carbon (DIC) concentrations. For instance, a process that removes 90 percent of the DIC while maintaining alkalinity would result in an aqueous CO_2 concentration on the order of pmol/kg and a CO_2 partial pressure <1 nanoatmosphere. Low DIC levels are possibly detrimental to autotrophic organisms (Hansen et al., 2007) and may be impacted at the point of effluent addition. This water will eventually mix into the surface ocean and promote CO_2 drawdown from the atmosphere. It may be possible to design discharge systems (active sparging or passive cascades) to promote gas exchange and limit these effects.

Co-benefits

In all of the approaches outlined above, hydrogen is a likely product of reactions at the cathode. Currently the United States consumes approximately 10 Mt of hydrogen per year, derived largely from steam methane reforming of natural gas or coal gasification, at costs of \$1–\$2.5/kg H_2, with an emission intensity of 40–60 kg CO_2/GJ (DOE, 2020). In the system proposed by Rau et al. (2013), 23 kg H_2 is produced for every tonne of CO_2 removed (or an emission intensity of −300 kg CO_2/GJ, assuming no emissions associated with the process of its energy supply). The baseline shared socioeconomic pathways (Riahi et al., 2017) suggest hydrogen may constitute 2.5 to 4.2 EJ/yr by 2050, rising to possibly tens of exajoules by 2100.

Processes that neutralize excess acid by reaction with crushed silicate rocks or minerals will create a brine composed of dissolved elements from the rock. This brine is likely to contain high concentrations of silica (which is present in some basic silicate rock ~40–50 mass % as SiO_2), aluminum (~15 mass % Al_2O_3), and iron (5–10 mass % iron oxide). Except for nickel and chromium in ultrabasic rocks, other elements are unlikely to be sufficiently concentrated to facilitate recovery.

TABLE 8.3 Emissions to Water from 1 Tonne of Cl_2 Production from Best Available Chlor-Alkali Electrochemical Technology

Water	Up to 2.7 m^3
Free chlorine	0.001–3.8 g
Chlorate	0.9–3,500 g
Bromate	50–300 mg
Chloride	0.6–1,060 kg
Sulfate	0.07–7.4 kg
Organic carbon (as total organic carbon)	2.5–34 g
Metals	Cadmium, chromium, copper, iron, nickel, lead, and zinc (derived from impurities in salt or brine)

SOURCE: Brinkmann et al., 2014. Licensed by Creative Commons CC BY 4.0.

Some processes also promote mineral carbonate formation. Niche high-value markets exist for these precipitated minerals (i.e., fine precipitated silica as a pozzolan in concrete manufacturing, precipitated calcium carbonate as a paint filler). However, use as construction products may be the more scalable destination.

Energy

Electricity consumption in the electrochemical cell dominates the energy balance of all described systems, and much of the reported values are normalized to mass of CO_2 removed from the atmosphere, rather than net removal (which would account for CO_2 emissions from electricity generation). The carbon intensity of the U.S. electricity grid is between 1.5 and 9.5 MWh/t CO_2,[8] so a system that consumes 1 MWh/t CO_2 removed, may require between 1.1 and 3.0 MWh/net t CO_2 removed. Processes that consume more electricity than the carbon intensity of the grid will emit more CO_2 than they remove.

Eisaman et al. (2012) suggest that approximately 1.5 MWh are required per ton of CO_2 extracted from their acid bipolar membrane electrodialysis process. Their system uses a stack of nine cells operating at a current density of ~0.4–0.5 (mA/cm^2)/Lpm to produce 0.1–0.4 liters per minute of CO_2 (at standard temperature and pressure) from 3–6 Lpm of seawater. A technoeconomic assessment of an upscaled process (7 kt CO_2/yr) suggests an energy requirement of 3.1 MWh for the acid process and 4.4 MWh for the base process per ton of CO_2 removed.[9]

There is no comprehensive technoeconomic assessment for processes that effect CDR and use electrochemistry to increase ocean alkalinity. House et al. (2007) consider electrolysis of seawater and suggest energy requirements between 0.8 and 2.5 MWh/t CO_2 removed, which is consistent with Rau (2008), Rau et al. (2013), Davies et al. (2018), and La Plante et al. (2021), who calculate 1.5–2.3 MWh/t CO_2 for similar systems.

Cost

Sodium hydroxide prices are sensitive to both energy costs and volatility in chlorine supply and demand. However, a typical price may be on the order of $300–$400 per tonne, typically delivered

[8] See https://www.epa.gov/egrid/data-explorer.

[9] Note that the final use or geological storage of the CO_2 was not included within the assessment.

as a 50 percent solution (HIS Markit, 2017). Table 8.4 presents a technoeconomic overview of chlor-alkali production. Although simplified, the levelized cost of NaOH is consistent with market prices. It suggests that using NaOH from current chlor-alkali methods may be on the order of \$500–\$700/t CO_2 removed (without revenue from H_2 and Cl_2 sale) or \$450–\$600/t CO_2 removed (with revenue). This does not account for acid disposal/neutralization, mineral extraction, valorization of precipitated products, seawater monitoring, or the emissions associated with the process or energy supply.

While chlor-alkali systems are not directly comparable to those considered in this chapter, it demonstrates that electrochemical reaction processes are capital and energy intensive. Although it is possible that many of the components will be different (e.g., no, or smaller, gas handling equipment if the sale or processing of Cl_2 or H_2 is not a priority; a fuel cell for Cl_2 or H_2 reaction; or less storage if deployment is integrated). However, the largest single capital component is the electrolyzer (i.e., the reaction half cells and electrodes, equating to 30–50 percent of the capital cost), which is essential for all electrochemical reaction systems, although the relative cost of membranes for electrodialysis systems may be greater and electrode material costs lower. The bare module cost (i.e., without other system components, installation costs, owner costs, or contingency) of an aqueous electrolyzer is shown in Figure 8.12. Those costs, incorporated into typical chlor-alkali process costs (Table 8.4) suggest that using a chlor-alkali system to produce NaOH for ocean alkalinity enhancement may cost on the order of \$533–\$668/t CO_2, without including costs associated with acid disposal. If additional revenue was generated from the sale of Cl_2 and H_2 gas, then the costs for

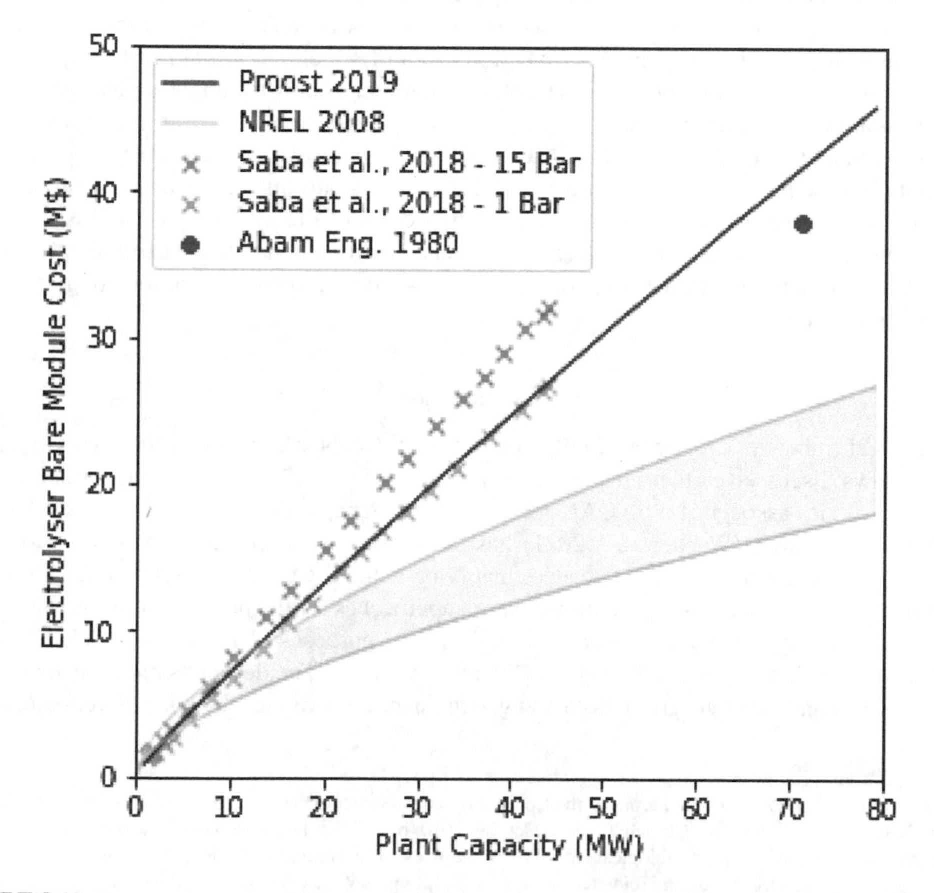

FIGURE 8.12 Bare module costs for an aqueous electrolyzer for a given power capacity.

CO_2 removal may be 10–15 percent lower. The energy requirements of these systems are approximately 8.8 MWh/t CO_2, which is much greater than ~2 MWh/t CO_2 estimated by Rau (2008), Rau et al. (2013), and La Plante et al. (2021) (see Figure 8.13). Using this energy requirement together with lower capital requirements (removing rectifiers, boilers, and storage) and reducing raw material costs (e.g., by using seawater as feedstock brine), then the overall cost for OAE by seawater electrolysis could be further reduced to ~$150/t CO_2, which is consistent with the preliminary assessments by Rau (2008), Rau et al. (2013), and La Plante et al. (2021). In the absence of cost improvements on conventional electrochemical systems, the cost of ocean alkalinity enhancement through electrochemical methods may be between $150 and $700/t CO_2 removed. These estimates do not account for CO_2 emissions from the process or supply chain. Powering an electrochemical system with ~2 MWh/t CO_2 removed using current unabated U.S. grid average emissions (400 kg/MWh) would result in 0.8 t CO_2 emitted (or a 5× cost increase for net removal).

The technoeconomic assessment of removing CO_2 as a gas from seawater through (acid and base) bipolar membrane electrodialysis was undertaken by Eisaman et al. (2018), who estimated costs on the order of $373–$2,355/t CO_2 removed. This included an acid stand-alone process ($1,839–2,355/t CO_2), an acid process colocated with seawater cooling ($727–$1,076/t CO_2), or an acid process colocated with desalination ($436–$717/t CO_2). Costs were also included for a base stand-alone process ($1,076–$1,349/t CO_2), a base process colocated with seawater cooling ($583–$899/t CO_2), or a base process colocated with desalination ($373–$604/t CO_2).

The above costs (summarized in Figure 8.13) do not include those associated with monitoring, verification, and reporting. For those processes that remove CO_2 from seawater for injection and storage into geological reservoirs, costs are incurred for seismic baseline monitoring and the installation of monitoring wells prior to injection (both of which are initial capital expenditures). Additional geophysical monitoring, water column seawater chemistry analysis, and groundwater monitoring are required during and following (~30 years) injection. However, the estimated costs of this are relatively small <$1/t CO_2 (IPCC, 2005). A monitoring program for well-mixed effluent quality will be independent of volume, and daily water sampling and analysis may cost on the order of $50–$100K/yr (<1 percent of the operating costs of the systems in Table 8.4). A more comprehensive environmental monitoring program and environmental impact assessment may form the basis of obtaining a permit for operating a plant within national jurisdictions (e.g., discussed in Chapters 2 and 9).

Governance

The legal framework for ocean CDR is discussed in Chapter 2. Many of the international and domestic laws discussed in that chapter could apply to ocean CDR electrochemical engineering, and specific issues associated with OAE are discussed in Chapter 7.

Previous research (Webb et al., 2021) has considered the application of international and domestic law to electrochemical approaches involving alkalinity creation. Webb et al. (2021) concluded that projects that employ alkalinity creation approaches for the purpose of mitigating climate change would likely constitute "geoengineering" for the purposes of Decisions X/33, XI/20, and XIII/4 under the Convention on Biological Diversity (CBD).[10] The decisions recommend that parties to the CBD and other governments avoid geoengineering activities that may affect biodiversity,

[10] Report of the Conference of the Parties to the Convention on Biological Diversity on the Work of its Tenth Meeting, Decision X/33 on Biodiversity and Climate Change, Oct. 29, 2010 (hereinafter Decision X/33)]; Report of the Conference of the Parties to the Convention on Biological Diversity on the Work of its Eleventh Meeting, Decision XI/20 on Climate-related Geoengineering, Dec. 5, 2012 (hereinafter Decision XI/20); Report of the Conference of the Parties to the Convention on Biological Diversity on the Work of its Thirteenth Meeting, Decision XIII/4, Dec. 10, 2016 (hereinafter Decision XIII/4).

TABLE 8.4 A Simplified Techno-Economic Overview of Diaphragm and Membrane Chlor-alkali Systems, and Then the Application of the Produced NaOH for CDR by Ocean Alkalinity Enhancement

	Diaphragm	Membrane
System size (kt/yr)		
Chlorine production	179	
Sodium hydroxide production	197	
Hydrogen production	5	
Maximum CO_2 removal	286	
Power (MW)	71	79
Capital costs (M$)		
Cells	38	73
Rectifiers	8	7
Boiler	9	3
Storage	12	12
Sulfate removal	1	2
Cell renewal	5	5
Initial materials	61	49
Total bare module costs[b]	134	152
Total production costs[b]	154	175
Total overnight costs[b]	172	195
Operating Costs (M$/yr)		
Energy costs[c]	118	83
Raw material costs[c]	14	15
Fixed running costs[c]	10	11
Cost summary		
Levelized cost of CO_2 removed ($/t CO_2)[d]	**668**	**533**
Possible H_2 and Cl_2 revenue (M$/yr)[e]	14	16
Levelized cost of CO_2 removed ($/t CO_2) including revenue	**608**	**466**
Levelized cost of NaOH ($/t) as 50% mass solution, including revenue	**442**	**339**

[a] Bare module costs were summarized from Abam Engineers (1980) and converted into current prices using the Chemical Engineering Plant Cost Index (https://www.chemengonline.com). Mass flows were taken from Abam Engineers (1980).

[b] Assuming 10% engineering procurement and contractor costs, 5% project contingency, 12% owner's costs.

[c] Cell power of 7,267–8,105 Mcal/ECU, motor power of 703–735 Mcal/ECU, produced hydrogen used in the boiler. In the diaphragm process, 625 Mcal/ECU of additional fuel is required. Raw material costs include salt ($50/t), sulfuric acid ($100/t), soda ash ($150/t), calcium chloride ($150/t). Labor costs include 40 shift and 25 day workers on an annual salary of $35K; 15% of labor as supervision; 35% of labor as administration and overheads; 2.5%, 1%, and 1% of production costs as maintenance, taxes, and insurance, respectively.

[d] Assuming a capital charge factor of 7.5%, a plant capacity factor of 90%, an escalation rate of 2%, a discount rate of 8%, and a levelization period of 30 years.

[e] Cl_2 at $80/t and H_2 at $800/t.

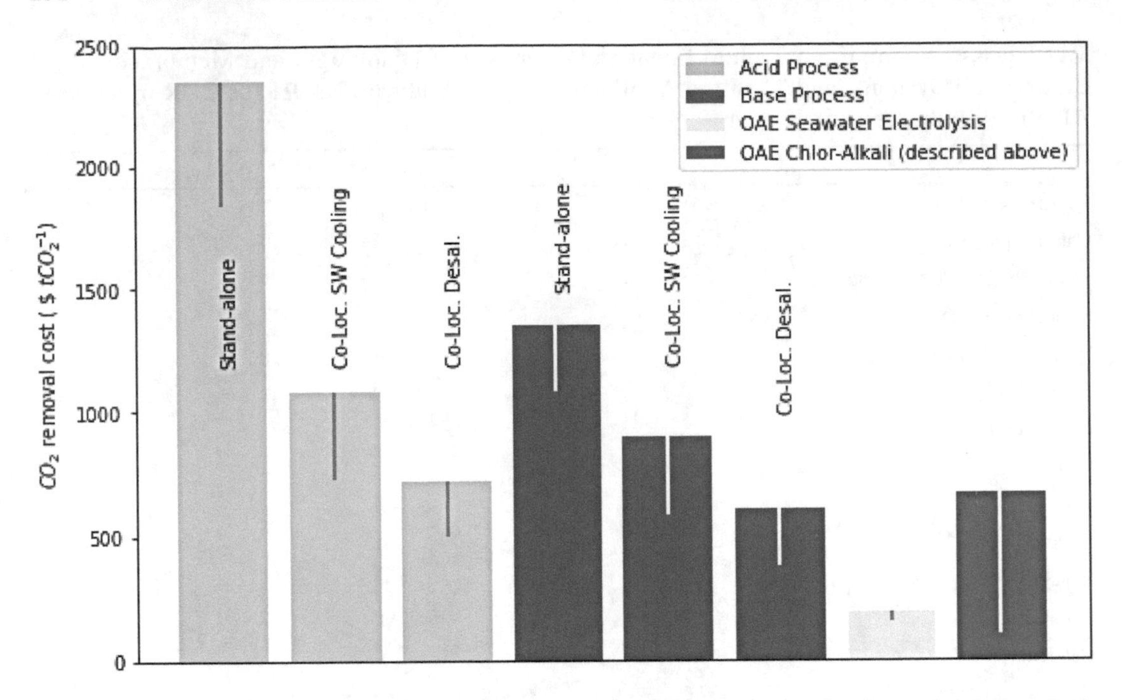

FIGURE 8.13 Summary of cost estimates for CDR from the atmosphere using the electrochemical methods described above. Downward lines show the "best-case" estimates for the processes.

except for "small scale scientific research studies . . . conducted in a controlled setting."[11] The decisions are not legally binding, however. The CBD itself arguably does not prevent countries from undertaking or authorizing electrochemical engineering projects, provided that they comply with all applicable consultation and other requirements imposed by the Convention (Webb et al., 2021).

Webb et al. (2021) further concluded that electrochemical engineering approaches that involve discharging materials into ocean waters could, in some circumstances, be considered "pollution" of the marine environment under the United Nations Convention on the Law of the Sea (UNCLOS) and/or marine "dumping" under the London Convention and Protocol. As discussed in Chapter 2, the parties to the London Protocol have agreed to an amendment, dealing specifically with the discharge of materials into ocean waters for the purposes of "marine geoengineering."[12] The amendment, which has not yet entered into force, establishes a framework under which parties to the London Convention may approve certain marine geoengineering activities involving ocean discharges.[13] While electrochemical engineering approaches could be considered a form of "marine geoengineering," at the time of writing, the framework only applied to activities relating to ocean fertilization. It could be expanded to apply to other techniques in the future (Brent et al., 2019; Webb et al., 2021). However, even if that occurred and the amendment entered into force, it would not be legally binding on the United States, which is not a party to the London Protocol (Webb et al., 2021).

The United States is a party to the London Convention. That instrument requires ocean dumping to be permitted and imposes restrictions on when permits can be issued. Webb et al. (2021)

[11] Para. 8(w), Decision X/33; Para. 1, Decision XI/20; Preamble, Decision XIII/4.

[12] Resolution LP.4(8), Amendment to the 1996 Protocol to the Convention on the Prevention of Marine Pollution by Dumping of Wastes and Other Matter, 1972 to Regulate Marine Geoengineering, Oct. 18, 2013.

[13] Annex 4, Resolution LP.4(8).

concluded that parties to the London Convention could issue permits for ocean CDR electrochemical engineering approaches involving the production of alkalinity, at least in some circumstances.

Webb et al. (2021) also reviewed the U.S. environmental laws potentially applicable to ocean CDR electrochemical engineering approaches involving the production of alkalinity. They concluded that the applicable laws will depend on the precise nature and location of each electrochemical engineering project. For example, projects involving the installation of structures in U.S. waters may require approvals from the U.S. Department of the Interior Bureau of Ocean Energy Management, U.S. Army Corps of Engineers, and/or other government agencies. Additionally, permits from the U.S. Environmental Protection Agency or state environmental agencies may be required for projects involving the discharge of materials into ocean waters, depending on where they occur. Projects in which CO_2 is injected into the sub-seabed could present additional legal challenges (see, e.g., Webb and Gerrard, 2018, 2019).

8.6 SUMMARY OF CARBON DIOXIDE REMOVAL POTENTIAL

The criteria for assessing the potential of electrochemical processes as a feasible approach to ocean CDR, described in Sections 8.2–8.5, is summarized in Table 8.5.

8.6 RESEARCH AGENDA

Components

Ocean-based electrochemical CDR approaches are promising in that (1) technologies that already exist (e.g., for alkaline electrolysis, in the chlor-alkali industry, and to a much smaller extent, involving electodialysis applications) could be adapted for the purpose, (2) existing technologies have previously been scaled to ~0.1 Gt/yr of products over several decades suggests scalability of new technologies, and (3) the environmental impact of the approaches that discharge effluent to ocean can be relatively easily monitored. However, electrochemical approaches present considerable challenges. For most proposed approaches, even best-case cost estimates are greater than other CDR approaches, which are primarily a result of capital expenditure and energy requirements. Given the energy requirements, these processes are unlikely to be feasible for CDR without exploiting low-carbon electricity. Some proposals generate chlorine and/or acids, which require treatment or disposal. An upscaled industry will likely require the movement of large volumes of seawater through the reaction system. Some of these proposals increase ocean alkalinity, and will share a research agenda that considers the wider social, environmental, and governance issues (Chapter 7). Table 8.6 presents the research agenda for advancing understanding of the feasibility of using electrochemical approaches to durably remove atmospheric CO_2.

Cost Discovery and Demonstration

Most processes have yet to be explored beyond bench-scale systems and/or early-stage technoeconomic analysis. While useful in scoping the possible range of costs, accurate assessment of first-of-a-kind systems requires the development of pilot and demonstration facilities approaching scales on the order of tonnes per day. For accurate cost discovery, exploration of these at a scale of, at minimum, hundreds of tonnes per year is required.

TABLE 8.5 CDR Potential of Electrochemical Processes

Knowledge base What is known about the system (low, mostly theoretical, few in situ experiments; medium, lab and some fieldwork, few carbon dioxide removal (CDR) publications; high, multiple in situ studies, growing body of literature)	**Low–Medium** Processes are based on well-understood chemistry with a long history of commercial deployment, but is yet to be adapted for CDR by ocean alkalinity enhancement (OAE) beyond benchtop scale.
Efficacy What is the confidence level that this approach will remove atmospheric CO_2 and lead to net increase in ocean carbon storage (low, medium, high)	**High Confidence** Monitoring within an enclosed engineered system, CO_2 stored either as increased alkalinity, solid carbonate, or aqueous CO_2 species. Additionality possible with the utilization of by-products to reduce carbon intensity.
Durability Will it remove CO_2 durably away from surface ocean and atmosphere (low, <10 years; medium, >10 years and <100 years; high, >100 years), and what is the confidence (low, medium, high)	**Medium–High** >100 years Dynamics similar to OAE.
Scalability What is the potential scalability at some future date with global-scale implementation (low, <0.1 Gt CO_2/yr; medium, >0.1 Gt CO_2/yr and <1.0 Gt CO_2/yr; high, >1.0 Gt CO_2/yr), and what is the confidence level (low, medium, high)	**Medium–High** Potential C removal >0.1–1.0 Gt CO_2/yr (medium confidence) Energy and water requirements may limit scale. For climate relevancy, the scale will need to be doubled to an order of magnitude greater than the current chlor-alkali industry.
Environmental risk Intended and unintended undesirable consequences at scale (unknown, low, medium, high), and what is the confidence level (low, medium, high)	**Medium–High** (low confidence) Impact on the ocean is possibly constrained to the point of effluent discharge. Poorly-known possible ecosystem impacts similar to alkalinity enhancement. Excess acid (or gases, particularly chlorine) will need to be treated and safely disposed. Provision of sufficient electrical power will likely have remote impacts.
Social considerations Encompass use conflicts, governance-readiness, opportunities for livelihoods, etc.	Similar to OAE and to any industrial site. Substantial electrical power demand may generate social impacts.
Co-benefits How significant are the co-benefits as compared to the main goal of CDR and how confident is that assessment	**Medium–High** (medium confidence) Mitigation of ocean acidification; production of H_2, Cl_2, silica.
Cost of scale-up Estimated costs in dollars per metric ton CO_2 for future deployment at scale; does not include all of monitoring and verification costs needed for smaller deployments during R&D phases (low, <\$50/t CO_2; medium, ~\$100/t CO_2; high, >>\$150/t CO_2) and confidence in estimate (low, medium, high)	**High** >\$150/t CO_2 (medium confidence) Gross current estimates \$150–\$2,500/t CO_2 removed. With further R&D, it may be possible to reduce this to <\$100/t CO_2.
Cost and challenges of carbon accounting Relative cost and scientific challenge associated with transparent and quantifiable carbon tracking (low, medium, high)	**Low–Medium**

TABLE 8.5 Continued

Cost of environmental monitoring Need to track impacts beyond carbon cycle on marine ecosystems (low, medium, high)	**Medium** (medium–high confidence) All CDR will require monitoring for intended and unintended consequences both locally and downstream of CDR site, and these monitoring costs may be substantial fraction of overall costs during R&D and demonstration-scale field projects.
Additional resources needed Relative low, medium, high to primary costs of scale-up	**Medium–High** High energy requirements (1–2.5 MWh/t CO_2 removed) and build-out of industrial CDR.

Electrode and Membrane Materials

Novel electrode materials may be able to substantially reduce the generation of Cl_2 gas from the anode favoring the production of O_2. A research agenda may be able to significantly reduce the cost and improve the technical feasibility of electrochemical approaches by exploring durable, long-lasting, and cost-effective electrode materials and their optimum design and configuration with either membrane-based or membraneless reaction systems.

Acid Management and Environmental Impact Assessment

At scale, the generation of large volumes of Cl_2 gas or acids may be unavoidable. If electrochemical approaches are to be deployed at scale, then an important component of a research strategy includes how these wastes can be safely disposed of. Waste management strategies are most effective when they are part of a comprehensive environmental impact assessments or life-cycle assessments that consider not only the impact of the process, but also its supply chain. This work could assess what impacts are possible to mitigate, what operating protocols are needed, and what impacts are likely even in highly controlled systems. It would also be useful to explore worst-case impact scenarios.

Coupling to Whole-Rock Dissolution Systems

Optimum system design will likely require careful coupling between the electrochemical component and that which facilitates rock dissolution. Furthermore, large-scale electrochemical approaches would use a range of local rock resources, and research to assess the impact of real-world rock chemistries on system performance and/or effluent quality is needed, including the potential of using waste rock (e.g., mine tailings, quarry fines, etc.) or alkaline materials (e.g., slag from the steel industry, concrete and demolition waste, etc.).

Hybrid Approaches

Processes that combine CDR and alkalinity increase are sparse in the literature but could be particularly useful in maximizing removal for small changes in system operation.

Resource Mapping and Pathways to Upscaling

It is not yet clear whether or how rapid a large-scale deployment of electrochemical approaches could be achieved. Research is needed to explore pathways and incentivization strategies for scale-up. These pathways could include integration within existing industries (e.g., desalination, power station cooling water, etc.), which may provide the most feasible, albeit niche, deployment for first-of-a-kind systems. It could also be possible to integrate products from these electrochemical systems into existing markets (e.g., hydrogen, chlorine, oxygen, silica) to lower the cost of initial systems. It is possible that the interaction between a growing electrochemical industry (and its supply chain) and existing industry, economies, and markets may have unintended consequences. A resource mapping exercise that identifies the colocation of low-carbon electricity, concentrated brine feedstocks, rock extraction, seawater, CO_2 disposal, and product utilization would be essential in designing upscaling pathways. This activity should be undertaken with stakeholders in the supply chain.

Cost and Time Frame of Research Agenda

Demonstration projects could be commissioned relatively quickly (~24 months) and would require a similar operational time in which to test the systems. These would be most effective if they were underpinned by smaller-scale laboratory-based experimentation and materials and systems development. Multiple rounds of demonstrations over a 5- to 10-year time frame would allow for iterative development between stages. Within the initial 36 months, an assessment of resources and upscaling pathways will be critical in assessing the viability of the approaches. Research exploring the environmental impact, and acid management or reduction strategies would support the demonstration research (in the first 60 months), but may not be necessary thereafter. In total the research agenda would cost ~$475 million over a 10-year period

Environmental Impacts of Research Agenda

The research described above could be undertaken with minimal environmental impact, and almost no impact on the ocean. Seawater may need to be collected as a feedstock for the processes (although artificial seawater could also be appropriate in some experiments). Depending on local regulations and the scale of the experiment, process effluents (i.e., brine, acids, and bases) could be neutralized and or diluted prior to disposal in municipal sewer systems. Volumes of solid wastes would be small and relatively inert. The research agenda described above considers the systems required to remove CO_2 from seawater, or to increase its alkalinity. The wider ocean impacts from ocean alkalinity enhancement are included in Chapter 7.

8.7 SUMMARY

The research agenda for electrochemical processes focuses on activities that reduce the cost and environmental impact of the approaches. Presently, only a small number of system configurations have been investigated for either CDR from seawater or as methods for increasing ocean alkalinity. These should be broadened to also consider novel designs of electrolyzers and electrochemical reactors, electrode materials that can minimize the production of unwanted by-products (e.g., Cl_2 gas), electrochemical reactor architectures, and hybrid approaches that both remove CO_2 and increase ocean alkalinity. Wider impacts on ocean ecosystems are included in the research agenda for Ocean Alkalinity Enhancement and are not repeated here. We recommend that meaningfully large demonstration-scale projects (>1,000 kg/day, and approaching the scale of several to tens of tonnes per day) be central to the research agenda. While part of this research agenda may focus on the development of specific technologies or materials (e.g., electrolyzer design or novel electrode

TABLE 8.6 Research and Development Needs: Electrochemical Processes for Ocean Alkalinity Enhancement

#	Recommended Research	Gap Filled	Environmental Impact of Research	Estimated Research Budget ($M/yr)	Time Frame (years)
8.1	**Demonstration projects including CDR verification and environmental monitoring**	**What is the likely upscaled cost of the approach?**	Low	**30**	**5**
8.2	**Development and assessment of novel and improved electrode and membrane materials**	**What is the minimal production possible of Cl_2 (or other waste gases) from electrochemical systems? Can membranes be optimized for the unique constraints encountered in electrochemical ocean CDR approaches?**	Low	**10**	**5**
8.3	Assessment of environmental impact and acid management strategies	What is the life-cycle environmental impact of approaches?	Low	7.5	10
8.4	**Coupling whole rock dissolution to electrochemical reactors and systems**	**Is it possible to scale electrochemical approaches while managing waste acid production?**	Low	**7.5**	**10**
8.5	**Development of hybrid approaches**	**Can approaches be optimized or combined for cost reduction?**	Low	**7.5**	**10**
8.6	Resource mapping and pathway assessment	Do we have sufficient resources in the right locations to upscale production?	None	10	5

NOTE: Bold type identifies priorities for taking the next step to advance understanding of electrochemical processes as an ocean CDR approach.

materials), it should also focus on systems integration (particularly with rock dissolution processes) and scale-up strategies that allow for meaningful process cost reductions to be discovered.

The processes explored by the research agenda will need to demonstrate pathways to large-scale deployment. As such, the research agenda should also include provision for resource mapping, stakeholder engagement, and wider economic impact from an ocean-based electrochemical CDR industry operating at a climate-relevant scale. The outcomes of the research should be transparently disseminated such that independent assessment of cost and environmental impact can be made. Social and governance considerations should also be addressed in the research.

9

Synthesis and Research Strategy

9.1 A GENERAL FRAMEWORK FOR OCEAN-BASED CARBON DIOXIDE REMOVAL STRATEGIES

By acting to remove carbon dioxide (CO_2) from the atmosphere and upper ocean and then store the excess carbon either in marine or geological reservoirs for some period of time, ocean CO_2 removal (CDR) approaches could complement CO_2 emission reductions and contribute to the portfolio of climate response strategies needed to limit climate change and surface ocean acidification over coming decades and centuries. While rapid and extensive decarbonization and abatement of other greenhouse gases in the United States and global economies are the primary action required to meet international climate goals, ocean and other CDR approaches can help balance difficult-to-mitigate human CO_2 emissions and contribute to mid-century to late-century net-zero carbon dioxide emissions targets.

The present state of knowledge on many ocean CDR approaches is inadequate, based in many cases only on laboratory-scale experiments, conceptual theory, and/or numerical models. Given the inadequate knowledge base and lack of time to develop climate solutions, it will be important to pursue research on multiple approaches in parallel to progress understanding of ocean CDR as a contributor to climate targets. Expanded research including field research is needed to assess the techniques' potential efficacy in removing and sequestering excess carbon away from the atmosphere and the permanence or durability of the carbon sequestration on timescales relevant to societal policy decisions. Research is also needed to identify and quantify environmental impacts, risks, benefits, and co-benefits as well as other factors governing possible decisions on deployment such as technological readiness, development timelines, energy and resource needs, economic costs, and potential social, policy, legal, and regulatory considerations. Additionally, **research on ocean CDR would greatly benefit from targeted studies on the interactions and trade-offs between ocean CDR, terrestrial CDR, greenhouse gas abatement and mitigation, and climate adaptation, including the potential of mitigation deterrence.**

Implementation of any ocean CDR strategy would require decision makers to make difficult trade-offs. Gathering and synthesizing as much information as possible can help to anticipate what those trade-offs will be. For example, it may seem premature to think about the social and livelihood implications of some of these techniques, or to assess their co-benefits to particular groups, when they are just in laboratory or demonstration stages. However, interest in ocean CDR approaches could move quickly as countries and companies seek to build roadmaps to the net-zero targets they have set. Pilot projects are sites for both demonstrating new technologies and for social learning, and an integrated research strategy would involve social scientists working together with geoscientists and engineers early in the research process to provide a broader evidence base around what the trade-offs for different communities might be.

Core Principles for Developing a Research Strategy

Research on societal responses to climate change has been approached through a number of conceptual and science policy frameworks. This report has not prescribed a single approach for policy makers, researchers, and stakeholders, but it has drawn from several complementary frameworks (see Box 9.1).

Recommendation 1: Ocean CDR Research Program Goals. To inform future societal decisions on a broader climate response mitigation portfolio, a research program for ocean CDR should be implemented, in parallel across multiple approaches, to address current knowledge gaps. The research program should not advocate for or lock in future ocean CDR deployments but rather provide an improved and unbiased knowledge base for the public, stakeholders, and policy makers. Funding for this research could come from both the public and private sectors, and collaboration between the two is encouraged. The integrated research program should include the following elements:

1. Assessment of whether the approach removes atmospheric CO_2, in net, and of the durability of the CDR, as a primary goal.
2. Assessment of intended and unintended environmental impacts beyond CDR.
3. Assessment of social and livelihood impacts, examining both potential harms and benefits.
4. Integration of research on social, legal, regulatory, policy, and economic questions relevant to ocean CDR research and possible future deployment with the natural science, engineering, and technological aspects.
5. Systematic examination of the biophysical and social interactions, synergies, and tensions between ocean CDR, terrestrial CDR, mitigation, and adaptation.

9.2 COMMON COMPONENTS OF ANY RESEARCH IMPLEMENTATION

The suite of six ocean CDR approaches, defined broadly and addressed in detail in the report chapters, covers a wide range from ecosystem recovery and alteration of marine ecosystems to more industrial-based techniques. The knowledge base and readiness levels for the approaches differ substantially as do the carbon sequestration potential, environmental impacts, and human dimensions. As such, the development of a comprehensive research strategy will require different emphasis for different approaches—no single research framework will be adequate for all CDR approaches. There are, however, several common components that are relevant to research into any ocean CDR approach.

Research Code of Conduct

The lack of a comprehensive international or domestic legal framework specific to ocean CDR research creates a risk that ill-considered projects, including projects that do little to

BOX 9.1
Frameworks for Research on Ocean CDR

Responsible research and innovation is an approach that aims to have multiactor and public engagement throughout the research process to ensure that science is guided by and serves the needs, values, and expectations of society. It has been adopted by various funding agencies and programs, such as the European Union's Horizon 2020 program and many of the United Kingdom's Research Councils. Responsible research and innovation involves not just public engagement, but also anticipating social and environmental impacts of research, reflecting on the motivations and unknowns in the research, and applying these insights into the conduct of the research.

Example Research Question:
- How do communities and publics want to be engaged in the development of ocean CDR?

Example of Research Activities:
- Deliberative engagements to explore social and ethical concerns as part of assessment processes
- Co-producing research and deployment scenarios with communities

Sociotechnical systems is a perspective used in sustainability studies, innovation studies, and transition studies that analyzes both the social and technical elements of a system together, as mutually co-constitutive: the technological components (such as artifacts) and social components (e.g., networks, institutions, norms, regulations, and knowledge) shape each other. For an ocean CDR strategy, this could be everything from biological responses in the ocean and physical aspects of the infrastructure needed to the policies, institutions, supply chains, and communities involved in the approach.

Example Research Questions:
- How do the life-cycle analyses of ocean CDR techniques change under different deployment scenarios?
- How do the perspectives, preferences, and norms of ocean CDR researchers and entrepreneurs influence the artifacts that are produced?
- How can society avoid lock-in of particular CDR techniques and continue to innovate once vested interests have developed?

Example of Research Activities:
- Mapping out relationships between parts of the sociotechnical system
- Ethnographies and critical discourse analyses of research practice
- Case studies of analog sociotechnical systems
- Collaborative research to understand the impacts of ocean CDR on other complex systems, such as food, water, energy, and material use

Environmental justice is a movement, a goal, and a field of scholarship, but it can also be a lens through which to evaluate research and development activities. That is, we can assess projects on their potential contributions to and incorporation of environmental justice, including dimensions of distributive, procedural, recognitive, and intergenerational justice.

Example Research Questions:
- Who bears the risks and reaps the benefits of ocean CDR technologies under different policy roadmaps and deployment scenarios?
- What are the barriers to community involvement in decision making, and how might they be overcome?
- What are the public health implications of increased manufacturing or new infrastructure for some ocean CDR techniques, and who bears any public health burdens?

Example of Research Activities:
- Deliberative workshops with communities to understand risks, benefits, and barriers to involvement in decision making
- Interdisciplinary, spatially explicit analysis of new infrastructures and practices related to mining, manufacturing, transportation, etc.

advance scientific knowledge and/or present significant risks, will be pursued. Developing a code of conduct for ocean CDR research may help to mitigate that risk. To maximize participation by researchers, compliance with the code of conduct could be made a condition of government or private funding for ocean CDR research.

Codes of conduct for geoengineering research have already been developed by academic researchers, in some cases with input from stakeholders (e.g., Leinen, 2008; Hubert and Reichwein, 2015; Hubert, 2017; Chhetri, 2018), but have not been widely adopted by the research community (NASEM, 2021b). Although not a full code of conduct, the so-called "Oxford Principles" for geoengineering do provide some guidance to researchers, and have been more widely accepted (see Box 9.2).[1]

Scholars have also suggested four principles for CDR research in just climate policy: emission cuts must remain in the center of climate policy; social, economic, and environmental impacts matter; CDR projects and approaches should be assessed individually; and climate policy needs

BOX 9.2
The Oxford Principles of Geoengineering Research

The Oxford Principles were developed by academic researchers at the University of Oxford, University College London, and the University of Cardiff in 2009. Later that year, the UK House of Commons Science and Technology Select Committee on the Regulation of Geoengineering endorsed the principles as "provid[ing] a basis to begin the discussion of . . . the regulation of geoengineering," and recommended that they be further developed (UK House of Commons Science and Technology Committee, 2010). Support for the principles was also expressed by geoengineering researchers at the International Conference on Climate Intervention Technologies in 2010. The principles are:

Principle 1: Geoengineering to be regulated as a public good. While the involvement of private actors in the delivery of a geoengineering technique should not be prohibited, and may indeed be encouraged to ensure that deployment of a suitable technique can be effected in a timely and efficient manner, regulation of such techniques should be undertaken in the public interest by the appropriate bodies at the state and/or international levels.

Principle 2: Public participation in geoengineering decision making. Wherever possible, those conducting geoengineering research should be required to notify, consult, and ideally obtain the prior informed consent of those affected by the research activities. . . .

Principle 3: Disclosure of geoengineering research and open publication of results. There should be complete disclosure of research plans and open publication of results in order to facilitate better understanding of the risks and to reassure the public as to the integrity of the process . . . .

Principle 4: Independent assessment of impacts. An assessment of the impacts of geoengineering research should be conducted by a body independent of those undertaking the research; where techniques are likely to have transboundary impact, such assessment should be carried out through the appropriate regional and/or international bodies. Assessments should address both the environmental and socioeconomic impacts of research, including mitigating the risks of lock-in to particular technologies or vested interests.

Principle 5: Governance before deployment. Any decisions with respect to deployment should only be taken with robust governance structures already in place, using existing rules and institutions wherever possible.

[1] See http://www.geoengineering.ox.ac.uk/www.geoengineering.ox.ac.uk/oxford-principles/principles/index.html.

to be resilient against unexpected outcomes (Morrow et al., 2020). However, neither the Oxford Principles nor the geoengineering codes of conduct specifically address ocean CDR research. Given the unique nature of such research, development of a specific code of conduct is likely to be useful. Some nongovernment organizations (NGOs) have recently advocated for the adoption of, and identified key principles that should be included in an ocean CDR-specific code of conduct.[2]

Ideally, a code of conduct for ocean CDR research would be developed and maintained by an international body, with input from the research community and other stakeholders. Key issues to be addressed in the code of conduct would include (among other things):

- the need for transparency in research planning, including ex ante review and disclosure of potential environmental and other impacts, and stakeholder and public engagement;
- use of independent review panels to evaluate research plans and outcomes;
- compliance with all applicable domestic and international laws and associated guidelines (even if not legally binding) in conducting research;
- use of open data systems with adequate metadata to evaluate data quality;
- use of peer-reviewed literature for reporting results as well as open presentations to stakeholders and the public;
- use of accepted carbon accounting methodologies, where available, for constraining CO_2 removal, permanence of CO_2 sequestration, and net greenhouse gas emissions;
- consideration and disclosure of other environmental and social impacts, both within and outside of the study area;
- consideration of full life-cycle impacts and costs; and
- responsibilities for post-research activities and site closure.

Permitting of Research

Significant permitting issues could arise in connection with ocean CDR research, particularly field trials and other in situ research. As discussed in Chapter 2, while there is no domestic legal framework specific to ocean CDR, research projects may be subject to a number of domestic environmental and other laws, which impose permitting requirements. **Further study is needed to identify and analyze the full range of potentially applicable laws, explore gaps in and barriers created by the application of those laws to ocean CDR, and evaluate possible alternative approaches to regulation.**

Requiring ocean CDR research projects to be permitted ensures some level of government oversight, which should, in turn, help to promote responsible research and minimize the risk of negative environmental and other outcomes. At the same time, however, unnecessarily burdensome or overly complex permitting requirements could create barriers to ocean CDR research.

Some in situ research projects may require multiple permits from multiple federal, state, and/or local regulatory agencies. Due to the novelty of the research, regulatory agencies may lack established processes for evaluating projects and issuing permits, which could lead to delays and other costs. Members of the scientific community may lack the time, resources, and expertise to navigate the permitting process and ensure compliance with other regulatory requirements.

The following actions can be taken by the research community, and by those funding or otherwise supporting research, to begin to address these complex permitting issues:

[2] See e.g., https://oceanvisions.org/roadmaps/growing-public-support/first-order-priorities/#developacodeofconducttoguidescientificexperimentationandfieldtrials and https://www.youtube.com/watch?v=2K_J2uaX--I&t=372s.

1. **Encourage interdisciplinary research.** See description in the section, Research Collaboration.
2. **Share knowledge.** It is imperative that government agency staff and other stakeholders, including the general public, have a solid understanding of ocean CDR techniques and the known and unknown benefits and risks that their use presents. To help build knowledge, researchers and other groups could offer briefings to agency staff and other stakeholders, or share information with them in other ways.
3. **Engage early.** When a field trial or other in situ research is being considered, the researchers should engage with relevant government agencies and other stakeholders early in the project design process wherever possible, and continue to share information during and after the project as appropriate. A clear data management plan should also be articulated early on and complied with throughout the project.

Government agencies can also play an important role, most notably by clarifying and, where possible, simplifying permitting processes. It is, however, important that any action taken not compromise existing safeguards designed to ensure that research is conducted responsibly and minimize the potential for negative environmental and other outcomes.

Research Collaboration

Co-production has been emphasized in marine sustainability science as a way of developing solutions that are adapted to local ecological, economic, and cultural contexts. Through projects that work with communities and discuss the changing ocean, blue economy, and energy and conservation, we can learn what other issues are informing how people think about ocean CDR and understand their priorities. **Having communities participate from the outset and guide the research can increase the likelihood of ocean CDR implementations that are compatible with environmental justice,** and avoid ocean CDR implementations that would exacerbate environmental injustice. Engagement with stakeholders from local government, business, NGOs, and other stakeholders as identified through stakeholder assessment will also be important.

Traditional ecological knowledge and traditional land and resource management systems would have many insights for some of the techniques discussed in this report, such as ecological restoration and seaweed cultivation. However, reversing the damage to marine ecosystems needs to be done with the permission, guidance, and collaboration of Indigenous peoples (Turner and Neis, 2020). Research conducted without consent and outside of a collaborative framework can be a form of colonialism that can potentially harm Indigenous peoples (Ban et al., 2018). Some of the challenges to doing this kind of research include the history of extractivist research by settler-colonial scholars; the fact that co-produced research takes more time, which can be in tension with the urgency of environmental and climate action; the fact that Indigenous resource offices are often under-resourced and swamped with other priority items; and the lack of scientists who are trained in co-production because it is at odds with publish-or-perish career incentives within science. However, **funders can address at least some of these challenges by funding co-produced research, consulting with Indigenous communities on what resources they would need to undertake such collaborations according to their existing research protocols, and developing a pipeline of researchers who have the capacity and skills to carry out such work.**

As well as being co-produced with communities, research into ocean CDR should also be interdisciplinary. Research into the scientific, social, legal, and other dimensions of ocean CDR tends to be highly siloed. Scientific research projects are, therefore, often designed without consideration of social or legal issues. Early engagement with those issues can avoid surprises and lead to improved project design. Examples of successful integration of legal and scientific research in the ocean CDR space include:

- CDRmare,[3] a project of the Germany Marine Research Alliance, which aims to determine whether and to what extent the oceans can be used to remove and store CO_2 from the atmosphere. The project is analyzing "individual actions that aim to enhance marine carbon sinks," with a "[s]trong emphasis on non-natural science aspects," including "legal, social, and ethical aspects."
- Solid Carbon,[4] a project funded by the Pacific Institute for Climate Solutions, which is exploring the feasibility of capturing and storing CO_2 in the sub-seabed. The project combines a study of key scientific and engineering questions with an analysis of potential regulatory and other challenges (Webb and Gerrard, 2021).

With oceans as a global commons, **it is critical that research in this area be international**—for social legitimacy, for research that is applicable to multiple cultural and geographic contexts, and for research that addresses the priorities of communities where ocean CDR may be used. Both prioritizing or mandating international collaborations and support of early-career researchers to build capacity can work toward the goal of supporting an international research community of ocean CDR research.

Recommendation 2: Common Components of an Ocean CDR Research Program. Implementation of the research program in Recommendation 1 should include several key common components:

1. The development and adherence to a common research code of conduct that emphasizes transparency and open public data access, verification of carbon sequestration, monitoring for intended and unintended environmental and other impacts, and stakeholder and public engagement.
2. Full consideration of, and compliance with, permitting and other regulatory requirements. Regulatory agencies should establish clear processes and criteria for permitting ocean CDR research, with input from funding entities and other stakeholders.
3. Co-production of knowledge and design of experiments with communities, Indigenous collaborators, and other key stakeholders.
4. Promotion of international cooperation in scientific research and issues relating to the governance of ocean CDR research, through prioritizing international research collaborations and enhancement of international oversight of projects (e.g., by establishing an independent expert review board with international representation).
5. Capacity building among researchers in the United States and other countries, including fellowships for early-career researchers in climate-vulnerable communities and underrepresented groups, including from Indigenous populations and the Global South.

9.3 SUMMARY OF ASSESSED OCEAN-BASED CARBON DIOXIDE REMOVAL STRATEGIES

Chapters 3 through 8 include detailed assessments of the current state of understanding of each of the six ocean-based CDR approaches examined within this report, as defined by the com-

---

[3] See https://www.cdrmare.de/en/die-mission/.

[4] See https://pics.uvic.ca/projects/solid-carbon-negative-emissions-technology-feasibility-study.

mittee's statement of task (Box 1.1). The committee assessed the potential of each approach as a viable mechanism for CDR according to common criteria described in Chapter 1, and listed here:

- **Knowledge base:** What do we know about the approach?
- **Efficacy:** Can effective CO_2 removal from the atmosphere be demonstrated?
- **Durability:** How long would the excess carbon be stored?
- **Scale:** How much CO_2 could potentially be removed and what are the temporal and geographic constraints?
- **Monitoring and verification:** What would be needed to monitor for both environmental impact and for carbon accounting?
- **Viability and barriers:** What is the potential viability based on technical, scientific, economic, safety, and sociopolitical factors? What are the intended and unintended environmental impacts, and what are the co-benefits?
- **Governance and social dimensions:** What is the governance landscape for both research on and possible future deployment of the approach?
- **R&D opportunities:** What needs to be done to fill the important gaps in understanding?

As previously stated, the knowledge base and readiness levels for the six ocean CDR approaches examined differ substantially as do the carbon sequestration potential, environmental impacts, and human dimensions. The committee's assessment on potential of scale-up for each approach is described in the sections that follow and is also summarized in Table 9.1 (see Box 9.3 for specifics on estimating costs). This assessment is the result of literature review, a series of public workshops and meetings convened by the committee, as well as committee member expertise and judgment, where needed. Although the report addresses in detail only a subset of all possible ocean CDR approaches, the criteria above and the research program elements in Recommendations 1 and 2 provide a general framework applicable more broadly to other ocean CDR approaches.

The following text describes research priorities to focus on, in parallel, as next steps to increasing understanding of each of the six approaches as a viable and responsible mechanism for durably removing CO_2 from the atmosphere. Tables 9.2 and 9.3 then summarize the research needed, including rough time frame and cost estimates. Table 9.2 summarizes the foundational research identified in Chapters 2 and 9 as research priorities common across ocean CDR approaches on potential social, policy, legal, and regulatory considerations. The research included in Table 9.2 is meant to inform the framework for any future ocean-based CDR effort. Table 9.3 then summarizes research needs identified within Chapters 3–8, with bolded text indicating research needed as next steps to better understanding the feasibility of that particular approach. The research costs listed in Tables 9.2 and 9.3 are approximate and were compiled by the committee based on publicly available information, experience, and judgment (see Box 9.3). Some caveats are appropriate, particularly when comparing across different research fields from ocean science and social science to more engineering and technological research. Further, these cost estimates are primarily for research and early development work, and costs could grow for some demonstration-phase projects.

Social and regulatory acceptability is likely to be a barrier to many ocean CDR approaches, both at the research and possible deployment stages, particularly ones requiring industrial infrastructure either in the ocean or on the coasts or ones involving restructuring of marine ecosystems. There will be both project-specific and approach-specific social, political, and regulatory discussions, as well as the possibility for diverging perspectives around the role of CDR broadly. Field-scale trials are likely to be a site of wider debates within the scientific community and in society around decarbonization and climate response strategies.

Ocean CDR approaches are already being discussed widely, and in some cases promoted, by some scientists, NGOs, and entrepreneurs as potential climate response strategies. At present,

BOX 9.3
Estimating Research Costs of Scale-up of Ocean CDR Approaches

When looking across ocean CDR methods, policy makers and investors may be looking for the lowest costs and minimal risk. Huesemann (2008) suggests, however, that it is "intrinsically impossible to conduct a truly objective, comparative cost-benefit analyses of different climate change mitigation technologies" due not only to the lack of data on the consequences, but also the political nature and different value judgments of the stakeholders. In any case, more basic research studies and assessments of the engineering costs will be needed to compare different ocean CDR approaches. Such assessments need to include not only deployment costs, but also costs associated with monitoring for verification of long-term storage costs and fair accounting of costs to track carbon. Do we know, for example, if these C accounting costs would be similar for ocean fertilization and upwelling tubes? And any CDR approach would need to assess the full range of ecological impacts, especially when done at scale. At this stage it is easier to consider the relative cost differences between these CDR approaches without assigning a specific dollar-per-ton value; however, the committee used publicly available information, experience, and judgment to come up with the costs included in Table 9.1. It should be noted that these numbers are rough estimates and should not be taken for anything more.

Despite the uncertainty in costs associated with deployment of ocean CDR at scale, a qualitative comparison can be made given the proposed effort and consequences of each approach. Most cost assessments that are reported on a per-ton CO_2 basis consider primarily the costs for deployment only (i.e., operational and capital costs of materials, energy, and structures needed to implement a given amount of CDR). This often extends across the ocean–land boundary as minerals and energy production and distribution are largely land-based activities needed to support ocean CDR. In general, the abiotic or more heavily engineered CDR approaches have higher deployment costs (Table 9.1, row 8). However, a full accounting of any CDR approach when brought to implementation will need to include costs for transparent and verifiable carbon accounting (Table 9.1, row 9). These costs are likely higher for biotic approaches, particularly during the research and early pilot and demonstration phases. The carbon accounting costs could potentially decline later if accepted standard methodologies are developed and agreed upon.

All ocean CDR approaches will also need to consider the costs associated with monitoring environmental and social impacts including assessment of nonlocal downstream effects on the environment and the ocean carbon cycle. In this regard, a research program on ocean CDR will benefit greatly from existing and expanding ocean observational capabilities being conducted for other purposes. Substantial leveraging opportunities exist, but only if these other ocean observing systems remain robust and adequately funded into the future. Examples include ocean climate, carbon cycle, and biogeochemistry observing programs (e.g., GO-SHIP, Bio-Argo, surface underway CO_2 observations, etc.), the Integrated Ocean Observing System (IOOS), satellite remote sensing of ocean color and biology, National Marine Fisheries Service ecosystem surveys, and emerging Marine Biodiversity Observing Networks.

society and policy makers lack sufficient knowledge to fully evaluate ocean CDR outcomes and weigh the trade-offs with other climate response approaches and with environmental and sustainable development goals. Research on ocean CDR, therefore, is needed whether or not society moves ahead with deployment, and/or to assess at what scales and locations the consequences would be acceptable. Research would also provide a framework for evaluating under what conditions various ocean CDR projects may be acceptable and thereby enable the development and implementation of an effective regulatory framework for deployment (if any).

Ocean Nutrient Fertilization

Ocean fertilization has received the most attention of any ocean CDR approach due to fieldwork that began more than 25 years ago focused on iron and its unique role as a micronutrient that is limiting phytoplankton growth, and hence CO_2 uptake, in large parts of the ocean. Those experiments were not large enough or long enough (<5 tons of iron, weeks to just over a month of observations) to observe the full cycle of carbon uptake and loss to the deep sea via the biological pump. Because of these limitations, carbon sequestration efficiencies are still poorly constrained, and concerns about unintended consequences, such as harmful algal blooms or production of other greenhouse gases such as nitrous oxide, along with models demonstrating downstream impacts and permanence issues led to a cessation of follow-on open-ocean iron fertilization experiments. However, studies of naturally iron-rich systems, for example, around Southern Ocean islands, and the biological response to episodic iron inputs, such as volcanic eruptions, suggest higher sequestration efficiencies and little harmful side effects with the possible co-benefit of enhanced fisheries.

With this background in mind, continued research is warranted, building on these earlier results to address key remaining issues related to how carbon uptake and sequestration efficiencies could be enhanced, and how carbon transport could be more readily tracked and thus accounted for. Some of this would be laboratory and mesocosm based, but a more complete understanding will require demonstration-scale field experiments—>100 tons of iron added over >10,000 km^2 and full seasonal and annual sampling of impacts. Monitoring for a full suite of geochemical and ecological shifts is needed using advanced autonomous platforms with sensors and samplers that would allow a continuous presence to observe changes. These demonstration-scale studies would need to be conducted in several ocean regimes. In addition to high-nutrient, low-chlorophyll areas, studies should be carried out in low-nutrient settings and in regions where the depth of permanent (>100-year) sequestration is shallower. Remote sensing and complementary measurements from ships will be needed to fully track the consequences. Models to plan out the best strategies for inducing blooms and measurement strategies for quantifying impacts are needed, as well as three-dimensional ocean ecosystem and Earth system models to extrapolate the regional impacts to areas outside of the immediate study site.

In part due to early and unsupported optimism by commercial interests in this CDR approach, many in the public and scientific community have urged caution on proceeding with ocean iron fertilization research. One response to these concerns was the development of a framework for evaluating and permitting research under the London Convention and Protocol. This is a necessary step in moving ahead and could possibly provide a model for regulating some other types of ocean CDR research. However, a broader investment in gathering stakeholders and regulatory agencies together is needed to assess the implications of CDR research and deployment at scale. The development of frameworks to assess CDR magnitude and durability as well as environmental impacts would need to be site- and scale-specific and include a delineation of responsibilities before, during, and after deployment for those deploying ocean fertilization and those regulating it. This is complicated in that those receiving the most benefit in terms of carbon credits or markets may not be the ones facing the greatest impact. Providing information for policy makers and the public about the trade-offs between ocean fertilization and other land- and ocean-based CDR approaches and the consequences of doing nothing for negative C emissions also needs attention.

The largest cost for research would be the demonstration-scale field experiments, which if upscaled from prior studies would be on the order of $25M/yr for 10 years to complete work at multiple sites. In parallel, laboratory/mesocosm studies to optimize conditions and methods to improve carbon accounting and the technologies to do so would total $18M/yr, which along with modeling ($5M/yr) and social and governance aspects ($4M/yr) would round out a comprehensive ocean fertilization research agenda.

Artificial Upwelling and Downwelling

The vertical movement of water in the ocean, termed upwelling and downwelling, acts to transfer heat, salt, nutrients, and energy between the well-lit surface ocean and the dark, nutrient-, and CO_2-rich deep ocean. Since the 1950s, researchers have sought to artificially simulate these physical transport processes to geoengineer localized regions of the ocean using a variety of technologies including wave pumps, airlift pumps, and salt fountains, each of which has varying pumping mechanisms, efficiencies, and energy requirements. Several field deployments have proven that at least some of these technologies can indeed deliver deep water to the surface ocean, and in some cases there is a measurable biological response. To date, however, in situ experiments with artificial upwelling have been short (generally less than weeks), limited in scale (a small number of pumps in smaller-than-kilometer regions, mostly coastal), and have not measured and/or demonstrated any net enhancement of carbon flux even though mesocosm studies and small-scale enrichment experiments do confirm that nutrient delivery via deep water can stimulate primary production.

In principle, as a CDR approach, artificial upwelling provides a means to supply growth-limiting nutrients to the upper ocean and generate increased primary production and net carbon sequestration. For this reason, it has been proposed as a potentially effective component of a portfolio of CDR approaches, either stand-alone or in concert with aquaculture. The important caveat for artificial upwelling is that to achieve the durable drawdown of atmospheric CO_2, the net enhancement of the biological production of carbon must exceed the delivery of dissolved inorganic carbon from the upwelled source water. This is an important distinction, because deep water contains relatively high levels of CO_2 that have been generated over time as sinking organic matter is remineralized, and so any stimulation of carbon export must exceed upwelled CO_2. Determination of where in the ocean this is feasible requires knowledge of the ratio (stoichiometry) of elements (carbon:nitrogen:phosphorus:trace elements) in source waters relative to the elemental needs of biological communities. This knowledge is not well constrained in the global ocean, and hence there is significant uncertainty as to where and when upwelling could generate net carbon sequestration. Moreover, natural analogs where upwelling occurs are generally net sources of CO_2 to the atmosphere (Takahashi et al., 1997) versus net sinks. In the particularly well-studied site of the Hawaii Ocean Time-series in the oligotrophic North Pacific Subtropical Gyre, Karl and Letelier (2008) estimate that artificial upwelling could only generate net carbon sequestration if the process of biological nitrogen gas fixation were stimulated to draw down residual nutrients not consumed by non-nitrogen gas-fixing phytoplankton. This hypothesis has not been proven or disproven, but it does illustrate that careful consideration must be paid to carbon accounting between deep-water sources and the long-term (> weeks to years) response of biological communities. In practice, as opposed to ocean iron fertilization, there exists no proof-of-concept sea trials that have demonstrated that artificial upwelling could act to sequester carbon below the ocean pycnocline. Lacking field demonstrations of the CDR efficacy (or lack thereof) of artificial upwelling, continued research is warranted to address key remaining issues related to regions where net CO_2 uptake could occur, how sequestration efficiencies could be enhanced, and how carbon transport could or should be tracked and thus accounted for.

In silico, the potential efficacy of artificial upwelling as a means of CDR has been assessed via simulations of the biological impact of large-scale (millions of "ocean pipes") global deployments operating efficiently over years to decades. As a composite, these models suggest that artificial upwelling would be an ineffective means for large-scale carbon sequestration (Dutreuil et al., 2009; Yool et al., 2009; Oschlies et al., 2010b; Keller et al., 2014) and would require a persistent and effective deployment of tens of million to hundreds of million functional pumps across the global ocean (Yool et al., 2009) operating at upwelling velocities generally in excess of what has been demonstrated in limited sea trials (Figure 4.3). Redistributing large amounts of deep ocean water is also expected to affect density and pressure fields in the surface ocean. If artificial upwelling (and the concomitant downwelling that will take place) is to be scaled up, research will need to investigate possible long-term changes in ocean circulation as well as dynamical termination effects once or if the millions of pumps are turned off.

Only targeted, regulated, and transparent field studies can help to minimize current uncertainties and determine whether this strategy could be an effective component of an ocean CDR portfolio. The largest cost for research would be the demonstration-scale field experiments in addition to the cost of materials, fabrication, and maintenance for pumping systems. Using ocean iron fertilization cost estimates as an analogy, field studies at multiple sites could be on the order of $25M/yr for 10 years to complete work at multiple sites, whereas current commercial estimates of instrumented wave pumps are on the order of $60K per pump.[5] "Cradle to grave" carbon-based life-cycle analyses for the pumping technology used would of course also be needed. A robust research agenda would also include parallel laboratory/mesocosm studies to address potential biological responses to deep-water additions of varying stoichiometry as well as modeling efforts to refine deployment scaling in space and time. Social and governance aspects would also be added as per any comprehensive ocean CDR research agenda.

Seaweed Cultivation

Seaweed cultivation and sequestration appears to be a compelling ocean CDR strategy. There is a good understanding of the underlying biology of seaweeds, their interactions with the ocean and its biogeochemistry, and decades of experience in farming seaweed. Seaweed cultivation has advantages over some other biotic CDR strategies as the fixed carbon biomass could be, in principle, conveyed to a known durable reservoir, which will increase the permanence of the sequestered carbon and in turn should simplify the carbon accounting. However, there remains uncertainty about how much productivity and carbon export that seaweed cultivation would displace as well as the durability of the sequestered carbon when conveyed to depth or the seafloor. Scaling to CDR-worthy levels will be challenging due to the large amount of farmed area required (many millions of hectares). However, much has been learned already from ongoing U.S. agency research programs, and the scale of the engineering and logistic efforts are similar to the many marine engineering accomplishments made by the marine engineering and oil and gas industries. The costs and energy expenditures should be small relative to some other CDR strategies. On the other hand, there will be environmental impacts where the farming occurs and where the seaweed biomass will be sequestered that are potentially detrimental, yet uncertain. There are both positive and negative social impacts from seaweed cultivation and sequestration CDR. If conducted at scale, seaweed cultivation will clearly enhance aspects of the blue economy associated with macroalagal farming (while perhaps diminishing other marine economic activities if there is competition for coastal space and resources), and it is an open question about the extent of net benefit for both coastal communities and many marine industries. There may also be several co-benefits from placing farms adjacent to other uses (e.g., fish farming, etc.) that could help mitigate some of the potential environmental damages conducted by these practices. On the negative side, the vast farms could represent hazards to navigation, and they may displace fishing and other uses of the ocean due to the implementation of farm infrastructure or the reduction in planktonic productivity and trophic exchanges that large-scale cultivation of seaweed biomass would create.

A decadal-scale research agenda for seaweed cultivation CDR should be focused on reducing the uncertainties outlined here as well as deploying demonstration-scale cultivation (several square kilometers in scale) and sequestration infrastructure. The accomplishment of these two elements will involve developing a predictive understanding of the entire process from farming on demonstration

[5] See https://ocean-based.com/frequently-asked-questions-faqs-about-our-autonomous-upwelling-pumps-aups/.

scales (many square kilometers) and its environmental and social impacts, to engineering conveyances to bring farmed biomass to depth, to understanding the biogeochemical fates, permanence, and environmental impacts of the farmed biomass and its by-products, to validating the end-to-end performance. In situ, in the laboratory, mesocosm experimentation and, in particular, numerical modeling will be required with the goal of creating modeling systems that can assess the potential of and impacts on local to global scales. Social and governance considerations would also need to be addressed in any research program. This research agenda should logically occur in phases (as mentioned in Section 4.6) and, based on present investment levels, will require approximately $25M per year.

Ecosystem Recovery

The research agenda for ecosystem recovery includes basic scientific research on the carbon removal potential and permanence of different organisms, ecosystems, and processes; an examination of the expected outcomes from different policy tools; and the socioeconomic and governance aspects of managing marine ecosystems and organisms for carbon removal. As with seaweed cultivation, there remains uncertainty in how much net carbon sequestration would result from protecting and restoring marine ecosystems. There are trade-offs to marine conservation efforts, with potential winners and losers if the focus of the ocean economy shifts from extraction to rebuilding marine life, with an emphasis on CDR and other services. Understanding of net carbon sequestration could be gained if carbon accounting and research became a central aspect of the monitoring of marine protected areas in the coming decades, including the examination of multiple systems, with geographic, ecological, and taxonomic differences, and varying levels of anthropogenic pressure.

For each of the systems discussed in Chapter 6, such as macroalgae reforestation, benthic algae and sediment protection, and animals in the carbon cycle, it is essential to understand the impact of human perturbation, which requires estimating the current and historical contribution of different species across the size spectrum in terms of annual carbon flux. Note that this agenda only focuses on the restoration of degraded ecosystems and recovery of depleted species to determine the respective CDR potential. In the absence of effective management, human activities will continue to threaten marine habitats, and any loss in ecosystem function and services via further marine degradation could have large consequences for the uptake of carbon by the oceans.

A logical next step to this work would be analysis of the best tools that could enhance marine ecosystem CDR, such as marine protected areas and habitat restoration, fish and fisheries management, and the restoration of food webs and large marine organisms. Kelp forest restoration, in particular, has been promoted for its CDR potential. All of these activities could have biodiversity and research co-benefits, such as improving the status of endangered or depleted species and enhancing the monitoring of marine mammals, fish, and other species. Research on the vulnerabilities of marine species and ecosystems, and their CDR potential, to rising sea-surface temperatures and reduced sea ice would also be valuable. Social and governance considerations would also need to be integrated into the research agenda. A budget of at least $20M per year would be required to comprehensively examine the carbon impacts of marine ecosystem-based recovery in the United States. Unlike several of the other approaches in this report, however, individual tools, such as macroalgal reforestation or marine protected areas, could be funded individually.

Ocean Alkalinity Enhancement

Approaches to increase ocean alkalinity have been proposed since the mid-1990s but remain at a relatively early stage of development. Ocean alkalinity enhancement attempts to mimic natural weathering processes either by adding crushed minerals directly to the ocean, to coastal, or to terrestrial environments. Given that the surface ocean is supersaturated with common carbonate minerals, they cannot be added directly to the ocean. To facilitate mineral dissolution, the reaction of carbonate minerals with elevated CO_2 has been suggested, initially for reducing fossil fuel emissions, but potentially for CDR by coupling with direct air capture or biomass energy carbon capture and storage. Others have proposed converting carbonate minerals into more reactive forms (e.g., lime or hydrated lime) for addition to the ocean. Finally, electrochemical approaches may be used to create basic and acidic solutions at each electrode; the former could be used for ocean alkalinity enhancement while the latter would need to be neutralized through reaction with silicate minerals.

The two key mechanisms by which ocean alkalinity enhancement could impact the environment are (1) elevated alkalinity that is "unequilibrated," that is, high-pH, low-CO_2 concentrations that may be more acute around the point of addition; and (2) the addition of other biologically active elements (i.e., iron, silica, as in ocean fertilization, or nickel, chromium, or other trace metals). Research is needed to assess the ecological response to ocean alkalinity enhancement. Although much can still be learned from laboratory-based and "contained" (mesocosm) experiments, including exploring the impacts of ocean alkalinity enhancement on the physiology and functionality of organisms and communities, pathways to responsible deployment will require field trials. Such trials will be essential in assessing how euphotic and benthic biogeochemical processes are affected, the response from complex communities, the indirect effects of ocean alkalinity enhancement, and optimal environments and treatment methods. Coupling this research to a rigorous monitoring program will be essential in accounting for CDR and the environmental impact of the experiments, but would help scope the requirements for monitoring of scaled-up ocean alkalinity enhancement.

The environmental impact of ocean alkalinity enhancement would be closely related to the methods or technologies that promote alkalinity changes (e.g., the choice of rock or mineral and the mechanisms by which it is dissolved). An experimental program that considers the environmental response to ocean alkalinity enhancement will be most effective if it were constrained by what might be practical. Research and development is required to explore and improve the technical feasibility and readiness level of ocean alkalinity enhancement approaches (including the development of pilot-scale facilities). Research on social and governance considerations associated with contained and pilot-scale experiments and deployment (if any) is also required.

Electrochemical Processes

The research agenda for electrochemical processes focuses on activities that reduce the cost and environmental impact of the approaches. Presently, only a small number of system configurations have been investigated for either CDR from seawater or as methods for increasing ocean alkalinity. These should be broadened to also consider novel designs of electrolyzers and electrochemical reactors, electrode materials that can minimize the production of unwanted by-products (e.g., chlorine gas), electrochemical reactor architectures, and hybrid approaches that both remove CO_2 and increase ocean alkalinity. Wider impacts on ocean ecosystems are included in the research agenda for ocean alkalinity enhancement and are not repeated here. Meaningfully large demonstration-scale projects (>1,000 kg per day, and approaching the scale of several or tens of tonnes per day) are central to the research agenda. While part of this research agenda may focus on the development of specific technologies or materials (e.g., electrolyzer design or novel electrode materials), it also focuses on systems integration (particularly with rock dissolution processes) and scale-up strategies that allow for process cost reductions to be discovered.

The processes explored by the research agenda will need to demonstrate pathways to large-scale deployment and thus include provisions for resource mapping, stakeholder engagement, and wider economic impact from an ocean-based electrochemical CDR industry operating at a climate-relevant scale. Transparent dissemination of the research outcomes would allow independent assessment of cost and environmental impact. Social and governance considerations are also important components of the research program.

TABLE 9.1 Summary of Scale-Up Potential of Ocean-Based CDR Approaches (summarized from Chapters 3–8)

	Ocean Nutrient Fertilization	Artificial Upwelling/ Downwelling	Seaweed Cultivation	Ecosystem Recovery	Ocean Alkalinity Enhancement	Electrochemical Processes
Knowledge base What is known about the system (low, mostly theoretical, few in situ experiments; medium, lab and some fieldwork, few carbon dioxide removal (CDR) publications; high, multiple in situ studies, growing body of literature)	**Medium–High** Considerable experience relative to any other ocean CDR approach with strong science on phytoplankton growth in response to iron, less experience on fate of carbon and unintended consequences. Natural iron-rich analogs provide valuable insight on larger temporal and spatial scales.	**Low–Medium** Various technologies have been demonstrated for artificial upwelling (AU), although primarily in coastal regimes for short duration. Uncertainty is high and confidence is low for CDR efficacy due to upwelling of CO_2, which may counteract any stimulation of the biological carbon pump (BCP).	**Medium–High** Science of macrophyte biology and ecology is mature; many mariculture facilities are in place globally. Less is known about the fate of macrophyte organic carbon and methods for transport to deep ocean or sediments.	**Low–Medium** There is abundant evidence that marine ecosystems can uptake large amounts of carbon and that anthropogenic impacts are widespread, but quantifying the collective impact of these changes and the CDR benefits of reversing them is complex and difficult.	**Low–Medium** Seawater CO_2 system and alkalinity thermodynamics are well understood. Need for empirical data on alkalinity enhancement; currently, knowledge is based on modeling work. Uncertainty is high for possible impacts.	**Low–Medium** Processes are based on well-understood chemistry with a long history of commercial deployment, but is yet to be adapted for CO_2 removal by ocean alkalinity enhancement (OAE) beyond benchtop scale.
Efficacy What is the confidence level that this approach will remove atmospheric CO_2 and lead to net increase in ocean carbon storage (low, medium, high)	**Medium–High Confidence** BCP known to work and productivity enhancement evident. Natural systems have higher rates of carbon sequestration in response to iron but low efficiencies seen thus far would limit effectiveness for CDR.	**Low Confidence** Upwelling of deep water also brings a source of CO_2 that can be exchanged with the atmosphere. Modeling studies generally predict that large-scale AU would not be effective for CDR.	**Medium Confidence** The growth and sequestration of seaweed crops should lead to net CDR. Uncertainties about how much existing net primary production (NPP) and carbon export downstream would be reduced due to large-scale farming.	**Low–Medium Confidence** Given the diversity of approaches and ecosystems, CDR efficacy is likely to vary considerably. Kelp forest restoration, marine protected areas, fisheries management, and restoring marine vertebrate carbon are promising tools.	**High Confidence** Need to conduct field deployments to assess CDR, alterations of ocean chemistry (carbon but also metals), how organic matter can impact aggregation, etc.	**High Confidence** Monitoring within an enclosed engineered system, CO_2 stored either as increased alkalinity, solid carbonate, or aqueous CO_2 species. Additionality possible with the utilization of by-products to reduce carbon intensity.

continued

TABLE 9.1 Continued

	Ocean Nutrient Fertilization	Artificial Upwelling/ Downwelling	Seaweed Cultivation	Ecosystem Recovery	Ocean Alkalinity Enhancement	Electrochemical Processes
Durability Will it remove CO_2 durably away from surface ocean and atmosphere (low, <10 years; medium, >10 years and <100 years; high, >100 years), and what is the confidence (low, medium, high)	**Medium** 10–100 years Depends highly on location and BCP efficiencies, with some fraction of carbon flux recycled faster or at shallower ocean depths; however, some carbon will reach the deep ocean with >100-year horizons for return of excess CO_2 to surface ocean.	**Low–Medium** <10–100 years As with ocean iron fertilization (OIF), dependent on the efficiency of the BCP to transport carbon to deep ocean.	**Medium–High** >10–100 years Dependent on whether the sequestered biomass is conveyed to appropriate sites (e.g., deep ocean with slow return time of waters to surface ocean).	**Medium** 10–100 years The durability of ecosystem recovery ranges from biomass in macroalgae to deep-sea whale falls expected to last >100 years.	**Medium–High** >100 years Processes for removing added alkalinity from seawater generally quite slow; durability not dependent simply on return time of waters with excess CO_2 to ocean surface.	**Medium–High** >100 years Dynamics similar to OAE.
Scalability Potential scalability at some future date with global-scale implementation (low, <0.1 Gt CO_2/yr; medium, >0.1 Gt CO_2/yr and <1.0 Gt CO_2/yr; high, >1.0 Gt CO_2/yr), and what is the confidence level (low, medium, high)	**Medium–High** Potential C removal >0.1–1.0 Gt CO_2/yr (medium confidence) Large areas of ocean have high-nutrient, low-chlorophyll conditions suitable to sequester >1 Gt CO_2/yr. Co-limitation of macronutrients and ecological impacts at large scales are likely. Low-nutrient, low-chlorophyll areas have not been explored to increase areas of possible deployment. (Medium confidence based on 13 field experiments).	**Medium** Potential C removal >0.1 Gt CO_2/yr and <1.0 Gt CO_2/yr (low confidence) Could be coupled with aquaculture efforts. Would require pilot trials to test materials durability for open ocean and assess CDR potential. Current model predictions would require deployment of tens of millions to hundreds of millions of pumps to enhance carbon sequestration. (Low confidence that this large-scale deployment would lead to permanent and durable CDR).	**Medium** Potential C removal >0.1 Gt CO_2/yr and <1.0 Gt CO_2/yr (medium confidence) Farms need to be many million hectares, which creates many logistic and cost issues. Uncertainties about nutrient availability and durability of sequestration, seasonality will limit sites, etc.	**Low–Medium** Potential C removal <0.1–1.0 Gt CO_2/yr (low–medium confidence) Given the widespread degradation of much of the coastal ocean, there are plenty of opportunities to restore ecosystems and depleted species. However, ecosystems and trophic interactions are complex and changing and research will be necessary to explore upper limits.	**Medium–High** Potential C removal >0.1–1.0 Gt CO_2/yr (medium confidence) Potential for sequestering >1 Gt CO_2/yr if applied globally. High uncertainty coming from potential aggregation and export to depth of added minerals and unintended chemical impacts of alkalinity addition.	**Medium–High** Potential C removal >0.1–1.0 Gt CO_2/yr (medium confidence) Energy and water requirements may limit scale. For climate relevancy, the scale will be double to an order of magnitude greater than the current chlor-alkali industry.

Environmental risk Intended and unintended undesirable consequences at scale (unknown, low, medium, high), and what is the confidence level (low, medium, high)	**Medium** (low to medium confidence) Intended environmental impacts increase NPP and carbon sequestration due to changes in surface ocean biology. If effective, there are deep-ocean impacts and concern for undesirable geochemical and ecological consequences. Impacts at scale uncertain.	**Medium–High** (low confidence) Similar impacts to OIF but upwelling also affects the ocean's density field and sea-surface temperature and brings likely ecological shifts due to bringing colder, inorganic carbon, and nutrient-rich waters to surface.	**Medium–High** (low confidence) Environmental impacts are potentially detrimental especially on local scales where seaweeds are farmed (i.e., nutrient removal due to farming will reduce NPP, carbon export, and trophic transfers) and in the deep ocean where the biomass is sequestered (leading to increases in acidification, hypoxia, eutrophication, and organic carbon inputs). The scale and nature of these impacts are highly uncertain.	**Low** (medium–high confidence) Environmental impacts would be generally viewed as positive. Restoration efforts are intended to provide measurable benefits to biodiversity across a diversity of marine ecosystems and taxa.	**Medium** (low confidence) Possible toxic effect of nickel and other leachates of olivine on biota, bio-optical impacts, removal of particles by grazers, unknown responses to increased alkalinity on functional diversity and community composition. Effects also from expanded mining activities (on land) on local pollution, CO_2 emissions.	**Medium–High** (low confidence) Impact on the ocean is possibly constrained to the point of effluent discharge. Poorly-known possible ecosystem impacts similar to alkalinity enhancement. Excess acid (or gases, particularly chlorine) will need to be treated and safely disposed. Provision of sufficient electrical power will likely have remote impacts.
Social considerations Encompass use conflicts, governance-readiness, opportunities for livelihoods, etc.	Potential conflicts with other uses of high seas and protections; downstream effects from displaced nutrients will need to be considered; legal uncertainties; potential for public acceptability and governance challenges (i.e., perception of "dumping").	Potential conflicts with other uses (shipping, marine protected areas, fishing, recreation); potential for public acceptability and governance challenges (i.e., perception of dumping).	Possibility for jobs and livelihoods in seaweed cultivation; potential conflicts with other marine uses. Downstream effects from displaced nutrients will need to be considered.	Trade-offs in marine uses to enhance ecosystem protection and recovery. Social and governance challenges may be less significant than with other approaches.	Expansion of mining production, with public health and economic implications; general public's potential for public acceptability and governance challenges (e.g., if perceived as "dumping").	Similar to OAE and to any industrial site. Substantial electrical power demand may generate social impacts.

continued

TABLE 9.1 Continued

	Ocean Nutrient Fertilization	Artificial Upwelling/ Downwelling	Seaweed Cultivation	Ecosystem Recovery	Ocean Alkalinity Enhancement	Electrochemical Processes
Co-benefits How significant are the co-benefits as compared to the main goal of CDR and how confident is that assessment	**Medium** (low confidence) Enhanced fisheries possible but not shown and difficult to attribute. Seawater dimethyl sulfide increase seen in some field studies that could enhance climate cooling impacts. Surface ocean decrease in ocean acidity possible.	**Medium–High** (low confidence) May be used as a tool in coordination with localized enhancement of aquaculture and fisheries.	**Medium–High** (medium confidence) Placing cultivation facilities near fish or shellfish aquaculture facilities could help alleviate environmental damages from these activities. Bio-fuels also possible.	**High** (medium–high confidence) Enhanced biodiversity conservation and the restoration of many ecological functions and ecosystem services damaged by human activities. Existence, spiritual, and other non-use values. Potential to enhance marine stewardship and tourism.	**Medium** (low confidence) Mitigation of ocean acidification; positive impact on fisheries.	**Medium–High** (medium confidence) Mitigation of ocean acidification; production of H_2, Cl_2, silica.
Cost of scale-up Estimated costs in dollars per metric ton CO_2 for future deployment at scale; does not include all of monitoring and verification costs needed for smaller deployments during R&D phases (low, <\$50/t CO_2; medium, ~\$100/t CO_2; high, >>\$150/t CO_2) and confidence in estimate (low, medium, high)	**Low** <\$50/t CO_2 (low–medium confidence) Deployment of <\$25/t CO_2 sequestered for deployment at scale are possible, but need to be demonstrated at scale	**Medium–High**. >\$100–\$150/t CO_2 (low confidence) Development of a robust monitoring program is the likely largest cost and would be of similar magnitude as OIF. Materials costs for pump assembly could be moderate for large-scale persistent deployments. Estimates for a kilometer-scale deployment are in the tens of million dollars.	**Medium** ~\$100/t CO_2 (medium confidence) Costs should be less than \$100/t CO_2. No direct energy used to fix CO_2.	**Low** <\$50/t CO_2 (medium confidence) Varies, but direct costs would largely be for management and opportunity costs for restricting uses of marine species and the environment. No direct energy used.	**Medium–High** >\$100–\$150/t CO_2 (low–medium confidence) Cost estimates range between tens of dollars and \$160/t CO_2. Need for expansion of mining, transportation, and ocean transport fleet.	**High** >\$150/t CO_2 (medium confidence) Gross current estimates \$150–\$2,500/t CO_2 removed. With further R&D, it may be possible to reduce this to <\$100/t CO_2.

Cost and challenges of carbon accounting Relative cost and scientific challenge associated with transparent and quantifiable carbon tracking (low, medium, high)	**Medium** Challenges tracking additional local carbon sequestration and impacts on carbon fluxes outside of boundaries of CDR application (additionality).	**High** Local and additionality monitoring needed for carbon accounting similar to OIF.	**Low–Medium** The amount of harvested and sequestered carbon will be known. However, an accounting of the carbon cycle impacts of the displaced nutrients will be required (additionality).	**High** Monitoring net effect on carbon sequestration is challenging.	**Low–Medium** Accounting more difficult for addition of minerals and non-equilibrated addition of alkalinity, than equilibrated addition.	**Low–Medium**
Cost of environmental monitoring Need to track impacts beyond carbon cycle on marine ecosystems (low, medium, high)	**Medium** (medium–high confidence) All CDR will require monitoring for intended and unintended consequences both locally and downstream of CDR site, and these monitoring costs may be substantial fraction of overall costs during R&D and demonstration-scale field projects. This cost of monitoring for ecosystem recovery may be lower.					
Additional resources needed Relative low, medium, high to primary costs of scale-up	**Low–Medium** Cost of material: iron is low and energy is sunlight.	**Medium–High** Materials, deployment, and potential recovery costs.	**Medium** Farms will require large amounts of ocean (many million hectares) to achieve CDR at scale.	**Low** Most recovery efforts will likely require few materials and little energy, though enforcement could be an issue. Active restoration of kelp and other ecosystems would require more resources.	**Medium–High** Adaptation and likely expansion of existing fleet for deployment; infrastructure for storage at ports. Infrastructure support for expansion of mineral extraction, processing, transportation, and deployment.	**Medium–High** High energy requirements (1–2.5 MWh/t CO_2 removed) and build-out of industrial CDR.

TABLE 9.2 Foundational Research Priorities Common to all Ocean-Based CDR

Research Priority	Estimated Budget	Duration (years)	Total
Model international governance framework for ocean CDR research	$2M–$3M/yr	2–4	$4M–$12M
Application of domestic laws to ocean CDR research	$1M/yr	1–2	$1M–$2M
Assessment of need for domestic legal framework specific to ocean CDR	$1M/yr	2–4	$2M–$4M
Development of domestic legal framework specific to ocean CDR			
Mixed methods, multisited research to understand community priorities and assessment of benefits and risks for ocean CDR as a strategy	$5M/yr	4	$20M
Interactions and trade-offs between ocean CDR, terrestrial CDR, adaptation, and mitigation, including the potential of mitigation deterrence	$2M/yr	4	$8M
Cross-sectoral research analyzing food system, energy, sustainable development goals, and other systems in their interaction with ocean CDR approaches	$1M/yr	4	$4M
Capacity-building research fellowship for diverse early-career scholars in ocean CDR	$1.5M/yr	2	$3M
Transparent, publicly accessible system for monitoring impacts from projects	$0.25M/yr	4	$1M
Research on how user communities (companies buying and selling CDR, nongovernmental organizations, practitioners, policy makers) view and use monitoring data, including certification	$0.5M/yr	4	$2M
Analysis of policy mechanisms and innovation pathways, including on the economics of scale-up	$1M–2M/yr	2	$2M–$4M
Development of standardized environmental monitoring and carbon accounting methods for ocean CDR	$0.2M/yr	3	$0.6M
Development of a coordinated research infrastructure to promote transparent research	$2M/yr	3–4	$6M–$8M
Development of a publicly accessible data management strategy for ocean CDR research	$2M–3M/yr	2	$4M–$6M
Development of a coordinated plan for science communication and public engagement of ocean CDR research in the context of decarbonization and climate response	$5M/yr	10	$50M
Development of a common code of conduct for ocean CDR research	$1M/yr	2	$2M
Total Estimated Research Budget (Assumes all six CDR approaches moving ahead)	**~$29M/yr**	**2–10**	**~$125M**

TABLE 9.3 Research Needed to Advance Understanding of Each Ocean CDR Approach

	Estimated Budget	Duration (years)	Total Budget
Ocean Fertilization			
Carbon sequestration delivery and bioavailability	**$5M/yr**	**5**	**~$25M**
Tracking carbon sequestration	**$3M/yr**	**5**	**~$15M**
In field experiments- >100 t Fe and >1,000 km^2 initial patch size followed over annual cycles	**$25M/yr**	**10**	**~$250M**
Monitoring carbon and ecological shifts	$10M/yr	10	~$100M
Experimental planning and extrapolation to global scales	$5M/yr	10	~$50M
Total Estimated Research Budget	$48M/yr	5–10	$440M
Estimated Budget of Research Priorities	**$33M/yr**	**5–10**	**$290M**
Artificial Upwelling and Downwelling			
Technological readiness: Limited and controlled open-ocean trials to determine durability and operability of artificial upwelling technologies (~100 pumps tested in various conditions)	**$5M/yr**	**5**	**$25M**
Feasibility studies	$1M/yr	1	$1M
Tracking carbon sequestration	$3M/yr	5	$15M
Modeling of carbon sequestration based upon achievable upwelling velocities and known stoichiometry of deep-water sources. Parallel mesocosm and laboratory experiments to assess potential biological responses to deep water of varying sources	$5M/yr	5	$25M
Planning and implementation of demonstration-scale in situ experimentation (>1 year, >1,000 km) in region sited-based input from modeling and preliminary experiments	$25M/yr	10	$250M
Monitoring carbon and ecological shifts	$10M/yr	10	$100M
Experimental planning and extrapolation to global scales (early for planning and later for impact assessments)	$5M/yr	10	$50M
Total Estimated Research Budget	~$54/yr	5–10	$466M
Estimated Budget of Research Priorities	**$5M/yr**	**5–10**	**$25M**
Seaweed Cultivation			
Technologies for efficient large-scale farming and harvesting of seaweed biomass	**$15M/yr**	**10**	**$150M**
Engineering studies focused on the conveying of harvested biomass to durable oceanic reservoir with minimal losses of carbon	**$2M/yr**	**10**	**$20M**
Assessment of long-term fates of seaweed biomass and by-products	**$5M/yr**	**5**	**$25M**
Implementation and deployment of a demonstration-scale seaweed cultivation and sequestration system	$10M/yr	10	$100M
Validation and monitoring the CDR performance of a demonstration-scale seaweed cultivation and sequestration system	$5M/yr	10	$50M
Evaluation of the environmental impacts of large-scale seaweed farming and sequestration	**$4M/yr**	**10**	**$40M**
Total Estimated Research Budget	$41M/yr	5–10	$385M
Estimated Budget of Research Priorities	**$26M/yr**	**5**	**$235M**

continued

TABLE 9.3 Continued

	Estimated Budget	Duration (years)	Total Budget
Ecosystem Recovery			
Restoration ecology and carbon	**$8M/yr**	**5**	**$40M**
Marine protected areas: Do ecosystem-level protection and restoration scale for marine CDR?	**$8M/yr**	**10**	**$80M**
Macroalgae: Carbon measurements, global range, and levers of protection	**$5M/yr**	**10**	**$50M**
Benthic communities: disturbance and restoration	$5M/yr	5	$25M
Marine animals and CO_2 removal	**$5M/yr**	**10**	**$50M**
Animal nutrient cycling	$5M/yr	5	$25M
Commercial fisheries and marine carbon	$5M/yr	5	$25M
Total Estimated Research Budget	$41M/yr	5–10	$295M
Estimated Budget of Research Priorities	**$26M/yr**	**5–10**	**$220M**
Ocean Alkalinity Enhancement			
Research and development to explore and improve the technical feasibility and readiness level of ocean alkalinity enhancement approaches (including the development of pilot-scale facilities)	$10M/yr	5	$50M
Laboratory and mesocosm experiments to explore impacts on physiology and functionality of organisms and communities	**$10M/yr**	**5**	**$50M**
Field experiments	**$15M/yr**	**5–10**	**$75M–$150M**
Research into the development of appropriate monitoring and accounting schemes, covering CDR potential and possible side effects	$10	5–10	$50M–$100M
Total Estimated Research Budget	$45M/yr	5–10	$180M–$350M
Estimated Budget of Research Priorities	**$25M/yr**	**5–10**	**$125M–$200M**
Electrochemical Processes			
Demonstration projects including CDR verification and environmental monitoring	**$30M/yr**	**5**	**$150M**
Development and assessment of novel and improved electrode and membrane materials	**$10M/yr**	**5**	**$50M**
Assessment of environmental impact and acid management strategies	$7.5M/yr	10	$75M
Coupling whole-rock dissolution to electrochemical reactors and systems	**$7.5M/yr**	**10**	**$75M**
Development of hybrid approaches	**$7.5M/yr**	**10**	**$75M**
Resource mapping and pathway assessment	$10M/yr	5	$50M
Total Estimated Research Budget	$73M/yr	5–10	$475M
Estimated Budget of Research Priorities	**$55M/yr**	**5–10**	**$350M**

NOTE: Bold type identifies priorities for taking the next step to advance understanding of each particular approach.

9.4 PROPOSED RESEARCH AGENDA

As summarized in this report, the state of knowledge and uncertainty surrounding ocean CDR varies greatly among and within different CDR approaches, with knowledge gaps remaining in determining carbon sequestration efficiencies, scaling, and permanence, as well as environmental and social impacts and costs. **At present, the current state of scientific understanding precludes choosing among different ocean CDR approaches, and there are opportunities and benefits from pursuing research along multiple pathways in parallel.** The resulting increased understanding and confidence will allow for more informed choices on the viability of ocean CDR relative to other climate change response options, given the ever-increasing and visible impacts of climate change.

Based on the present state of knowledge, there are substantial uncertainties in all of the ocean CDR approaches evaluated in this report. The knowledge gaps differ among the CDR approaches as highlighted by the research elements and priorities (marked in bold) in Table 9.2. **The best approach for reducing knowledge gaps will involve a diversified research investment strategy that includes both crosscutting, common components (Table 9.2) and coordination across multiple individual CDR approaches (Table 9.3) in parallel (Figure 9.1).** The development of a robust research portfolio will reflect a balance among several factors: common elements and infrastructure versus targeted studies on specific approaches; biotic versus abiotic CDR approaches; and more established versus emerging CDR approaches. The research cost values in Table 9.2 and 9.3 are only rough estimates, and as an ocean CDR research program develops and priorities are refined with time, improved cost estimates will likely become clearer.

Crosscutting foundation research priorities listed in Table 9.2 include research on international governance and the domestic legal framework of ocean CDR research. Other priorities include the development of a common code of conduct for ocean CDR research and coordinated research infrastructure including components on standardized environmental monitoring and carbon accounting methods, publicly accessible data management, and science communication and public engagement.

The research priorities in Table 9.3 for each of the four biotic ocean CDR approaches (Chapters 3–6) differ based on the current knowledge base, extent of previous research, and distinctions in the underlying biological processes. As discussed above, evaluation of research needs across CDR approaches is more challenging, suggesting some investment in all methods; however, a first-order attempt at prioritization can be constructed based on current knowledge using the CDR criteria outlined at the beginning of Section 9.3. **Among the biotic approaches, research on ocean iron fertilization and seaweed cultivation offer the greatest opportunities for evaluating the viability of possible biotic ocean CDR approaches; research on the potential CDR and sequestration permanence for ecosystem recovery would also be beneficial in the context of ongoing marine conservation efforts.**

For abiotic ocean CDR approaches (Chapters 7 and 8), the research agenda (Table 9.2) will be most impactful if it combines a thorough understanding of potential environmental impacts alongside technology development and upscaling efforts. Based on present understanding, there is considerable CDR potential for ocean alkalinity enhancement, which spans a number of approaches including, but not restricted to, ocean liming, accelerated rock weathering, and electrochemical methods for alkalinity enhancement, among others. Next steps for alkalinity enhancement research offer large opportunities for closing knowledge gaps but include the complexity of undertaking large-scale experimentation to assess whole-ecosystem responses across the range of technologies and approaches for increasing alkalinity. Therefore, **among the abiotic approaches, research on ocean alkalinity enhancement, including electrochemical alkalinity enhancement, have priority over electrochemical approaches that only seek to achieve CDR from seawater (also known as carbon dioxide stripping).**

Early research findings might indicate a low viability for particular approaches. The approach advocated in the research agenda below is to be adaptive, meaning that decisions on future investments in research activities will need to take into account new findings on the efficacy and durability of a technique, whether the social and environmental impacts outweigh benefits, or face social and governance challenges. This is in line with a responsible research and innovation approach (Box 9.1). Generally speaking, we can anticipate that there may be showstoppers for some approaches from factors both internal and external to the research. Internal showstoppers include findings that indicate that the viability is so low as to not warrant further research investments. For example, there are questions about whether artificial upwelling/downwelling technologies will be operable at sufficient scales; this is why the technique has a priority item to better evaluate technological readiness (Table 9.3), recommending limited and controlled open ocean trials to determine durability and operability of artificial upwelling technologies.

There may also be external showstoppers to the research, such as lack of social license or governance actions, which preclude further investigation. Given that climate change governance and politics are dynamic, it is impossible to predict what these showstoppers will be. However, there are examples of research that was deemed too risky or unnecessary by social stakeholders. For example, in the early 2000s, scientists were researching injecting CO_2 directly into the mid-depth ocean using small-scale experiments. In 1997 in Kyoto, during UNFCCC COP-3, an international project agreement was signed for the study of direct CO_2 injection, with sponsors from the U.S. Department of Energy, the New Energy and Industrial Technology Development Organization of Japan, and the Norwegian Research Council; researchers from Australia, Canada, and Switzerland also joined. However, the research faced criticism from local civil society organizations as well as larger organizations such as Greenpeace, and planned experiments off the coasts of Hawaii and Norway were halted, with the Norwegian Environmental Minister stating that using deep marine areas as future storage places for CO_2 required more international discussion and the clarification of legal implications (de Figueiredo, 2003). Within the scientific community, there was discussion of what research could be done to identify potential harms to deep-sea biology, with some recommending very aggressive research to provide information on impacts on deep-sea organisms within a relevant time frame (Seibel and Walsh, 2003). In other words, direct deep-sea CO_2 injection faced external social showstoppers and may have also faced potential scientific, internal showstoppers, should it have proceeded further. Scientists must be alert and responsive to potential showstoppers. **Early investment in public engagement and governance activities may help advance understanding of what external showstoppers may be, and structuring the research agenda to focus on priority items can help advance understanding of internal showstoppers**.

Recommendation 3: Ocean CDR Research Program Priorities. A research program should move forward integrating studies, in parallel, on multiple aspects of different ocean CDR approaches, recognizing the different stages of the knowledge base and technological readiness of specific ocean CDR approaches. Priorities for the research program should include development of:

1. Overarching implementation plan for the next decade adhering to the crosscutting strategy elements in Recommendation 1 and incorporating from its onset the common research components in Recommendation 2 and Table 9.2. Achieving progress on these common research components is essential to lay a foundation for all other recommended research.
2. Tailored implementation planning for specific ocean CDR approaches focused on reducing critical knowledge gaps by moving sequentially from laboratory-scale to pilot-scale field experiments, as appropriate, with adequate environmental and social risk reduction measures and transparent decision-making processes (priority components bolded in Table 9.3).

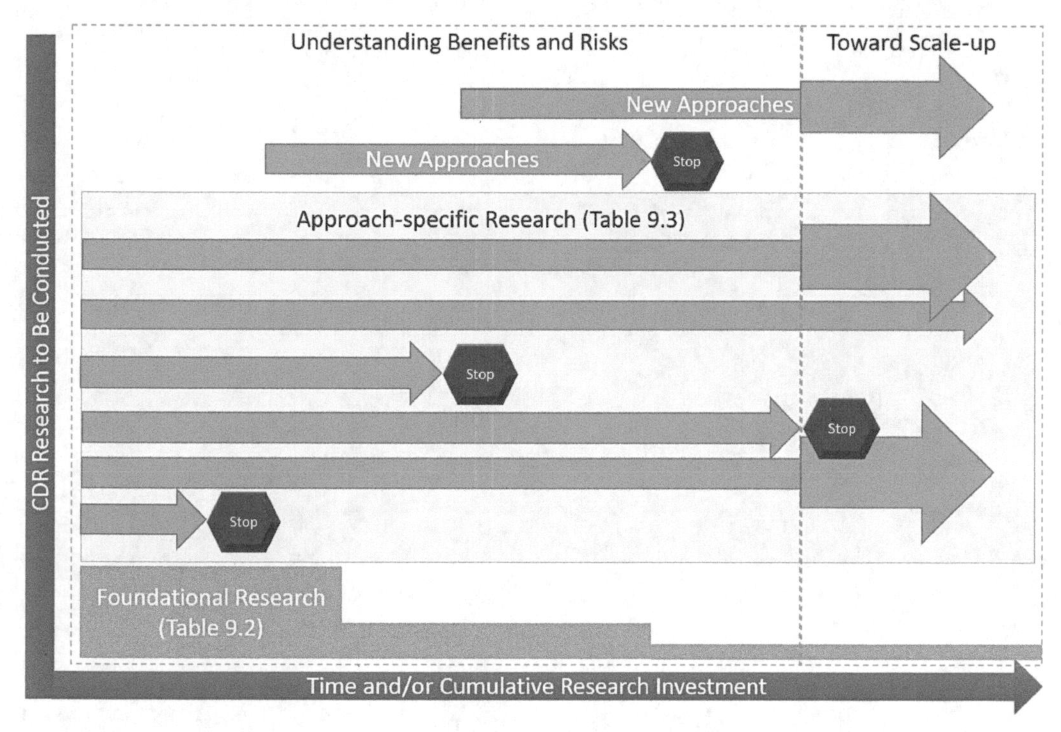

FIGURE 9.1 Conceptual timeline of ocean-based CDR research based on Tables 9.2 and 9.3. Stops included on the diagram represent possible internal and external showstoppers or barriers to a particular approach.

3. Common framework for intercomparing the viability of ocean CDR approaches with each other and with other climate response measures using standard criteria for efficacy, permanence, costs, environmental and social impacts, and governance and social dimensions.
4. Research framework including program-wide components for experimental planning and public engagement, monitoring and verification (carbon accounting), and open publicly accessible data management.
5. Strategy and implementation for engaging and communicating with stakeholders, policy makers, and publics.
6. Research agenda that emphasizes advancing understanding of ocean fertilization, seaweed cultivation, and ocean alkalinity enhancement.

References

Aagaard, P., and H. C. Helgeson. 1982. Thermodynamic and kinetic constraints on reaction rates among minerals and aqueous solutions. I. Theoretical considerations. *American Journal of Science* 282(3):237-285. doi:10.2475/ajs.282.3.237.

Abam Engineers. 1980. *Process Engineering and Economic Evaluations of Diaphragm and Membrane Chlorine Cell Technologies*. Illinois: University of North Texas Libraries, UNT Digital Library.

Abate, R. S., and A. B. Greenlee. 2009. Sowing seeds uncertain: Ocean iron fertilization, climate change, and the international environmental law framework. *Pace Environmental Law Review* 27:555.

Abraham, E. R., C. S. Law, P. W. Boyd, S. J. Lavender, M. T. Maldonado, and A. R. Bowie. 2000. Importance of stirring in the development of an iron-fertilized phytoplankton bloom. *Nature* 407(6805):727-30. doi:10.1038/35037555.

Aguzzi, J., E. Fanelli, T. Ciuffardi, A. Schirone, F. C. De Leo, C. Doya, M. Kawato, M. Miyazaki, Y. Furushima, C. Costa, and Y. Fujiwara. 2018. Faunal activity rhythms influencing early community succession of an implanted whale carcass offshore Sagami Bay, Japan. *Scientific Reports* 8(1):11163. doi:10.1038/s41598-018-29431-5.

Agyeman, J., D. Schlosberg, L. Craven, and C. Matthews. 2016. Trends and directions in environmental justice: From inequity to everyday life, community, and just sustainabilities. *Annual Review of Environment and Resources* 41 (1):321-340. doi:10.1146/annurev-environ-110615-090052.

Alexander, S. M., J. F. Provencher, D. A. Henri, J. J. Taylor, J. I. Lloren, L. Nanayakkara, J. T. Johnson, and S. J. Cooke. 2019. Bridging Indigenous and science-based knowledge in coastal and marine research, monitoring, and management in Canada. *Environmental Evidence* 8(1). doi:10.1186/s13750-019-0181-3.

Alldredge, A. L., and M. W. Silver. 1988. Characteristics, dynamics and significance of marine snow. *Progress in Oceanography* 20(1):41-82. doi:10.1016/0079-6611(88)90053-5.

Allen, K. A., B. Hönisch, S. M. Eggins, L. L. Haynes, Y. Rosenthal, and J. Yu. 2016. Trace element proxies for surface ocean conditions: A synthesis of culture calibrations with planktic foraminifera. *Geochimica et Cosmochimica Acta* 193:197-221. doi:10.1016/j.gca.2016.08.015.

Allen, M. R., O. P. Dube, W. Solecki, F. Aragón-Durand, W. Cramer, S. Humphreys, M. Kainuma, J. Kala, N. Mahowald, and Y. Mulugetta. 2018. Framing and context. Pp. 41-91 in *Global Warming of 1.5 C: An IPCC Special Report*. Intergovernmental Panel on Climate Change. In Press.

Allsopp, M., D. Santillo, and P. Johnston. 2007. *A Scientific Critique of Oceanic Iron Fertilization as a Climate Change Mitigation Strategy*. GRL-TN-07-2007. Exeter, UK: Greenpeace Research Laboratories. https://www.whoi.edu/fileserver.do?id=27223&pt=2&p=28442.

Amelung, D., and J. Funke. 2014. Laypeople's risky decisions in the climate change context: Climate engineering as a risk-defusing strategy? *Human and Ecological Risk Assessment: An International Journal* 21 (2):533-559. doi:10.1080/10807039.2014.932203.

Anagnostou, E., R. M. Sherrell, A. Gagnon, M. LaVigne, M. P. Field, and W. F. McDonough. 2011. Seawater nutrient and carbonate ion concentrations recorded as P/Ca, Ba/Ca, and U/Ca in the deep-sea coral *Desmophyllum dianthus*. *Geochimica et Cosmochimica Acta* 75(9):2529-2543. doi:10.1016/j.gca.2011.02.019.

Andersson, A. J. 2014. The oceanic $CaCO_3$ cycle. Pp. 519-542 in *Treatise on Geochemistry*, H., D. Holland and K, K. Turekian, eds. Palisades, NY: Elsevier Science.

Andrew, R. M. 2018. Global CO_2 emissions from cement production. *Earth System Science Data* 10(1):195-217.

Angelsen, A., M. Brockhaus, M. Kanninen, E. Sills, W. D. Sunderlin, S. Wertz-Kanounnikoff, T. Dokken, and E. A. Johnson, eds. 2009. *Realising REDD+: National Strategy and Policy Options*. Bogor, Indonesia: Center for International Forestry Research.

Arafeh-Dalmau, N., K. C. Cavanaugh, H. P. Possingham, A. Munguia-Vega, G. Montaño-Moctezuma, T. W. Bell, K. Cavanaugh, and F. Micheli. 2021. Southward decrease in the protection of persistent giant kelp forests in the northeast Pacific. *Communications Earth & Environment* 2(1):1-7.

Archer, D., M. Eby, V. Brovkin, A. Ridgwell, L. Cao, U. Mikolajewicz, K. Caldeira, K. Matsumoto, G. Munhoven, A. Montenegro, and K. Tokos. 2009. Atmospheric lifetime of fossil fuel carbon dioxide. *Annual Review of Earth and Planetary Sciences* 37(1):117-134. doi:10.1146/annurev.earth.031208.100206.

Archibald, K. M., D. A. Siegel, and S. C. Doney. 2019. Modeling the impact of zooplankton diel vertical migration on the carbon export flux of the biological pump. *Global Biogeochemical Cycles* 33(2):181-199. doi:10.1029/2018GB005983.

Armstrong, R. A., C. Lee, J. I. Hedges, S. Honjo, and S. G. Wakeham. 2001. A new, mechanistic model for organic carbon fluxes in the ocean based on the quantitative association of POC with ballast minerals. *Deep-Sea Research Part II: Topical Studies in Oceanography* 49(1-3):219-236. doi:10.1016/S0967-0645(01)00101-1.

Arnstein, S. R. 1969. A ladder of citizen participation. *Journal of the American Institute of Planners* 35 (4):216-224. doi:10.1080/01944366908977225.

ARPA-e (Advanced Research Projects Agency-e). 2021a. MARINER Annual Review 2021. https://arpa-e.energy.gov/mariner-annual-review-2021.

ARPA-e. 2021b. Ocean Era (formerly Kampachi Farms)—Blue Fields. https://arpa-e.energy.gov/mariner-annual-review-2021/ocean-era-blue-fields.

Arribére, M. A., S. Ribeiro Guevara, R. S. Sánchez, M. I. Gil, G. Román Ross, L. E. Daurade, V. Fajon, M. Horvat, R. Alcalde, and A. J. Kestelman. 2003. Heavy metals in the vicinity of a chlor-alkali factory in the upper Negro River ecosystem, Northern Patagonia, Argentina. *Science of the Total Environment* 301(1-3):187-203. doi:10.1016/s0048-9697(02)00301-7.

Atwood, T. B., R. M. Connolly, E. G. Ritchie, C. E. Lovelock, M. R. Heithaus, G. C. Hays, J. W. Fourqurean, and P. I. Macreadie. 2015. Predators help protect carbon stocks in blue carbon ecosystems. *Nature Climate Change* 5(12):1038-1045. doi:10.1038/nclimate2763.

Atwood, T. B., E. M. P. Madin, A. R. Harborne, E. Hammill, O. J. Luiz, Q. R. Ollivier, C. M. Roelfsema, P. I. Macreadie, and C. E. Lovelock. 2018. Predators shape sedimentary organic carbon storage in a coral reef ecosystem. *Frontiers in Ecology and Evolution* 6:110. doi:10.3389/fevo.2018.00110.

Aumont, O., and L. Bopp. 2006. Globalizing results from ocean in situ iron fertilization studies. *Global Biogeochemical Cycles* 20(2):GB2017. doi:10.1029/2005GB002591.

Aure, J., Ø. Strand, S. R. Erga, and T. Strohmeier. 2007. Primary production enhancement by artificial upwelling in a western Norwegian fjord. *Marine Ecology Progress Series* 352:39-52. doi:10.3354/meps07139.

Avery, W. H., and C. Wu. 1994. *Renewable Energy from the Ocean: A Guide to OTEC*: New York, NY: Oxford University Press.

Azevedo, I. C., P. M. Duarte, G. S. Marinho, F. Neumann, and I. Sousa-Pinto. 2019. Growth of *Saccharina latissima* (Laminariales, Phaeophyceae) cultivated offshore under exposed conditions. *Phycologia* 58(5):504-515. doi:10.1080/00318884.2019.1625610.

Bach, L. T., S. J. Gill, R. E. M. Rickaby, S. Gore, and P. Renforth. 2019. CO_2 removal with enhanced weathering and ocean alkalinity enhancement: Potential risks and co-benefits for marine pelagic ecosystems. 1:7. doi:10.3389/fclim.2019.00007.

Bach, L. T., V. Tamsitt, J. Gower, C. L. Hurd, J. A. Raven, and P. W. Boyd. 2021. Testing the climate intervention potential of ocean afforestation using the Great Atlantic Sargassum Belt. *Nature Communications* 12(1). doi:10.1038/s41467-021-22837-2.

Baduini, C. L., K. D. Hyrenbach, K. O. Coyle, A. Pinchuk, V. Mendenhall, and G. L. Hunt Jr. 2001. Mass mortality of short-tailed shearwaters in the south-eastern Bering Sea during summer 1997. *Fisheries Oceanography* 10(1):117-130. doi:10.1046/j.1365-2419.2001.00156.x.

Bak, U. G., A. Mols-Mortensen, and O. Gregersen. 2018. Production method and cost of commercial-scale offshore cultivation of kelp in the Faroe Islands using multiple partial harvesting. *Algal Research* 33:36-47. doi:10.1016/j.algal.2018.05.001.

Bak, U. G., Ó. Gregersen, and J. Infante. 2020. Technical challenges for offshore cultivation of kelp species: Lessons learned and future directions. *Botanica Marina* 63(4):341-353. doi:10.1515/bot-2019-0005.

Bakker, D. C. E., B. Pfeil, C. S. Landa, N. Metzl, K. M. O'Brien, A. Olsen, K. Smith, C. Cosca, S. Harasawa, S. D. Jones, S. Nakaoka, Y. Nojiri, U. Schuster, T. Steinhoff, C. Sweeney, T. Takahashi, B. Tilbrook, C. Wada, R. Wanninkhof, S. R. Alin, C. F. Balestrini, L. Barbero, N. R. Bates, A. A. Bianchi, F. Bonou, J. Boutin, Y. Bozec, E. F. Burger, W. J. Cai, R. D. Castle, L. Chen, M. Chierici, K. Currie, W. Evans, C. Featherstone, R. A. Feely, A. Fransson, C. Goyet, N. Greenwood, L. Gregor, S. Hankin, N. J. Hardman-Mountford, J. Harlay, J. Hauck, M. Hoppema, M. P. Humphreys, C. W. Hunt, B. Huss, J. S. P. Ibánhez, T. Johannessen, R. Keeling, V. Kitidis, A. Körtzinger, A. Kozyr, E. Krasakopoulou, A. Kuwata, P. Landschützer, S. K. Lauvset, N. Lefèvre, C. Lo Monaco, A. Manke, J. T. Mathis, L. Merlivat, F. J. Millero, P. M. S. Monteiro, D. R. Munro, A. Murata, T. Newberger, A. M. Omar, T. Ono, K. Paterson, D. Pearce, D. Pierrot, L. L. Robbins, S. Saito, J. Salisbury, R. Schlitzer, B. Schneider, R. Schweitzer, R. Sieger, I. Skjelvan, K. F. Sullivan, S. C. Sutherland, A. J. Sutton, K. Tadokoro, M. Telszewski, M. Tuma, S. M. A. C. van Heuven, D. Vandemark, B. Ward, A. J. Watson, and S. Xu. 2016. A multi-decade record of high-quality fCO_2 data in version 3 of the Surface Ocean CO_2 Atlas (SOCAT). *Earth System Science Data* 8(2):383-413. doi:10.5194/essd-8-383-2016.

Balch, W. M., K. A. Kilpatrick, P. Holligan, D. Harbour, and E. Fernandez. 1996. The 1991 coccolithophore bloom in the central North Atlantic. 2. Relating optics to coccolith concentration. *Limnology and Oceanography* 41(8):1684-1696.

Balch, W. M., A. J. Plueddeman, B. C. Bowler, and D. T. Drapeau. 2009. Chalk-Ex—Fate of $CaCO_3$ particles in the mixed layer: Evolution of patch optical properties. *Journal of Geophysical Research Oceans*. 114(C7):C07020. doi:https://doi.org/10.1029/2008JC004902.

Ban, N. C., A. Frid, M. Reid, B. Edgar, D. Shaw, and P. Siwallace. 2018. Incorporate Indigenous perspectives for impactful research and effective management. *Nature Ecology & Evolution* 2(11):1680-1683. doi:10.1038/s41559-018-0706-0.

Banse, K. 1994. Grazing and zooplankton production as key controls of phytoplankton production in the open ocean. *Oceanography* 7(1):13-20.

Bard, A. J., and L. Faulkner. 2001. Fundamentals and applications. *Electrochemical Methods* 2(482):580-632.

Bar-On, Y. M., R. Phillips, and R. Milo. 2018. The biomass distribution on Earth. *Proceedings of the National Academy of Sciences of the United States of America* 115(25):6506-6511. doi:10.1073/pnas.1711842115.

Bass, R., A. Mudaliar, and H. Dithrich. 2018. Annual Impact Investor Survey 2018. Global Impact Investing Network.

Batel, Susana. 2020. Research on the social acceptance of renewable energy technologies: Past, present and future. *Energy Research & Social Science* 68. doi:10.1016/j.erss.2020.101544.

Bates, N. R., Y. M. Astor, M. J. Church, K. Currie, J. E. Dore, M. González-Dávila, L. Lorenzoni, F. Muller-Karger, J. Olafsson, and J. M. Santana-Casiano. 2014. A time-series view of changing surface ocean chemistry due to ocean uptake of anthropogenic CO_2 and ocean acidification. *Oceanography* 27(1):126-141. doi:10.5670/oceanog.2014.16.

Batten, S. D., and J. F. R. Gower. 2014. Did the iron fertilization near Haida Gwaii in 2012 affect the pelagic lower trophic level ecosystem? *Journal of Plankton Research* 36(4):925-932. doi:10.1093/plankt/fbu049.

Baumann, M., J. Taucher, A. J. Paul, M. Heinemann, M. Vanharanta, L. T. Bach, K. Spilling, J. Ortiz, J. Arístegui, N. Hernández-Hernández, I. Baños, and U. Riebesell. 2021. Effect of intensity and mode of artificial upwelling on particle flux and carbon export. *Frontiers in Marine Science* 8(1579). doi:10.3389/fmars.2021.742142.

Bednar, J., M. Obersteiner, and F. Wagner. 2019. On the financial viability of negative emissions. *Nature Communications* 10(1):1783. doi:10.1038/s41467-019-09782-x.

Behrenfeld, M. J., E. Boss, D. A. Siegel, and D. M. Shea. 2005. Carbon-based ocean productivity and phytoplankton physiology from space. *Global Biogeochemical Cycles* 19(1). doi:https://doi.org/10.1029/2004GB002299.

Behrenfeld, M. J., T. K. Westberry, E. S. Boss, R. T. O'Malley, D. A. Siegel, J. D. Wiggert, B. A. Franz, C. R. McClain, G. C. Feldman, S. C. Doney, J. K. Moore, G. Dall'Olmo, A. J. Milligan, I. Lima, and N. Mahowald. 2009. Satellite-detected fluorescence reveals global physiology of ocean phytoplankton. *Biogeosciences* 6(5):779-794. doi:10.5194/bg-6-779-2009.

Bell, T. W., K. C. Cavanaugh, D. C. Reed, and D. A. Siegel. 2015a. Geographical variability in the controls of giant kelp biomass dynamics. *Journal of Biogeography* 42(10):2010-2021. doi:10.1111/jbi.12550.

Bell, T. W., K. C. Cavanaugh, and D. A. Siegel. 2015b. Remote monitoring of giant kelp biomass and physiological condition: An evaluation of the potential for the Hyperspectral Infrared Imager (HyspIRI) mission. *Remote Sensing of Environment* 167:218-228. doi:10.1016/j.rse.2015.05.003.

Bell, T. W., N. J. Nidzieko, D. A. Siegel, R. J. Miller, K. C. Cavanaugh, N. B. Nelson, D. C. Reed, D. Fedorov, C. Moran, J. N. Snyder, K. C. Cavanaugh, C. E. Yorke, and M. Griffith. 2020. The utility of satellites and autonomous remote sensing platforms for monitoring offshore aquaculture farms: A case study for canopy forming kelps. *Frontiers in Marine Science* 7:1083. doi:10.3389/fmars.2020.520223.

Bellamy, R. 2018. Incentivize negative emissions responsibly. *Nature Energy* 3(7):532-534. doi:10.1038/s41560-018-0156-6.

Bellamy, R., J. Lezaun, and J. Palmer. 2017. Public perceptions of geoengineering research governance: An experimental deliberative approach. *Global Environmental Change* 45:194-202. doi:10.1016/j.gloenvcha.2017.06.004.

Bellamy, R., and S. Osaka. 2020. Unnatural climate solutions? *Nature Climate Change* 10(2):98-99. doi:10.1038/s41558-019-0661-z.

Bennett, J. 1980. Electrodes for generation of hydrogen and oxygen from seawater. *International Journal of Hydrogen Energy* 5(4):401-408. doi:10.1016/0360-3199(80)90021-x.

Bergstrom, E., J. Silva, C. Martins, and P. Horta. 2019. Seagrass can mitigate negative ocean acidification effects on calcifying algae. *Scientific Reports* 9 (1):1932. doi:10.1038/s41598-018-35670-3.

Bernardino, A. F., C. R. Smith, A. Baco, I. Altamira, and P. Y. G. Sumida. 2010. Macrofaunal succession in sediments around kelp and wood falls in the deep NE Pacific and community overlap with other reducing habitats. *Deep-Sea Research Part I: Oceanographic Research Papers* 57(5):708-723. doi:10.1016/j.dsr.2010.03.004.

Bertram, C., and C. Merk. 2020. Public perceptions of ocean-based carbon dioxide removal: The nature-engineering divide? *Frontiers in Climate* 2:31. doi:10.3389/fclim.2020.594194.

Bianchi, D., D. A. Carozza, E. D. Galbraith, J. Guiet, and T. DeVries. 2021. Estimating global biomass and biogeochemical cycling of marine fish with and without fishing. *Science Advances* 7(41). doi:10.1126/sciadv.abd7554.

Bidwell, D. 2017. Ocean beliefs and support for an offshore wind energy project. *Ocean & Coastal Management* 146:99-108. doi:10.1016/j.ocecoaman.2017.06.012.

Biello, D. 2012. Can controversial ocean iron fertilization save salmon? *Scientific American,* October 24.

Billing, S.-L., J. Rostan, P. Tett, and A. Macleod. 2021. Is social license to operate relevant for seaweed cultivation in Europe? *Aquaculture* 534:736203. doi:10.1016/j.aquaculture.2020.736203.

Bischoff, W. D., F. T. Mackenzie, and F. C. Bishop. 1987. Stabilities of synthetic magnesian calcites in aqueous solution: Comparison with biogenic materials. *Geochimica et Cosmochimica Acta* 51(6):1413-1423. doi:10.1016/0016-7037(87)90325-5.

Black, E. E., S. S. Kienast, N. Lemaitre, P. J. Lam, R. F. Anderson, H. Planquette, F. Planchon, and K. O. Buesseler. 2020. Ironing out Fe residence time in the dynamic upper ocean. *Global Biogeochemical Cycles* 34(9):e2020GB006592. doi:10.1029/2020gb006592.

Blain, S., B. Quéguiner, and T. Trull. 2008. The natural iron fertilization experiment KEOPS (KErguelen Ocean and Plateau compared Study): An overview. *Deep Sea Research Part II: Topical Studies in Oceanography* 55(5-7):559-565. doi:10.1016/j.dsr2.2008.01.002.

Blewett, T. A., and C. M. Wood. 2015a. Low salinity enhances Ni-mediated oxidative stress and sub-lethal toxicity to the green shore crab (*Carcinus maenas*). *Ecotoxicology and Environmental Safety* 122:159-170.

Blewett, T. A., and C. M. Wood. 2015b. Salinity-dependent nickel accumulation and oxidative stress responses in the euryhaline killifish (*Fundulus heteroclitus*). *Archives of Environmental Contamination and Toxicology* 68(2):382-394. doi:10.1007/s00244-014-0115-6.

Blewett, T. A., C. N. Glover, S. Fehsenfeld, M. J. Lawrence, S. Niyogi, G. G. Goss, and C. M. Wood. 2015. Making sense of nickel accumulation and sub-lethal toxic effects in saline waters: Fate and effects of nickel in the green crab. *Carcinus maenas. Aquatic Toxicology* 164:23-33. doi:10.1016/j.aquatox.2015.04.010.

Böckmann, S., F. Koch, B. Meyer, F. Pausch, M. Iversen, R. Driscoll, L. M. Laglera, C. Hassler, and S. Trimborn. 2021. Salp fecal pellets release more bioavailable iron to Southern Ocean phytoplankton than krill fecal pellets. *Current Biology* 31(13):2737-2746 e3. doi:10.1016/j.cub.2021.02.033.

Bodansky, D. 2011. Governing climate engineering: Scenarios for analysis. *SSRN Electronic Journal*. doi:10.2139/ssrn.1963397.

Bonan, G. B., and S. C. Doney. 2018. Climate, ecosystems, and planetary futures: The challenge to predict life in Earth system models. *Science* 359 (6375):eaam8328. doi:10.1126/science.aam8328.

Bonino, G., E. Di Lorenzo, S. Masina, and D. Iovino. 2019. Interannual to decadal variability within and across the major eastern boundary upwelling systems. *Scientific Reports* 9(1):19949. doi:10.1038/s41598-019-56514-8.

Boscolo-Galazzo, F., K. A. Crichton, A. Ridgwell, E. M. Mawbey, B. S. Wade, and P. N. Pearson. 2021. Temperature controls carbon cycling and biological evolution in the ocean twilight zone. *Science* 371(6534):1148-1152. doi:10.1126/science.abb6643.

Boumans, R., J. Roman, I. Altman, and L. Kaufman. 2015. The Multiscale Integrated Model of Ecosystem Services (MIMES): Simulating the interactions of coupled human and natural systems. *Ecosystem Services* 12:30-41. doi:10.1016/j.ecoser.2015.01.004.

Boyce, D. G., M. R. Lewis, and B. Worm. 2010. Global phytoplankton decline over the past century. *Nature* 466(7306):591-596. doi:10.1038/nature09268.

Boyd, P. 2008. Implications of large-scale iron fertilization of the oceans. *Marine Ecology Progress Series* 364:213-218. doi:10.3354/meps07541.

Boyd, P. W., H. Claustre, M. Levy, D. A. Siegel, and T. Weber. 2019. Multi-faceted particle pumps drive carbon sequestration in the ocean. *Nature* 568(7752):327-335. doi:10.1038/s41586-019-1098-2.

Boyd, P. W., A. J. Watson, C. S. Law, E. R. Abraham, T. Trull, R. Murdoch, D. C. E. Bakker, A. R. Bowie, K. O. Buesseler, H. Chang, M. Charette, P. Croot, K. Downing, R. Frew, M. Gall, M. Hadfield, J. Hall, M. Harvey, G. Jameson, J. LaRoche, M. Liddicoat, R. Ling, M. T. Maldonado, R. M. McKay, S. Nodder, S. Pickmere, R. Pridmore, S. Rintoul, K. Safi, P. Sutton, R. Strzepek, K. Tanneberger, S. Turner, A. Waite, and J. Zeldis. 2000. A mesoscale phytoplankton bloom in the polar Southern Ocean stimulated by iron fertilization. *Nature* 407(6805):695-702. doi:10.1038/35037500.

Boyd, P. W., and S. C. Doney. 2003. The impact of climate change and feedback processes on the ocean carbon cycle. Pp. 157-193 in *Ocean Biogeochemistry*. Springer, Berlin, Heidelberg.

Boyd, P. W., T. Jickells, C. S. Law, S. Blain, E. A. Boyle, K. O. Buesseler, K. H. Coale, J. J. Cullen, H. J. W. de Baar, M. Follows, M. Harvey, C. Lancelot, M. Levasseur, N. P. J. Owens, R. Pollard, R. B. Rivkin, J. Sarmiento, V. Schoemann, V. Smetacek, S. Takeda, A. Tsuda, S. Turner, and A. J. Watson. 2007. Mesoscale iron enrichment experiments 1993-2005: Synthesis and future directions. *Science* 315(5812):612-617. doi:10.1126/science.1131669.

Boyle, A. 2012. Law of the sea perspectives on climate change. *International Journal of Marine and Coastal Law* 27(4):831-838. doi:10.1163/15718085-12341244.

Boyle, E. A. 1988. The role of vertical chemical fractionation in controlling late Quaternary atmospheric carbon dioxide. *Journal of Geophysical Research* 93(C12):15701-15714. doi:10.1029/JC093iC12p15701.

Brantley, S. L. 2008. Kinetics of mineral dissolution. Pp. 151-210 in *Kinetics of Water-Rock Interaction*, S. Brantley, J. Kubicki, and A. White, eds. New York: Springer.

Breitburg, D., L. A. Levin, A. Oschlies, M. Gregoire, F. P. Chavez, D. J. Conley, V. Garcon, D. Gilbert, D. Gutierrez, K. Isensee, G. S. Jacinto, K. E. Limburg, I. Montes, S. W. A. Naqvi, G. C. Pitcher, N. N. Rabalais, M. R. Roman, K. A. Rose, B. A. Seibel, M. Telszewski, M. Yasuhara, and J. Zhang. 2018. Declining oxygen in the global ocean and coastal waters. *Science* 359(6371):eeam7240. doi:10.1126/science.aam7240.

Brent, K., J. McGee, J. McDonald, and E. J. Rohling. 2018. International law poses problems for negative emissions research. *Nature Climate Change* 8(6):451-453. doi:10.1038/s41558-018-0181-2.

Brent, K., W. Burns, and J. McGee. 2019. *Governance of Marine Geoengineering*. Waterloo, ON: Centre for International Governance Innovation.

Bressac, M., C. Guieu, D. Doxaran, F. Bourrin, G. Obolensky, and J. M. Grisoni. 2012. A mesocosm experiment coupled with optical measurements to assess the fate and sinking of atmospheric particles in clear oligotrophic waters. *Geo-Marine Letters* 32(2):153-164. doi:10.1007/s00367-011-0269-4.

Bressac, M., C. Guieu, M. J. Ellwood, A. Tagliabue, T. Wagener, E. C. Laurenceau-Cornec, H. Whitby, G. Sarthou, and P. W. Boyd. 2019. Resupply of mesopelagic dissolved iron controlled by particulate iron composition. *Nature Geoscience* 12(12):995-1000. doi:10.1038/s41561-019-0476-6.

Brinkmann, T., G. G. Santonja, F. Schorcht, S. Roudier, and L. D. Sancho. 2014. Best Available Techniques (BAT) Reference Document for the Production of Chlor-Alkali: Industrial Emissions Directive 2010/75/EU. Luxembourg: European Union.

Britten, G. L., M. Dowd, and B. Worm. 2016. Changing recruitment capacity in global fish stocks. *Proceedings of the National Academy of Sciences of the United States of America* 113(1):134-139. doi:10.1073/pnas.1504709112.

Broecker, W. S. 1991. Keeping global change honest. *Global Biogeochemical Cycles* 5(3):191-192. doi:10.1029/91gb01421.

Broecker, W. S., and T. H. Peng. 1982. *Tracers in the Sea*. Palisades, NY: Eldigio Press.

Browning, T. J., H. A. Bouman, G. M. Henderson, T. A. Mather, D. M. Pyle, C. Schlosser, E. M. S. Woodward, and C. M. Moore. 2014. Strong responses of Southern Ocean phytoplankton communities to volcanic ash. *Geophysical Research Letters* 41(8):2851-2857. doi:10.1002/2014gl059364.

Bruckner, T., I. A. Bashmakov, Y. Mulugetta, H. Chum, A. de la Vega Navarro, J. Edmonds, A. Faaij, B. Fungtammasan, A. Garg, E. Hertwich, D. Honnery, D. Infield, M. Kainuma, S. Khennas, S. Kim, H. B. Nimir, K. Riahi, N. Strachan, R. Wiser, and X. Zhang. 2014. Energy systems. In *Climate Change 2014: Mitigation of Climate Change. Contribution of Working Group III to the Fifth Assessment Report of the Intergovernmental Panel on Climate Change,* O. Edenhofer, R. Pichs-Madruga, Y. Sokona, E. Farahani, S. Kadner, K. Seyboth, A. Adler, I. Baum, S. Brunner, P. Eickemeier, B. Kriemann, J. Savolainen, S. Schlömer, C. von Stechow, T. Zwickel, and J. C. Minx, eds. Cambridge, UK and New York: Cambridge University Press.

Brzezinski, M. A. 2004. The Si:C:N ratio of marine diatoms: Interspecific variability and the effect of some environmental variables. *Journal of Phycology* 21(3):347-357. doi:10.1111/j.0022-3646.1985.00347.x.

Buck, H. J. 2018. The politics of negative emissions technologies and decarbonization in rural communities. *Global Sustainability* 1. doi:10.1017/sus.2018.2.

Budinis, S., S. Krevor, N. M. Dowell, N. Brandon, and A. Hawkes. 2018. An assessment of CCS costs, barriers and potential. *Energy Strategy Reviews* 22:61-81. doi:10.1016/j.esr.2018.08.003.

Buesseler, K. O., and P. W. Boyd. 2003. Will ocean fertilization work? *Science* 300(5616):67-68.

Buesseler, K. O., C. H. Lamborg, P. W. Boyd, P. J. Lam, T. W. Trull, R. R. Bidigare, J. K. B. Bishop, K. L. Casciotti, F. Dehairs, M. Elskens, M. Honda, D. M. Karl, D. A. Siegel, M. W. Silver, D. K. Steinberg, J. Valdes, B. Van Mooy, and S. Wilson. 2007. Revisiting carbon flux through the ocean's twilight zone. *Science* 316(5824):567-570. doi:10.1126/science.1137959.

Buesseler, K. O., P. W. Boyd, E. E. Black, and D. A. Siegel. 2020. Metrics that matter for assessing the ocean biological carbon pump. *Proceedings of the National Academy of Sciences of the United States of America* 117(18):9679-9687. doi:10.1073/pnas.1918114117.

Buesseler, K. O., S. C. Doney, D. M. Karl, P. W. Boyd, K. Caldeira, F. Chai, K. H. Coale, H. J. de Baar, P. G. Falkowski, K. S. Johnson, R. S. Lampitt, A. F. Michaels, S. W. Naqvi, V. Smetacek, S. Takeda, and A. J. Watson. 2008. Environment. Ocean iron fertilization—moving forward in a sea of uncertainty. *Science* 319(5860):162. doi:10.1126/science.1154305.

Burns, E., and V. Suarez. 2020. Everything you need to know about federal funding for carbon removal: Big money and big CDR wins. *Carbon180.* August 31. https://carbon180.medium.com/everything-you-need-to-know-about-federal-funding-for-carbon-removal-bb2548595b41.

Burton, J. D., M. Althaus, G. E. Millward, A. W. Morris, P. J. Statham, A. D. Tappin, and A. Turner. 1994. Processes influencing the fate of trace metals in the North Sea. Pp. 179-190 in *Understanding the North Sea System.* Dordrecht, Netherlands: Springer.

Busch, D. S., M. O'Donnell, C. Hauri, K. Mach, M. Poach, S. Doney, and S. Signorini. 2015. Understanding, characterizing, and communicating responses to ocean acidification: Challenges and uncertainties. *Oceanography* 25(2):30-39. doi:10.5670/oceanog.2015.29.

Bushinsky, S. M., P. Landschützer, C. Rödenbeck, A. R. Gray, D. Baker, M. R. Mazloff, L. Resplandy, K. S. Johnson, and J. L. Sarmiento. 2019. Reassessing Southern Ocean air-sea CO_2 flux estimates with the addition of biogeochemical float observations. *Global Biogeochemical Cycles* 33(11):1370-1388. doi:https://doi.org/10.1029/2019GB006176.

Butenschön, M., T. Lovato, S. Masina, S. Caserini, and M. Grosso. 2021. Alkalinization scenarios in the Mediterranean Sea for efficient removal of atmospheric CO_2 and the mitigation of ocean acidification. *Frontiers in Climate* 3:14. doi:10.3389/fclim.2021.614537.

Caldeira, K., and G. H. Rau. 2000. Accelerating carbonate dissolution to sequester carbon dioxide in the ocean: Geochemical implications. *Geophysical Research Letters* 27(2):225-228. doi:10.1029/1999GL002364.

Caldeira, K., M. Akai, P. Brewer, B. Chen, P. Haugan, T. Iwama, P. Johnston, H. Kheshgi, Q. Li, T. Ohsumi, H. Pörtner, C. Sabine, Y. Shirayama, J. Thomson, J. Barry, and L. Hansen. 2018. Ocean storage. In *IPCC Special Report on Carbon dioxide Capture and Storage*, B. De Young and F. Joos, eds. New York, NY: IPCC.

Callaway, E. 2015. Lab staple agar hit by seaweed shortage. *Nature* 528(7581):171-172. doi:10.1038/528171a.

Campbell-Arvai, V., P. S. Hart, K. T. Raimi, and K. S. Wolske. 2017. The influence of learning about carbon dioxide removal (CDR) on support for mitigation policies. *Climatic Change* 143(3-4):321-336. doi:10.1007/s10584-017-2005-1.

Camus, C., and A. H. Buschmann. 2017. *Macrocystis pyrifera* aquafarming: Production optimization of rope-seeded juvenile sporophytes. *Aquaculture* 468:107-114. doi:10.1016/j.aquaculture.2016.10.010.

Camus, C., J. Infante, and A. H. Buschmann. 2018. Overview of 3-year precommercial seafarming of *Macrocystis pyrifera* along the Chilean coast. *Reviews in Aquaculture* 10(3):543-559. doi:10.1111/raq.12185.

Canadell, J. G., P. M. S. Monteiro, M. H. Costa, L. Cotrim da Cunha, P. M. Cox, A. V. Eliseev, S. Henson, M. Ishii, S. Jaccard, C. Koven, A. Lohila, P. K. Patra, S. Piao, J. Rogelj, S. Syampungani, S. Zaehle, and K. Zickfeld. 2021. Global carbon and other biogeochemical cycles and feedbacks. In *Climate Change 2021: The Physical Science Basis. Contribution of Working Group I to the Sixth Assessment Report of the Intergovernmental Panel on Climate Change,* V. Masson-Delmotte, P. Zhai, A. Pirani, S. L. Connors, C. Péan, S. Berger, N. Caud, Y. Chen, L. Goldfarb, M. I. Gomis, M. Huang, K. Leitzell, E. Lonnoy, J. B. R. Matthews, T. K. Maycock, T. Waterfield, O. Yelekçi, R. Yu, and B. Zhou, eds.. Cambridge University Press.

Candeias, C., P. Ávila, P. Coelho, and J. P. Teixeira. 2019. Mining activities: Health impacts. Pp. 415-435 in *Encyclopedia of Environmental Health*, 2nd ed., J. O. Nriagu, ed. Cambridge, MA: Elsevier.

Cao, L., and K. Caldeira. 2010. Can ocean iron fertilization mitigate ocean acidification? *Climatic Change* 99(1):303-311.

Cao, Z., R. J. Myers, R. C. Lupton, H. Duan, R. Sacchi, N. Zhou, T. Reed Miller, J. M. Cullen, Q. Ge, and G. Liu. 2020. The sponge effect and carbon emission mitigation potentials of the global cement cycle. *Nature Communications* 11(1). doi:10.1038/s41467-020-17583-w.

Carter, B. R., R. A. Feely, S. K. Lauvset, A. Olsen, T. DeVries, and R. Sonnerup. 2021. Preformed properties for marine organic matter and carbonate mineral cycling quantification. *Global Biogeochemical Cycles* 35(1):e2020GB006623. doi:10.1029/2020GB006623.

Carton, W., J. F. Lund, and K. Dooley. 2021. Undoing equivalence: Rethinking carbon accounting for just carbon removal. *Frontiers in Climate* 3. doi:10.3389/fclim.2021.664130.

Caserini, S., D. Pagano, F. Campo, A. Abbà, S. De Marco, D. Righi, P. Renforth, and M. Grosso. 2021. Potential of maritime transport for ocean liming and atmospheric CO_2 removal. *Frontiers in Climate* 3:22. doi:10.3389/fclim.2021.575900.

Cattau, M. E., C. Wessman, A. Mahood, J. K. Balch, and B. Poulter. 2020. Anthropogenic and lightning-started fires are becoming larger and more frequent over a longer season length in the U.S.A. *Global Ecology and Biogeography* 29(4):668-681. doi:10.1111/geb.13058.

Cavan, E. L., E. C. Laurenceau-Cornec, M. Bressac, and P. W. Boyd. 2019. Exploring the ecology of the mesopelagic biological pump. *Progress in Oceanography* 176:102125. doi:10.1016/j.pocean.2019.102125.

Cavanaugh, K. C., K. C. Cavanaugh, T. W. Bell, and E. G. Hockridge. 2021. An automated method for mapping giant kelp canopy dynamics from UAV. *Frontiers in Environmental Science* 8:301. doi:10.3389/fenvs.2020.587354.

Charles, M. H., inventor. 1889. Process of reducing aluminium from its fluoride salts by electrolysis. U.S. Patent 400,664A.

Chen, G. 2004. Electrochemical technologies in wastewater treatment. *Separation and Purification Technology* 38(1):11-41. doi:10.1016/j.seppur.2003.10.006.

Chen, S., E. F. M. Littley, J. W. B. Rae, C. D. Charles, and J. F. Adkins. 2021. Uranium distribution and incorporation mechanism in deep-sea corals: Implications for seawater CO_3^{2-} proxies. *Frontiers in Earth Science* 9:159. doi:10.3389/feart.2021.641327.

Cheung, W. W. L., G. Reygondeau, and T. L. Frölicher. 2016. Large benefits to marine fisheries of meeting the 1.5°C global warming target. *Science* 354(6319):1591-1594. doi:10.1126/science.aag2331.

Chhetri, N., D. Chong, K. Conca, R. Falk, A. Gillespie, A. Gupta, S. Jinnah, P. Kashwan, M. Lahsen, and A. Light. 2018. *Governing Solar Radiation Management.* Report from the Academic Working Group on Climate Engineering Governance, Forum for Climate Engineering Assessment. Washington, DC: American University. https://dra.american.edu/islandora/object/auislandora%3A78266.

Childs, J. 2019. Greening the blue? Corporate strategies for legitimising deep sea mining. *Political Geography* 74. doi:10.1016/j.polgeo.2019.102060.

Childs, J. 2020. Performing 'blue degrowth': critiquing seabed mining in Papua New Guinea through creative practice. *Sustainability Science* 15(1):117-129. doi:10.1007/s11625-019-00752-2.

Chilvers, J., H. Pallett, and T. Hargreaves. 2018. Ecologies of participation in socio-technical change: The case of energy system transitions. *Energy Research & Social Science* 42:199-210. doi:10.1016/j.erss.2018.03.020.

Chisholm, S. W., P. G. Falkowski, and J. J. Cullen. 2001. Oceans: Discrediting ocean fertilization. *Science* 294(5541):309-310. doi:10.1126/science.1065349.

Chlistunoff, J. 2005. *Advanced Chlor-Alkali Technology.* Los Alamos, NM: Los Alamos National Laboratory.

Christensen, V., M. Coll, C. Piroddi, J. Steenbeek, J. Buszowski, and D. Pauly. 2014. A century of fish biomass decline in the ocean. *Marine Ecology Progress Series* 512:155-166. doi:10.3354/meps10946.

Chukwuma, J. S., H. Pullin, and P. Renforth. 2021. Assessing the carbon capture capacity of South Wales' legacy iron and steel slag. *Minerals Engineering* 173:1107232. doi:10.1016/j.mineng.2021.107232.

Chung, A. E., L. M. Wedding, A. L. Green, A. M. Friedlander, G. Goldberg, A. Meadows, and M. A. Hixon. 2019. Building coral reef resilience through spatial herbivore management. *Frontiers in Marine Science* 6:98. doi:10.3389/fmars.2019.00098.

Chung, I. K., J. Beardall, S. Mehta, D. Sahoo, and S. Stojkovic. 2010. Using marine macroalgae for carbon sequestration: A critical appraisal. *Journal of Applied Phycology* 23(5):877-886. doi:10.1007/s10811-010-9604-9.

Ciais, P., C. Sabine, G. Bala, L. Bopp, V. Brovkin, J. Canadell, A. Chhabra, R. DeFries, J. Galloway, and M. Heimann. 2013. Carbon and other biogeochemical cycles. Pp. 465-570 in *Climate Change 2013: The Physical Science Basis. Contribution of Working Group I to the Fifth Assessment Report of the Intergovernmental Panel on Climate Change.* Cambridge University Press. In Press.

Cisneros-Montemayor, A. M., M. Moreno-Báez, M. Voyer, E. H. Allison, W. W. L. Cheung, M. Hessing-Lewis, M. A. Oyinlola, G. G. Singh, W. Swartz, and Y. Ota. 2019. Social equity and benefits as the nexus of a transformative Blue Economy: A sectoral review of implications. *Marine Policy* 109. doi:10.1016/j.marpol.2019.103702.

Claudet, J., L. Bopp, W. W. L. Cheung, R. Devillers, E. Escobar-Briones, P. Haugan, J. J. Heymans, V. Masson-Delmotte, N. Matz-Lück, P. Miloslavich, L. Mullineaux, M. Visbeck, R. Watson, A. M. Zivian, I. Ansorge, M. Araujo, S. Aricò, D. Bailly, J. Barbière, C. Barnerias, C. Bowler, V. Brun, A. Cazenave, C. Diver, A. Euzen, A. T. Gaye, N. Hilmi, F. Ménard, C. Moulin, N. P. Muñoz, R. Parmentier, A. Pebayle, H.-O. Pörtner, S. Osvaldina, P. Ricard, R. Serrão Santos, M.-A. Sicre, S. Thiébault, T. Thiele, R. Troublé, A. Turra, J. Uku, and F. Gaill. 2020. A roadmap for using the UN decade of pcean science for sustainable development in support of science, policy, and action. *One Earth* 2 (1):34-42. doi:10.1016/j.oneear.2019.10.012.

Closek, C. J., J. A. Santora, H. A. Starks, I. D. Schroeder, E. A. Andruszkiewicz, K. M. Sakuma, S. J. Bograd, E. L. Hazen, J. C. Field, and A. B. Boehm. 2019. Marine vertebrate biodiversity and distribution within the Central California Current using environmental DNA (eDNA) metabarcoding and ecosystem surveys. *Frontiers in Marine Science* 6:00732. doi:10.3389/fmars.2019.00732.

Corner, A., N. Pidgeon, and K. Parkhill. 2012. Perceptions of geoengineering: public attitudes, stakeholder perspectives, and the challenge of 'upstream' engagement. *Wiley Interdisciplinary Reviews: Climate Change* 3(5):451-466. doi:10.1002/wcc.176.

Cox, E. M., N. Pidgeon, E. Spence, and G. Thomas. 2018. Blurred lines: The ethics and policy of greenhouse gas removal at scale. *Frontiers in Environmental Science* 6. doi:10.3389/fenvs.2018.00038.

Cox, E., and N. R. Edwards. 2019. Beyond carbon pricing: policy levers for negative emissions technologies. *Climate Policy* 19(9):1144-1156. doi:10.1080/14693062.2019.1634509.

Cox, E., E. Spence, and N. Pidgeon. 2020. Public perceptions of carbon dioxide removal in the United States and the United Kingdom. *Nature Climate Change* 10(8):744-749. doi:10.1038/s41558-020-0823-z.

Cox, E., M. Boettcher, E. Spence, and R. Bellamy. 2021. Casting a wider net on ocean NETs. *Frontiers in Climate* 3. doi:10.3389/fclim.2021.576294.

Craik, A. N., and W. C. G. Burns. 2016. Climate engineering under the Paris Agreement: A legal and policy primer: Special Report. In *Center for International Governance and Innovation White Paper*. Waterloo: Centre for International Governance Innovation.

Crippa, M., E. Solazzo, D. Guizzardi, F. Monforti-Ferrario, F. N. Tubiello, and A. Leip. 2021. Food systems are responsible for a third of global anthropogenic GHG emissions. *Nature Food* 2(3):198-209. doi:10.1038/s43016-021-00225-9.

Csanady, G. T. 1986. Mass transfer to and from small particles in the sea. *Limnology and Oceanography* 31(2):237-248. doi:10.4319/lo.1986.31.2.0237.

Dalsøren, S. B., B. H. Samset, G. Myhre, J. J. Corbett, R. Minjares, D. Lack, and J. S. Fuglestvedt. 2013. Environmental impacts of shipping in 2030 with a particular focus on the Arctic region. *Atmospheric Chemistry and Physics* 13(4):1941-1955. doi:10.5194/acp-13-1941-2013.

Daly, A., and P. Zannetti. 2007. Air pollution modeling—An overview. Pp. 15-28 in *Ambient Air Pollution*, P. Zannetti, D. Al-Ajmi, and S. Al-Rashied, eds. Arab School for Science and Technology and EnviroComp Institute.

Danovaro, R., E. Fanelli, J. Aguzzi, D. Billett, L. Carugati, C. Corinaldesi, A. Dell'Anno, K. Gjerde, A. J. Jamieson, S. Kark, C. McClain, L. Levin, N. Levin, E. Ramirez-Llodra, H. Ruhl, C. R. Smith, P. V. R. Snelgrove, L. Thomsen, C. L. Van Dover, and M. Yasuhara. 2020. Ecological variables for developing a global deep-ocean monitoring and conservation strategy. *Nature Ecoogy & Evolution* 4(2):181-192. doi:10.1038/s41559-019-1091-z.

Dasgupta, P. 2021. *The Economics of Biodiversity: The Dasgupta Review*. London: HM Treasury.

Datta, S., M. P. Henry, Y. P. J. Lin, A. T. Fracaro, C. S. Millard, S. W. Snyder, R. L. Stiles, J. Shah, J. Yuan, L. Wesoloski, R. W. Dorner, and W. M. Carlson. 2013. Electrochemical CO_2 capture using resin-wafer electrodeionization. *Industrial & Engineering Chemistry Research* 52(43):15177-15186. doi:10.1021/ie402538d.

Davies, P. A. 2015. Solar thermal decomposition of desalination reject brine for carbon dioxide removal and neutralisation of ocean acidity. *Environmental Science: Water Research and Technology* 1(2):131-137. doi:10.1039/c4ew00058g.

Davies, P. A., Q. Yuan, and R. de Richter 2018. Desalination as a negative emissions technology. *Environmental Science: Water Research & Technology* 4(6):839-850. doi.org/10.1039/C7EW00502D.

de Baar, H. J. W., P. W. Boyd, K. H. Coale, M. R. Landry, A. Tsuda, P. Assmy, D. C. E. Bakker, Y. Bozec, R. T. Barber, M. A. Brzezinski, K. O. Buesseler, M. Boyé, P. L. Croot, F. Gervais, M. Y. Gorbunov, P. J. Harrison, W. T. Hiscock, P. Laan, C. Lancelot, C. S. Law, M. Levasseur, A. Marchetti, F. J. Millero, J. Nishioka, Y. Nojiri, T. van Oijen, U. Riebesell, M. J. A. Rijkenberg, H. Saito, S. Takeda, K. R. Timmermans, M. J. W. Veldhuis, A. M. Waite, and C. S. Wong. 2005. Synthesis of iron fertilization experiments: From the Iron Age in the Age of Enlightenment. *Journal of Geophysical Research C: Oceans* 110(9):1-24. doi:10.1029/2004JC002601.

de Baar, H. J. W., L. J. A. Gerringa, P. Laan, and K. R. Timmermans. 2008. Efficiency of carbon removal per added iron in ocean iron fertilization. *Marine Ecology Progress Series* 364:269-282. doi:10.3354/meps07548.

de Figueiredo, M. A. 2003. *The Hawaii Carbon Dioxide Ocean Sequestration Field Experiment: A Case Study in Public Perceptions and Institutional Effectiveness.* Graduate Thesis, Massachusetts Institute of Technology.

de Lannoy, C. F., D. Jassby, D. D. Davis, and M. R. Wiesner. 2012. A highly electrically conductive polymer–multiwalled carbon nanotube nanocomposite membrane. *Journal of Membrane Science* 415-416:718-724. doi: https://doi.org/10.1016/j.memsci.2012.05.061.

de Lannoy, C.-F., M. D. Eisaman, A. Jose, S. D. Karnitz, R. W. DeVaul, K. Hannun, and J. L. B. Rivest. 2018. Indirect ocean capture of atmospheric CO_2: Part I. Prototype of a negative emissions technology. *International Journal of Greenhouse Gas Control* 70:243-253. doi:10.1016/j.ijggc.2017.10.007.

Deer, W. A., R. A. Howie, and J. Zussmann. 2013. *An Introduction to the Rock-Forming Minerals,* 3rd ed. London: Mineralogical Society.

Devine-Wright, P., and B. Wiersma. 2020. Understanding community acceptance of a potential offshore wind energy project in different locations: An island-based analysis of 'place-technology fit'. *Energy Policy* 137. doi:10.1016/j.enpol.2019.111086.

Devol, A. H., and H. E. Hartnett. 2001. Role of the oxygen-deficient zone in transfer of organic carbon to the deep ocean. *Limnology and Oceanography* 46(7):1684-1690. doi:10.4319/lo.2001.46.7.1684.

DeVries, T., and T. Weber. 2017. The export and fate of organic matter in the ocean: New constraints from combining satellite and oceanographic tracer observations. *Global Biogeochemical Cycles* 31(3):535-555. doi:10.1002/2016GB005551.

Dickson, A. G. 1981. An exact definition of total alkalinity and a procedure for the estimation of alkalinity and total inorganic carbon from titration data. *Deep Sea Research Part A: Oceanographic Research Papers* 28(6):609-623. doi:10.1016/0198-0149(81)90121-7.

DOE (U.S. Department of Energy). 2020. *Hydrogen Strategy: Enabling a Low-Carbon Economy.* Washington, DC: DOE.

Doney, S. C., D. S. Busch, S. R. Cooley, and K. J. Kroeker. 2020. The impacts of ocean acidification on marine ecosystems and reliant human communities. *Annual Review of Environment and Resources* 45(1):83-112. doi:10.1146/annurev-environ-012320-083019.

Dooley, K., C. Holz, S. Kartha, S. Klinsky, J. T. Roberts, H. Shue, H. Winkler, T. Athanasiou, S. Caney, E. Cripps, N. K. Dubash, G. Hall, P. G. Harris, B. Lahn, D. Moellendorf, B. Müller, A. Sagar, and P. Singer. 2021. Ethical choices behind quantifications of fair contributions under the Paris Agreement. *Nature Climate Change* 11 (4):300-305. doi:10.1038/s41558-021-01015-8.

Drazen, J. C., C. R. Smith, K. M. Gjerde, S. H. D. Haddock, G. S. Carter, C. A. Choy, M. R. Clark, P. Dutrieux, E. Goetze, and C. Hauton. 2020. Opinion: Midwater ecosystems must be considered when evaluating environmental risks of deep-sea mining. *Proceedings of the National Academy of Sciences of the United States of America* 117(30):17455-17460.

Duarte, C. M., I. J. Losada, I. E. Hendriks, I. Mazarrasa, and N. Marbà. 2013. The role of coastal plant communities for climate change mitigation and adaptation. *Nature Climate Change* 3(11):961-968. doi:10.1038/nclimate1970.

Duarte, C. M., S. Agusti, E. Barbier, G. L. Britten, J. C. Castilla, J. P. Gattuso, R. W. Fulweiler, T. P. Hughes, N. Knowlton, C. E. Lovelock, H. K. Lotze, M. Predragovic, E. Poloczanska, C. Roberts, and B. Worm. 2020. Rebuilding marine life. *Nature* 580(7801):39-51. doi:10.1038/s41586-020-2146-7.

Ducklow, H. W., D. K. Steinberg, and K. O. Buesseler. 2001. Upper ocean carbon export and the biological pump. *Oceanography* 14(4):50-58. doi:10.5670/oceanog.2001.06.

Duffy, J. E., L. Benedetti-Cecchi, J. Trinanes, F. E. Muller-Karger, R. Ambo-Rappe, C. Boström, A. H. Buschmann, J. Byrnes, R. G. Coles, J. Creed, L. C. Cullen-Unsworth, G. Diaz-Pulido, C. M. Duarte, G. J. Edgar, M. Fortes, G. Goni, C. Hu, X. Huang, C. L. Hurd, C. Johnson, B. Konar, D. Krause-Jensen, K. Krumhansl, P. Macreadie, H. Marsh, L. J. McKenzie, N. Mieszkowska, P. Miloslavich, E. Montes, M. Nakaoka, K. M. Norderhaug, L. M. Norlund, R. J. Orth, A. Prathep, N. F. Putman, J. Samper-Villarreal, E. A. Serrao, F. Short, I. S. Pinto, P. Steinberg, R. Stuart-Smith, R. K. F. Unsworth, M. van Keulen, B. I. van Tussenbroek, M. Wang, M. Waycott, L. V. Weatherdon, T. Wernberg, and S. M. Yaakub. 2019. Toward a coordinated global observing system for seagrasses and marine macroalgae. *Frontiers in Marine Science* 6:317. doi:10.3389/fmars.2019.00317.

Duggen, S., P. Croot, U. Schacht, and L. Hoffmann. 2007. Subduction zone volcanic ash can fertilize the surface ocean and stimulate phytoplankton growth: Evidence from biogeochemical experiments and satellite data. *Geophysical Research Letters* 34(1):L01612. doi:10.1029/2006GL027522.

Durakovic, A. 2020. REF: — Norther to double as seaweed farm: Wind farm update. offshoreWIND.biz, July 15. https://www.offshorewind.biz/2020/07/15/norther-to-double-as-seaweed-farm/.

Dutreuil, S., L. Bopp, and A. Tagliabue. 2009. Impact of enhanced vertical mixing on marine biogeochemistry: Lessons for geo-engineering and natural variability. *Biogeosciences* 6(5):901-912. doi:10.5194/bg-6-901-2009.

Egenhofer, C., L. Schrefler, V. Rizos, A. Marcu, F. Genoese, A. Renda, J. Wieczorkiewicz, S. Roth, F. Infelise, and G. Luchetta. 2014. *The Composition and Drivers of Energy Prices and Costs in Energy-Intensive Industries: The Case of Ceramics, Glass and Chemicals.* CEPS Special Report No. 85. Brussels: Centre for European Policy Studies.

Eger, A. M., A. Vergés, C. G. Choi, H. Christie, M. A. Coleman, C. W. Fagerli, D. Fujita, M. Hasegawa, J. H. Kim, M. Mayer-Pinto, D. C. Reed, P. D. Steinberg, and E. M. Marzinelli. 2020. Financial and institutional support are important for large-scale kelp forest restoration. *Frontiers in Marine Science* 7:811. doi:10.3389/fmars.2020.535277.

EIA (U.S. Energy Information Administration). 2021. *Annual Energy Outlook 2021.* https://www.eia.gov/outlooks/aeo/.

Eisaman, M. D. 2020. Negative emissions technologies: The tradeoffs of air-capture economics. *Joule* 4(3):516-520. doi:10.1016/j.joule.2020.02.007.

Eisaman, M. D., K. Parajuly, A. Tuganov, C. Eldershaw, N. Chang, and K. A. Littau. 2012. CO_2 extraction from seawater using bipolar membrane electrodialysis. *Energy & Environmental Science* 5(6):7346-7352. doi:10.1039/c2ee03393c.

Eisaman, M. D., J. L. B. Rivest, S. D. Karnitz, C.-F. de Lannoy, A. Jose, R. W. DeVaul, and K. Hannun. 2018. Indirect ocean capture of atmospheric CO_2: Part II. Understanding the cost of negative emissions. *International Journal of Greenhouse Gas Control* 70:254-261. doi:10.1016/j.ijggc.2018.02.020.

Eisner, L. B., E. V. Farley, J. M. Murphy, and J. H. Helle. 2005. Distributions of oceanographic variables, juvenile sockeye salmon and age-0 walleye pollock in the southeastern Bering Sea during fall 2000–2003. Pp. 16-18 in *NPAFC Technical Report No. 6, BASIS 2004: Salmon and Marine Ecosystems in the Bering Sea and Adjacent Waters*, T. Azumaya, R. Beamish, J. Helle, S. Kang, V. Karpenko, C. S. Lee, K. Myers, T. Nagasawa, O. Temnykh, and M. Trudel, eds. Vancouver, Canada: North Pacific Anadromous Fish Commission.

Eke, J., A. Yusuf, A. Giwa, and A. Sodiq. 2020. The global status of desalination: An assessment of current desalination technologies, plants and capacity. *Desalination* 495:114633. doi:10.1016/j.desal.2020.114633.

Engel, A. 2000. The role of transparent exopolymer particles (TEP) in the increase in apparent particle stickiness (α) during the decline of a diatom bloom. *Journal of Plankton Research* 22(3):485-497. doi:10.1093/plankt/22.3.485.

Estes, J. A., J. Terborgh, J. S. Brashares, M. E. Power, J. Berger, W. J. Bond, S. R. Carpenter, T. E. Essington, R. D. Holt, J. B. C. Jackson, R. J. Marquis, L. Oksanen, T. Oksanen, R. T. Paine, E. K. Pikitch, W. J. Ripple, S. A. Sandin, M. Scheffer, T. W. Schoener, J. B. Shurin, A. R. E. Sinclair, M. E. Soulé, R. Virtanen, and D. A. Wardle. 2011. Trophic downgrading of planet Earth. *Science* 333(6040):301-306. doi:10.1126/science.1205106.

Fajardy, M., P. Patrizio, H. A. Daggash, and N. Mac Dowell. 2019. Negative emissions: Priorities for research and policy design. 1(6). doi:10.3389/fclim.2019.00006.

Falkowski, P. G., R. T. Barber, and V. Smetacek. 1998. Biogeochemical controls and feedbacks on ocean primary production. *Science* 281(5374):200-206. doi:10.1126/science.281.5374.200.

Falkowski, P., R. J. Scholes, E. Boyle, J. Canadell, D. Canfield, J. Elser, N. Gruber, K. Hibbard, P. Högberg, S. Linder, F. T. Mackenzie, B. Moore III, T. Pedersen, Y. Rosenthal, S. Seitzinger, V. Smetacek, and W. Steffen. 2000. The global carbon cycle: A test of our knowledge of earth as a system. 290 (5490):291-296. doi:10.1126/science.290.5490.291.

Fan, W., Z. Zhang, Z. Yao, C. Xiao, Y. Zhang, Y. Zhang, J. Liu, Y. Di, Y. Chen, and Y. Pan. 2020. A sea trial of enhancing carbon removal from Chinese coastal waters by stimulating seaweed cultivation through artificial upwelling. *Applied Ocean Research* 101:102260. doi:10.1016/j.apor.2020.102260.

FAO (Food and Agriculture Organization of the United Nations). 2016. *The State of World Fisheries and Aquaculture 2016: Contributing to Food Security and Nutrition for All.* Rome.

Fassbender, A. J., C. L. Sabine, N. Lawrence-Slavas, E. H. De Carlo, C. Meinig, and S. Maenner Jones. 2015. Robust sensor for extended autonomous measurements of surface ocean dissolved inorganic carbon. *Environmental Science & Technology* 49(6):3628-3635. doi:10.1021/es5047183.

Feehan, C. J., K. Filbee-Dexter, and T. Wernberg. 2021. Embrace kelp forests in the coming decade. 373(6557):863-863. doi:10.1126/science.abl3984.

Felthoven, Ronald, and Stephen Kasperski. 2013. Socioeconomic indicators for United States fisheries and fishing communities. *PICES Press* 21(2):20.

Feng, E. Y., D. P. Keller, W. Koeve, and A. Oschlies. 2016. Could artificial ocean alkalinization protect tropical coral ecosystems from ocean acidification? *Environmental Research Letters* 11(7):074008. doi:10.1088/1748-9326/11/7/074008.

Feng, E. Y., W. Koeve, D. P. Keller, and A. Oschlies. 2017. Model-based assessment of the CO_2 sequestration potential of coastal ocean alkalinization. *Earth's Future* 5(12):1252-1266. doi:10.1002/2017ef000659.

Feng, E. Y., B. Su, and A. Oschlies. 2020. Geoengineered ocean vertical water exchange can accelerate global deoxygenation. *Geophysical Research Letters* 47(16):e2020GL088263. doi:10.1029/2020GL088263.

Fennel, K. 2008. Widespread implementation of controlled upwelling in the North Pacific Subtropical Gyre would counteract diazotrophic N_2 fixation. *Marine Ecology Progress Series* 371:301-303.

Ferrari, R., M. F. Jansen, J. F. Adkins, A. Burke, A. L. Stewart, and A. F. Thompson. 2014. Antarctic sea ice control on ocean circulation in present and glacial climates. *Proceedings of the National Academy of Sciences of the United States of America* 111(24):8753-8758. doi:10.1073/pnas.1323922111.

Ferretti, F., B. Worm, G. L. Britten, M. R. Heithaus, and H. K. Lotze. 2010. Patterns and ecosystem consequences of shark declines in the ocean. *Ecology Letters* 13(8):1055-1071. doi:10.1111/j.1461-0248.2010.01489.x.

Filer, C., and J. Gabriel. 2018. How could Nautilus Minerals get a social licence to operate the world's first deep sea mine? *Marine Policy* 95:394-400. doi:10.1016/j.marpol.2016.12.001.

Fiorino, D. J. 2016. Citizen participation and environmental risk: A survey of institutional mechanisms. *Science, Technology, & Human Values* 15(2):226-243. doi:10.1177/016224399001500204.

Fodrie, F. J., A. B. Rodriguez, R. K. Gittman, J. H. Grabowski, N. L. Lindquist, C. H. Peterson, M. F. Piehler, and J. T. Ridge. 2017. Oyster reefs as carbon sources and sinks. *Proceedings of the Royal Society B: Biological Sciences* 284(1859):20170891. doi:10.1098/rspb.2017.0891.

Forbord, S., S. Matsson, G. E. Brodahl, B. A. Bluhm, O. J. Broch, A. Handå, A. Metaxas, J. Skjermo, K. B. Steinhovden, and Y. Olsen. 2020. Latitudinal, seasonal and depth-dependent variation in growth, chemical composition and biofouling of cultivated *Saccharina latissima* (Phaeophyceae) along the Norwegian coast. *Journal of Applied Phycology* 32(4):2215-2232. doi:10.1007/s10811-020-02038-y.

Foster, G. L. 2008. Seawater pH, pCO_2 and $[CO_2^{-3}]$ variations in the Caribbean Sea over the last 130 kyr: A boron isotope and B/Ca study of planktic foraminifera. *Earth and Planetary Science Letters* 271(1-4):254-266.

Francois, R., S. Honjo, R. Krishfield, and S. Manganini. 2002. Factors controlling the flux of organic carbon to the bathypelagic zone of the ocean. *Global Biogeochemical Cycles* 16(4):34-1-34-20. doi:10.1029/2001gb001722.

Franks, D. M., D. Brereton, and C. J. Moran. 2011. Cumulative social impacts. In *New Directions in Social Impact Assessment*. Northampton, MA: Edward Elgar Publishing.

Fredriksen, S., K. Filbee-Dexter, K. M. Norderhaug, H. Steen, T. Bodvin, M. A. Coleman, F. Moy, and T. Wernberg. 2020. Green gravel: A novel restoration tool to combat kelp forest decline. *Scientific Reports* 10(1):3983. doi:10.1038/s41598-020-60553-x.

Freestone, D., and R. Rayfuse. 2008. Ocean iron fertilization and international law. *Marine Ecology Progress Series* 364:227-233. doi:10.3354/meps07543.

Fricko, O., P. Havlik, J. Rogelj, Z. Klimont, M. Gusti, N. Johnson, P. Kolp, M. Strubegger, H. Valin, M. Amann, T. Ermolieva, N. Forsell, M. Herrero, C. Heyes, G. Kindermann, V. Krey, D. L. McCollum, M. Obersteiner, S. Pachauri, S. Rao, E. Schmid, W. Schoepp, and K. Riahi. 2017. The marker quantification of the Shared Socioeconomic Pathway 2: A middle-of-the-road scenario for the 21st century. *Global Environmental Change* 42:251-267. doi:https://doi.org/10.1016/j.gloenvcha.2016.06.004.

Friedlingstein, P., M. O'Sullivan, M. W. Jones, R. M. Andrew, J. Hauck, A. Olsen, G. P. Peters, W. Peters, J. Pongratz, S. Sitch, C. Le Quéré, J. G. Canadell, P. Ciais, R. B. Jackson, S. Alin, L. E. O. C. Aragão, A. Arneth, V. Arora, N. R. Bates, M. Becker, A. Benoit-Cattin, H. C. Bittig, L. Bopp, S. Bultan, N. Chandra, F. Chevallier, L. P. Chini, W. Evans, L. Florentie, P. M. Forster, T. Gasser, M. Gehlen, D. Gilfillan, T. Gkritzalis, L. Gregor, N. Gruber, I. Harris, K. Hartung, V. Haverd, R. A. Houghton, T. Ilyina, A. K. Jain, E. Joetzjer, K. Kadono, E. Kato, V. Kitidis, J. I. Korsbakken, P. Landschützer, N. Lefèvre, A. Lenton, S. Lienert, Z. Liu, D. Lombardozzi, G. Marland, N. Metzl, D. R. Munro, J. E. M. S. Nabel, S. I. Nakaoka, Y. Niwa, K. O'Brien, T. Ono, P. I. Palmer, D. Pierrot, B. Poulter, L. Resplandy, E. Robertson, C. Rödenbeck, J. Schwinger, R. Séférian, I. Skjelvan, A. J. P. Smith, A. J. Sutton, T. Tanhua, P. P. Tans, H. Tian, B. Tilbrook, G. Van Der Werf, N. Vuichard, A. P. Walker, R. Wanninkhof, A. J. Watson, D. Willis, A. J. Wiltshire, W. Yuan, X. Yue, and S. Zaehle. 2020. Global carbon budget 2020. *Earth System Science Data* 12(4):3269-3340. doi:10.5194/essd-12-3269-2020.

Fuhrman, J., H. McJeon, S. C. Doney, W. Shobe, and A. F. Clarens. 2019. From zero to hero? Why integrated assessment modeling of negative emissions technologies is hard and how we can do better." *Frontiers in Climate* 1:11. doi:10.3389/fclim.2019.00011.

Fuhrman, J., H. McJeon, P. Patel, S. C. Doney, W. M. Shobe, and A. F. Clarens. 2020. Food–energy–water implications of negative emissions technologies in a +1.5 °C future. *Nature Climate Change* 10(10):920-927. doi:10.1038/s41558-020-0876-z.

Fuhrman, J., A. Clarens, K. Calvin, S. C. Doney, J. A. Edmonds, P. O'Rourke, P. Patel, S. Pradhan, W. Shobe, and H. McJeon. 2021. The role of direct air capture and negative emissions technologies in the shared socioeconomic pathways towards +1.5°C and +2°C futures. *Environmental Research Letters* 16(11):114012. doi:10.1088/1748-9326/ac2db0.

Fujii, M., A. C. Y. Yeung, and T. D. Waite. 2015. Competitive effects of calcium and magnesium ions on the photochemical transformation and associated cellular uptake of iron by the freshwater cyanobacterial phytoplankton *Microcystis aeruginosa. Environmental Science & Technology* 49(15):9133-9142. doi:10.1021/acs.est.5b01583.

Gallo, N. D., D. G. Victor, and L. A. Levin. 2017. Ocean commitments under the Paris Agreement. *Nature Climate Change* 7(11):833-838. doi:10.1038/nclimate3422.

Galvani, L. 1841. *Opere Edite ed Inedite del Professore Luigi Galvani*. Dall'Olmo.

Gangal, M., A. B. Gafoor, E. D'Souza, N. Kelkar, R. Karkarey, N. Marbà, R. Arthur, and T. Alcoverro. 2021. Sequential overgrazing by green turtles causes archipelago-wide functional extinctions of seagrass meadows. *Biological Conservation* 260(1):109195. doi:10.1016/j.biocon.2021.109195.

Garcia-Herrero, I., M. Margallo, R. Onandía, R. Aldaco, and A. Irabien. 2017. Life cycle assessment model for the chlor-alkali process: A comprehensive review of resources and available technologies. *Sustainable Production and Consumption* 12:44-58. doi:10.1016/j.spc.2017.05.001.

Gartner, N., T. Kosec, and A. Legat. 2020. Monitoring the corrosion of steel in concrete exposed to a marine environment. *Materials* 13(2):407. doi:10.3390/ma13020407.

Gattuso, J.-P., A. K. Magnan, L. Bopp, W. W. L. Cheung, C. M. Duarte, J. Hinkel, E. McLeod, F. Micheli, A. Oschlies, P. Williamson, R. Billé, V. I. Chalastani, R. D. Gates, J. O. Irisson, J. J. Middelburg, H. O. Pörtner, and G. H. Rau. 2018. Ocean solutions to address climate change and its effects on marine ecosystems. *Frontiers in Marine Science* 5:337. doi:10.3389/fmars.2018.00337.

Gattuso, J.-P., P. Williamson, C. M. Duarte, and A. K. Magnan. 2021. The potential for ocean-based climate action: Negative emissions technologies and beyond. *Frontiers in Climate* 2:37. doi:10.3389/fclim.2020.575716.

Geden, O., and F. Schenuit. 2020. Unconventional mitigation: carbon dioxide removal as a new approach in EU climate policy. In *SWP Research Paper, 8/2020*. Berlin.

Gee, K. 2019. The ocean perspective. Pp. 23-45, in *Maritime Spatial Planning*. J. Zaucha, K. Gee, eds. Cham, Switzerland: Springer Nature.

Geraldi, N. R., A. Ortega, O. Serrano, P. I. Macreadie, C. E. Lovelock, D. Krause-Jensen, H. Kennedy, P. S. Lavery, M. L. Pace, J. Kaal, and C. M. Duarte. 2019. Fingerprinting blue carbon: Rationale and tools to determine the source of organic carbon in marine depositional environments. *Frontiers in Marine Science* 6:263. doi:10.3389/fmars.2019.00263.

Gerard, V. A. 1982. Growth and utilization of internal nitrogen reserves by the giant kelp *Macrocystis pyrifera* in a low-nitrogen environment. *Marine Biology* 66(1):27-35. doi:10.1007/BF00397251.

Gertner, J. 2021. The dream of carbon air capture edges toward reality. *Yale Environment 360,* August 25. https://e360.yale.edu/features/the-dream-of-co2-air-capture-edges-toward-reality.

GESAMP (Joint Group of Experts on the Scientific Aspects of Marine Environmental Protection). 2019. *High Level Review of a Wide Range of Proposed Marine Geoengineering Techniques*, P. W. Boyd and C. M. G. Vivian, eds. GESAMP Reports & Studies Series. London: International Maritime Organization.

Giering, S. L. C., E. L. Cavan, S. L. Basedow, N. Briggs, A. B. Burd, L. J. Darroch, L. Guidi, J.-O. Irisson, M. H. Iversen, R. Kiko, D. Lindsay, C. R. Marcolin, A. M. P. McDonnell, K. O. Möller, U. Passow, S. Thomalla, T. W. Trull, and A. M. Waite. 2020. Sinking organic particles in the ocean—flux estimates from in situ optical devices. *Frontiers in Marine Science* 6:00834. doi:10.3389/fmars.2019.00834.

Gifford, L. 2020. "You can't value what you can't measure": a critical look at forest carbon accounting. *Climatic Change* 161(2):291-306. doi:10.1007/s10584-020-02653-1.

Gilman, E., A. Perez Roda, T. Huntington, S. J. Kennelly, P. Suuronen, M. Chaloupka, and P. A. H. Medley. 2020. Benchmarking global fisheries discards. *Scientific Reports* 10(1):14017. doi:10.1038/s41598-020-71021-x.

Global Carbon Project. 2021. Global Carbon Budget 2021. https://www.globalcarbonproject.org/carbonbudget/index.htm.

GML (Global Monitoring Laboratory). 2021. Trends in Atmospheric Carbon Dioxide. National Oceanic and Atmospheric Administration. https://gml.noaa.gov/ccgg/trends/.

Gnanadesikan, A., J. L. Sarmiento, and R. D. Slater. 2003. Effects of patchy ocean fertilization on atmospheric carbon dioxide and biological production. *Global Biogeochemical Cycles* 17(2):1050. doi:10.1029/2002gb001940.

Goldman, J. G. 2019. Haida Gwaii's kelp forests disappeared. Here's how they're being brought back to life. *bioGraphic*, October 15.

Gollan, N., and K. Barclay. 2020. "It's not just about fish": Assessing the social impacts of marine protected areas on the well-being of coastal communities in New South Wales. *PLoS ONE* 15(12):e0244605. doi:10.1371/journal.pone.0244605.

Golubev, S. V., O. S. Pokrovsky, and J. Schott. 2005. Experimental determination of the effect of dissolved CO_2 on the dissolution kinetics of Mg and Ca silicates at 25°C. *Chemical Geology* 217(3-4):227-238. doi:10.1016/j.chemgeo.2004.12.011.

González, M. F., and T. Ilyina. 2016. Impacts of artificial ocean alkalinization on the carbon cycle and climate in Earth system simulations. *Geophysical Research Letters* 43(12):6493-6502. doi:10.1002/2016GL068576.

Gore, S., P. Renforth, and R. Perkins. 2019. The potential environmental response to increasing ocean alkalinity for negative emissions. *Mitigation and Adaptation Strategies for Global Change* 24(7):1191-1211. doi:10.1007/s11027-018-9830-z.

Gough, C., R. Cunningham, and S. Mander. 2018. Understanding key elements in establishing a social license for CCS: An empirical approach. *International Journal of Greenhouse Gas Control* 68:16-25. doi:10.1016/j.ijggc.2017.11.003.

Grandstaff, D. E. 1978. Changes in surface area and morphology and the mechanism of forsterite dissolution. *Geochimica et Cosmochimica Acta* 42(12):1899-1901. doi:10.1016/0016-7037(78)90245-4.

Gray, L. A., A. G. Bisonó León, F. E. Rojas, S. S. Veroneau, and A. H. Slocum. 2021. Caribbean-wide, negative emissions solution to *Sargassum* spp. low-cost collection device and sustainable disposal method. *Phycology* 1(1):49-75. doi:10.3390/phycology1010004.

Gregr, E. J., V. Christensen, L. Nichol, R. G. Martone, R. W. Markel, J. C. Watson, C. D. G. Harley, E. A. Pakhomov, J. B. Shurin, and K. M. A. Chan. 2020. Cascading social-ecological costs and benefits triggered by a recovering keystone predator. *Science* 368(6496):1243-1247. doi:10.1126/science.aay5342.

Griffioen, J. 2017. Enhanced weathering of olivine in seawater: The efficiency as revealed by thermodynamic scenario analysis. *Science of the Total Environment* 575:536-544. doi:10.1016/j.scitotenv.2016.09.008.

Griscom, B. W., J. Adams, P. W. Ellis, R. A. Houghton, G. Lomax, D. A. Miteva, W. H. Schlesinger, D. Shoch, J. V. Siikamaki, P. Smith, P. Woodbury, C. Zganjar, A. Blackman, J. Campari, R. T. Conant, C. Delgado, P. Elias, T. Gopalakrishna, M. R. Hamsik, M. Herrero, J. Kiesecker, E. Landis, L. Laestadius, S. M. Leavitt, S. Minnemeyer, S. Polasky, P. Potapov, F. E. Putz, J. Sanderman, M. Silvius, E. Wollenberg, and J. Fargione. 2017. Natural climate solutions. *Proceedings of the National Academy of Sciences of the United States of America* 114(44):11645-11650. doi:10.1073/pnas.1710465114.

Grove, W. R. 1839. XXIV. On voltaic series and the combination of gases by platinum. *London, Edinburgh, and Dublin Philosophical Magazine and Journal of Science* 14(86-87):127-130. doi:10.1080/14786443908649684.

Gruber, N. 2011. Warming up, turning sour, losing breath: Ocean biogeochemistry under global change. *Philosophical Transactions of the Royal Society A: Mathematical, Physical and Engineering Sciences* 369(1943):1980-1996. doi:10.1098/rsta.2011.0003.

Gruber, N., M. Gloor, S. E. Mikaloff Fletcher, S. C. Doney, S. Dutkiewicz, M. J. Follows, M. Gerber, A. R. Jacobson, F. Joos, K. Lindsay, D. Menemenlis, A. Mouchet, S. A. Müller, J. L. Sarmiento, and T. Takahashi. 2009. Oceanic sources, sinks, and transport of atmospheric CO_2. *Global Biogeochemical Cycles* 23(1):GB1005. doi:10.1029/2008GB003349.

Gruber, N., D. Clement, B. R. Carter, R. A. Feely, S. van Heuven, M. Hoppema, M. Ishii, R. M. Key, A. Kozyr, S. K. Lauvset, C. Lo Monaco, J. T. Mathis, A. Murata, A. Olsen, F. F. Perez, C. L. Sabine, T. Tanhua, and R. Wanninkhof. 2019. The oceanic sink for anthropogenic CO_2 from 1994 to 2007. *Science* 363(6432):1193-1199. doi:10.1126/science.aau5153.

Gunn, K., and C. Stock-Williams. 2012. Quantifying the global wave power resource. *Renewable Energy* 44:296-304. doi:10.1016/j.renene.2012.01.101.

Guo, J., F. Wang, R. D. Vogt, Y. Zhang, and C. Q. Liu. 2015. Anthropogenically enhanced chemical weathering and carbon evasion in the Yangtze Basin. *Scientific Reports* 5:11941. doi:10.1038/srep11941.

Haggett, C. 2008. Over the sea and far away? A consideration of the planning, politics and public perception of offshore wind farms. *Journal of Environmental Policy & Planning* 10(3):289-306. doi:10.1080/15239080802242787.

Hamm, C. E. 2002. Interactive aggregation and sedimentation of diatoms and clay-sized lithogenic material. *Limnology and Oceanography* 47(6):1790-1795. doi:10.4319/lo.2002.47.6.1790.

Hamme, R. C., P. W. Webley, W. R. Crawford, F. A. Whitney, M. D. DeGrandpre, S. R. Emerson, C. C. E., K. E. Giesbrecht, J. F. R. G., M. T. Kavanaugh, M. A. Peña, C. L. Sabine, S. D. Batten, L. A. Coogan, D. S. Grundle, and D. Lockwood. 2010. Volcanic ash fuels anomalous plankton bloom in subarctic northeast Pacific. *Geophysical Research Letters* 37(19):L19604. doi:10.1029/2010gl044629.

Hanak, D. P., B. G. Jenkins, T. Kruger, and V. Manovic. 2017. High-efficiency negative-carbon emission power generation from integrated solid-oxide fuel cell and calciner. *Applied Energy* 205:1189-1201. doi:10.1016/j.apenergy.2017.08.090.

Handå, A., T. A. McClimans, K. I. Reitan, Ø. Knutsen, K. Tangen, and Y. Olsen. 2013. Artificial upwelling to stimulate growth of non-toxic algae in a habitat for mussel farming. *Aquaculture Research* 45:1798-1809. https://doi.org/10.1111/are.12127.

Hangx, S. J. T., and C. J. Spiers. 2009. Coastal spreading of olivine to control atmospheric CO_2 concentrations: A critical analysis of viability. *International Journal of Greenhouse Gas Control* 3(6):757-767. doi:10.1016/j.ijggc.2009.07.001.

Hansell, D. A. 2013. Recalcitrant dissolved organic carbon fractions. *Annual Review of Marine Science* 5:421-445. doi:10.1146/annurev-marine-120710-100757.

Hansen, J., M. Sato, P. Kharecha, K. von Schuckmann, D. J. Beerling, J. Cao, S. Marcott, V. Masson-Delmotte, M. J. Prather, E. J. Rohling, J. Shakun, and P. Smith. 2017. Young people's burden: Requirement of negative CO_2 emissions. *Earth System Dynamics*. doi:10.5194/esd-2016-42.

Hansen, P. J., N. Lundholm, and B. Rost. 2007. Growth limitation in marine red-tide dinoflagellates: Effects of pH versus inorganic carbon availability. *Marine Ecology Progress Series* 334:63-71.

Hanssen, S. V., V. Daioglou, Z. J. N. Steinmann, J. C. Doelman, D. P. Van Vuuren, and M. A. J. Huijbregts. 2020. The climate change mitigation potential of bioenergy with carbon capture and storage. *Nature Climate Change* 10(11):1023-1029. doi:10.1038/s41558-020-0885-y.

Harland, A.D. and B. Brown. 1989. Metal tolerance in the scleractinian coral Porites lutea. *Marine Pollution Bulletin*. 20: 353-357. 10.1016/0025-326X(89)90159-8.

Harris, R. 2012. Can adding iron to oceans slow global warming? *NPR*. https://www.npr.org/2012/07/18/156976147/can-adding-iron-to-oceans-slow-global-warming.

Harrison, D. P. 2017. Global negative emissions capacity of ocean macronutrient fertilization. *Environmental Research Letters* 12(3):035001. doi:10.1088/1748-9326/aa5ef5.

Harrold, C., K. Light, and S. Lisin. 1998. Organic enrichment of submarine-canyon and continental-shelf benthic communities by macroalgal drift imported from nearshore kelp forests. *Limnology and Oceanography* 43(4):669-678. doi:10.4319/lo.1998.43.4.0669.

Hartmann, J., A. J. West, P. Renforth, P. Köhler, C. L. De La Rocha, D. A. Wolf-Gladrow, H. H. Dürr, and J. Scheffran. 2013. Enhanced chemical weathering as a geoengineering strategy to reduce atmospheric carbon dioxide, supply nutrients, and mitigate ocean acidification. *Reviews of Geophysics* 51(2):113-149. doi:10.1002/rog.20004.

Harvey, L. D. D. 2008. Mitigating the atmospheric CO_2 increase and ocean acidification by adding limestone powder to upwelling regions. *Journal of Geophysical Research: Oceans* 113(4):C04028. doi:10.1029/2007JC004373.

Haupin, W. E. 1983. Electrochemistry of the Hall-Heroult process for aluminum smelting. *Journal of Chemical Education* 60(4):279. doi:10.1021/ed060p279.

Hay, B. J. 1988. Sediment accumulation in the central western Black Sea over the past 5100 years. *Paleoceanography* 3(4):491-508. doi:10.1029/PA003i004p00491.

Hayes, D. J., R. Vargas, S. R. Alin, R. T. Conant, L. R. Hutyra, A. R. Jacobson, W. A. Kurz, S. Liu, A. D. McGuire, B. Poulter, and C. W. Woodall. 2018. The North American carbon budget. Pp. 71-108 in *Second State of the Carbon Cycle Report (SOCCR2): A Sustained Assessment Report*, N. Cavallaro, G. Shrestha, R. Birdsey, M. A. Mayes, R. G. Najjar, S. C. Reed, P. Romero-Lankao, and Z. Zhu, eds. Washington, DC: U.S. Global Change Research Program.

Helgeson, H. C., W. M. Murphy, and P. Aagaard. 1984. Thermodynamic and kinetic constraints on reaction rates among minerals and aqueous solutions. II. Rate constants, effective surface area, and the hydrolysis of feldspar. *Geochimica et Cosmochimica Acta* 48(12):2405-2432. doi:10.1016/0016-7037(84)90294-1.

Henderson, G. M., R. E. M. Rickaby, and H. Bouman. 2008. *Decreasing Atmosphere CO_2 by Increasing Ocean Alkalinity*. University of Oxford and James Martin 21st Century Ocean Institute. http://www.earth.ox.ac.uk/~gideonh/reports/Cquestrate_report.pdf.

Hernández-Romero, I. M., L. F. Fuentes-Cortés, R. Mukherjee, M. M. El-Halwagi, M. Serna-González, and F. Nápoles-Rivera. 2019. Multi-scenario model for optimal design of seawater air-conditioning systems under demand uncertainty. *Journal of Cleaner Production* 238:117863. doi:10.1016/j.jclepro.2019.117863.

Herr, D., J. Blum, A. Himes-Cornell, and A. Sutton-Grier. 2019. An analysis of the potential positive and negative livelihood impacts of coastal carbon offset projects. *Journal of Environmental Management* 235:463-479. doi:10.1016/j.jenvman.2019.01.067.

Ho, T. Y. 2013. Nickel limitation of nitrogen fixation in *Trichodesmium*. *Limnology and Oceanography* 58(1):112-120. doi:10.4319/lo.2013.58.1.0112.

Hoegh-Guldberg, O., E. Northrop, and J. Lubchenco. 2019. The ocean is key to achieving climate and societal goals. *Science* 365(6460):1372-1374. doi:10.1126/science.aaz4390.

Holifield, Ryan. 2013. Defining environmental justice and environmental racism. *Urban Geography* 22(1):78-90. doi:10.2747/0272-3638.22.1.78.

Holzer, M., B. Pasquier, T. DeVries, and M. Brzezinski. 2019. Diatom physiology controls silicic acid leakage in response to iron fertilization. *Global Biogeochemical Cycles* 33(12):1631-1653. doi:10.1029/2019GB006460.

Honegger, M., and D. Reiner. 2018. The political economy of negative emissions technologies: consequences for international policy design. *Climate Policy* 18 (3):306-321. doi:10.1080/14693062.2017.1413322.

House, K. Z., C. H. House, D. P. Schrag, and M. J. Aziz. 2007. Electrochemical acceleration of chemical weathering as an energetically feasible approach to mitigating anthropogenic climate change. *Environmental Science & Technology* 41(24):8464-8470. doi:10.1021/es0701816.

Howard, J., E. Babij, R. Griffis, B. Helmuth, A. Himes-Cornell, P. Niemier, M. Orbach, L. Petes, S. Allen, G. Auad, C. Auer, R. Beard, M. Boatman, N. Bond, T. Boyer, D. Brown, P. Clay, K. Crane, S. Cross, M. Dalton, J. Diamond, R. Diaz, Q. Dortch, E. Duffy, D. Fauquier, W. Fisher, M. Graham, B. Halpern, L. Hansen, B. Hayum, S. Herrick, A. Hollowed, D. Hutchins, E. Jewett, D. Jin, N. Knowlton, D. Kotowicz, T. Kristiansenl, P. Little, C. Lopez, P. Loring, R. Lumpkin, A. Mace, K. Mengerink, J. Ru Morrison, J. Murray, K. Norman, J. O'Donnell, J. Overland, R. Parsons, N. Pettigrew, L. Pfeiffer, E. Pidgeon, M. Plummer, J. Polovina, J. Quintrell, T. Rowles, J. Runge, M. Rust, E. Sanford, U. Send, M. Singer, C. Speir, D. Stanitski, C. Thornber, C. Wilson, and Y. Xue. 2013. Oceans and marine resources in a changing climate. In *Oceanography and Marine Biology: An Annual Review*, Vol. 51, R. N. Hughes, D. Hughes, and P. Smith, eds. Boca Raton, FL: CRC Press.

Howard, J., A. Sutton-Grier, D. Herr, J. Kleypas, E. Landis, E. McLeod, E. Pidgeon, and S. Simpson. 2017. Clarifying the role of coastal and marine systems in climate mitigation. *Frontiers in Ecology and the Environment* 15(1):42-50. doi:10.1002/fee.1451.

Hu, X., and W.-J. Cai. 2011. An assessment of ocean margin anaerobic processes on oceanic alkalinity budget. *Global Biogeochemical Cycles* 25(3):GB3003. doi:10.1029/2010gb003859.

Hubert, A.-M. 2017. A code of conduct for responsible geoengineering research. *Global Policy* 12:82-96. https://www.ce-conference.org/system/files/documents/revised_code_of_conduct_for_geoengineering_research_2017.pdf.

Hubert, A.-M., and D. Reichwein. 2015. *An Exploration of a Code of Conduct for Responsible Scientific Research Involving Geoengineering: Introduction, Draft Articles and Commentaries.* IASS Working Paper. Institute for Advanced Sustainability Studies. https://www.insis.ox.ac.uk/sites/default/files/insis/documents/media/an_exploration_of_a_code_of_conduct.pdf.

Huesemann, M. H. 2008. Ocean fertilization and other climate change mitigation strategies: An overview. *Marine Ecology Progress Series* 364:243-250. doi:10.3354/meps07545.

Huppert, H. E., and J. S. Turner. 1981. Double-diffusive convection. *Journal of Fluid Mechanics* 106:299-329. doi:10.1017/S0022112081001614.

IEA (International Energy Agency). 2020. Cement. Paris: IEA.

IEA. 2021a. *About CCUS: Playing an Important and Diverse Role in Meeting Global Energy and Climate Goals.* Technology Report. https://www.iea.org/reports/about-ccus.

IEA. 2021b. Is carbon capture too expensive? Paris: IEA. https://www.iea.org/commentaries/is-carbon-capture-too-expensive.

Iglesias-Rodríguez, M. Débora, C. W. Brown, S. C. Doney, J. Kleypas, D. Kolber, Z. Kolber, P. K. Hayes, and P. G. Falkowski. 2002. Representing key phytoplankton functional groups in ocean carbon cycle models: Coccolithophorids. *Global Biogeochemical Cycles* 16(4):47-1–47-20. doi:10.1029/2001gb001454.

IHS Markit. 2017. *The Economic Benefits of Sodium Hydroxide Chemistry in the Production of Organic Chemicals in the United States and Canada.* American Chemical Council.

Ilyina, T., D. Wolf-Gladrow, G. Munhoven, and C. Heinze. 2013. Assessing the potential of calcium-based artificial ocean alkalinization to mitigate rising atmospheric CO_2 and ocean acidification. *Geophysical Research Letters* 40(22):5909-5914. doi:10.1002/2013gl057981.

Indigenous Delegates at OceanObs'19. 2019. "Aha Honua Coastal Indigenous Peoples' Declaration at OceanObs'19."

InflationTool. 2021. Value of 1983 US Dollars Today. https://www.inflationtool.com/us-dollar/1983-to-present-value.

Inoue, M., R. Suwa, A. Suzuki, K. Sakai, and H. Kawahata. 2011. Effects of seawater pH on growth and skeletal U/Ca ratios of *Acropora digitifera* coral polyps. *Geophysical Research Letters* 38(12):L12809. doi:10.1029/2011gl047786.

IOOS (Integrated Ocean Observing System). 2018. U.S. IOOS Enterprise: Strategic Plan 2018-2022. Silverspring, MD: NOAA.

IPCC (Intergovernmental Panel on Climate Change). 2005. *IPCC Special Report on Carbon Dioxide Capture and Storage*, B. Metz, O. Davidson, H. de Coninck, M. Loos, and L. Meyer, eds. Cambridge, UK: Cambridge University Press.

IPCC (Intergovernmental Panel on Climate Change). 2018. *Global Warming of 1.5°C: An IPCC Special Report.* SR15. http://www.ipcc.ch/report/sr15.

IPCC. 2021. *Climate Change 2021: The Physical Science Basis. Contribution of Working Group I to the Sixth Assessment Report of the Intergovernmental Panel on Climate Change. Summary for Policymakers*, V. Masson-Delmotte, P. Zhai, A. Pirani, S. L. Connors, C. Péan, S. Berger, N. Caud, Y. Chen, L. Goldfarb, M. I. Gomis, M. Huang, K. Leitzell, E. Lonnoy, J. B. R. Matthews, T. K. Maycock, T. Waterfield, O. Yelekçi, Yu R. and B. Zhou, eds. Cambridge University Press.

IPCS (International Programme on Chemical Safety). 1991. Environmental Health Criteria No. 108: Nickel. Geneva, Switzerland: World Health Organization.

Isaacs, J. D., D. Castel, and G. L. Wick. 1976. Utilization of the energy in ocean waves. *Ocean Engineering* 3(4):175-187. doi:10.1016/0029-8018(76)90022-6.

Itsweire, E. C., J. R. Koseff, D. A. Briggs, and J. H. Ferziger. 1993. Turbulence in stratified shear flows: Implications for interpreting shear-induced mixing in the ocean. *Journal of Physical Oceanography* 23(7):1508-1522. doi: 10.1175/1520-0485(1993)023<1508:TISSFI>2.0.CO;2.

Iversen, M. H., E. A. Pakhomov, B. P. V. Hunt, H. van der Jagt, D. Wolf-Gladrow, and C. Klaas. 2017. Sinkers or floaters? Contribution from salp pellets to the export flux during a large bloom event in the Southern Ocean. *Deep Sea Research Part II: Topical Studies in Oceanography* 138:116-125. doi:10.1016/j.dsr2.2016.12.004.

Izumiya, K., E. Akiyama, H. Habazaki, N. Kumagai, A. Kawashima, and K. Hashimoto. 1997. Effects of additional elements on electrocatalytic properties of thermally decomposed manganese oxide electrodes for oxygen evolution from seawater. *Materials Transactions, JIM* 38(10):899-905. doi:10.2320/matertrans1989.38.899.

Jackson, G. A. 1977. Nutrients and production of giant kelp, *Macrocystis pyrifera,* off southern California. *Limnology and Oceanography* 22(6):979-995. doi:10.4319/lo.1977.22.6.0979.

Jackson, G. A. 1987. Modelling the growth and harvest yield of the giant kelp *Macrocystis pyrifera. Marine Biology* 95(4):611-624. doi:10.1007/BF00393105.

Jahediesfanjani, H., P. D. Warwick, and S. T. Anderson. 2018. Estimating the pressure-limited CO_2 injection and storage capacity of the United States saline formations: Effect of the presence of hydrocarbon reservoirs. *International Journal of Greenhouse Gas Control* 79:14-24. doi:10.1016/j.ijggc.2018.09.011.

Janasie, C., and A. Nichols. 2018. Navigating the kelp forest: Current legal issues surrounding seaweed wild harvest and aquaculture. *Natural Resources & Environment* 33(1):17-21.

Jebari, J., O. O. Táíwò, T. M. Andrews, V. Aquila, B. Beckage, M. Belaia, M. Clifford, J. Fuhrman, D. P. Keller, K. J. Mach, D. R. Morrow, K. T. Raimi, D. Visioni, S. Nicholson, and C. H. Trisos. 2021. From moral hazard to risk-response feedback. *Climate Risk Management* 33. doi:10.1016/j.crm.2021.100324.

Jiang, L.-Q., R. A. Feely, B. R. Carter, D. J. Greeley, D. K. Gledhill, and K. M. Arzayus. 2015. Climatological distribution of aragonite saturation state in the global oceans. *Global Biogeochemical Cycles* 29(10):1656-1673. doi:10.1002/2015gb005198.

Jin, X., and N. Gruber. 2003. Offsetting the radiative benefit of ocean iron fertilization by enhancing N_2O emissions. *Geophysical Research Letters* 30(24):2249. doi:10.1029/2003GL018458.

Jin, X., N. Gruber, H. Frenzel, S. C. Doney, and J. C. McWilliams. 2008. The impact on atmospheric CO_2 of iron fertilization induced changes in the ocean's biological pump. *Biogeosciences* 5(2):385-406. doi:10.5194/bg-5-385-2008.

Jobin, M., and M. Siegrist. 2020. Support for the deployment of climate engineering: A comparison of ten different technologies. *Risk Analysis* 40(5):1058-1078. doi:10.1111/risa.13462.

Johnson, D. H., and J. Decicco. 1983. *An Artificial Upwelling Driven by Salinity Differences in the Ocean.* U.S. Department of Energy.

Johnson, K. S., and D. M. Karl. 2002. Is ocean fertilization credible and creditable? (Letters). *Science* 296(5567):467+.

Jolly, W. M., M. A. Cochrane, P. H. Freeborn, Z. A. Holden, T. J. Brown, G. J. Williamson, and D. M. J. S. Bowman. 2015. Climate-induced variations in global wildfire danger from 1979 to 2013. *Nature Communications* 6(1):7537. doi:10.1038/ncomms8537.

Jones, I. S. F., and H. E. Young. 1997. Engineering a large sustainable world fishery. *Environmental Conservation* 24(2):99-104. doi:10.1017/S0376892997000167.

Jones, M. T., and S. R. Gislason. 2008. Rapid releases of metal salts and nutrients following the deposition of volcanic ash into aqueous environments. *Geochimica et Cosmochimica Acta* 72(15):3661-3680. doi:10.1016/j.gca.2008.05.030.

Joos, F., R. Roth, J. S. Fuglestvedt, G. P. Peters, I. G. Enting, W. von Bloh, V. Brovkin, E. J. Burke, M. Eby, N. R. Edwards, T. Friedrich, T. L. Frölicher, P. R. Halloran, P. B. Holden, C. Jones, T. Kleinen, F. T. Mackenzie, K. Matsumoto, M. Meinshausen, G. K. Plattner, A. Reisinger, J. Segschneider, G. Shaffer, M. Steinacher, K. Strassmann, K. Tanaka, A. Timmermann, and A. J. Weaver. 2013. Carbon dioxide and climate impulse response functions for the computation of greenhouse gas metrics: A multi-model analysis. *Atmospheric Chemistry and Physics* 13(5):2793-2825. doi:10.5194/acp-13-2793-2013.

Jouffray, J. B., R. Blasiak, A. V. Norström, H. Österblom, and M. Nyström. 2020. The blue acceleration: The trajectory of human expansion into the ocean. *One Earth* 2(1):43-54. doi:10.1016/j.oneear.2019.12.016.

Juniper, S. K., K. Thornborough, K. Douglas, and J. Hillier. 2019. Remote monitoring of a deep-sea marine protected area: The Endeavour Hydrothermal Vents. *Aquatic Conservation: Marine and Freshwater Ecosystems* 29(S2):84-102. doi:10.1002/aqc.3020.

Kachelreiss, D., M. Wegmann, M. Gollock, and N. Pettorelli. 2014. The application of remote sensing for marine protected area management. *Ecological Indicators* 36:169-177.

Kaiser, B. A., M. Hoeberechts, K. H. Maxwell, L. Eerkes-Medrano, N. Hilmi, A. Safa, C. Horbel, S. K. Juniper, M. Roughan, N. T. Lowen, K. Short, and D. Paruru. 2019. The importance of connected ocean monitoring knowledge systems and communities. *Frontiers in Marine Science* 6 (JUN). doi:10.3389/fmars.2019.00309.

Kaplan, M. B., and S. Solomon. 2016. A coming boom in commercial shipping? The potential for rapid growth of noise from commercial ships by 2030. *Marine Policy* 73:119-121. doi:10.1016/j.marpol.2016.07.024.

Karl, D. M., and R. M. Letelier. 2008. Nitrogen fixation-enhanced carbon sequestration in low nitrate, low chlorophyll seascapes. *Marine Ecology Progress Series* 364:257-268. doi:10.3354/meps07547.

Kelemen, P., S. M. Benson, H. Pilorgé, P. Psarras, and J. Wilcox. 2019. An overview of the status and challenges of CO_2 storage in minerals and geological formations. *Frontiers in Climate* 1:9. doi:10.3389/fclim.2019.00009.

Kelemen, P. B., N. McQueen, J. Wilcox, P. Renforth, G. Dipple, and A. P. Vankeuren. 2020. Engineered carbon mineralization in ultramafic rocks for CO_2 removal from air: Review and new insights. *Chemical Geology* 550:119628. doi:10.1016/j.chemgeo.2020.119628.

Keller, D. P., E. Y. Feng, and A. Oschlies. 2014. Potential climate engineering effectiveness and side effects during a high carbon dioxide-emission scenario. *Nature Communications* 5:3304. doi:10.1038/ncomms4304.

Kelly, L. W., G. J. Williams, K. L. Barott, C. A. Carlson, E. A. Dinsdale, R. A. Edwards, A. F. Haas, M. Haynes, Y. W. Lim, T. McDole, C. E. Nelson, E. Sala, S. A. Sandin, J. E. Smith, M. J. A. Vermeij, M. Youle, and F. Rohwer. 2014. Local genomic adaptation of coral reef-associated microbiomes to gradients of natural variability and anthropogenic stressors. *Proceedings of the National Academy of Sciences of the United States of America* 111(28):10227-10232. doi:10.1073/pnas.1403319111.

Kelly, R. L., X. Bian, S. J. F., K. L. Fornace, T. Gunderson, N. J. Hawco, H. Liang, J. Niggemann, S. E. Paulson, P. Pinedo Gonzalez, A. J. West, S. C. Yang, and S. G. John. 2021. Delivery of metals and dissolved black carbon to the Southern California coastal ocean via aerosols and floodwaters following the 2017 Thomas Fire. *Journal of Geophysical Research: Biogeosciences* 126(3):e2020JG006117. doi:10.1029/2020jg006117.

Keul, N., G. Langer, L. J. Nooijer, G. Nehrke, G. J. Reichart, and J. Bijma. 2013. Incorporation of uranium in benthic foraminiferal calcite reflects seawater carbonate ion concentration. *Geochemistry, Geophysics, Geosystems* 14(1):102-111. doi:10.1029/2012gc004330.

Khalil, M. A. K. 1999. Reactive chlorine compounds in the atmosphere. Pp. 45-79 in *Reactive Halogen Compounds in the Atmosphere*. Springer.

Kheshgi, H. S. 1995. Sequestering atmospheric carbon dioxide by increasing ocean alkalinity. *Energy* 20(9):915-922. doi:10.1016/0360-5442(95)00035-F.

Kim, A. S., and H.-J. Kim, eds. 2020. *Ocean Thermal Energy Conversion (OTEC): Past, Present, and Progress*. London: IntechOpen. doi:10.5772/intechopen.86591.

Kirke, B. 2003. Enhancing fish stocks with wave-powered artificial upwelling. *Ocean and Coastal Management* 46(9-10):901-915. doi:10.1016/S0964-5691(03)00067-X.

Klaas, C., and D. E. Archer. 2002. Association of sinking organic matter with various types of mineral ballast in the deep sea: Implications for the rain ratio. *Global Biogeochemical Cycles* 16(4):63-1-63-14. doi:10.1029/2001gb001765.

Kockel, A., N. C. Ban, M. Costa, and P. Dearden. 2020. Addressing distribution equity in spatial conservation prioritization for small-scale fisheries. *PLoS ONE* 15(5):e0233339. doi:10.1371/journal.pone.0233339.

Koeve, W., P. Kähler, and A. Oschlies. 2020. Does export production measure transient changes of the biological carbon pump's feedback to the atmosphere under global warming? *Geophysical Research Letters* 47(22):e2020GL089928. doi:10.1029/2020gl089928.

Köhler, P., J. Hartmann, and D. A. Wolf-Gladrow. 2010. Geoengineering potential of artificially enhanced silicate weathering of olivine. *Proceedings of the National Academy of Sciences of the United States of America* 107(47):20228-20233. doi:10.1073/pnas.1000545107.

Köhler, P., J. F. Abrams, C. Völker, J. Hauck, and D. A. Wolf-Gladrow. 2013. Geoengineering impact of open ocean dissolution of olivine on atmospheric CO_2, surface ocean pH and marine biology. *Environmental Research Letters* 8(1):014009. doi:10.1088/1748-9326/8/1/014009.

Kolotyrkin, Y. M., V. V. Losev, and A. N. Chemodanov. 1988. Relationship between corrosion processes and oxygen evolution on anodes made from noble metals and related metal oxide anodes. *Materials Chemistry and Physics* 19(1-2):1-95. doi:10.1016/0254-0584(88)90002-8.

Konishi, S., H. Ohno, H. Yoshida, and Y. Naruse. 1983. Decomposition of tritiated water with solid oxide electrolysis cell. *Nuclear Technology—Fusion* 3(2):195-198. doi:10.13182/fst83-a20839.

Kopelevich, O., V. Burenkov, S. Sheberstov, S. Vazyulya, M. Kravchishina, L. Pautova, V. Silkin, V. Artemiev, and A. Grigoriev. 2014. Satellite monitoring of coccolithophore blooms in the Black Sea from ocean color data. *Remote Sensing of Environment* 146:113-123. doi:10.1016/j.rse.2013.09.009.

Koppers, A. A. P., and R. Coggon. 2020. *Exploring Earth by Scientific Ocean Drilling: 2050 Science Framework*. UC San Diego Library Digital Collections. https://doi.org/10.6075/J0W66J9H.

Koschinsky, A., L. Heinrich, K. Boehnke, J. C. Cohrs, T. Markus, M. Shani, P. Singh, K. Smith Stegen, and W. Werner. 2018. Deep-sea mining: Interdisciplinary research on potential environmental, legal, economic, and societal implications. *Integrated Environmental Assessment and Management* 14(6):672-691. doi:10.1002/ieam.4071.

Kramer, S. J., K. M. Bisson, and A. D. Fischer. 2020. Observations of phytoplankton community composition in the Santa Barbara Channel during the Thomas Fire. *Journal of Geophysical Research: Oceans* 125(12):e2020JC016851. doi:10.1029/2020jc016851.

Kramer, S. J., D. A. Siegel, S. Maritorena, and D. Catlett. 2021. Modeling surface ocean phytoplankton pigments from hyperspectral remote sensing reflectance on global scales. *In review*.

Krause-Jensen, D., and C. M. Duarte. 2016. Substantial role of macroalgae in marine carbon sequestration. *Nature Geoscience* 9(10):737-742. doi:10.1038/ngeo2790.

Krause-Jensen, D., P. Lavery, O. Serrano, N. Marba, P. Masque, and C. M. Duarte. 2018. Sequestration of macroalgal carbon: The elephant in the blue carbon room. *Biology Letters* 14(6):20180236. doi:10.1098/rsbl.2018.0236.

Kroschwitz, J. 1991. In *Kirk-Othmer Encyclopedia of Chemical Technology, Vol 1: A to Alkali*. New York: John Wiley & Sons.

Krueger, A. D., G. R. Parsons, and J. Firestone. 2011. Valuing the visual disamenity of offshore wind power projects at varying distances from the shore: An application on the Delaware shoreline. *Land Economics* 87(2):268-283. doi:10.3368/le.87.2.268.

Krumhansl, K. A., and R. E. Scheibling. 2012. Production and fate of kelp detritus. *Marine Ecology Progress Series* 467:281-302. doi:10.3354/meps09940.

Krumhansl, K. A., D. K. Okamoto, A. Rassweiler, M. Novak, J. J. Bolton, K. C. Cavanaugh, S. D. Connell, C. R. Johnson, B. Konar, S. D. Ling, F. Micheli, K. M. Norderhaug, A. Pérez-Matus, I. Sousa-Pinto, D. C. Reed, A. K. Salomon, N. T. Shears, T. Wernberg, R. J. Anderson, N. S. Barrett, A. H. Buschmann, M. H. Carr, J. E. Caselle, S. Derrien-Courtel, G. J. Edgar, M. Edwards, J. A. Estes, C. Goodwin, M. C. Kenner, D. J. Kushner, F. E. Moy, J. Nunn, R. S. Steneck, J. Vásquez, J. Watson, J. D. Witman, and J. E. K. Byrnes. 2016. Global patterns of kelp forest change over the past half-century. *Proceedings of the National Academy of Sciences of the United States of America* 113(48):13785-13790. doi:10.1073/pnas.1606102113.

Krumins, J. A., D. van Oevelen, T. M. Bezemer, G. B. De Deyn, W. H. G. Hol, E. van Donk, W. de Boer, P. C. de Ruiter, J. J. Middelburg, F. Monroy, K. Soetaert, E. Thébault, J. van de Koppel, J. A. van Veen, M. Viketoft and W. H. van der Putten. 2013, Soil and freshwater and marine sediment food webs: Their structure and function. *BioScience* 63(1):35-42. doi.org/10.1525/bio.2013.63.1.8.

Kunze, E. 2019. Biologically generated mixing in the ocean. *Annual Review of Marine Science* 11(1):215-226. doi:10.1146/annurev-marine-010318-095047.

Kuokkanen, T., and Y. Yamineva. 2013. Regulating geoengineering in international environmental law. *Carbon & Climate Law Review* 7(3):161-167. doi:10.21552/cclr/2013/3/261.

Kwon, E. Y., F. Primeau, and J. L. Sarmiento. 2009. The impact of remineralization depth on the air–sea carbon balance. *Nature Geoscience* 2(9):630-635. doi:10.1038/ngeo612.

La Plante, E. C., D. A. Simonetti, J. Wang, A. Al-Turki, X. Chen, D. Jassby, and G. N. Sant. 2021. Saline water-based mineralization pathway for gigatonne-scale CO_2 Management. *ACS Sustainable Chemistry and Engineering* 9(3):1073-1089. doi:10.1021/acssuschemeng.0c08561.

Lackner, K. S., C. H. Wendt, D. P. Butt, E. L. Joyce, and D. H. Sharp. 1995. Carbon dioxide disposal in carbonate minerals. *Energy* 20(11):1153-1170. doi:10.1016/0360-5442(95)00071-n.

Lamborg, C. H., K. O. Buesseler, and P. J. Lam. 2008. Sinking fluxes of minor and trace elements in the North Pacific Ocean measured during the VERTIGO program. *Deep-Sea Research Part II: Topical Studies in Oceanography* 55(14-15):1564-1577. doi:10.1016/j.dsr2.2008.04.012.

Langer, G., G. Nehrke, I. Probert, J. Ly, and P. Ziveri. 2009. Strain-specific responses of *Emiliania huxleyi* to changing seawater carbonate chemistry. *Biogeosciences* 6(11):2637-2646. doi:10.5194/bg-6-2637-2009.

Lau, J. D., G. G. Gurney, and J. Cinner. 2021. Environmental justice in coastal systems: Perspectives from communities confronting change. *Global Environmental Change* 66. doi:10.1016/j.gloenvcha.2020.102208.

Lau, W. W. Y. 2013. Beyond carbon: Conceptualizing payments for ecosystem services in blue forests on carbon and other marine and coastal ecosystem services. *Ocean & Coastal Management* 83:5-14. doi:10.1016/j.ocecoaman.2012.03.011.

Laufkötter, C., J. Zscheischler, and T. L. Frolicher. 2020. High-impact marine heatwaves attributable to human-induced global warming. *Science* 369(6511):1621-1625. doi:10.1126/science.aba0690.

Lavery, T. J., B. Roudnew, P. Gill, J. Seymour, L. Seuront, G. Johnson, J. G. Mitchell, and V. Smetacek. 2010. Iron defecation by sperm whales stimulates carbon export in the Southern Ocean. *Proceedings of the Royal Society B: Biological Sciences* 277(1699):3527-3531. doi:10.1098/rspb.2010.0863.

Lavery, T. J., B. Roudnew, J. Seymour, J. G. Mitchell, V. Smetacek, and S. Nicol. 2014. Whales sustain fisheries: Blue whales stimulate primary production in the Southern Ocean. *Marine Mammal Science* 30(3):888-904. doi:10.1111/mms.12108.

Law, C. S. 2008. Predicting and monitoring the effects of large-scale ocean iron fertilization on marine trace gas emissions. *Marine Ecology Progress Series* 364:283-288. doi:10.3354/meps07549.

Law, C. S., and R. D. Ling. 2001. Nitrous oxide flux and response to increased iron availability in the Antarctic Circumpolar Current. *Deep-Sea Research Part II: Topical Studies in Oceanography* 48(11-12):2509-2527. doi:10.1016/S0967-0645(01)00006-6.

Lawford-Smith, H., and A. Currie. 2017. Accelerating the carbon cycle: the ethics of enhanced weathering. *Biology Letters* 13 (4). doi:10.1098/rsbl.2016.0859.

Lawrence, M. W. 2014. Efficiency of carbon sequestration by added reactive nitrogen in ocean fertilisation. *International Journal of Global Warming* 6(1):15-33. doi:10.1504/ijgw.2014.058754.

Lebrato, M., M. Pahlow, J. R. Frost, M. Küter, P. J. Mendes, J.-C. Molinero, and A. Oschlies. 2019. Sinking of gelatinous zooplankton biomass increases deep carbon transfer efficiency globally. *Global Biogeochemical Cycles* 33 (12):1764-1783. doi:10.1029/2019gb006265.

Leinen, M. 2008. Building relationships between scientists and business in ocean iron fertilization. *Marine Ecology Progress Series* 364:251-256.

Lenton, T. M., and N. E. Vaughan. 2009. The radiative forcing potential of different climate geoengineering options. *Atmospheric Chemistry and Physics* 9(15):5539-5561. doi:10.5194/acp-9-5539-2009.

Lenzi, D. 2021. On the permissibility (or otherwise) of negative emissions. *Ethics, Policy & Environment*:1-14. doi:10.1080/21550085.2021.1885249.

Lepofsky, D., and M. Caldwell. 2013. Indigenous marine resource management on the Northwest Coast of North America. *Ecological Processes* 2(1). doi:10.1186/2192-1709-2-12.

Li, H., Z. Tang, N. Li, L. Cui, and X. Mao. 2020. Mechanism and process study on steel slag enhancement for CO_2 capture by seawater. *Applied Energy* 276. doi:10.1016/j.apenergy.2020.115515.

Liang, N., and H. Peng. 2005. A study of air-lift artificial upwelling. *Ocean Engineering* 32:731-745.

Liittge, A., and R. S. Arvidson. 2008. The mineral-water interface. Pp. 73-107 in *Kinetics of Water-Rock Interaction*, S. Brantley, J. Kubicki, and A. White, eds. New York: Springer.

Liu, C. C. K., and Q. Jin. 1995. Artificial upwelling in regular and random waves. *Ocean Engineering* 22(4):337-350. doi:10.1016/0029-8018(94)00019-4.

Liu, C. C. K., H. H. Chen, and L. C. Sun. 1989. Artificial upwelling and mixing. In *Proceedings of the International Workshop on Artificial Upwelling and Mixing in Coastal Waters.*

Liu, H., C. Consoli, and A. Zapantis. 2018. Overview of carbon capture and storage (CCS) facilities globally. Presented at 14th Greenhouse Gas Control Technologies Conference, Melbourne.

Longman, J., M. R. Palmer, and T. M. Gernon. 2020. Viability of greenhouse gas removal via artificial addition of volcanic ash to the ocean. *Anthropocene* 32:100264. doi:10.1016/j.ancene.2020.100264.

López-Martínez, S., C. Morales-Caselles, J. Kadar, and M. L. Rivas. 2021. Overview of global status of plastic presence in marine vertebrates. *Global Change Biology* 27(4):728-737. doi:10.1111/gcb.15416.

Lovelock, C. E., D. A. Friess, J. B. Kauffman, and J. W. Fourqurean. 2018. Human impacts on blue carbon ecosystems. Pp. 17-24 in *A Blue Carbon Primer,* L. Windham-Myers, S. Crooks, and T. G. Troxler, eds.. Boca Raton, FL: CRC Press.

Lovvorn, J. R., C. L. Baduini, and G. L. Hunt, Jr. 2001. Modeling underwater visual and filter feeding by planktivorous shearwaters in unusual sea conditions. *Ecology* 82(8):2342-2356. doi:10.1890/0012-9658(2001)082[2342:MUVAFF]2.0.CO;2.

Lüthi, D., M. Le Floch, B. Bereiter, T. Blunier, J.-M. Barnola, U. Siegenthaler, D. Raynaud, J. Jouzel, H. Fischer, K. Kawamura, and T. F. Stocker. 2008. High-resolution carbon dioxide concentration record 650,000-800,000 years before present. *Nature*, 453: 379-382.

Luo, J. Y., R. H. Condon, C. A. Stock, C. M. Duarte, C. H. Lucas, K. A. Pitt, and R. K. Cowen. 2020. Gelatinous zooplankton-mediated carbon flows in the global oceans: A data-driven modeling study. *Global Biogeochemical Cycles* 34(9):e2020GB006704. doi:10.1029/2020GB006704.

Mabon, L., S. Shackley, and N. Bower-Bir. 2014. Perceptions of sub-seabed carbon dioxide storage in Scotland and implications for policy: A qualitative study. *Marine Policy* 45:9-15. doi:10.1016/j.marpol.2013.11.011.

MacArthur, J. L. 2015. Challenging public engagement: participation, deliberation and power in renewable energy policy. *Journal of Environmental Studies and Sciences* 6(3):631-640. doi:10.1007/s13412-015-0328-7.

Magera, A. M., J. E. Mills Flemming, K. Kaschner, L. B. Christensen, and H. K. Lotze. 2013. Recovery trends in marine mammal populations. *PLoS ONE* 8(10):e77908. doi:10.1371/journal.pone.0077908.

Maia, M. R. G., A. J. M. Fonseca, H. M. Oliveira, C. Mendonça, and A. R. J. Cabrita. 2016. The potential role of seaweeds in the natural manipulation of rumen fermentation and methane production. *Scientific Reports* 6:32321. doi:10.1038/srep32321.

Maier-Reimer, E., U. Mikolajewicz, and A. Winguth. 1996. Future ocean uptake of CO_2: Interaction between ocean circulation and biology. *Climate Dynamics* 12(10):711-722.

Manning, D. A. C. 2010. Mineral sources of potassium for plant nutrition. A review. *Agronomy for Sustainable Development* 30(2):281-294. doi:10.1051/agro/2009023.

Mao, J., H. L. Burdett, R. A .R. McGill, J. Newton, P. Gulliver, and N. A. Kamenos. 2020. Carbon burial over the last four millennia is regulated by both climatic and land use change. *Global Change Biology* 26(4):2496-2504.

Marchetti, A., M. S. Parker, L. P. Moccia, E. O. Lin, A. L. Arrieta, F. Ribalet, M. E. P. Murphy, M. T. Maldonado, and E. V. Armbrust. 2009. Ferritin is used for iron storage in bloom-forming marine pennate diatoms. *Nature* 457(7228):467-470. doi:10.1038/nature07539.

Mariani, G., W. W. L. Cheung, A. Lyet, E. Sala, J. Mayorga, L. Velez, S. D. Gaines, T. Dejean, M. Troussellier, and D. Mouillot. 2020. Let more big fish sink: Fisheries prevent blue carbon sequestration–half in unprofitable areas. *Science Advances* 6(44):eabb4848. doi:10.1126/sciadv.abb4848.

Marinov, I., A. Gnanadesikan, J. R. Toggweiler, and J. L. Sarmiento. 2006. The Southern Ocean biogeochemical divide. *Nature* 441(7096):964-967. doi:10.1038/nature04883.

Markels, M., Jr., and R. Barber. 2002. Sequestration of CO_2 by ocean fertilization. Pp. 118-131 in *Environmental Challenges and Greenhouse Gas Control for Fossil Fuel Utilization in the 21st Century*, M. M. Maroto-Valer, C. Song, and Y. Soong, eds. Boston: Springer. https://doi.org/10.1007/978-1-4615-0773-4_9.

Markusson, N., D. McLaren, and D. Tyfield. 2018. Towards a cultural political economy of mitigation deterrence by negative emissions technologies (NETs). *Global Sustainability* 1. doi:10.1017/sus.2018.10.

Marshall, J. B. 2017. Geoengineering: A promising weapon or an unregulated disaster in the fight against climate change. *Journal of Land Use Law* 33:183.

Martin, A., P. Boyd, K. Buesseler, I. Cetinic, H. Claustre, S. Giering, S. Henson, X. Irigoien, I. Kriest, L. Memery, C. Robinson, G. Saba, R. Sanders, D. Siegel, M. Villa-Alfageme, and L. Guidi. 2020. The oceans' twilight zone must be studied now, before it is too late. *Nature* 580(7801):26-28. doi:10.1038/d41586-020-00915-7.

Martin, A. H., H. C. Pearson, G. K. Saba, and E. M. Olsen. 2021. Integral functions of marine vertebrates in the ocean carbon cycle and climate change mitigation. *One Earth* 4(5):680-693. doi:10.1016/j.oneear.2021.04.019.

Martin, J. H., and S. E. Fitzwater. 1988. Iron deficiency limits phytoplankton growth in the north-east Pacific subarctic. *Nature* 331(6154):341-343. doi:10.1038/331341a0.

Martin, J. H., G. A. Knauer, D. M. Karl, and W. W. Broenkow. 1987. VERTEX: Carbon cycling in the northeast Pacific. *Deep Sea Research Part A: Oceanographic Research Papers* 34(2):267-285. doi:10.1016/0198-0149(87)90086-0.

Martin, J. H., R. M. Gordon, and S. E. Fitzwater. 1990. Iron in Antarctic waters. *Nature* 345(6271):156-158. doi:10.1038/345156a0.

Martin, S. L., L. T. Ballance, and T. Groves. 2016. An ecosystem services perspective for the oceanic eastern tropical Pacific: commercial fisheries, carbon storage, recreational fishing, and biodiversity. 3(50). doi:10.3389/fmars.2016.00050.

Martz, T. R., J. G. Connery, and K. S. Johnson. 2010. Testing the Honeywell Durafet® for seawater pH applications. *Limnoogy and Oceanography Methods* 8(5):172-184. doi:https://doi.org/10.4319/lom.2010.8.172.

Martz, T., K. McLaughlin, and S. B. Weisberg. 2015. *Best Practices for Autonomous Measurement of Seawater pH with the Honeywell Durafet pH Sensor.* California Current Acidification Network (C-CAN).

Maruyama, S., T. Yabuki, T. Sato, K. Tsubaki, A. Komiya, M. Watanabe, H. Kawamura, and K. Tsukamoto. 2011. Evidences of increasing primary production in the ocean by Stommel's perpetual salt fountain. *Deep Sea Research Part I: Oceanographic Research Papers* 58(5):567-574. doi:10.1016/j.dsr.2011.02.012.

Mascia, M. B., C. A. Claus, and R. Naidoo. 2010. Impacts of marine protected areas on fishing communities. *Conservation Biology* 24(5):1424-1429. doi:10.1111/j.1523-1739.2010.01523.x.

Mason, C., G. Paxton, J. Parr, and N. Boughen. 2010. Charting the territory: Exploring stakeholder reactions to the prospect of seafloor exploration and mining in Australia. *Marine Policy* 34 (6):1374-1380. doi:10.1016/j.marpol.2010.06.012.

Masuda, T., K. Furuya, N. Kohashi, M. Sato, S. Takeda, M. Uchiyama, N. Horimoto, and T. Ishimaru. 2010. Lagrangian observation of phytoplankton dynamics at an artificially enriched subsurface water in Sagami Bay, Japan. *Journal of Oceanography* 66(6):801-813. doi:10.1007/s10872-010-0065-1.

Matabos, M., M. Hoeberechts, C. Doya, J. Aguzzi, J. Nephin, T. E. Reimchen, S. Leaver, R. M. Marx, A. Branzan Albu, R. Fier, U. Fernandez-Arcaya, and S. K. Juniper. 2017. Expert, crowd, students or algorithm: Who holds the key to deep-sea imagery "big data" processing? *Methods in Ecology and Evolution* 8(8):996-1004. doi:10.1111/2041-210X.12746.

Matsson, S., A. Metaxas, S. Forbord, S. Kristiansen, A. Handå, and B. A. Bluhm. 2021. Effects of outplanting time on growth, shedding and quality of *Saccharina latissima* (Phaeophyceae) in its northern distribution range. *Journal of Applied Phycology* 33:2415-2431. doi:10.1007/s10811-021-02441-z.

Matsuda, F., T. Tsurutani, J. P. Szyper, and P. Takahashi. 1998. Ultimate ocean ranch. Pp. 971-976 in *OCEANS'98. Conference Proceedings,* Vol. 2. IEEE Oceanic Engineering Society. doi:10.1109/OCEANS.1998.724382.

Mayer, L. M., L. L. Schick, R. F. L. Self, P. A. Jumars, R. H. Findlay, Z. Chen, and S. Sampson. 1997. Digestive environments of benthic macroinvertebrate guts: Enzymes, surfactants and dissolved organic matter. *Journal of Marine Research* 55(4):785-812. doi:10.1357/0022240973224247.

McClimans, T. A. 2008. Improved efficiency of bubble curtains. In *Proceedings, Coastal Technology Workshop Trondheim, Norway.*

McClimans, T. A., A. Handå, A. Fredheim, E. Lien, and K. I. Reitan. 2010. Controlled artificial upwelling in a fjord to stimulate non-toxic algae. *Aquacultural Engineering* 42(3):140-147. doi:10.1016/j.aquaeng.2010.02.002.

McDonnell, A. M. P., P. J. Lam, C. H. Lamborg, K. O. Buesseler, R. Sanders, J. S. Riley, C. Marsay, H. E. K. Smith, E. C. Sargent, R. S. Lampitt, and J. K. B. Bishop. 2015. The oceanographic toolbox for the collection of sinking and suspended marine particles. *Progress in Oceanography* 133:17-31. doi:10.1016/j.pocean.2015.01.007.

McGee, J., K. Brent, and W. Burns. 2017. Geoengineering the oceans: an emerging frontier in international climate change governance. *Australian Journal of Maritime & Ocean Affairs* 10(1):67-80. doi:10.1080/18366503.2017.1400899.

McKinnell, S. 2013. Challenges for the Kasatoshi volcano hypothesis as the cause of a large return of sockeye salmon (*Oncorhynchus nerka*) to the Fraser River in 2010. *Fisheries Oceanography* 22(4):337-344. doi:10.1111/fog.12023.

McLaren, D., K. A. Parkhill, A. Corner, N. E. Vaughan, and N. F. Pidgeon. 2016. Public conceptions of justice in climate engineering: Evidence from secondary analysis of public deliberation. *Global Environmental Change* 41:64-73. doi:10.1016/j.gloenvcha.2016.09.002.

McLaren, D. P., D. P. Tyfield, R. Willis, B. Szerszynski, and N. O. Markusson. 2019. Beyond "Net-Zero": A Case for Separate Targets for Emissions Reduction and Negative Emissions. 1 (4). doi:10.3389/fclim.2019.00004.

McLaren, D. 2020. Quantifying the potential scale of mitigation deterrence from greenhouse gas removal techniques. *Climatic Change* 162(4):2411-2428. doi:10.1007/s10584-020-02732-3.

Merk, C., G. Klaus, J. Pohlers, A. Ernst, K. Ott, and K. Rehdanz. 2019. Public perceptions of climate engineering: Laypersons' acceptance at different levels of knowledge and intensities of deliberation. *GAIA - Ecological Perspectives for Science and Society* 28(4):348-355. doi:10.14512/gaia.28.4.6.

Merriam-Webster. n.d. NIMBY.

Meysman, F. J., and F. Montserrat. 2017. Negative CO_2 emissions via enhanced silicate weathering in coastal environments. *Biology Letters* 13(4):20160905. doi:10.1098/rsbl.2016.0905.

Meysman, F. J., J. J. Middelburg, and C. H. Heip. 2006. Bioturbation: A fresh look at Darwin's last idea. *Trends in Ecology and Evolution* 21(12):688-695. doi:10.1016/j.tree.2006.08.002.

Meysman, F. J. R., N. Risgaard-Petersen, S. Y. Malkin, and L. P. Nielsen. 2015. The geochemical fingerprint of microbial long-distance electron transport in the seafloor. *Geochimica et Cosmochimica Acta* 152:122-142. doi:10.1016/j.gca.2014.12.014.

Mihnea, P. E. 1997. Major shifts in the phytoplankton community (1980-1994) in the Romanian Black Sea. *Oceanologica Acta* 20(1):119-129.

Milledge, J. J., B. Smith, P. W. Dyer, and P. Harvey. 2014. Macroalgae-derived biofuel: A review of methods of energy extraction from seaweed biomass. *Energies* 7(11):7194-7222. doi:10.3390/en7117194.

Millero, F. J. 2007. The marine inorganic carbon cycle. *Chemical Reviews* 107(2):308-341. doi:10.1021/cr0503557.

Millward, G. E., S. Kadam, and A. N. Jha. 2012. Tissue-specific assimilation, depuration and toxicity of nickel in *Mytilus edulis*. *Environmental Pollution* 162:406-412. doi:10.1016/j.envpol.2011.11.034.

Mizumukai, K., S. T., S. Tabeta, and D. Kitazawa. 2008. Numerical studies on ecological effects of artificial mixing of surface and bottom waters in density stratification in semi-enclosed bay and open sea. *Ecological Modelling* 214:251-270.

Mohan, A., O. Geden, M. Fridahl, H. J. Buck, and G. P. Peters. 2021. UNFCCC must confront the political economy of net-negative emissions. *One Earth*. doi:https://doi.org/10.1016/j.oneear.2021.10.001.

Montserrat, F., P. Renforth, J. Hartmann, M. Leermakers, P. Knops, and F. J. Meysman. 2017. Olivine dissolution in seawater: Implications for CO_2 sequestration through enhanced weathering in coastal environments. *Environmental Science and Technology* 51(7):3960-3972. doi:10.1021/acs.est.6b05942.

Moore, J. K., W. Fu, F. Primeau, G. L. Britten, K. Lindsay, M. Long, S. C. Doney, N. Mahowald, F. Hoffman, and J. T. Randerson. 2018. Sustained climate warming drives declining marine biological productivity. *Science* 359(6380):1139-1143.

Morel, F. M., and N. M. Price. 2003. The biogeochemical cycles of trace metals in the oceans. *Science* 300(5621):944-947. doi:10.1126/science.1083545.

Morrow, D. R., M. S. Thompson, A. Anderson, M. Batres, H. J. Buck, K. Dooley, O. Geden, A. Ghosh, S. Low, A. Njamnshi, J. Noël, O. O. Táíwò, S. Talati, and J. Wilcox. 2020. Principles for thinking about carbon dioxide removal in just climate policy. *One Earth* 3(2):150-153. doi:https://doi.org/10.1016/j.oneear.2020.07.015.

Muirhead, J. R., M. S. Minton, W. A. Miller, and G. M. Ruiz. 2015. Projected effects of the Panama Canal expansion on shipping traffic and biological invasions. *Diversity and Distributions* 21(1):75-87. doi:10.1111/ddi.12260.

Müller, J. D., B. Schneider, and G. Rehder. 2016. Long-term alkalinity trends in the Baltic Sea and their implications for CO_2-induced acidification. *Limnology and Oceanography* 61(6):1984-2002. doi:10.1002/lno.10349.

NASEM (National Academies of Sciences, Engineering, and Medicine). 2019. *Negative Emissions Technologies and Reliable Sequestration: A Research Agenda*. Washington, DC: The National Academies Press.

NASEM. 2021a. *Accelerating Decarbonization of the U.S. Energy System*. Washington, DC: The National Academies Press.

NASEM. 2021b. *Global Change Research Needs and Opportunities for 2022-2031*. Washington, DC: The National Academies Press.

NASEM. 2021c. *Reflecting Sunlight: Recommendations for Solar Geoengineering Research and Governance*. Washington, DC: The National Academies Press.

National Climate Task Force. 2021. *Conserving and Restoring America the Beautiful*. https://www.doi.gov/sites/doi.gov/files/report-conserving-and-restoring-america-the-beautiful-2021.pdf.

Navarrete, I. A, D. Y. Kim, C. Wilcox, D. C. Reed, D. W. Ginsburg, J. M. Dutton, J. Heidelberg, Y. Raut, and B. H. Wilcox. 2021. Effects of depth-cycling on nutrient uptake and biomass production in the giant kelp *Macrocystis pyrifera*. *Renewable and Sustainable Energy Reviews* 141:110747.

Nayak, A. R., and M. S. Twardowski. 2020. Breaking news for the ocean's carbon budget. *Science* 367(6479):738-739. doi:10.1126/science.aba7109.

NCEI (National Centers for Environmental Information). 2020. *Global Climate Report—Annual 2020*. National Oceanic and Atmospheric Administration. https://www.ncdc.noaa.gov/sotc/global/202013.

Needham, S. J., R. H. Worden, and D. McIlroy. 2004. Animal-sediment interactions: The effect of ingestion and excretion by worms on mineralogy. *Biogeosciences* 1(2):113-121. doi:10.5194/bg-1-113-2004.

Nemet, G. F., M. W. Callaghan, F. Creutzig, S. Fuss, J. Hartmann, J. Hilaire, W. F. Lamb, J. C. Minx, S. Rogers, and P. Smith. 2018. Negative emissions—Part 3: Innovation and upscaling. *Environmental Research Letters* 13(6):063003. doi:10.1088/1748-9326/aabff4.

Neori, A., T. Chopin, M. Troell, A. H. Buschmann, G. P. Kraemer, C. Halling, M. Shpigel, and C. Yarish. 2004. Integrated aquaculture: Rationale, evolution and state of the art emphasizing seaweed biofiltration in modern mariculture. *Aquaculture* 231(1-4):361-391.

NOAA (National Oceanic and Atmospheric Administration). 2015. Introduction to Stakeholder Participation. In *Social Science Tools for Coastal Programs*. National Oceanic and Atmospheric Administration.

NOAA. 2021. What is the EEZ?. Last Modified 02/26/21. https://oceanservice.noaa.gov/facts/eez.html.

Nørskov, J. K., J. Rossmeisl, A. Logadottir, L. Lindqvist, J. R. Kitchin, T. Bligaard, and H. Jónsson. 2004. Origin of the overpotential for oxygen reduction at a fuel-cell cathode. *Journal of Physical Chemistry B* 108(46):17886-17892. doi:10.1021/jp047349j.

NRC (National Research Council). 2008. *Public Participation in Environmental Assessment and Decision Making*. T. Dietz and P. C. Stern, eds. Washington, DC: The National Academies Press.

NRC. 2015a. *Climate Intervention: Carbon Dioxide Removal and Reliable Sequestration*. Washington, DC: The National Academies Press.

NRC. 2015b. *Climate Intervention: Reflecting Sunlight to Cool Earth*. Washington, DC: The National Academies Press. https://doi.org/10.17226/18988.

O'Brien, T. F., T. V. Bommaraju, and F. Hine. 2005. *Handbook of Chlor-Alkali Technology: Volume I: Fundamentals*, Vol. 1. New York, NY: Springer Science & Business Media.

Ocean artUp. 2021. FAQ. https://ocean-artup.eu/faq/.

Ocean Frontier Institute, Dillon Consulting. 2021. Ocean Frontier Institute's Indigenous Engagement Guide.

Oelkers, E. H., J. Declercq, G. D. Saldi, S. R. Gislason, and J. Schott. 2018. Olivine dissolution rates: A critical review. *Chemical Geology* 500:1-19. doi:10.1016/j.chemgeo.2018.10.008.

Okada, T., H. Abe, A. Murakami, T. Shimizu, K. Fujii, T. Wakabayashi, and M. Nakayama. 2020. A bilayer structure composed of Mg|Co-MnO_2 deposited on a $Co(OH)_2$ film to realize selective oxygen evolution from chloride-containing water. *Langmuir* 36(19):5227-5235. doi:10.1021/acs.langmuir.0c00547.

Olgun, N., S. Duggen, B. Langmann, M. Hort, C. F. Waythomas, L. Hoffmann, and P. Croot. 2013. Geochemical evidence of oceanic iron fertilization by the Kasatochi volcanic eruption in 2008 and the potential impacts on Pacific sockeye salmon. *Marine Ecology Progress Series* 488:81-88. doi:10.3354/meps10403.

Oloye, O., and A. P. O'Mullane. 2021. Electrochemical capture and storage of CO_2 as calcium carbonate. *ChemSusChem* 14(7):1767-1775. doi:10.1002/cssc.202100134.

Olsen, A. A., and J. D. Rimstidt. 2008. Oxalate-promoted forsterite dissolution at low pH. *Geochimica et Cosmochimica Acta* 72(7):1758-1766. doi:10.1016/j.gca.2007.12.026.

Opdyke, B. N., and J. C. G. Walker. 1992. Return of the coral reef hypothesis: Basin to shelf partitioning of $CaCO_3$ and its effect on atmospheric CO_2. *Geology* 20(8):733-736.

Oreska, M. P. J., K. J. McGlathery, L. R. Aoki, A. C. Berger, P. Berg, and L. Mullins. 2020. The greenhouse gas offset potential from seagrass restoration. *Scientific Reports* 10(1):7325. doi:10.1038/s41598-020-64094-1.

Oschlies, A. 2009. Impact of atmospheric and terrestrial CO_2 feedbacks on fertilization-induced marine carbon uptake. *Biogeosciences* 6(8):1603-1613. doi:10.5194/bg-6-1603-2009.

Oschlies, A., W. Koeve, W. Rickels, and K. Rehdanz. 2010a. Side effects and accounting aspects of hypothetical large-scale Southern Ocean iron fertilization. *Biogeosciences* 7(12):4014-4035. doi:10.5194/bg-7-4017-2010.

Oschlies, A., M. Pahlow, A. Yool, and R. J. Matear. 2010b. Climate engineering by artificial ocean upwelling: Channelling the sorcerer's apprentice. *Geophysical Research Letters* 37(4):L04701. doi:10.1029/2009GL041961.

Ouchi, K., K. Otsuka, and H. Omura. 2005. Recent advances of ocean nutrient enhancer "TAKUMI" project. Paper presented at the Sixth ISOPE Ocean Mining Symposium, Changsha, Hunan, China.

Pacala, S., and R. Socolow. 2004. Stabilization wedges: Solving the climate problem for the next 50 years with current technologies. *Science* 305(5686):968-972.

Page, B., G. Turan, A. Zapantis, J. Burrows, C. Consoli, J. Erikson, I. Havercroft, D. Kearns, H. Liu, and D. Rassool. 2020. *The Global Status of CCS 2020: Vital to Achieve Net Zero*. Global CCS Institute. https://www.globalccsinstitute.com/wp-content/uploads/2020/11/Global-Status-of-CCS-Report-2020_FINAL.pdf.

Paidar, M., V. Fateev, and K. Bouzek. 2016. Membrane electrolysis—History, current status and perspective. *Electrochimica Acta* 209:737-756. doi:10.1016/j.electacta.2016.05.209.

Paine, E. R., M. Schmid, P. W. Boyd, G. Diaz-Pulido, and C. L. Hurd. 2021. Rate and fate of dissolved organic carbon release by seaweeds: A missing link in the coastal ocean carbon cycle. *Journal of Phycology* 57(5):1375-1391. doi:10.1111/jpy.13198.

Pan, Y. W., W. Fan, T.-H. Huang, S.-L. Wang, and C.-T. A. Chen. 2015. Evaluation of the sinks and sources of atmospheric CO2 by artificial upwelling. *Science of the Total Environment* 511:692-702. doi:https://doi.org/10.1016/j.scitotenv.2014.11.060.

Pan, Y. W., W. Fan, D. H. Zhang, J. W. Chen, H. C. Huang, S. X. Liu, Z. P. Jiang, Y. N. Di, M. M. Tong, and Y. Chen. 2016. Research progress in artificial upwelling and its potential environmental effects. *Science China Earth Sciences* 59(2):236-248. doi:10.1007/s11430-015-5195-2.

Paquay, F. S., and R. E. Zeebe. 2013. Assessing possible consequences of ocean liming on ocean pH, atmospheric CO_2 concentration and associated costs. *International Journal of Greenhouse Gas Control* 17:183-188. doi:10.1016/j.ijggc.2013.05.005.

Parkhurst, D. L., and C. A. J. Appelo. 1999. *User's Guide to PHREEQC (Ver. 2): A Computer Program for Speciation, Batch-Reaction, One-Dimensional Transport, and Inverse Geochemical Calculations.* Water-Resources Investigations Report 99-4259. Reston, VA: U.S. Geological Survey.

Pärnamäe, R., S. Mareev, V. Nikonenko, S. Melnikov, N. Sheldeshov, V. Zabolotskii, H. V. M. Hamelers, and M. Tedesco. 2021. Bipolar membranes: A review on principles, latest developments, and applications. *Journal of Membrane Science* 617:118538. doi:10.1016/j.memsci.2020.118538.

Parsons, T. R., and F. A. Whitney. 2012. Did volcanic ash from Mt. Kasatoshi in 2008 contribute to a phenomenal increase in Fraser River sockeye salmon (*Oncorhynchus nerka*) in 2010? *Fisheries Oceanography* 21(5):374-377.

Passow, U., and C. A. Carlson. 2012. The biological pump in a high CO_2 world. *Marine Ecology Progress Series* 470:249-271. doi:10.3354/meps09985.

Pearson, H. C. 2019. *Assessment of Oceanic Blue Carbon in the UAE: Biomass Carbon Audit.* Arendal, Norway: AGEDI and GRID-Arendal.

Pedersen, M. F., K. Filbee-Dexter, N. L. Frisk, Z. Sárossy, and T. Wernberg. 2021. Carbon sequestration potential increased by incomplete anaerobic decomposition of kelp detritus. *Marine Ecology Progress Series* 660:53-67. doi:10.3354/meps13613.

Pershing, A. J., L. B. Christensen, N. R. Record, G. D. Sherwood, and P. B. Stetson. 2010. The impact of whaling on the ocean carbon cycle: Why bigger was better. *PLoS ONE* 5(8):e12444. doi:10.1371/journal.pone.0012444.

Pershing, A. J., R. B. Griffis, E. B. Jewett, C. T. Armstrong, J. F. Bruno, D. S. Busch, A. C. Haynie, S. A. Siedlecki, and D. Tommasi. 2018. Oceans and marine resources. Pp. 353-390 in *Impacts, Risks, and Adaptation in the United States: Fourth National Climate Assessment*, Vol. II, D. R. Reidmiller, C. W. Avery, D. R. Easterling, K. E. Kunkel, K. L. M. Lewis, T. K. Maycock and B. C. Stewart, eds. Washington, DC: U.S. Global Change Research Program.

Pessarrodona, A., K. Filbee-Dexter, T. Alcoverro, J. Boada, C. J. Feehan, S. Fredriksen, S. P. Grace, Y. Nakamura, C. A. Narvaez, K. M. Norderhaug, and T. Wernberg. 2021. Homogenization and miniaturization of habitat structure in temperate marine forests. *Global Change Biology* 27 (20):5262-5275. doi:10.1111/gcb.15759.

Peteiro, C., and Ó. Freire. 2013. Biomass yield and morphological features of the seaweed *Saccharina latissima* cultivated at two different sites in a coastal bay in the Atlantic coast of Spain. *Journal of Applied Phycology* 25(1):205-213. doi:10.1007/s10811-012-9854-9.

Petrova, S., and D. Marinova. 2013. Social impacts of mining: Changes within the local social landscape. *Rural Society* 22(2):153-165. doi:10.5172/rsj.2013.22.2.153.

Petrykin, V., K. Macounova, O. A. Shlyakhtin, and P. Krtil. 2010. Tailoring the selectivity for electrocatalytic oxygen evolution on ruthenium oxides by zinc substitution. *Angewandte Chemie International Edition in English* 49(28):4813-4815. doi:10.1002/anie.200907128.

Pfeffer, C., S. Larsen, J. Song, M. Dong, F. Besenbacher, R. L. Meyer, K. U. Kjeldsen, L. Schreiber, Y. A. Gorby, M. Y. El-Naggar, K. M. Leung, A. Schramm, N. Risgaard-Petersen, and L. P. Nielsen. 2012. Filamentous bacteria transport electrons over centimetre distances. *Nature* 491(7423):218-221. doi:10.1038/nature11586.

Pilorgé, H., N. McQueen, D. Maynard, P. Psarras, J. He, T. Rufael, and J. Wilcox. 2020. Cost analysis of carbon capture and sequestration of process emissions from the US industrial sector. *Environmental Science & Technology* 54(12):7524-7532.

Pilson, M. E. Q. 1998. Chemical extraction of useful substance from sea water. Pp. 341-350 in *Introduction to the Chemistry of the Sea.* Cambridge: Prentice Hall.

Pinti, J., T. DeVries, T. Norin, C. Serra-Pompei, R. Proud, D. A. Siegel, T. Kiørboe, C. M. Petrik, K. H. Andersen, and A. S. Brierley. 2021. Metazoans, migrations, and the ocean's biological carbon pump. *bioRxiv*. doi:10.1101/2021.03.22.436489.

Platonov, N. G., I. N. Mordvintsev, and V. V. Rozhnov. 2013. The possibility of using high resolution satellite images for detection of marine mammals. *Biology Bulletin* 40(2):197-205. doi:10.1134/S1062359013020106.

Platt, D., M. Workman, and S. Hall. 2018. A novel approach to assessing the commercial opportunities for greenhouse gas removal technology value chains: Developing the case for a negative emissions credit in the UK. *Journal of Cleaner Production* 203:1003-1018. doi:https://doi.org/10.1016/j.jclepro.2018.08.291.

Pollard, R. T., I. Salter, R. J. Sanders, M. I. Lucas, C. M. Moore, R. A. Mills, P. J. Statham, J. T. Allen, A. R. Baker, D. C. E. Bakker, M. A. Charette, S. Fielding, G. R. Fones, M. French, A. E. Hickman, R. J. Holland, J. A. Hughes, T. D. Jickells, R. S. Lampitt, P. J. Morris, F. H. Nédélec, M. Nielsdóttir, H. Planquette, E. E. Popova, A. J. Poulton, J. F. Read, S. Seeyave, T. Smith, M. Stinchcombe, S. Taylor, S. Thomalla, H. J. Venables, R. Williamson, and M. V. Zubkov. 2009. Southern Ocean deep-water carbon export enhanced by natural iron fertilization. *Nature* 457(7229):577-580. doi:10.1038/nature07716.

Pörtner, H. O., R. J. Scholes, J. Agard, E. Archer, A. Arneth, X. Bai, D. Barnes, M. Burrows, L. Chan, W. L. Cheung, S. Diamond, C. Donatti, C. Duarte, N. Eisenhauer, W. Foden, M. A. Gasalla, C. Handa, T. Hickler, O. Hoegh-Guldberg, K. Ichii, U. Jacob, G. Insarov, W. Kiessling, P. Leadley, R. Leemans, L. Levin, M. Lim, S. Maharaj, S. Managi, P. A. Marquet, P. McElwee, G. Midgley, T. Oberdorff, D. Obura, E. Osman, R. Pandit, U. Pascual, A. P. F. Pires, A. Popp, V. Reyes-García, M. Sankaran, J. Settele, Y. J. Shin, D. W. Sintayehu, P. Smith, N. Steiner, B. Strassburg, R. Sukumar, C. Trisos, A. L. Val, J. Wu, E. Aldrian, C. Parmesan, R. Pichs-Madruga, D. C. Roberts, A. D. Rogers, S. Díaz, M. Fischer, S. Hashimoto, S. Lavorel, N. Wu, and H. T. Ngo. 2021. *Biodiversity and Climate Change: Workshop Report.* Intergovernmental Science-Policy Platform for Biodiversity and Ecosystem Services and Intergovernmental Panel on Climate Chang.

Pozo, C., Á. Galán-Martín, D. M. Reiner, N. Mac Dowell, and G. Guillén-Gosálbez. 2020. Equity in allocating carbon dioxide removal quotas. *Nature Climate Change* 10(7):640-646. doi:10.1038/s41558-020-0802-4.

Prall, E. 2021. Is seaweed a fish? US seaweed fisheries and climate change. *Natural Resources & Environment* 35 (4):14-18.

Primeau, F. 2005. Characterizing transport between the surface mixed layer and the ocean interior with a forward and adjoint global ocean transport model. *Journal of Physical Oceanography* 35(4):545-564. doi:10.1175/JPO2699.1.

Proelss, A. 2012. "International Legal Challenges Concerning Marine Scientific Research in the Era of Climate Change." LOSI-KIOST Conference on Securing the Ocean for the Next Generation, Seoul, Korea.

Proelss, A., and C. Hong. 2012. Ocean upwelling and international law. *Ocean Development and International Law* 43 (4):371-385. doi:10.1080/00908320.2012.726843.

Queirós, A. M., N. Stephens, S. Widdicombe, K. Tait, S. J. McCoy, J. Ingels, S. Rühl, R. Airs, A. Beesley, G. Carnovale, P. Cazenave, S. Dashfield, E. Hua, M. Jones, P. Lindeque, C. L. McNeill, J. Nunes, H. Parry, C. Pascoe, C. Widdicombe, T. Smyth, A. Atkinson, D. Krause-Jensen, and P. J. Somerfield. 2019. Connected macroalgal-sediment systems: Blue carbon and food webs in the deep coastal ocean. *Ecological Monographs* 89(3):e01366. doi:10.1002/ecm.1366.

Raddatz, J., A. Rüggeberg, S. Flögel, E. C. Hathorne, V. Liebetrau, A. Eisenhauer, and W. Chr Dullo. 2014. The influence of seawater pH on U/Ca ratios in the scleractinian cold-water coral *Lophelia pertusa. Biogeosciences* 11(7):1863-1871. doi:10.5194/bg-11-1863-2014.

Rädecker, N., C. Pogoreutz, M. Ziegler, A. Ashok, M. M. Barreto, V. Chaidez, C. G. B. Grupstra, Y. M. Ng, G. Perna, M. Aranda, and C. R. Voolstra. 2017. Assessing the effects of iron enrichment across holobiont compartments reveals reduced microbial nitrogen fixation in the Red Sea coral *Pocillopora verrucosa. Ecology and Evolution* 7(16):6614-6621. doi:10.1002/ece3.3293.

Raitzsch, M., H. Kuhnert, E. C. Hathorne, J. Groeneveld, and T. Bickert. 2011. U/Ca in benthic foraminifers: A proxy for the deep-sea carbonate saturation. *Geochemistry, Geophysics, Geosystems* 12(6):Q06019. doi:10.1029/2010gc003344.

Rao, A. M. F., L. Polerecky, D. Ionescu, F. J. R. Meysman, and D. de Beer. 2012. The influence of pore-water advection, benthic photosynthesis, and respiration on calcium carbonate dynamics in reef sands. *Limnology and Oceanography* 57(3):809-825. doi:10.4319/lo.2012.57.3.0809.

Rao, C. K., and V. K. Indusekhar. 1987. Carbon, nitrogen and phosphorus ratios in seawater and seaweeds of Saurashtra, northwest coast of India. *Indian Journal of Geo-Marine Sciences* 16(2):117-121.

Rassweiler, A., D. C. Reed, S. L. Harrer, and J. C. Nelson. 2018. Improved estimates of net primary production, growth, and standing crop of *Macrocystis pyrifera* in Southern California. *Ecology* 99(9):2132. doi:10.1002/ecy.2440.

Ratnarajah, L., J. Melbourne-Thomas, M. P. Marzloff, D. Lannuzel, K. M. Meiners, F. Chever, S. Nicol, and A. R. Bowie. 2016. A preliminary model of iron fertilisation by baleen whales and Antarctic krill in the Southern Ocean: Sensitivity of primary productivity estimates to parameter uncertainty. *Ecological Modelling* 320:203-212. doi:10.1016/j.ecolmodel.2015.10.007.

Ratnarajah, L., S. Nicol, and A. R. Bowie. 2018. Pelagic iron recycling in the Southern Ocean: Exploring the contribution of marine animals. *Frontiers in Marine Science* 5:109. doi:10.3389/fmars.2018.00109.

Rau, G. H. 2008. Electrochemical splitting of calcium carbonate to increase solution alkalinity: Implications for mitigation of carbon dioxide and ocean acidity. *Environmental Science & Technology* 42(23):8935-8940. doi.org/10.1021/es800366q.

Rau, G. H. 2011. CO_2 mitigation via capture and chemical conversion in seawater. *Environmental Science & Technology* 45(3):1088-1092. doi:10.1021/es102671x.

Rau, G. H., and K. Caldeira. 1999. Enhanced carbonate dissolution: A means of sequestering waste CO_2 as ocean bicarbonate. *Energy Conversion and Management* 40(17):1803-1813. doi:10.1016/S0196-8904(99)00071-0.

Rau, G. H., S. A. Carroll, W. L. Bourcier, M. J. Singleton, M. M. Smith, and R. D. Aines. 2013. Direct electrolytic dissolution of silicate minerals for air CO_2 mitigation and carbon-negative H_2 production. *Proceedings of the National Academy of Sciences of the United States of America* 110(25):10095-10100. doi:10.1073/pnas.1222358110.

Rau, G. H., H. D. Willauer, and Z. J. Ren. 2018. The global potential for converting renewable electricity to negative-CO_2-emissions hydrogen. *Nature Climate Change* 8(7):621-625. doi:10.1038/s41558-018-0203-0.

Read, A. J., P. Drinker, and S. Northridge. 2006. Bycatch of marine mammals in U.S. and global fisheries. *Conservation Biology* 20(1):163-169. doi:10.1111/j.1523-1739.2006.00338.x.

Reed, D. C., A. Rassweiler, and K. K. Arkema. 2008. Biomass rather than growth rate determines variation in net primary production by giant kelp. *Ecology* 89(9):2493-2505.

Reed, D. C., C. A. Carlson, E. R. Halewood, J. Cl. Nelson, S. L. Harrer, A. Rassweiler, and R. J. Miller. 2015. Patterns and controls of reef-scale production of dissolved organic carbon by giant kelp *Macrocystis pyrifera*. *Limnology and Oceanography* 60(6):1996-2008. doi:10.1002/lno.10154.

Regnier, P., P. Friedlingstein, P. Ciais, F. T. Mackenzie, N. Gruber, I. A. Janssens, G. G. Laruelle, R. Lauerwald, S. Luyssaert, A. J. Andersson, S. Arndt, C. Arnosti, A. V. Borges, A. W. Dale, A. Gallego-Sala, Y. Goddéris, N. Goossens, J. Hartmann, C. Heinze, T. Ilyina, F. Joos, D. E. LaRowe, J. Leifeld, F. J. R. Meysman, G. Munhoven, P. A. Raymond, R. Spahni, P. Suntharalingam, and M. Thullner. 2013. Anthropogenic perturbation of the carbon fluxes from land to ocean. *Nature Geoscience* 6(8):597-607. doi:10.1038/ngeo1830.

Reith, F., W. Koeve, D. P. Keller, J. Getzlaff, and A. Oschlies. 2019. Meeting climate targets by direct CO2 injections: what price would the ocean have to pay? *Earth System Dynamics* 10(4):711-727. doi:10.5194/esd-10-711-2019.

Renforth, P. 2012. The potential of enhanced weathering in the UK. *International Journal of Greenhouse Gas Control* 10:229-243. doi:10.1016/j.ijggc.2012.06.011.

Renforth, P. 2019. The negative emission potential of alkaline materials. *Nature Communications* 10(1):1401. doi:10.1038/s41467-019-09475-5.

Renforth, P., and G. Henderson. 2017. Assessing ocean alkalinity for carbon sequestration. *Reviews of Geophysics* 55(3):636-674. doi:10.1002/2016rg000533.

Renforth, P., B. G. Jenkins, and T. Kruger. 2013. Engineering challenges of ocean liming. *Energy* 60:442-452. doi:10.1016/j.energy.2013.08.006.

Reynolds, J. 2015. The international legal framework for climate engineering. *SSRN Electronic Journal*. doi:10.2139/ssrn.2586927.

Reynolds, J. L. 2018a. Governing Experimental Responses: Negative Emissions Technologies and Solar Climate Engineering. Pp. 285-302 in *Governing Climate Change: Polycentricity in Action?*, A. Jordan, D. Huitema, H. van Asselt and J. Forster, eds., 285-302. Cambridge: Cambridge University Press.

Reynolds, J. L. 2018b. "International Law." In *Climate Engineering and the Law: Regulation and Liability for Solar Radiation Management and Carbon Dioxide Removal*, edited by Michael B. Gerrard and Tracy Hester, 57-153. Cambridge: Cambridge University Press.

Riahi, K., D. P. van Vuuren, E. Kriegler, J. Edmonds, B. C. O'Neill, S. Fujimori, N. Bauer, K. Calvin, R. Dellink, O. Fricko, W. Lutz, A. Popp, J. C. Cuaresma, S. Kc, M. Leimbach, L. Jiang, T. Kram, S. Rao, J. Emmerling, K. Ebi, T. Hasegawa, P. Havlik, F. Humpenöder, L. A. Da Silva, S. Smith, E. Stehfest, V. Bosetti, J. Eom, D. Gernaat, T. Masui, J. Rogelj, J. Strefler, L. Drouet, V. Krey, G. Luderer, M. Harmsen, K. Takahashi, L. Baumstark, J. C. Doelman, M. Kainuma, Z. Klimont, G. Marangoni, H. Lotze-Campen, M. Obersteiner, A. Tabeau, and M. Tavoni. 2017. The shared socioeconomic pathways and their energy, land use, and greenhouse gas emissions implications: An overview. *Global Environmental Change* 42:153-168. doi:https://doi.org/10.1016/j.gloenvcha.2016.05.009.

Ricke, K., D. Ivanova, T. McKie, and M. Rugenstein. 2021. Reversing Sahelian droughts. *Geophysical Research Letters* 48(14). doi:10.1029/2021GL093129.

Rickels, W, A. Proelß, O. Geden, J. Burhenne, and M. Fridahl. 2020. The future of (negative) emissions trading in the European Union. Kiel Working Paper.

Riebesell, U., A. K. Rtzinger, and A. Oschlies. 2009. Sensitivities of marine carbon fluxes to ocean change. *Proceedings of the National Academy of Sciences of the United States of America* 106(49):20602-20609. doi:10.1073/pnas.0813291106.

Ries, J. B., A. L. Cohen, and D. C. McCorkle. 2009. Marine calcifiers exhibit mixed responses to CO_2-induced ocean acidification. *Geology* 37(12):1131-1134. doi:10.1130/G30210A.1.

Ries, J. B., M. N. Ghazaleh, B. Connolly, I. Westfield, and K. D Castillo. 2016. Impacts of seawater saturation state (ΩA = 0.4–4.6) and temperature (10, 25 C) on the dissolution kinetics of whole-shell biogenic carbonates. *Geochimica et Cosmochimica Acta* 192:318-337.

Rimstidt, J. D., S. L. Brantley, and A. A. Olsen. 2012. Systematic review of forsterite dissolution rate data. *Geochimica et Cosmochimica Acta* 99:159-178. doi:10.1016/j.gca.2012.09.019.

Riosmena-Rodríguez, R., W. Nelson, and J. Aguirre. 2017. *Rhodolith/Maërl Beds: A Global Perspective*. Switzerland: Springer.

Risgaard-Petersen, N., A. Revil, P. Meister, and L. P. Nielsen. 2012. Sulfur, iron-, and calcium cycling associated with natural electric currents running through marine sediment. *Geochimica et Cosmochimica Acta* 92:1-13. doi:10.1016/j.gca.2012.05.036.

Roberts, C. M., B. C. O'Leary, D. J. McCauley, P. M. Cury, C. M. Duarte, J. Lubchenco, D. Pauly, A. Saenz-Arroyo, U. R. Sumaila, R. W. Wilson, B. Worm, and J. C. Castilla. 2017. Marine reserves can mitigate and promote adaptation to climate change. *Proceedings of the National Academy of Sciences of the United States of America* 114(24):6167-6175. doi:10.1073/pnas.1701262114.

Rockwood, R. C., J. Calambokidis, and J. Jahncke. 2017. High mortality of blue, humpback and fin whales from modeling of vessel collisions on the US West Coast suggests population impacts and insufficient protection. *PLoS ONE* 12(8):e0183052.

Rogelj, J., O. Geden, A. Cowie, and A. Reisinger. 2021. Three ways to improve net-zero emissions targets. *Nature* 591 (7850):365-368.

Roman, J., J. A. Estes, L. Morissette, C. Smith, D. Costa, J. McCarthy, J. B. Nation, S. Nicol, A. Pershing, and V. Smetacek. 2014. Whales as marine ecosystem engineers. *Frontiers in Ecology and the Environment* 12(7):377-385. doi:10.1890/130220.

Roman, J., M. M. Dunphy-Daly, D. W. Johnston, and A. J. Read. 2015. Lifting baselines to address the consequences of conservation success. *Trends in Ecology and Evolution* 30(6):299-302. doi:10.1016/j.tree.2015.04.003.

Roman, J., V. DeLauer, I. Altman, B. Fisher, R. Boumans, and L. Kaufman. 2018. Stranded capital: Environmental stewardship is part of the economy, too. *Frontiers in Ecology and the Environment* 16(3):169-175. doi:10.1002/fee.1780.

Roman, J., J. J. Kiszka, H. Pearson, M. S. Savoca, and C. Smith. 2021. *Ecological Roles and Impacts of Large Cetaceans in Marine Ecosystems*. International Whaling Committee.

Russell, A. D., B. Hönisch, H. J. Spero, and D. W. Lea. 2004. Effects of seawater carbonate ion concentration and temperature on shell U, Mg, and Sr in cultured planktonic foraminifera. *Geochimica et Cosmochimica Acta* 68(21):4347-4361. doi:10.1016/j.gca.2004.03.013.

Saba, G. K., A. B. Burd, J. P. Dunne, S. Hernández-León, A. H. Martin, K. A. Rose, J. Salisbury, D. K. Steinberg, C. N. Trueman, R. W. Wilson, and S. E. Wilson. 2021. Toward a better understanding of fish-based contribution to ocean carbon flux. *Limnology and Oceanography* 66(5):1639-1664. doi:10.1002/lno.11709.

Sabine, C. L., and T. Tanhua. 2010. Estimation of anthropogenic CO_2 inventories in the ocean. *Annual Review of Marine Science* 2(1):175-198. doi:10.1146/annurev-marine-120308-080947.

Sabine, C. L., R. A. Feely, N. Gruber, R. M. Key, K. Lee, J. L. Bullister, R. Wanninkhof, C. S. Wong, D. W. R. Wallace, B. Tilbrook, F. J. Millero, T. H. Peng, A. Kozyr, T. Ono, and A. F. Rios. 2004. The oceanic sink for anthropogenic CO_2. *Science* 305(5682):367-371. doi:10.1126/science.1097403.

Sakamoto, T., and D. A. Bryant. 2001. Requirement of nickel as an essential micronutrient for the utilization of urea in the marine cyanobacterium *Synechococcus* sp. PCC 7002. *Microbes and Environments* 16(3):177-184. doi:10.1264/jsme2.2001.177.

Sala, E., J. Mayorga, D. Bradley, R. B. Cabral, T. B. Atwood, A. Auber, W. Cheung, C. Costello, F. Ferretti, A. M. Friedlander, S. D. Gaines, C. Garilao, W. Goodell, B. S. Halpern, A. Hinson, K. Kaschner, K. Kesner-Reyes, F. Leprieur, J. McGowan, L. E. Morgan, D. Mouillot, J. Palacios-Abrantes, H. P. Possingham, K. D. Rechberger, B. Worm, and J. Lubchenco. 2021. Protecting the global ocean for biodiversity, food and climate. *Nature* 592(7854):397-402. doi:10.1038/s41586-021-03371-z.

Sarmiento, J. L., and J. R. Toggweiler. 1984. A new model for the role of the oceans in determining atmospheric P CO_2. *Nature* 308 (5960):621-624. doi:10.1038/308621a0.

Sarmiento, J. L., and N. Gruber. 2002. Sinks for anthropogenic carbon. *Physics Today* 55(8):30. doi:10.1063/1.1510279.

Sarmiento, J. L., and N. Gruber. 2013. *Ocean Biogeochemical Dynamics*. Princeton, Woodstock: Princeton University Press.

Sarmiento, J. L., N. Gruber, M. A. Brzezinski, and J. P. Dunne. 2004. High-latitude controls of thermocline nutrients and low latitude biological productivity. *Nature* 427(6969):56-60. doi:10.1038/nature02127.

Sarmiento, J. L., R. D. Slater, J. Dunne, A. Gnanadesikan, and M. R. Hiscock. 2010. Efficiency of small scale carbon mitigation by patch iron fertilization. *Biogeosciences* 7(11):3593-3624. doi:10.5194/bg-7-3593-2010.

Sawall, Y., M. Harris, Ma. Lebrato, M. Wall, and E. Y. Feng. 2020. Discrete pulses of cooler deep water can decelerate coral bleaching during thermal stress: Implications for artificial upwelling during heat stress events. *Frontiers in Marine Science* 7:00720. doi:10.3389/fmars.2020.00720.

Schenuit, F., R. Colvin, M. Fridahl, B. McMullin, A. Reisinger, D. L. Sanchez, S. M. Smith, A. Torvanger, A. Wreford, and O. Geden. 2021. Carbon dioxide removal policy in the making: Assessing developments in 9 OECD cases. 3 (7). doi:10.3389/fclim.2021.638805.

Schlosberg, D., and L. B. Collins. 2014. From environmental to climate justice: climate change and the discourse of environmental justice. *WIREs Climate Change* 5 (3):359-374. doi:10.1002/wcc.275.

Schmitz, O. J., C. C. Wilmers, S. J. Leroux, C. E. Doughty, T. B. Atwood, M. Galetti, A. B. Davies, and S. J. Goetz. 2018. Animals and the zoogeochemistry of the carbon cycle. *Science* 362(6419):eaar3213. doi:10.1126/science.aar3213.

Schroeder, R. E., A. L. Green, E. E. DeMartini, and J. C. Kenyon. 2008. Long-term effects of a ship-grounding on coral reef fish assemblages at Rose Atoll, American Samoa. *Bulletin of Marine Science* 82(3):345-364.

Schuiling, R. D., and P. Krijgsman. 2006. Enhanced weathering: An effective and cheap tool to sequester CO_2. *Climatic Change* 74:349-354.

Schuiling, R. D., and O. Tickell. 2010. Enhanced weathering of olivine to capture CO_2. *Journal of Applied Geochemistry* 12(4):510-519.

Schuiling, R. D., and P. L. de Boer. 2011. Rolling stones; fast weathering of olivine in shallow seas for cost-effective CO_2 capture and mitigation of global warming and ocean acidification. *Earth System Dynamics Discussions* 2:551-568. doi:10.5194/esdd-2-551-2011.

Scott, A. L., P. H. York, and M. A. Rasheed. 2021. Herbivory has a major influence on structure and condition of a Great Barrier Reef subtropical seagrass meadow. *Estuaries and Coasts* 44(2):506-521. doi:10.1007/s12237-020-00868-0.

Scott, K. N. 2013. Regulating ocean fertilization under international law: The risks. *Carbon & Climate Law Review* 7(2):108-116.

Seddon, N., A. Chausson, P. Berry, C. A. J. Girardin, A. Smith, and B. Turner. 2020. Understanding the value and limits of nature-based solutions to climate change and other global challenges. *Philosophical Transactions of the Royal Society B: Biological Sciences* 375(1794):20190120. doi:10.1098/rstb.2019.0120.

Seddon, N., A. Smith, P. Smith, I. Key, A. Chausson, C. Girardin, J. House, S. Srivastava, and B. Turner. 2021. Getting the message right on nature-based solutions to climate change. *Global Change Biology* 27(8):1518-1546. doi:10.1111/gcb.15513.

Seddon, N., B. Turner, P. Berry, A. Chausson, and C. A. J. Girardin. 2019. Grounding nature-based climate solutions in sound biodiversity science. *Nature Climate Change* 9(2):84-87. doi:10.1038/s41558-019-0405-0.

Seibel, B. A., and P. J. Walsh. 2001. Carbon cycle: Potential impacts of CO_2 injection on deep-sea biota. *Science* 294(5541):319-320. doi:10.1126/science.1065301.

Seibel, B. A., and P. J. Walsh. 2003. Biological impacts of deep-sea carbon dioxide injection inferred from indices of physiological performance. *Journal of Experimental Biology* 206(4):641-650. doi:10.1242/jeb.00141.

Seiffert, F., N. Bandow, J. Bouchez, F. Von Blanckenburg, and A. A. Gorbushina. 2014. Microbial colonization of bare rocks: Laboratory biofilm enhances mineral weathering. *Procedia Earth and Planetary Science* 10:123-129.

Sengupta, M. 2021. *Environmental Impacts of Mining: Monitoring, Restoration, and Control*, 2nd ed. Boca raton, FL: CRC Press.

Sharma, R., S. E. Swearer, R. L. Morris, and E. M. A. Strain. 2021. Testing the efficacy of sea urchin exclusion methods for restoring kelp. *Marine Environmental Research* 170:105439. doi:https://doi.org/10.1016/j.marenvres.2021.105439.

Shepherd, J. G. 2009. *Geoengineering the Climate: Science, Governance and Uncertainty*. London: The Royal Society.

Shirokova, L. S., P. Bénézeth, O. S. Pokrovsky, E. Gerard, B. Ménez, and H. Alfredsson. 2012. Effect of the heterotrophic bacterium *Pseudomonas reactans* on olivine dissolution kinetics and implications for CO_2 storage in basalts. *Geochimica et Cosmochimica Acta* 80:30-50.

Siegel, D. A., E. Fields, and K. O. Buesseler. 2008. A bottom-up view of the biological pump: Modeling source funnels above ocean sediment traps. *Deep-Sea Research Part I: Oceanographic Research Papers* 55(1):108-127. doi:10.1016/j.dsr.2007.10.006.

Siegel, D. A., K. O. Buesseler, S. C. Doney, S. F. Sailley, M. J. Behrenfeld, and P. W. Boyd. 2014. Global assessment of ocean carbon export by combining satellite observations and food-web models. *Global Biogeochemical Cycles* 28(3):181-196. doi:10.1002/2013gb004743.

Siegel, D. A., K. O. Buesseler, M. J. Behrenfeld, C. R. Benitez-Nelson, E. Boss, M. A. Brzezinski, A. Burd, C. A. Carlson, E. A. D'Asaro, S. C. Doney, M. J. Perry, R. H. R. Stanley, and D. K. Steinberg. 2016. Prediction of the export and fate of global ocean net primary production: The EXPORTS Science Plan. *Frontiers in Marine Science* 3:22. doi:10.3389/fmars.2016.00022.

Siegel, D. A., T. DeVries, S. Doney, and T. Bell. 2021a. Assessing the sequestration time scales of some ocean-based carbon dioxide reduction strategies. *Environmental Research Letters* 16(10):104003.

Siegel, D. A., I. Cetinić, J. R. Graff, C. M. Lee, N. Nelson, M. J. Perry, I. S. Ramos, D. K. Steinberg, K. Buesseler, R. Hamme, A. J. Fassbender, D. Nicholson, M. M. Omand, M. Robert, A. Thompson, V. Amaral, M. Behrenfeld, C. Benitez-Nelson, K. Bisson, E. Boss, P. W. Boyd, M. Brzezinski, K. Buck, A. Burd, S. Burns, S. Caprara, C. Carlson, N. Cassar, H. C., E. D'Asaro, C. Durkin, Z. Erickson, M. L. Estapa, E. Fields, J. Fox, S. Freeman, S. Gifford, W. Gong, D. Gray, L. Guidi, N. Haëntjens, K. Halsey, Y. Huot, D. Hansell, B. Jenkins, L. Karp-Boss, S. Kramer, P. Lam, J.-M. Lee, A. Maas, O. Marchal, A. Marchetti, A. McDonnell, H. McNair, S. Menden-Deuer, F. Morison, A. K. Niebergall, U. Passow, B. Popp, G. Potvin, L. Resplandy, M. Roca-Martí, C. Roesler, T. Rynearson, S. Traylor, A. Santoro, K. D. Seraphin, H. M. Sosik, K. Stamieszkin, B. Stephens, W. Tang, B. Van Mooy, Y.g Xiong, and X. Zhang. 2021b. An operational overview of the EXport Processes in the Ocean from RemoTe Sensing (EXPORTS) Northeast Pacific field deployment. *Elementa: Science of the Anthropocene* 9(1):00107. doi:10.1525/elementa.2020.00107.

Sigman, D. M., M. P. Hain, and G. H. Haug. 2010. The polar ocean and glacial cycles in atmospheric CO_2 concentration. *Nature* 466(7302):47-55. doi:10.1038/nature09149.

Silver, M. W., S. Bargu, S. L. Coale, C. R. Benitez-Nelson, A. C. Garcia, K. J. Roberts, E. Sekula-Wood, K. W. Bruland, and K. H. Coale. 2010. Toxic diatoms and domoic acid in natural and iron enriched waters of the oceanic Pacific. *Proceedings of the National Academy of Sciences of the United States of America* 107(48):20762-7. doi:10.1073/pnas.1006968107.

Sincovich, A., T. Gregory, A. Wilson, and S. Brinkman. 2018. The social impacts of mining on local communities in Australia. *Rural Society* 27(1):18-34. doi:10.1080/10371656.2018.1443725.

Singh, A., L. T. Bach, C. R. Löscher, A. J. Paul, N. Ojha, and U. Riebesell. 2021. Impact of increasing carbon dioxide on dinitrogen and carbon fixation rates under oligotrophic conditions and simulated upwelling. *Limnology and Oceanography*. doi:10.1002/lno.11795.

Smetacek, V., and S. W. A. Naqvi. 2010. *The Expedition of the Research Vessel "Polarstern" to the Antarctic in 2009 (ANT-XXV/3-LOHAFEX)*. Berichte zur Polar-und Meeresforschung No. 613. Bremerhaven, Germany: Alfred Wegener Institute for Polar and Marine Research.

Smith, C. R. l. 1983. *Enrichment, Disturbance and Deep-Sea Community Structure: The Significance of Large Organic Falls to Bathyal Benthos in Santa Catalina Basin (California, Polychaetes)*. Ph.D. Dissertation, University of California, San Diego.

Smith, C. R., J. Roman, and J. B. Nation. 2019b. A metapopulation model for whale-fall specialists: The largest whales are essential to prevent species extinctions. *Journal of Marine Research* 77:283-302. doi:10.1357/002224019828474250.

Smith, R. J., J. C. Benavides, S. Jovan, M. Amacher, and B. McCune. 2015. A rapid method for landscape assessment of carbon storage and ecosystem function in moss and lichen ground layers. *Bryologist* 118(1):32-45. doi:10.1639/0007-2745-118.1.032.

Smith, S. R., G. Alory, A. Andersson, W. Asher, A. Baker, D. I. Berry, K. Drushka, D. Figurskey, E. Freeman, P. Holthus, T. Jickells, H. Kleta, E. C. Kent, N. Kolodziejczyk, M. Kramp, Z. Loh, P. Poli, U. Schuster, E. Steventon, S. Swart, O. Tarasova, L. P. De La Villéon, and N. V. Shiffer. 2019a. Ship-based contributions to global ocean, weather, and climate observing systems. *Frontiers in Marine Science* 6:434. doi:10.3389/fmars.2019.00434.

Smith, S. V. 1981. Marine macrophytes as a global carbon sink. *Science* 211(4484):838-840. doi:10.1126/science.211.4484.838.

Solan, M., E. M. Bennett, P. J. Mumby, J. Leyland, and J. A. Godbold. 2020. Benthic-based contributions to climate change mitigation and adaptation. *Philosophical Transactions of the Royal Society B: Biological Sciences* 375(1794):20190107. doi:10.1098/rstb.2019.0107.

Soma, K., and C. Haggett. 2015. Enhancing social acceptance in marine governance in Europe. *Ocean & Coastal Management* 117:61-69. doi:10.1016/j.ocecoaman.2015.11.001.

Song, A. M., W. H. Dressler, P. Satizábal, and M. Fabinyi. 2021. From conversion to conservation to carbon: The changing policy discourse on mangrove governance and use in the Philippines. *Journal of Rural Studies* 82:184-195. doi:10.1016/j.jrurstud.2021.01.008.

Spence, E., N. Pidgeon, and P. Pearson. 2018. UK public perceptions of ocean acidification—the importance of place and environmental identity. *Marine Policy* 97:287-293. doi:10.1016/j.marpol.2018.04.006.

Spiers, E. K. A., R. Stafford, M. Ramirez, D. F. V. Izurieta, M. Cornejo, and J. Chavarria. 2016. Potential role of predators on carbon dynamics of marine ecosystems as assessed by a Bayesian belief network. *Ecological Informatics* 36:77-83. doi:10.1016/j.ecoinf.2016.10.003.

Stafford, R. 2019. Sustainability: A flawed concept for fisheries management? *Elementa: Science of the Anthropocene* 7(1). doi:10.1525/elementa.346.

Stafford, R., M. Ashley, L. Clavey, L. S. Esteve, N. Hicks, A. Jones, P. Leonard, T. Luisetti, A. Martin, R. Parker, S. Rees, M. Schratzberger, and R. K. F. Unsworth. 2021. Coastal and marine systems. In *Nature-Based Solutions for Climate Change in the UK: A Report by the British Ecological Society*, R. Stafford, B. Chamberlain, L. Clavey, P. K. Gillingham, S. McKain, M. D. Morecroft, C. Morrison-Bell, and O. Watts, eds. London: British Ecological Society.

Stewart, H. L., J. P. Fram, D. C. Reed, S. L. Williams, M. A. Brzezinski, S. MacIntyreb, and B. Gaylord. 2009. Differences in growth, morphology and tissue carbon and nitrogen of *Macrocystis pyrifera* within and at the outer edge of a giant kelp forest in California, USA. *Marine Ecology Progress Series* 375:101-112. doi:10.3354/meps07752.

Stigebrandt, A., and B. Liljebladh. 2011. Oxygenation of large volumes of natural waters by geo-engineering: With particular reference to a pilot experiment in Byfjorden. Pp. 303-315 in *Macro-engineering Seawater in Unique Environments*, V. Badescu and R. B. Cathcart, eds. Heidelberg: Springer.

Stigebrandt, A., B. Liljebladh, L. de Brabandere, M. Forth, A. Granmo, P. Hall, J. Hammar, D. Hansson, M. Kononets, M. Magnusson, F. Noren, L. Rahm, A. H. Treusch, and L. Viktorsson. 2015. An experiment with forced oxygenation of the deepwater of the anoxic By Fjord, western Sweden. *Ambio* 44(1):42-54. doi:10.1007/s13280-014-0524-9.

Stilgoe, J., R. Owen, and P. Macnaghten. 2013. Developing a framework for responsible innovation. *Research Policy* 42:1568-1580. doi:https://doi.org/10.1016/j.respol.2013.05.008.

Stommel, H. 1956. An oceanographic curiosity: The perpetual salt fountain. *Deep-Sea Research* 3:152-153.

Strathmann, H. 2004. *Ion-Exchange Membrane Separation Processes.* Elsevier.

Strong, A., S. Chisholm, C. Miller, and J. Cullen. 2009. Ocean fertilization: Time to move on. *Nature* 461(7262):347-348. doi:10.1038/461347a.

Sugiyama, M., and T. Sugiyama. 2010. Interpretation of CBD COP10 decision on geoengineering. *SERC Discussion.*

Sumaila, U. R., and T. C. Tai. 2020. End overfishing and increase the resilience of the ocean to climate change. *Frontiers in Marine Science* 7:523. doi:10.3389/fmars.2020.00523.

Sunda, W. G., and S. A. Huntsman. 1995. Iron uptake and growth limitation in oceanic and coastal phytoplankton. *Marine Chemistry* 50(1-4):189-206. doi:10.1016/0304-4203(95)00035-p.

Sundarambal, P., R. Balasubramanian, P. Tkalich, and J. He. 2010. Impact of biomass burning on ocean water quality in Southeast Asia through atmospheric deposition: Field observations. *Atmospheric Chemistry and Physics* 10(23):11323-11336. doi:10.5194/acp-10-11323-2010.

Sutton, A. J., R. A. Feely, S. Maenner-Jones, S. Musielwicz, J. Osborne, C. Dietrich, N. Monacci, J. Cross, R. Bott, A. Kozyr, A. J. Andersson, N. R. Bates, W. J. Cai, M. F. Cronin, E. H. De Carlo, B. Hales, S. D. Howden, C. M. Lee, D. P. Manzello, M. J. McPhaden, M. Meléndez, J. B. Mickett, J. A. Newton, S. E. Noakes, J. H. Noh, S. R. Olafsdottir, J. E. Salisbury, U. Send, T. W. Trull, D. C. Vandemark, and R. A. Weller. 2019. Autonomous seawater pCO2 and pH time series from 40 surface buoys and the emergence of anthropogenic trends. *Earth System Science Data* 11(1):421-439. doi:10.5194/essd-11-421-2019.

Tagliabue, A., O. Aumont, R. DeAth, J. P. Dunne, St. Dutkiewicz, E. Galbraith, K. Misumi, J. K. Moore, A. Ridgwell, E. Sherman, C. Stock, M. Vichi, C. Völker, and A. Yool. 2016. How well do global ocean biogeochemistry models simulate dissolved iron distributions? *Global Biogeochemical Cycles* 30(2):149-174. doi:10.1002/2015gb005289.

Takahashi, T. 2004. The fate of industrial carbon dioxide. *Science* 305(5682):352-353.

Takahashi, T., R. A. Feely, R. F. Weiss, R. H. Wanninkhof, D. W. Chipman, S. C. Sutherland, and T. T. Takahashi. 1997. Global air-sea flux of CO_2: An estimate based on measurements of sea-air pCO_2 difference. *Proceedings of the National Academy of Sciences of the United States of America* 94(16):8292-8299. doi:10.1073/pnas.94.16.8292.

Takahashi, T., S. C. Sutherland, R. Wanninkhof, C. Sweeney, R. A. Feely, D. W. Chipman, B. Hales, G. Friederich, F. Chavez, C. Sabine, A. Watson, D. C. E. Bakker, U. Schuster, N. Metzl, H. Yoshikawa-Inoue, M. Ishii, T. Midorikawa, Y. Nojiri, A. Körtzinger, T. Steinhoff, M. Hoppema, J. Olafsson, T. S. Arnarson, B. Tilbrook, T. Johannessen, A. Olsen, R. Bellerby, C. S. Wong, B. Delille, N. R. Bates, and H. J. W. de Baar. 2009. Climatological mean and decadal change in surface ocean pCO_2, and net sea-air CO_2 flux over the global oceans. *Deep-Sea Research Part II: Topical Studies in Oceanography* 56(8-10):554-577. doi:10.1016/j.dsr2.2008.12.009.

Talley, L. D., R. A. Feely, B. M. Sloyan, R. Wanninkhof, M. O. Baringer, J. L. Bullister, C. A. Carlson, S. C. Doney, R. A. Fine, E. Firing, N. Gruber, D. A. Hansell, M. Ishii, G. C. Johnson, K. Katsumata, R. M. Key, M. Kramp, C. Langdon, A. M. MacDonald, J. T. Mathis, E. L. McDonagh, S. Mecking, F. J. Millero, C. W. Mordy, T. Nakano, C. L. Sabine, W. M. Smethie, J. H. Swift, T. Tanhua, A. M. Thurnherr, M. J. Warner, and J. Z. Zhang. 2016a. Changes in ocean heat, carbon content, and ventilation: A review of the first decade of GO-SHIP global repeat hydrography. *Annual Review of Marine Science* 8:185-215. doi:10.1146/annurev-marine-052915-100829.

Talley, L. D., S. Riser, K. S. Johnson, J. Wang, I. V. Kamenkovich, I. Rosso, M. R. Mazloff, S. Ogle, and J. L. Sarmiento. 2016b. *SOCCOM Biogeochemical Profiling Floats: Representativeness and Deployment Strategies Utilizing GO-SHIP/Argo Observations and SOSE/Hycom Model Output.* Presented at American Geophysical Union Fall Meeting, December 15.

Tan, R. R., K. B. Aviso, D. C. Y. Foo, J.-Y. Lee, and A. T. Ubando. 2019. Optimal synthesis of negative emissions polygeneration systems with desalination. *Energy* 187:115953. doi:https://doi.org/10.1016/j.energy.2019.115953.

Taucher, J., L. T. Bach, T. Boxhammer, A. Nauendorf, E. P. Achterberg, M. Algueró-Muñiz, J. Arístegui, J. Czerny, M. Esposito, W. Guan, M. Haunost, H. G. Horn, A. Ludwig, J. Meyer, C. Spisla, M. Sswat, P. Stange, and U. Riebesell. 2017. Influence of ocean acidification and deep water upwelling on oligotrophic plankton communities in the subtropical North Atlantic: Insights from an in situ mesocosm study. *Frontiers in Marine Science* 4:00085. doi:10.3389/fmars.2017.00085.

Tavares, D. C., J. F. Moura, E. Acevedo-Trejos, and A. Merico. 2019. Traits shared by marine megafauna and their relationships with ecosystem functions and services. *Frontiers in Marine Science* 6:262. doi:10.3389/fmars.2019.262.

Thingstad, T. F., M. D. Krom, R. F. C. Mantoura, C. A. F. Flaten, S. Groom, B. Herut, N. Kress, C. S. Law, A. Pasternak, P. Pitta, S. Psarra, F. Rassoulzadegan, T. Tanaka, A. Tselepides, P. Wassmann, E. M. S. Woodward, C. W. Riser, C. Zodiatis, and T. Zohary. 2005. Nature of phosphorus limitation in the ultraoligotrophic eastern Mediterranean. *Science* 309(5737):1068-1071. doi:10.1126/science.1112632.

Thomas, H., L. S. Schiettecatte, K. Suykens, Y. J. M. Koné, E. H. Shadwick, A. E. F. Prowe, Y. Bozec, H. J. W. De Baar, and A. V. Borges. 2009. Enhanced ocean carbon storage from anaerobic alkalinity generation in coastal sediments. *Biogeosciences* 6(2):267-274. doi:10.5194/bg-6-267-2009.

Thomson, D. J. M., and D. R. Barclay. 2020. Real-time observations of the impact of COVID-19 on underwater noise. *Journal of the Acoustical Society of America* 147(5):3390-3396. doi:10.1121/10.0001271.

Timothy, D. A., C. S. Wong, J. E. Barwell-Clarke, J. S. Page, L. A. White, and R. W. Macdonald. 2013. Climatology of sediment flux and composition in the subarctic northeast Pacific Ocean with biogeochemical implications. *Progress in Oceanography* 116:95-129. doi:10.1016/j.pocean.2013.06.017.

Toggweiler, J. R., J. L. Russell, and S. R. Carson. 2006. Midlatitude westerlies, atmospheric CO_2, and climate change during the ice ages. *Paleoceanography* 21(2):PA2005. doi:10.1029/2005PA001154.

Tollefson, J. 2012. Ocean-fertilization project off Canada sparks furore. *Nature* 490(7421):458-459. doi:10.1038/490458a.

Tollefson, J. 2017. Iron-dumping ocean experiment sparks controversy. *Nature* 545(7655):393-394. doi:10.1038/545393a.

Townsend, A., and A. Gillespie. 2020. *Scaling up the CCS Market to Deliver Net-Zero Emissions.* Docklands, Australia: Global CCS Institute.

Trick, C. G., B. D. Bill, W. P. Cochlan, M. L. Wells, V. L. Trainer, and L. D. Pickell. 2010. Iron enrichment stimulates toxic diatom production in high-nitrate, low-chlorophyll areas. *Proceedings of the National Academy of Sciences of the United States of America* 107(13):5887-5892. doi:10.1073/pnas.0910579107.

Tsubaki, K., S. Maruyama, A. Komiya, and H. Mitsugashira. 2007. Continuous measurement of an artificial upwelling of deep sea water induced by the perpetual salt fountain. *Deep-Sea Research Part I: Oceanographic Research Papers* 54:75-84.

Tulloch, V. J. D., É. E. Plagányi, C. Brown, A. J. Richardson, and R. Matear. 2019. Future recovery of baleen whales is imperiled by climate change. *Global Change Biology* 25(4):1263-1281. doi:10.1111/gcb.14573.

Tunnicliffe, V., K. T. A. Davies, D. A. Butterfield, R. W. Embley, J. M. Rose, and W. W. Chadwick, Jr. 2009. Survival of mussels in extremely acidic waters on a submarine volcano. *Nature Geoscience* 2(5):344-348. doi:10.1038/ngeo500.

Turetsky, M. R. 2003. New frontiers in bryology and lichenology: The role of bryophytes in carbon and nitrogen cycling. *Bryologist* 106(3):395-409. doi:10.1639/05.

Turner, N. J., and B. Neis. 2020. From "taking" to "tending": Learning about Indigenous land and resource management on the Pacific Northwest Coast of North America. *ICES Journal of Marine Science* 77(7-8):2472-2482. doi:10.1093/icesjms/fsaa095.

U.S. EPA. 2021. Public Participation Guide. Last Modified 07/12/2021. https://www.epa.gov/international-cooperation/public-participation-guide.

Uhlig, H. H., and R. W. Revie. 1985. *Corrosion and Corrosion Control. An Introduction to Corrosion Science and Engineering,* 3rd ed. Hoboken, NJ: John Wiley & Sons.

Uitz, J., D. Stramski, R. A. Reynolds, and J. Dubranna. 2015. Assessing phytoplankton community composition from hyperspectral measurements of phytoplankton absorption coefficient and remote-sensing reflectance in open-ocean environments. *Remote Sensing of Environment* 171:58-74. doi:10.1016/j.rse.2015.09.027.

UK House of Commons Science and Technology Committee. 2010. *The Regulation of Geoengineering: Fifth Report of Session 2009-10.* http://www.geoengineering.ox.ac.uk/publications.parliament.uk/pa/cm200910/cmselect/cmsctech/221/221.pdf.

UNEP (United Nations Environmental Programme). 2017. *The Emissions Gap Report 2017.* Nairobi: UNEP.

USGCRP (U.S. Global Change Research Program). 2017. *Climate Science Special Report: Fourth National Climate Assessment,* Vol. I, D. J. Wuebbles, D. W. Fahey, K. A. Hibbard, D. J. Dokken, B. C. Stewart and T. K. Maycock, eds. Washington, DC: USGCRP.

USGCRP. 2018. *Impacts, Risks, and Adaptation in the United States: Fourth National Climate Assessment,* Vol. II. D. R. Reidmiller, C. W. Avery, D. R. Easterling, K. E. Kunkel, K. L. M. Lewis, T. K. Maycock and B. C. Stewart, eds. Washington, DC: U.S. Global Change Research Program.

USGS (U.S. Geological Survey). 2021. Aluminum. Pp. 20-21 in *Mineral Commodity Summaries 2021.* Reston, VA: USGS.

van Bets, Linde K. J., Jan P. M. van Tatenhove, and Arthur P. J. Mol. 2016. Liquefied natural gas production at Hammerfest: A transforming marine community. *Marine Policy* 69:52-61. doi:10.1016/j.marpol.2016.03.020.

van der Heijden, L. H., and N. A. Kamenos. 2015. Calculating the global contribution of coralline algae to carbon burial. *Biogeosciences* 12(10):7845-7877.

van der Jagt, H., C. Friese, J. B. W. Stuut, G. Fischer, and M. H. Iversen. 2018. The ballasting effect of Saharan dust deposition on aggregate dynamics and carbon export: Aggregation, settling, and scavenging potential of marine snow. *Limnology and Oceanography* 63(3):1386-1394. doi:10.1002/lno.10779.

Van Mooy, B. A. S., R. G. Keil, and A. H. Devol. 2002. Impact of suboxia on sinking particulate organic carbon: Enhanced carbon flux and preferential degradation of amino acids via denitrification. *Geochimica et Cosmochimica Acta* 66(3):457-465. doi:10.1016/S0016-7037(01)00787-6.

van Son, T. C., N. Nikolioudakis, H. Steen, J. Albretsen, B. R. Furevik, S. Elvenes, F. Moy, and K. M. Norderhaug. 2020. Achieving reliable estimates of the spatial distribution of kelp biomass. *Frontiers in Marine Science* 7:107. doi:10.3389/fmars.2020.00107.

van Weelden, C., J. R. Towers, and T. Bosker. 2021. Impacts of climate change on cetacean distribution, habitat and migration. *Climate Change Ecology* 1:100009. doi:https://doi.org/10.1016/j.ecochg.2021.100009.

Vanclay, F. 2019. Reflections on social impact assessment in the 21st century. *Impact Assessment and Project Appraisal* 38(2):126-131. doi:10.1080/14615517.2019.1685807.

Verlaan, P. 2009. Geo-engineering, the Law of the Sea, and Climate Change. *Carbon & Climate Law Review* 3(4). doi:10.21552/cclr/2009/4/115.

Vershinsky, N. V., B. P. Psenichnyy, and A. V. Solovyev. 1987. Artificial upwelling using the energy of surface waves. Oceanology 27 (3):400-2.

Visch, W., M. Kononets, P. O. J. Hall, G. M. Nylund, and H.Pavia. 2020. Environmental impact of kelp (*Saccharina latissima*) aquaculture. *Marine Pollution Bulletin* 155:110962.

Vos, J. G., and M. T. M. Koper. 2018. Measurement of competition between oxygen evolution and chlorine evolution using rotating ring-disk electrode voltammetry. *Journal of Electroanalytical Chemistry* 819:260-268. doi:10.1016/j.jelechem.2017.10.058.

Vos, J. G., T. A. Wezendonk, A. W. Jeremiasse, and M. T. M. Koper. 2018. MnOx/IrOx as selective oxygen evolution electrocatalyst in acidic chloride solution. *Journal of the American Chemical Society* 140(32):10270-10281. doi:10.1021/jacs.8b05382.

Vos, J. G., Z. Liu, F. D. Speck, N. Perini, W. Fu, S. Cherevko, and M. T. M. Koper. 2019. Selectivity trends between oxygen evolution and chlorine evolution on iridium-based double perovskites in acidic media. *ACS Catalysis* 9(9):8561-8574. doi:10.1021/acscatal.9b01159.

Voyer, M., and J. van Leeuwen. 2018. Social license to operate and the blue economy. In *A Report to the World Ocean Council*. Wollongong, Australia: Australian National Centre for Ocean Resources and Security.

Walker, B. J. A., B. Wiersma, and E. Bailey. 2014. Community benefits, framing and the social acceptance of offshore wind farms: An experimental study in England. *Energy Research & Social Science* 3:46-54. doi:10.1016/j.erss.2014.07.003.

Wang, F., and D. E. Giammar. 2013. Forsterite dissolution in saline water at elevated temperature and high CO_2 pressure. *Environmental Science & Technology* 47(1):168-173. doi:10.1021/es301231n.

Wanninkhof, R. 1992. Relationship between wind speed and gas exchange over the ocean. *Journal of Geophysical Research* 97(C5):7373-7382. doi:10.1029/92JC00188.

Waples, J. T., C. Benitez-Nelson, N. Savoye, M. R. van der Loeff, M. Baskaran, and O. Gustafsson. 2006. An introduction to the application and future use of ^{234}Th in aquatic systems. *Marine Chemistry* 100(3-4):166-189. doi:10.1016/j.marchem.2005.10.011.

Watanabe, K., G. Yoshida, M. Hori, Y. Umezawa, H. Moki, and T. Kuwae. 2020. Macroalgal metabolism and lateral carbon flows can create significant carbon sinks. *Biogeosciences* 17(9):2425-2440. doi:10.5194/bg-17-2425-2020.

Watson, A. J., U. Schuster, J. D. Shutler, T. Holding, I. G. C. Ashton, P. Landschützer, D. K. Woolf, and L. Goddijn-Murphy. 2020. Revised estimates of ocean-atmosphere CO_2 flux are consistent with ocean carbon inventory. *Nature Communications* 11(1):1-6.

Webb, R. M. 2020. *The Law of Enhanced Weathering for Carbon Dioxide Removal.* White Paper. New York: Sabin Center for Climate Change Law.

Webb, R. M., and M. B. Gerrard. 2019. Overcoming impediments to offshore CO_2 storage: Legal issues in the United States and Canada. *Environmental Law Reporter* 49:10634.

Webb, R., and M. B. Gerrard. 2018. Sequestering carbon dioxide undersea in the Atlantic: Legal problems and solutions. *UCLA Journal of Environmental Law and Policy* 36(1). doi:10.5070/l5361039900.

Webb, R., and M. Gerrard. 2021. *The Legal Framework for Offshore Carbon Capture and Storage in Canada.* White Paper. New York: Sabin Center for Climate Change Law. https://climate.law.columbia.edu/sites/default/files/content/Webb%20%26%20Gerrard%20-%20Offshore%20CCS%20in%20Canada.pdf.

Webb, R. M., K. G. Silverman-Roati, and M. B. Gerrard. 2021. *Removing Carbon Dioxide Through Ocean Alkalinity Enhancement and Seaweed Cultivation: Legal Challenges and Opportunities.* Columbia Public Law Research Paper. Sabin Center for Climate Change Law. https://scholarship.law.columbia.edu/faculty_scholarship/2739.

Wei, H.-Z., Y. Zhao, X. Liu, Y.-J. Wang, F. Lei, W.-Q. Wang, Y.-C. Li, and H.-Y. Lu. 2021. Evolution of paleo-climate and seawater pH from the late Permian to postindustrial periods recorded by boron isotopes and B/Ca in biogenic carbonates. *Earth-Science Reviews* 215:103546. doi:10.1016/j.earscirev.2021.103546.

Werdell, P. J., M. J. Behrenfeld, P. S. Bontempi, E. Boss, B. Cairns, G. T. Davis, B. A. Franz, U. B. Gliese, E. T. Gorman, O. Hasekamp, K. D. Knobelspiesse, A. Mannino, J. V. Martins, C. R. McClain, G. Meister, and L. A. Remer. 2019. The plankton, aerosol, cloud, ocean ecosystem mission: Status, science, advances. *Bulletin of the American Meteorological Society* 100(9):1775-1794. doi:10.1175/bams-d-18-0056.1.

Wernberg, T., and K. Filbee-Dexter. 2018. Grazers extend blue carbon transfer by slowing sinking speeds of kelp detritus. *Scientific Reports* 8(1):17180. doi:10.1038/s41598-018-34721-z.

Westberry, T. K., M. J. Behrenfeld, A. J. Milligan, and S. C. Doney. 2013. Retrospective satellite ocean color analysis of purposeful and natural ocean iron fertilization. *Deep Sea Research Part I: Oceanographic Research Papers* 73:1-16. doi:10.1016/j.dsr.2012.11.010.

Wheeler, P. A., and W. J. North. 1981. Nitrogen supply, tissue composition and frond growth rates for *Macrocystis pyrifera* off the coast of southern California. *Marine Biology* 64(1):59-69. doi:10.1007/BF00394081.

White, A. F., and M. L. Peterson. 1990. Role of reactive-surface-area characterization in geochemical kinetic models. Pp. 461-475 in *Chemical Modeling of Aqueous Systems II*, D. C. Melchoir and R. L. Bassett, eds. Washington, DC: American Chemical Society.

White, A. F., and S. L. Brantley. 1995. Chemical weathering rates of silicate minerals: An overview. Pp. 1-22 in *Chemical Weathering Rates of Silicate Minerals*, A. F. White and S. L. Brantley, eds. Berlin, Boston: De Gruyter.

White, A., K. Björkman, E. Grabowski, R. Letelier, S. Poulos, B. Watkins, and D. Karl. 2010. An open ocean trial of controlled upwelling using wave pump technology. *Journal of Atmospheric and Oceanic Technology* 27(2):385-396. doi:10.1175/2009JTECHO679.1.

Whyte, K. P. 2011. The Recognition Dimensions of Environmental Justice in Indian Country. *Environmental Justice* 4 (4):199-205. doi:10.1089/env.2011.0036.

Willauer, H. D., D. R. Hardy, M. K. Lewis, E. C. Ndubizu, and F. W. Williams. 2009. Recovery of CO_2 by phase transition from an aqueous bicarbonate system under pressure by means of multilayer gas permeable membranes. *Energy & Fuels* 23(3):1770-1774. doi:10.1021/ef8009298.

Willauer, H. D., F. Dimascio, D. R. Hardy, M. K. Lewis, and F. W. Williams. 2011. Development of an electrochemical acidification cell for the recovery of CO_2 and H_2 from seawater. *Industrial and Engineering Chemistry Research* 50(17):9876-9882. doi:10.1021/ie2008136.

Williamson, N., A. Komiya, S. Maruyama, M. Behnia, and S. W. Armfield. 2009. Nutrient transport from an artificial upwelling of deep sea water. *Journal of Oceanography* 65(3):349-359. doi:10.1007/s10872-009-0032-x.

Williamson, P., and C. Turley. 2012. Ocean acidification in a geoengineering context. *Philosophical Transactions of the Royal Society A: Mathematical Physical and Engineering Sciences* 370(1974):4317-4342. doi:10.1098/rsta.2012.0167.

Williamson, P., D. W. R. Wallace, C. S. Law, P. W. Boyd, Y. Collos, P. Croot, K. Denman, U. Riebesell, S. Takeda, and C. Vivian. 2012. Ocean fertilization for geoengineering: A review of effectiveness, environmental impacts and emerging governance. *Process Safety and Environmental Protection* 90(6):475-488. doi:10.1016/j.psep.2012.10.007.

Wilmers, C. C., J. A. Estes, M. Edwards, K. L. Laidre, and B. Konar. 2012. Do trophic cascades affect the storage and flux of atmospheric carbon? An analysis of sea otters and kelp forests. *Frontiers in Ecology and the Environment* 10(8):409-415.

Wilson, G. S. 2013. Murky waters: Ambiguous international law for ocean fertilization and other geoengineering. *SSRN Electronic Journal*. doi:10.2139/ssrn.2312755.

Wing, S. R., L. Jack, O. Shatova, J. J. Leichter, D. Barr, R. D. Frew, and M. Gault-Ringold. 2014. Seabirds and marine mammals redistribute bioavailable iron in the Southern Ocean. *Marine Ecology Progress Series* 510:1-13. doi:10.3354/meps10923.

Wogelius, R. A., and J. V. Walther. 1991. Olivine dissolution at 25°C: Effects of pH, CO_2, and organic acids. *Geochimica et Cosmochimica Acta* 55(4):943-954. doi:10.1016/0016-7037(91)90153-V.

Wogelius, R. A., and J. V. Walther. 1992. Olivine dissolution kinetics at near-surface conditions. *Chemical Geology* 97(1-2):101-112. doi:10.1016/0009-2541(92)90138-U.

Wolff, G. A., D. S. M. Billett, B. J. Bett, J. Holtvoeth, T. FitzGeorge-Balfour, E. H. Fisher, I. Cross, R. Shannon, I. Salter, B. Boorman, N. J. King, A. Jamieson, and F. Chaillan. 2011. The effects of natural iron fertilisation on deep-sea ecology: The Crozet Plateau, southern Indian Ocean. *PLoS ONE* 6(6):e20697. doi:10.1371/journal.pone.0020697.

Wolf-Gladrow, D. A., R. E. Zeebe, C. Klaas, A. Körtzinger, and A. G. Dickson. 2007. Total alkalinity: The explicit conservative expression and its application to biogeochemical processes. *Marine Chemistry* 106(1-2):287-300. doi:10.1016/j.marchem.2007.01.006.

Wolsink, M. 2019. Social acceptance, lost objects, and obsession with the 'public'—The pressing need for enhanced conceptual and methodological rigor. *Energy Research & Social Science* 48:269-276. doi:10.1016/j.erss.2018.12.006.

Wolske, K. S., K. T. Raimi, V. Campbell-Arvai, and P. S. Hart. 2019. "Public support for carbon dioxide removal strategies: the role of tampering with nature perceptions." *Climatic Change* 152 (3-4):345-361. doi:10.1007/s10584-019-02375-z.

Wong, C. S., F. A. Whitney, K. Iseki, J. S. Page, and J. Zeng. 1995. Analysis of trends in primary productivity and chlorophyll-a over two decades at Ocean Station P (50°N, 145°W) in the subarctic northeast Pacific Ocean. *Canadian Journal of Fisheries and Aquatic Sciences* 121:107-117.

Woodson, C. B., J. R. Schramski, and S. B. Joye. 2020. Food web complexity weakens size-based constraints on the pyramids of life. *Proceedings of the Royal Society B: Biological Sciences* 287(1934):20201500. doi:10.1098/rspb.2020.1500.

Worden, R. H., S. J. Needham, and J. Cuadros. 2006. The worm gut; a natural clay mineral factory and a possible cause of diagenetic grain coats in sandstones. *Journal of Geochemical Exploration* 89(1-3):428-431. doi:10.1016/j.gexplo.2005.12.011.

Work, T. M., G. S. Aeby, and J. E. Maragos. 2008. Phase shift from a coral to a corallimorph-dominated reef associated with a shipwreck on Palmyra Atoll. *PLoS ONE* 3(8). doi:10.1371/journal.pone.0002989.

World Chlorine Council. 2017. *Sustainable Progress*. https://worldchlorine.org/wp-content/uploads/2018/10/WCC_Sustainable-Progress_Version-3-2017.pdf.

WRI. 2020. "4 Charts Explain Greenhouse Gas Emissions by Countries and Sectors." https://www.wri.org/insights/4-charts-explain-greenhouse-gas-emissions-countries-and-sectors.

Wüstenhagen, R., M. Wolsink, and M. J. Bürer. 2007. "Social acceptance of renewable energy innovation: An introduction to the concept." *Energy Policy* 35(5):2683-2691. doi:10.1016/j.enpol.2006.12.001.

Wylie, L., A. E. Sutton-Grier, and A. Moore. 2016. "Keys to successful blue carbon projects: Lessons learned from global case studies." *Marine Policy* 65:76-84. doi:10.1016/j.marpol.2015.12.020.

Xi, H., M. Hieronymi, H. Krasemann, and R. Röttgers. 2017. Phytoplankton group identification using simulated and in situ hyperspectral remote sensing reflectance. *Frontiers in Marine Science* 4:272. doi:10.3389/fmars.2017.00272.

Xiao, X., S. Agusti, F. Lin, K. Li, Y. Pan, Y.n Yu, Y. Zheng, J. Wu, and C. M. Duarte. 2017. Nutrient removal from Chinese coastal waters by large-scale seaweed aquaculture. *Scientific Reports* 7(1):46613. doi:10.1038/srep46613.

Xiu, P., A. C. Thomas, and F. Chai. 2014. Satellite bio-optical and altimeter comparisons of phytoplankton blooms induced by natural and artificial iron addition in the Gulf of Alaska. *Remote Sensing of Environment* 145:38-46. doi:10.1016/j.rse.2014.02.004.

Yan, Z., J. L. Hitt, J. A. Turner, and T. E. Mallouk. 2020. Renewable electricity storage using electrolysis. *Proceedings of the National Academy of Sciences of the United States of America* 117(23):12558-12563. doi:10.1073/pnas.1821686116.

Yang, B., J. Fox, M. J. Behrenfeld, E. S. Boss, N. Haëntjens, K. H. Halsey, S. R. Emerson, and S. C. Doney. 2021. In situ estimates of net primary production in the western North Atlantic with Argo profiling floats. *Journal of Geophysical Research: Biogeosciences* 126(2):e2020JG006116. doi:10.1029/2020JG006116.

Yodzis, P. 2001. Must top predators be culled for the sake of fisheries? *Trends in Ecology and Evolution* 16(2):78-84.

Yool, A., J. G. Shepherd, H. L. Bryden, and A. Oschlies. 2009. Low efficiency of nutrient translocation for enhancing oceanic uptake of carbon dioxide. *Journal of Geophysical Research: Oceans* 114(8):C08009. doi:10.1029/2008JC004792.

Yoon, J.-E., K.-C. Yoo, A. M. Macdonald, H.-I. Yoon, K.-T. Park, E. J. Yang, H.-C. Kim, J. I. L., M. K. L., J. Jung, J. Park, J. Lee, S. Kim, S.-S. Kim, K. Kim, and I.-N. Kim. 2018. Reviews and syntheses: Ocean iron fertilization experiments—past, present, and future looking to a future Korean Iron Fertilization Experiment in the Southern Ocean (KIFES) project. *Biogeosciences* 15(19):5847-5889. doi:10.5194/bg-15-5847-2018.

Young, D. R., and T.-K. Jan. 1977. Fire fallout of metals off California. *Marine Pollution Bulletin* 8(5):109-112.

Yu, Z., D. W. Beilman, S. Frolking, G. M. MacDonald, N. T. Roulet, P. Camill, and D. J. Charman. 2011. Peatlands and their role in the global carbon cycle. *Eos* 92(12):97-98. doi:10.1029/2011EO120001.

Zahariev, K., J. R. Christian, and K. L. Denman. 2008. Preindustrial, historical, and fertilization simulations using a global ocean carbon model with new parameterizations of iron limitation, calcification, and N_2 fixation." *Progress in Oceanography* 77(1):56-82. doi:https://doi.org/10.1016/j.pocean.2008.01.007.

Zhai, S., X. Wang, J. R. McConnell, L. Geng, J. Cole-Dai, M. Sigl, N. Chellman, T. S., R. Pound, K. Fujita, S. Hattori, Jo. M. Moch, L. Zhu, M. Evans, M. Legrand, P. Liu, D. Pasteris, Y.-C. Chan, L. T. Murray, and B. Alexander. 2021. Anthropogenic impacts on tropospheric reactive chlorine since the preindustrial. *Geophysical Research Letters* 48(14):e2021GL093808. doi:10.1029/2021gl093808.

Zhang, G., T. N. Forland, E. Johnsen, G. Pedersen, and H. Dong. 2020. Measurements of underwater noise radiated by commercial ships at a cabled ocean observatory. *Marine Pollution Bulletin* 153:110948. doi:10.1016/j.marpolbul.2020.110948.

Zhang, R., T. Jiang, Y. Tian, S. Xie, L. Zhou, Q. Li, and N. Jiao. 2017. Volcanic ash stimulates growth of marine autotrophic and heterotrophic microorganisms. *Geology* 45(8):G38833.1. doi:10.1130/g38833.1.

Zhao, Y., J. Wang, Z. Ji, J. Liu, X. Guo, and J. Yuan. 2020. A novel technology of carbon dioxide adsorption and mineralization via seawater decalcification by bipolar membrane electrodialysis system with a crystallizer. *Chemical Engineering Journal* 381:1222542. doi:10.1016/j.cej.2019.122542.

Zhou, S., and P. C. Flynn. 2005. Geoengineering downwelling ocean currents: A cost assessment. *Climatic Change* 71(1-2):203-220. doi:10.1007/s10584-005-5933-0.

Zickfeld, K., D. Azevedo, S. Mathesius, and H. D. Matthews. 2021. Asymmetry in the climate–carbon cycle response to positive and negative CO_2 emissions. *Nature Climate Change* 11(7):613-617. doi:10.1038/s41558-021-01061-2.

Zimmerman, R. C., and J. N. Kremer. 1986. In situ growth and chemical composition of the giant kelp, *Macrocystis pyrifera*: Response to temporal changes in ambient nutrient availability. *Marine Ecology Progress Series* 27(2):277-285.

Acronyms and Abbreviations

AD	artificial downwelling
AIS	automatic identification system
ANSI	American National Standards Institute
aOIF	artificial ocean iron fertilization
AU	artificial upwelling
BBNJ	Biodiversity Beyond National Jurisdiction
BCP	biological carbon pump
C	carbon
$Ca(HCO_3)_2$	calcium bicarbonate
$Ca(OH)_2$	portlandite or calcium hydroxide
$CaAl_2Si_2O_8$	anorthite
CAB	coastal area of benefit
$CaCO_3$	calcium carbonate or calcite
$CaMg(CO_3)_2$	dolomite
CaO	quicklime or calcium oxide
CBD	Convention on Biological Diversity
CCBS	Climate, Community, and Biodiversity Standards
CCS	carbon capture and storage
CDR	carbon dioxide removal
CH_4	methane
CO_2	carbon dioxide
CO_2 (aq)	aqueous carbon dioxide
CO_3^{2-}	carbonate ion
CREATE	Carbon Removal, Efficient Agencies, Technology Expertise
CWA	Clean Water Act
CZMA	Coastal Zone Management Act

DA	domoic acid
DIC	dissolved inorganic carbon
DMS	dimethyl sulfide
DOC	dissolved organic carbon
DOE	Department of Energy
EEZ	Exclusive Economic Zone
EIA	environmental impact assessment
ENSO	El Niño–Southern Oscillation
EPA	Environmental Protection Agency
ESA	Endangered Species Act
EW	enhanced weathering
EZ	euphotic zone
Fe	iron
Fe_2SiO_4	fayalite
GHGs	greenhouse gases
GO-SHIP	Global Ocean Hydrographic Investigations Program
H^+	hydrogen ion
H_2CO_3	carbonic acid
H_2O	water
HABs	harmful algal blooms
HCO_3^-	bicarbonate
HNLC	high nutrient, low chlorophyll
IODP	International Ocean Discovery Program
IOOS	Integrated Ocean Observing System
IUCN	International Union for Conservation of Nature
LC/LP	London Convention and London Protocol
LNLC	low nutrient, low chlorophyll
$Mg(OH)_2$	brucite
Mg_2SiO_4	forsterite
$MgCO_3$	manganesite
MgO	periclase/magnesia
MMPA	Marine Mammal Protection Act
MPAs	marine protected areas
MPRSA	Marine Protection, Research, and Sanctuaries Act
MSFCMA	Magnuson-Stevens Fishery Conservation and Management Act
N	nitrogen
N_2O	nitrous oxide
NaOH	sodium hydroxide
NBS	nature-based solutions
NDC	nationally determined contribution
NEPA	National Environmental Policy Act

NETs	negative emissions technologies
NIMBY	not in my backyard
NMFS	National Marine Fisheries Service
NMSA	National Marine Sanctuaries Act
nOIF	natural ocean iron fertilization
NPP	net primary productivity
NPSG	North Pacific Subtropical Gyre
NRC	National Research Council
NSTC	National Science and Technology Council
O_2	oxygen
OA	ocean alkalinity
OAE	ocean alkalinity enhancement
OCCAM	Ocean Circulation and Climate Advanced Modeling
OCSLA	Outer Continental Shelf Lands Act
OF	ocean fertilization
OIF	ocean iron fertilization
OMF	ocean macronutrient fertilization
OOI	Ocean Observatories Initiative
OTEC	ocean thermal energy conversion
P	phosphorus
$p\mathrm{CO_2}$	partial pressure of carbon dioxide
PDO	Pacific Decadal Oscillation
POC	particulate organic carbon
REDD+	Reducing Emissions from Deforestation and Forest Degradation
RHA	Rivers and Harbors Act
Si	silica
SLO	social license to operate
SOD	superoxide dismutase
TZ	twilight zone
UNCLOS	United Nations Convention on the Law of the Sea
UNFCCC	United Nations Framework Convention on Climate Change
VCS	Verified Carbon Standard
WHOI	Woods Hole Oceanographic Institute

Appendix A

Committee Biographies

Scott Doney, *Chair*, is the inaugural Joe D. and Helen J. Kington Professor in Environmental Change at the University of Virginia. He was a postdoctoral fellow at the National Center for Atmospheric Research from 1991 to 1993, and he served as a scientist at the National Center for Atmospheric Research from 1993 to 2002 and then the Woods Hole Oceanographic Institution from 2002 to 2017 before moving to the University of Virginia. Dr. Doney's expertise spans oceanography, climate, and biogeochemistry, with particular emphasis on the application of numerical models and data analysis to global-scale questions. His research focuses on how the global carbon cycle and ocean ecology respond to natural and human-driven climate change and ocean acidification. His previous experience with the National Academies of Sciences, Engineering, and Medicine includes membership on a number of committees in association with the Space Studies Board, Board on Atmospheric Sciences and Climate, and Ocean Studies Board. He has also served as an external reviewer for several National Academies reports. Dr. Doney graduated with a B.A. in chemistry from the University of California, San Diego in 1986 and a Ph.D. in chemical oceanography from the Massachusetts Institute of Technology/Woods Hole Oceanographic Institution Joint Program in Oceanography in 1991.

Holly Buck is an assistant professor of Environment and Sustainability at the University at Buffalo. She is an interdisciplinary social scientist who works across rural sociology, human geography, and science and technology studies to understand the social and environmental dimensions of emerging technologies. Her Ph.D. in development sociology from Cornell University in 2017 focused on public engagement with emerging environmental technologies, with chapters on carbon removal, solar geoengineering, and emerging "blue revolution" marine technologies. Dr. Buck's recent and current research involves the social dimensions of technologies to remove carbon from the atmosphere. Specific to marine carbon removal, she has authored a book chapter on an ocean fertilization event off the coast of Haida Gwaii, based on stakeholder interviews she conducted in Haida Gwaii and Vancouver in 2013. Her book *After Geoengineering* (2019) also discusses marine bioenergy with carbon capture and storage and draws on interviews with people working in the field. She also has experience with public engagement on solar climate intervention strategies, and presented some of this research to a National Academies committee on solar climate intervention at a meeting at Stanford in 2019, as part of that committee's information gathering process. Dr. Buck's work relating to climate intervention more broadly has focused on themes such as the governance of stopgap measures (Nature Sustainability), public engagement in research (Geoforum), and possible harms to ocean life (chapter in edited volume *Blue Legalities: The Life and Laws of the Sea*).

Ken Buesseler is a marine radiochemist and member of the Ocean Twilight Zone Project at the Woods Hole Oceanographic Institution. He is best known for work using natural and manmade isotopes in the ocean to study processes such as the movement of carbon and iron from the surface to deep ocean, as well as studies of the fate and transport of radioactive contaminants in the ocean. Dr. Buesseler participated in two ocean iron fertilization (OIF) experiments, leading one of three research vessels during the last major U.S. OIF experiment off Antarctica. In 2009 he was elected a Fellow of the American Geophysical Union; in 2013 selected as foreign member of the Dutch Academy of Sciences; and in 2018 elected as a Fellow of the American Association for the Advancement of Science. He is author of more than 175 research publications with 10 papers focused on OIF. In 2011 he was noted as the top-cited ocean scientist by the Times Higher Education for the decade 2000-2010. Dr. Buesseler received his Ph.D. from Massachusetts Institute of Technology and Woods Hole Oceanographic Institute in 1986.

M. Debora Iglesias-Rodriguez is professor of biological oceanography and vice-chair of the Department of Ecology, Evolution, and Marine Biology at the University of California, Santa Barbara. Dr. Iglesias-Rodriguez has worked for 20 years on diversity and function in marine phytoplankton, combining molecular approaches, carbon physiology, and biogeochemistry in the lab and in the field. She has broad interest in mechanisms controlling diversity and function in marine biota and the effect of ocean acidification (OA) on marine plankton. She has contributed to several white papers on OA, was a speaker at the 2011 IPCC workshop on OA, and one of her papers (*Science* 320:336-340) was identified by Thomson Reuters as "fast breaking paper" and at the top 0.01 most cited papers in Geoscience in 2008. Dr. Iglesias-Rodriguez has a B.Sc. in biology and biochemistry (University of Santiago de Compostela, Spain) and a Ph.D. (1996) on carbon utilization in phytoplankton (Swansea University, UK).

Kathryn Moran joined the University of Victoria in September 2011 as a professor in the Faculty of Sciences and as director of NEPTUNE Canada. In 2012, she was promoted to the position of president and CEO, Ocean Networks Canada. Since then, she has led and grown the organization following the vision of enhancing life on Earth by providing knowledge and leadership that deliver solutions to science, society, and industry. Dr. Moran's interests include topics related to the Arctic, ocean drilling, ocean observing, and climate change including ocean carbon sequestration methods and offshore renewable energy. She previously served on three National Academies committees: the Committee on Emerging Research Questions in the Arctic; the Gulf Research Program Advisory Board; and the Standing Committee on Understanding Gulf Ocean Systems. Dr. Moran holds degrees in marine science and engineering from the University of Pittsburgh, the University of Rhode Island, and Dalhousie University.

Andreas Oschlies is of Marine Biogeochemical Modelling at GEOMAR and the University of Kiel, Germany. His research interests include the global carbon, nitrogen and oxygen cycles, their sensitivities to environmental change, and the development and quality assessment of numerical models appropriate to investigate these. He was head of the Collaborative Research Centre "Climate-Biogeochemistry Interactions in the Tropical Ocean" (SFB754) running from 2008 to 2019, is a founding member of the Global Ocean Oxygen Network (GO2NE), member of the GESAMP Working Group 41 on Marine Geoengineering, and currently leads the Priority Program on Climate Engineering: Risks, Challenges, Opportunities? (SPP1689) funded by the German Research Foundation. He has contributed to a number of assessment reports, such as *Large-Scale Intentional Interventions into the Climate System? Assessing the Climate Engineering Debate* for the German Ministry of Education and Research (2011), *The European Transdisciplinary Assessment of Climate Engineering (EuTRACE)* for the European Commission (2015), the GESAMP *High Level Review of a Wide Range of Proposed Marine Geoengineering Techniques* (2019), and the *IPCC Special Report on the Oceans and the Cryosphere* (2019). Dr. Oschlies studied theoretical physics at Heidelberg and Cambridge (P. Phil. 1990) and received his Ph.D. in Oceanography from the University of Kiel (1994).

Phil Renforth is an associate professor for the School of Engineering and Physical Sciences, Institute of Mechanical, Process and Energy Engineering at Heriot-Watt University, Edinburgh, UK. He is an engineer and geochemist interested in understanding how reacting carbon dioxide with rocks and minerals may be able to help prevent climate change. His research expertise and interests include enhanced weathering, negative emission technologies, and alkaline waste, and he is also interested in understanding geochemical carbon sequestration in the ocean by increasing ocean alkalinity. Dr. Renforth serves on the Scientific Committee for the international conference on negative CO_2 emissions, the Geological Society's Engineering Geology Committee, Scientific Council for the UK Carbon Capture and Storage Research Centre, and on several editorial boards. He presented to the National Academies Committee on Developing a Research Agenda for Carbon Dioxide Removal and Reliable Sequestration. Dr. Renforth earned his Ph.D. in geoenvironmental engineering from the University of Newcastle-upon-Tyne (2011).

Joe Roman is a fellow in conservation biology and marine ecology at the Gund Institute for Environment, University of Vermont. His research focuses on the ecological functions and services provided by whales and other marine mammals. His team's work shows that great whales, the largest animals ever to have lived, play important ecological roles in the ocean—from enhancing primary productivity to providing habitat for more than 80 endemic species when their carcasses sink to the deep sea. Dr. Roman is dedicated to science policy, scientific diplomacy, teaching, and research. He received his A.B. from Harvard College in visual and environmental studies (1985), his M.S. from the University of Florida in wildlife ecology and conservation (1999), and his Ph.D. in organismic and evolutionary biology from Harvard University (2003).

Gaurav N. Sant is a professor and Henry Samueli Fellow at the University of California, Los Angeles with appointments in the Departments of Civil and Environmental Engineering and Materials Science and Engineering, and a member of the California Nanosystems Institute and the director of the Institute for Carbon Management. Dr. Sant's research interests include interfacial solid–liquid, solid–vapor, and solid–liquid–vapor reactions including dissolution, precipitation, and electrochemical reactions with applications to (i) cement, concrete, and porous media; (ii) hard biological tissues; (iii) metals and alloys; (iv) natural and synthetic minerals; and (v) glasses. In his research, special focus is placed on decarbonizing construction, the development of carbon mitigation technologies, and promoting manufacturing disruptions in entrenched heavy-industry sectors. In 2021, CarbonBuilt, a team led by Dr. Sant, was selected as the Grand Prize Winner of the NRG COSIA Carbon XPRIZE, the first university team to win an XPRIZE, and he has participated on the National Academies committee, Gaseous Carbon Waste Streams Utilization: Status and Research Needs. Dr. Sant received his Ph.D. in civil engineering from Purdue University in 2009.

David A. Siegel is presently a Distinguished Professor in the Department of Geography and chair of the Interdepartmental Graduate Program in Marine Science at the University of California, Santa Barbara. His research focuses on aquatic ecosystems and their functioning on local to global scales. He has worked extensively in marine bio-optics, satellite ocean color remote sensing and oceanographic observations, and numerical modeling on a wide range of problems from assessing marine biodiversity, quantifying the ocean's biological carbon pump, measuring and modeling giant kelp spatial population dynamics, and understanding the efficacy of nearshore fisheries management scenarios. Dr. Siegel has served as a member of the National Academies Committee on Assessing Requirements for Sustained Ocean Color Research and Operations (2011), and most recently was a member on the Ecosystems Panel of the National Academies 2018 Committee on the Decadal Survey for Earth Science and Applications from Space. He is a fellow of both the American Geophysical Union and the American Association for the Advancement of Science. Dr. Siegel received a B.A. in chemistry and a B.S. in engineering sciences from the University of California, San Diego (1982) and M.S. (1986) and Ph.D. (1988) in geological sciences from the University of Southern California.

Romany Webb is an associate research scholar at Columbia Law School and senior fellow at the Sabin Center for Climate Change Law. Her research focuses on climate change mitigation, exploring how legal and policy tools can be used to drive reductions in greenhouse gas emissions and support efforts to remove greenhouse gases from the atmosphere. Much of Ms. Webb's work centers on legal issues associated with the use of oceans for carbon dioxide removal and storage. She is a member of the Pacific Institute for Climate Solutions' "Solid Carbon" research study, which is assessing the feasibility of removing carbon dioxide from the ambient air using direct air capture facilities located on offshore platforms, and injecting the captured carbon dioxide into sub-seabed basalt rock formations. Ms. Webb received an LL.M., with a certificate of specialization in environmental law, from the University of California, Berkeley in 2013. She also holds an LL.B., awarded with first class honors, from the University of New South Wales (Australia).

Angelicque White is an associate professor in the School of Ocean and Earth Science and Technology at the University of Hawaii. Her primary research interests involve understanding how specific organisms acquire the elements necessary for growth and how different nutrient sources impact primary productivity and particle export. She is also working on the development of stochastic optimization models that can allow for more realistic simulations of the taxonomic and biogeochemical diversity of the phytoplankton community in the upper water column of the North Pacific. Dr. White received her B.S. (1998) and M.S. (2001) in biology from the University of Alabama in Huntsville and her Ph.D. in biological oceanography (2006) from Oregon State University.

Appendix B

Workshop and Meeting Public Presentations to the Committee

Sponsor Briefings
Jan Mazurek, ClimateWorks Foundation
Antonius Gagern, CEA Consulting

Current Activities in the Ocean-Based CDR Space
Brad Ack, Ocean Visions
Joseph Hezir, Energy Futures Initiative
Gyami Shrestha, UGSCRP Carbon Cycle Interagency Working Group
Dwight Gledhill, National Oceanic and Atmospheric Administration
Marc von Keitz, Department of Energy
Shuchi Talati, Carbon180
Jill Hamilton, Legislative Assistant, Senator Sheldon Whitehouse (D-RI)
K. John Holmes, National Academies of Sciences, Engineering, and Medicine

Legal and Political Aspects of Ocean-Based CDR
William Burns, American University
Anna-Maria Hubert, University of Calgary
Lisa Suatoni, Natural Resources Defense Council
Oliver Geden, German Institute for International and Security Affairs

Social Acceptance and Ethical Considerations to Ocean-Based CDR
David Morrow, American University
Sarah Cooley, Ocean Conservancy
Terre Satterfield, University of British Columbia
Emily Cox, Cardiff University
Bill Collins, Cascadia Seaweed

Financial and Economic Considerations to Ocean-Based CDR
Juan Moreno Cruz, University of Waterloo
Ryan Orbuch, Stripe Climate
Barbara Haya, University of California, Berkeley
Julio Friedmann, Columbia University

Parallel Efforts in Advancement of Ocean-Based CDR Strategies
David Keller, GEOMAR
Mark Preston, Bellona Foundation
Phillip Williamson, University of East Anglia
Filip Meysman, University of Antwerp

Technological Approaches to Increase Alkalinity/Remove CO_2
Greg Rau, University of California, Santa Cruz
Heather Willauer, U.S. Naval Research Laboratory
Matt Eisaman, Stony Brook University
David Babson, Advanced Research Projects Agency–energy
Tim Kruger, Oxford Martin School
Jess Adkins, California Institute of Technology
Robert Zeller, Occidental Petroleum Corporation
Niall Mac Dowell, Imperial College
Natural Approaches to Alkalinity Enhancement
Ulf Riebesell, GEOMAR, Kiel
Rosalind Rickaby, Oxford University
Lennart Bach, University of Tasmania
George Waldbusser, Oregon State University
Bärbel Hönisch, Lamont-Doherty Earth Observatory
Andy Ridgwell, University of California, Riverside

Validation & Monitoring and Environmental Risk
Ellen Briggs, University of Hawai'i
Chris Sabine, University of Hawai'i
Andrew Dickson, Scripps Institution of Oceanography
Albert Plueddemann, Woods Hole Oceanographic Institution

Seaweed Cultivation: Opportunities and Challenges
Marc von Keitz, Advanced Research Projects Agency–energy
Halley E. Froehlich, University of California, Santa Barbara
Olavur Gregersen, Ocean Rainforests
Brian von Herzen, Climate Foundation
Dorte Krause-Jensen, Aarhus University
William Collins, Cascadia Seaweed

Ecosystem Recovery: Opportunities and Challenges
Carlos Duarte, King Abdullah University of Science and Technology
Trisha Atwood, Utah State University
Nick Kamenos, University of Glasgow
Andy Pershing, Climate Central
Xabier Irigoien, AZTI-Basque Research
Catherine Lovelock, University of Queensland

Nutrient Fertilization: Challenges and Opportunities
Francisco Chavez, Monterey Bay Aquarium Research Institute
Stephanie Henson, National Oceanographic Centre, Southampton
Fei Chai, University of Maine

Artificial Upwelling and Downwelling: Challenges and Opportunities
Ricardo Letelier, Oregon State University
David Koweek, Ocean Visions
Ulf Riebesell, GEOMAR, Kiel
Kate Ricke, University of California, San Diego
Ray Schmitt, Woods Hole Oceanographic Institution

Overarching Challenges and Opportunities
Joellen Russell, University of Arizona
Nicolas Gruber, ETH Zürich
Chris Vivian, Retired, ex-Cefas, Centre for Environment, Fisheries and Aquaculture Science